An Introduction to Logical Theory

Aladdin M. Yaqub

broadview press

An Introduction to Logical Theory

Aladdin M. Yaqub

broadview press

Library and Archives Canada Cataloguing in Publication

Yaqub, Aladdin Mahmud
An introduction to logical theory / Aladdin M. Yaqub.

Includes bibliographical references and index.
ISBN 978-1-55111-993-9

1. Logic—Textbooks. 2. Language and logic—Textbooks.
I. Title.

BC15.Y36 2013 160 C2012-907967-7

Broadview Press is an independent, international publishing house, incorporated in 1985.

We welcome comments and suggestions regarding any aspect of our publications—please feel free to contact us at the addresses below or at broadview@broadviewpress.com.

North America
Post Office Box 1243, Peterborough, Ontario, Canada K9J 7H5
2215 Kenmore Avenue, Buffalo, NY, USA 14207
Tel: (705) 743-8990; Fax: (705) 743-8353
email: customerservice@broadviewpress.com

UK, Europe, Central Asia, Middle East, Africa, India, and Southeast Asia
Eurospan Group, 3 Henrietta St., London WC2E 8LU, United Kingdom
Tel: 44 (0) 1767 604972; Fax: 44 (0) 1767 601640
email: eurospan@turpin-distribution.com

Australia and New Zealand
NewSouth Books
c/o TL Distribution, 15-23 Helles Ave., Moorebank, NSW, Australia 2170
Tel: (02) 8778 9999; Fax: (02) 8778 9944
email: orders@tldistribution.com.au

www.broadviewpress.com

Edited by Robert M. Martin

This book is printed on paper containing 100% post-consumer fibre.

Typesetting and assembly: True to Type Inc., Claremont, Canada.

PRINTED IN CANADA

To Mariam and Ranah,
who learned the art of
argumentation too early in life

Contents

Introduction and Instructor's Guide

I

I wrote this book for two main reasons: to reclaim logic as a branch of philosophy, and to give a self-contained and fairly complete introduction to the three central systems of classical logic. I labored for more than sixteen years in the classroom trying to achieve this goal and for almost six years writing this book. I initially wrote a concise version of this textbook, which I used as the textbook for my introductory symbolic logic courses. It consisted of the summary paragraphs and exercises, and during my lectures I unpacked those paragraphs into elaborate commentaries on them and I solved most of the exercises with the students' input. The sections and subsections of this book are essentially those lectures in print.

This book has several unique features. First, it is *truly* self-contained: no background in logic, philosophy, or mathematics is presupposed. Second, it gives fairly complete introductions to the systems of Term Logic (TL), Sentence Logic (SL), and Predicate Logic (PL). These introductions cover the syntax, semantics, and logical concepts of all these systems as well as the proof theory of Sentence and Predicate Logics. There are no shortcuts or artificial simplifications of any aspects of these systems. For instance, the whole syntax and semantics of Predicate Logic are presented, including function symbols and set-theoretic interpretations, and a complete diagrammatic semantics for Term Logic is described—not merely Venn diagrams that work for only two-premise syllogisms. Also, the proof-theory chapter includes almost all the traditional rules of inference, and not only introduction and elimination rules. However, due to the philosophical importance of the Gentzen introduction and elimination rules, they are discussed in a separate subsection.

Third, practical methods are preferred to more traditional, less practical methods without ignoring the traditional methods. For example, while the method of truth tables for Sentence Logic is described and its philosophical significance is discussed, the preferred method for solving exercises is what I term "truth analysis." It is similar to the tableau method but without the formalities. Truth analysis is much simpler, less tedious, less mechanical, and far more practical than either the truth table or the tableau method. It is also a method that engages the reader's understanding of the truth conditions of the sentential connectives and the logical concepts of Sentence Logic more directly and "creatively" than the other two methods. Throughout this book, wherever possible, "creative" methods are preferred to merely mechanical approaches.

Fourth, the central part of the exercises consists of natural-language arguments that are philosophically interesting and logically complex. These are not arguments cooked up solely because their deductive validity can be demonstrated by constructing simple truth tables and short derivations. These are serious arguments, albeit simplified and modified, which are taken from the philosophical literature. Their formalizations are beyond the *practical* reach of truth tables or truth trees, and their derivations are strategically demanding. There are arguments from Plato, Descartes, Spinoza, Hume, Frege, Russell, Wittgenstein, and other philosophers and thinkers. However, there are also "playful" arguments for the sake of practice.

Fifth, the exercises are divided into two groups: starred exercises, whose solutions are included in the book, and non-starred exercises, whose solutions are available to instructors only. This feature by itself is not unique to this book. Many textbooks include solved and unsolved exercises. The difference, however, is that some of the included solutions are quite elaborate when needed. Some of these solutions run into several passages. They cover important topics and they supply meaningful digressions. For example, the role of logically true premises are discussed with elaboration in the chapter on Sentence Logic, and the emergence of the concept of the infinite is demonstrated in detail in the chapter on Predicate Logic. The exercises of Chapter Four (Predicate Logic) call for the construction of many PL interpretations, which should solidify the reader's understanding of the semantics of PL. The chapter itself contains many examples of these interpretations, so that the reader may gain some expertise in this creative enterprise. English-to-PL and PL-to-English translations are emphasized in the chapter and in the exercises. Many of these translations are not trivial, and some of them are quite complex. The goal is to develop some "natural feel" for the symbolism of PL, and to prepare the way for symbolizing English arguments.

Sixth, the book is seriously philosophical. It covers almost all the philosophical issues that concern introductory symbolic logic. Every logical system begins with a discussion of the "worldview" on which the system is based. For the philosophically informed reader, these worldviews are the metaphysical foundations of these systems. Each of the chapters on Term, Sentence, and Predicate Logic ends with a discussion of how successful the semantics of the system and the system's worldview are in capturing the intuitive and informal notion of logical possibility. Between the beginning and end of a chapter, many philosophical issues are discussed, whether in the main text or in the notes—issues such as the decidability or undecidability of the system, non-referring singular and general terms, negative existentials, essentialism, logical versus metaphysical possibility, the relation between the meaning and the referent of a proper name, the conceptual content of an expression, the relation between logical possibilities and their formal representations, the status of the material conditional, and many more. One of the guiding principles in writing this book was to produce an intellectually honest introduction to classical logic. The book does not shy away from philosophical difficulties and controversies. It is as much an introduction to the three main systems of classical logic as it is an introduction to the philosophy of classical logic.

Seventh, the book is conceptually tight. It begins with a chapter on the informal theory of classical logic, introducing the notion of logical possibility and eight logical concepts: deductive validity (or logical consequence), deductive invalidity, logical truth, logical falsehood, contingency, logical equivalence, consistency, and inconsistency. Of course, these concepts are not independent of each other. Some of them are denials of others, and all of them could be defined in terms of logical consequence. However the book takes an "inflationary" attitude towards the logical concepts because they are discussed frequently and widely in the philosophical literature as separate notions. It is common to read in a philosophical paper a claim that such-and-such position is inconsistent, or consistent, or logically follows from another position, and so on. I want the reader to be directly aware of these concepts, rather than to spend the whole semester studying the notion of deductive validity, and consequently lose connection with the other notions. The first chapter makes it clear that in order to give precise definitions of these eight logical concepts, we need precise definitions of the notions of a declarative sentence, of a logical possibility, and of true-in-a-logical-possibility. Each of Chapters Two through Four develops a logical system, which offers precise characterizations of these three notions. Hence the eight logical concepts immediately acquire precise definitions in each logical system. The readers do not need to relearn these concepts. They simply transform the definitions introduced in Chapter One into the conceptual framework of the formal system developed in that chapter. This process is repeated three times: for Term, Sentence, and Predicate Logic. The readers, therefore, acquire complete familiarity and mastery of these concepts as they progress through the discussion in the book.

Eighth, Chapter Four, which is the longest chapter in the book, is devoted to a fairly complete introduction to the system of Predicate Logic. PL interpretations are introduced fully without shortcuts and oversimplifications, with one notable exception: the quantifiers are given substitutional, rather than objectual, truth conditions. This approach allows me to avoid a discussion of the technical notion of satisfaction, and to maintain that variables receive no semantical interpretation. Objectual quantification, however, is discussed and compared to substitutional quantification. The reader will be able to follow this philosophical discussion without actually learning the technicalities of the relation of satisfaction. Substitutional quantification, though simpler to learn, exacts a logical and philosophical price. The main objections to substitutional quantifications, and possible rebuttals, are discussed in Chapter Four. Near the end of this chapter, I present a brief and informal discussion of Gödel's first incompleteness theorem and Church's undecidability theorem. The discussion is meant to convince the reader that the eight logical concepts introduced in Chapter One are undecidable in Predicate Logic. There is also a very brief introduction to second-order quantification when the expressive limitation of Predicate Logic is discussed

Ninth, the final chapter of the book is motivated by a discussion of the notion of demonstrative proof and its formal representations. This approach requires the introduction of many rules of inference in order to mirror as closely as possible natural demonstrative reasoning. The discussion, however, is sensitive to the fact that the Gentzen introduction and elimination rules are philosophically and logically significant. Hence a whole subsection is devoted to the discussion of these rules and their philosophical significance. All of the rules of inference introduced in Chapter Five are given their traditional names as they appear in the philosophical and logical literature. Three strategies for constructing formal derivations are discussed in detail and illustrated through several examples. The method of constructing PL derivations is put into the service of proving the deductive validity of many philosophically interesting English arguments, and of showing the applicability of many semantical concepts, such as inconsistency and logical truth. This approach requires a clear link between the proof theory and the semantics of Predicate Logic. So, early in Chapter Five, the Soundness and Completeness Theorems for Sentence and Predicate Logics are introduced and several of their corollaries, including the Compactness and the Finite-Satisfiability Theorems, are proved rigorously.

II

This book is based on a certain point of view about classical logic: the two central tasks of classical logic are to give precise representations of logical possibilities and formal representations of demonstrative proofs. Of course, there are non-classical logics that may be described as aiming at the same tasks. For example, it could be correctly argued that many systems of modal logic are developed in order to give precise descriptions of different types of possibility, including logical possibility, and to capture demonstrative proofs that invoke modal statements, such as statements of the forms "it is necessary that X" and "it is possible that X." This is definitely true. Classical logic has no monopoly over these tasks. Classical logic may be characterized by certain semantical and proof-theoretic presuppositions and principles. For instance, every system of classical logic presupposes the semantical principle of bivalence, that is, every sentence is either true or false, but not both, on any interpretation for it; it does not presuppose relevance, that is, it does not require that there be relevance between the premises and conclusion of a deductively valid argument; classical proof theory validates the Law of Excluded Middle, which allows the inference of any sentence of the form "X or not-X"; and it validates the inference rule Double Negation, which licenses the inference of X form not-not-X. Since this is a book about classical logic, all the standard presuppositions and principles of classical logic are incorporated in the systems discussed in this book. The purpose of this book, however, is not to provide a characterization of classical logic. Rather, its purpose is to introduce the reader to logical theory through the development of three systems of classical logic and through an exposition of their philosophical foundations, aims, and significance. In other words, this book introduces logical theory by developing an important branch of it as a response to certain philosophical questions.

The three logical systems developed in this book are selected because of their historical, logical, and philosophical significance. These are the three traditional systems of classical logic that dominated most of logical theory for a very long time. Term logic, as described in this book, is a modern version of an ancient system of classical logic. The traditional name for Term Logic is "Syllogistic Logic." Aristotle (384 BCE–322 BCE) introduced Syllogistic Logic. Later the system was enriched with some sentential and modal syllogisms. It remained the main logical system until the nineteenth century. Gottlob Frege (1848–1925) introduced both Sentence and Predicate Logics in the last third of the nineteenth century. These two systems are the "core" logics of all systems of modern logic, in the sense that every system of modern logic is either an extension of one of these systems or an alternative to it. Hence understanding these systems is a prerequisite for a full understanding of the logical systems that aim to extend them or to replace them. These three logical systems, perhaps together with some systems of modal logic, are the most commonly invoked logical apparatus in the philosophical literature, whether this literature belongs to the ancient, medieval, or modern eras. In contemporary philosophy, the analytic tradition, which is the brand of philosophy mostly practiced by Anglo-American philosophers, makes frequent use of Predicate and Modal Logics. Thus understanding these three systems of classical logic is necessary for understanding a good deal of the philosophical literature.

I must make an important disclaimer. These systems are not selected because I believe that they are the correct logical systems, the most likely to be correct among all the logical systems, the best available systems, or even the most useful systems. This book is not a defense of classical logic in general or of any of the systems developed here in particular. These systems are selected mainly because of their historical significance, their central place in modern logic, the richness of their philosophy, and their relative simplicity. Hence they serve as a good introduction to logical theory. The study of logical theory has to start at some point; these logical systems seemed to me to be a natural starting point.

I am aware that there is an approach to the proof theory of Predicate Logic that conceives of PL derivations as a way of formalizing the semantical relation of logical consequence. This is not the approach that is adopted in this book. PL proof theory is introduced here as a way of formalizing informal demonstrative proofs. According to this approach, PL proof theory is a response to a philosophical problem that is different from the philosophical problem that motivates the semantics of PL. PL semantics is motivated by the philosophical problem of giving precise definitions of the notions of logical possibility and true-in-a-logical-possibility. According to the approach to proof theory adopted in this book, the Soundness Theorem is not surprising, but the Completeness Theorem is. The Soundness Theorem says that any PL sentence that can be derived from a PL set is a logical consequence of that set. This is expected if PL derivations are indeed formalizations of demonstrative proofs. An essential property of demonstrations is that they are truth-preserving, that is, the conclusion of a demonstration must be true if the premises invoked in the demonstration are true. Therefore, if PL derivations are successful representations of demonstrations, then they too must be truth-preserving. Since being truth-preserving and being sound amount to the same thing, it should be expected that PL derivations are sound. What is surprising is that PL proof theory is complete.

The Completeness Theorem says that any logical consequence of a PL set is formally derivable from that set. This is surprising because it is not obvious at all that there should always be a way of demonstrating that the conclusion of any deductively valid argument logically follows from its premises. To say that a PL argument is deductively valid is to say that its conclusion is true on every PL interpretation that makes all the premises of the argument true. This semantical notion is quite different from the formal notion of derivability. The fact that the conclusion of any PL valid argument is formally derivable from its premises is unexpected. In fact, when Frege introduced Predicate Logic in 1879, he did not offer a semantical theory for the new logic. Frege introduced proof-theoretic axioms for PL. The semantical theory of PL was developed later in the first half of the twentieth century. The Completeness Theorem was such an unexpected and important result that it constituted Kurt Gödel's (1906–1978) doctoral dissertation (1929). Its importance is further illustrated by the fact that in 1949 Leon Henkin (1921–2006) gave another proof of it that is also based on his PhD dissertation (1947). While these historical facts do not, on their own, vindicate our approach to classical proof theory, they are instructive in suggesting that the proof theory of PL was not developed to formalize the relation of logical consequence.

Of course, one might maintain that, in spite of the historical facts, the best way to conceive of PL proof theory is as a formalization of the semantical relation of logical consequence. A possible response to this assertion is that, in addition to the historical development of PL proof theory, the actual technical development of the formal rules of inference of this proof theory conforms to our approach more naturally. When a formal inference rule is introduced, it is typically motivated by pointing out that given the truth conditions of the logical vocabulary invoked by the rule, the rule is truth-preserving. No rule is introduced with the claim that it is complete. There is a reason for this: completeness is only meaningful as a property of a sound system of rules. In constructing rules of inference, one aims at introducing rules that allow the inference of *only* logical consequences of a set of premises. This is the standard motivation for the introduction of any inference rule. A rule of inference is hardly ever introduced because it allows the inference of *all* the logical consequences of a set of premises. This is a property of a sound system of inference rules or axioms—a property that is usually established after the fact. Add to this that there are many logical systems that have sound, but incomplete, proof theories. The rules of inference in these systems are meant to capture modes of demonstrative reasoning, even though it is known that not every deductively valid argument can be proved valid by the system's inference rules.

III

The material of the book is divided into five chapters. The first four chapters deal with the semantical task of defining the notions of logical possibility and true-in-a-logical-possibility. Since in this book the truth bearers, i.e., the things that are said to be true or false, are declarative sentences, these chapters must also accomplish the formal task of defining the class of the declarative sentences of each system. The book deviates from the standard organization, found in other textbooks, of developing each system semantically and proof-theoretically as a single unit. Given the philosophical motivations of this book, the book is divided into two main parts. The first part, which consists of the first four chapters, formulates the first philosophical question, "What is the nature of logical possibility and of true-in-a-logical-possibility?", and offers three responses to this question. These responses are represented by the development of the logical systems of Term Logic, Sentence Logic, and Predicate Logic. The second part of the book, which consists of Chapter Five, formulates the second philosophical question, "What is the nature of demonstrative proof?", and offers a response to this question. This response is represented by the development of formal derivations and the Natural Deduction System.

The first chapter is an informal discussion of the theory of classical logic. In this chapter, the notion of logical possibility is introduced and eight logical concepts are defined in terms of logical possibility and true-in-a-logical-possibility. As indicated previously, these eight logical concepts are not independent of each other: some of them are the denials of others, and all of them can be defined on the basis of the relation of logical consequence. The chapter addresses several philosophical issues and explains that the best informal definition of logical possibility is circular. This prepares the way for the project of developing the logical systems of the following three chapters. Some of the exercises of this chapter are challenging for readers without any background in constructing informal proofs. The solutions to the starred exercises include many such proofs to help the reader develop the necessary skills for proving theorems. The exercises also introduce the reader to the task of interpreting argumentative passages and schematizing their arguments. An example from Hume is discussed with elaboration because it is the first example of interpreting an argumentative passage and schematizing an argument.

Term Logic (also known as "Syllogistic Logic" and "Categorical Logic") is developed in Chapter Two. Like all the systems developed in this book, the first stage of the development is to explain the worldview on which the system is based. Every worldview specifies the basic objects that constitute the logical possibilities admitted by that worldview. These objects are called "the basic ingredients of reality." A philosophically informed reader will notice that the approach to logical possibility adopted in this book is what philosophers call *the combinatorial theory of possibility* (or simply, *combinatorialism*). This theory is interpreted in this book as asserting that every combination of basic ingredients of reality is a logical possibility. Combinatorialism is quite old, but in contemporary philosophy it is typically associated with the Australian philosopher D.M. Armstrong (b. 1926). However, there are very important differences between the combinatorialism of this book and the one defended by Armstrong. Armstrong is a nominalist, that is, he does not believe in the existence of abstract objects, such as numbers, vector spaces, and sets; and he only admits states of affairs as basic ingredients of reality. Our combinatorialism makes no such commitments. In fact, only the worldview of Sentence Logic takes states of affairs as basic ingredients of reality. The worldviews of Term and Predicate Logics have no states of affairs among their basic ingredients of reality. Furthermore, all these worldviews do not rule out the existence of abstract objects. They take a neutral stance on this issue. What sorts of individuals or states of affairs are to be admitted into the worldview of a system depend on the philosopher's convictions. If the philosopher is a nominalist, abstract objects would be ruled out as ingredients of reality, and if he or she is not a nominalist, abstract objects would be allowed into the worldview.

The worldview of Term Logic takes individuals and properties as the basic ingredients of reality. Hence any combination of these objects constitutes a logical possibility according to this worldview. Following the standard commitment of Aristotelian Syllogistic Logic, no property is allowed to have an empty extension, that is, every property must be instantiated by at least one individual. Thus while this worldview permits logical possibilities to consist solely of individuals, it does not admit possibilities that consist solely of properties without individuals. Term Logic makes for a good starting point. It is the least symbolic of the classical systems, its syntax is very simple and has a natural feel to it, its semantics is pictorial (and hence, more visual and less abstract), and it illustrates the use of quantifiers and identity, which paves the way to the more complex quantification of Predicate Logic. Several philosophical issues are discussed in this chapter, and it ends with a discussion of the decidability of the system, and of the extent to which its semantics is successful in capturing the informal notion of logical possibility. As is the case with Chapters Three and Five, the bulk of the exercises is made of English-language arguments, most of which are philosophically interesting and logically complex. The reader is asked to schematize these arguments, translate them into the symbolic system, and test for deductive validity. Special care is given to whether the status of the symbolic argument can be considered as indicative of the status of the English-language argument of which it is a translation.

Chapter Three introduces the system of Sentence Logic (also known as "Propositional Logic" and "Sentential Logic"). The worldview of Sentence Logic takes states of affairs as the basic ingredients of reality. It was the Austrian philosopher Ludwig Wittgenstein (1889–1951) who made, in the *Tractatus Logico-Philosophicus* (1921), this account of the worldview of Sentence Logic the standard approach. Wittgenstein also invented the method of truth tables. I argue in this book that another philosophical commitment of Sentence Logic is the *correspondence conception of truth*, according to which a declarative sentence is true if and only if it represents a state of affairs that obtains, i.e., that is actually the case. It is standard to develop Sentence Logic before introducing Predicate Logic. Sentence Logic has a simple *recursive* syntax that consists of sentence letters and sentential connectives. Hence quantifiers are introduced in Chapter Two and sentential connectives are introduced in Chapter Three. This approach prepares the reader for the syntax of Predicate Logic, which contains sentential connectives and quantifiers. Sentence Logic is also a decidable system, with simple semantics. The mechanical truth-table method is deemphasized in favor of a less mechanical and more practical method that I call "truth analysis." It is similar to, but less formal than, the tableau method. As usual, the chapter ends with a discussion of the decidability of Sentence Logic and of the extent to which its semantics is successful in capturing the informal notion of logical possibility.

Predicate Logic (also known as "First-Order Logic" and "Quantificational Logic") is developed in Chapter Four, which is the longest chapter in the book. The chapter begins with a discussion of the worldview of Predicate Logic. According to this worldview the basic ingredients of reality are individuals, properties, and relations. Unlike Term Logic, properties and relations in Predicate Logic may have empty extensions. Predicate Logic has relatively complex syntax and semantics, and the philosophical issues associated with it are far more interesting and demanding than the philosophies of Term Logic and Sentence Logic. So there is plenty of philosophical discussion in this chapter. The main focus of the chapter is PL-English and English-PL translations and the construction of PL interpretations. Many examples and exercises are worked into the chapter in order to help the reader understand the theoretical material. The chapter ends with an informal discussion of the undecidability of PL, its limitation in representing logical possibilities, and a philosophical overview of the three logical systems.

Chapter Five is devoted to classical proof theory—more precisely, to the proof theory of Sentence and Predicate Logics. The chapter begins with a discussion of the notion of demonstrative proof. The discussion explains that the most important property of demonstrative proofs is that they are truth-preserving, that is, a conclusion derived from true premises is true. Then formal derivations are introduced as formal representations of demonstrative proofs. Before discussing the inference rules of the Natural Deduction System, the proof theory of PL is linked to its semantics via the Soundness and Completeness theorems and some of their corollaries. The Soundness and Completeness Theorems are not proved but their corollaries are. After introducing and explaining the inference rules of the Natural Deduction Systems, the standard Gentzen introduction and elimination rules are discussed and their philosophical significance is explained. Three strategies for constructing formal derivations are discussed and illustrated with several examples.

IV

The book can be read and taught in a few different ways. The best way is to study the whole book in its specified order and to solve all the starred exercises and as many as possible of the non-starred exercises. I usually teach the book in this way in my introductory symbolic logic courses. I have at my disposal 150 minutes of teaching time per week and fourteen weeks per semester. I also quickly cover Chapters Four and Five as the essential background in the first five weeks of my other, more advanced logic courses, such as Topics in Philosophical Logic and Mathematical Logic.

In a quarter system, the instructor would need to be selective. There are several options. One of them is to cover all the chapters but to skip certain sections, such as the sections on decidability and representability of logical possibilities. Another approach is to skip Chapter Two on Term Logic and cover the other four chapters in their entirety. Some instructors might prefer a nontraditional approach. They might wish to skip Chapter Three on Sentence Logic and cover all of the other four chapters. A justification for this approach might be that Predicate Logic is an extension of Sentence Logic; the latter can be quickly and briefly introduced when the relation between Term Logic and Predicate Logic is discussed (Sentence Logic would be introduced as an example of a "real" sublogic of Predicate Logic). There is another approach to covering the entire book in a quarter. Given the fact that the book is designed for self-study as well as for classroom use, the instructor can assign large chunks of the book to the students and discuss in class only the main and more difficult topics and exercises. This is the approach I use when I teach Chapters Four and Five in five weeks. I could easily adopt the same approach, and cover the other three chapters if I had only nine or ten weeks available to me.

I offer below some suggested syllabi for a fourteen-week semester and a ten-week quarter. The first syllabus is based on actual syllabi, which I used for introductory symbolic logic courses through the years. Thus in the first syllabus, I will describe what I actually do in class.

Suggested syllabus for a fourteen-week semester
Given the length of some of the exercises, I suggest that the instructor teach two 75-minute sessions, instead of three 50-minute sessions, per week.

Chapter One: Informal Logic. 5 lectures (2.5 weeks). Although this is the shortest chapter in the book, I cover it in detail and solve all its exercises, since it furnishes the conceptual foundations for Chapters Two through Four.

Lecture 1. Section 1.1: taxonomy of arguments. I cover this section with elaboration.

Lecture 2. Section 1.2: the notion of logical possibility and the eight logical concepts. I take my time here, since this section will be revisited in Chapters Two through Four.

Lectures 3–5. Section 1.3: exercises. Many of these exercises ask for informal proofs. Students usually find constructing proofs quite hard. Almost all of them have no experience with proofs at all. Thus I cover these exercises with depth and elaboration. I also spend a reasonable amount of time discussing arguments, argument schematization, and paraphrasing. All of this material is totally foreign to almost all of the students. Thus covering it carefully pays well for their understanding of many aspects of the subsequent chapters. This is why I devote 5 lectures to this chapter, and 3 lectures solely to the exercises.

Chapter Two: Term Logic (TL). 4 Lectures (2 weeks). This chapter can be covered quickly, since its syntax is very simple and natural, and its semantics is diagrammatic, which makes it easy to comprehend and work with.

Lecture 1. Sections 2.1–2.2: the TL worldview (including the introduction to set theory) and the syntax of TL. I usually cover the translations (Section 2.3) when I discuss the argument exercises.

Lecture 2. Sections 2.4–2.5: the semantics of TL and logical concepts in TL. The logical concepts (Section 2.5) can be covered very quickly with most of the definitions being left to the students, since these definitions are covered in depth in Chapter One; in Section 2.5, those definitions are simply modified to fit TL. I spend some time discussing Subsection 2.5.12, which addresses the relation between TL diagrams and logical possibilities and the extent to which the status of a symbolic argument can be indicative of the status of the English-language argument of which it is a translation. This discussion initiates the students into the philosophy of logic.

Lectures 3–4. Section 2.6: exercises. I spend a fair amount of time symbolizing arguments, since these exercises are the students' first encounter with formalization.

Chapter Three: Sentence Logic (SL). 5 lectures (2.5 weeks). I spend very little time constructing truth tables and a considerable amount of time solving exercises using the truth analysis method. I also devote a fair amount of time discussing Subsection 3.5.12 on the scope and limits of SL.

Lecture 1. Sections 3.1–3.2 and parts of Section 3.4: the SL worldview, the syntax of SL, and the notion of an SL truth valuation without covering the truth conditions of the SL sentences. I leave matters of translation (Section 3.3) until we work on the translation exercises.

Lectures 2–3. The remainder of Section 3.4, Section 3.5, and parts of Section 3.6: the truth conditions of the SL sentences, logical concepts in SL, and some exercises. As usual, the logical concepts are covered very quickly—the students can study these definitions on their own. I pay close attention to Subsection 3.5.12, so the students may come to appreciate the philosophical aspects of Sentence Logic. I teach the methods of truth tables and truth analysis by actually working some of the exercises from Section 3.6.

Lectures 4–5. The remainder of Section 3.6: exercises. I devote a fair amount of time to symbolizing and analyzing arguments. Students typically find these exercises much more interesting than the purely symbolic exercises.

Chapter Four: Predicate Logic (PL). 7 lectures (3.5 weeks). Although this is the longest chapter of the book, I find 3.5 weeks to be adequate for covering all of its material. Section 4.6 does not contain too many exercises, since there is a fair amount of examples discussed in the theoretical part of the chapter. Also, due to the nature of PL methodology, students cannot prove deductive validity until Chapter Five, where the notion of formal derivation is introduced. Most of this chapter concerns matters of translation and the construction of PL interpretations. The chapter also contains a good deal of philosophy of logic.

Lecture 1. Sections 4.1–4.2: the PL worldview and the syntax of PL. Given the relative richness of the PL worldview and its syntax, I cover these two sections with elaborations. I also introduce some natural readings of PL symbolism. I find that students tend to have a better understanding of the formal syntax if they are able to interpret the symbolism in English. I spend a fair amount of time on singular terms, including definite descriptions and their formal counterparts.

Lecture 2. Section 4.3: translation. The language of PL is the most demanding formal language in this book. Learning how to translate PL into English and English into PL improves the students' ability to construct PL interpretations and to evaluate the truth values of PL sentences on those interpretations. Furthermore, it makes it possible for them to give faithful PL translations of English arguments—a skill they will need in Chapter Five. In this lecture I work in some of the translation exercises in Section 4.6.

Lectures 3–4. Sections 4.4–4.5: the semantics of PL and logical concepts in PL. I explain the notion of PL interpretation and the truth conditions of PL sentences by first working actual examples and then giving the formal definitions. I spend some time explaining the difference between substitutional and objectual quantification, and the importance of having names for all the individuals in the universe of discourse. As in the previous two chapters, very little time is spent on the logical concepts. I merely point out how to convert the definitions given in Chapter One into the system of PL and leave the rest to the students.

Lecture 5. I devote this lecture to philosophical issues that relate to PL. I discuss essentialism, the distinctions between metaphysical and logical possibility, the conceptual content of an expression, and the relation between PL interpretations and logical possibilities. All of these issues are discussed in the book with some elaboration. I usually ask the students to read these sections prior to the lectures, and conduct this lecture mostly as a discussion session.

Lectures 6–7. Section 4.6: exercises. Some of these exercises are already solved during the previous lectures as examples. During these lectures, I solve many exercises in an interactive fashion, which is my usual way of solving exercises.

Chapter Five: Classical Proof Theory. 7 lectures (3.5 weeks). Although this chapter is relatively short, I allocate to it 7 lectures. It is an exercise-driven unit. The theoretical material is limited, and the rules of inference can only be fully explained through actual derivation construction.

Lectures 1-2. Sections 5.1–5.2: the notion of demonstrative proof, the notion of formal derivations, and the Soundness and Completeness Theorems and their corollaries. I prove all the corollaries. This is a good place to explain how the eight logical concepts introduced in Chapter One can all be reduced to the relation of logical consequence. I give examples of demonstrative proofs and their formal representations as PL derivations. PL derivations can be introduced without a full coverage of the Natural Deduction System.

Lectures 3–7: Sections 5.3–5.4: the Natural Deduction System, proof strategies, exercises, and the Gentzen Deduction System and its philosophical significance. I teach all of these concepts by solving exercises. I usually introduce the Gentzen Deduction System after the students gain some experience with the Natural Deduction System. I spend a considerable amount of time translating English arguments into PL and constructing PL derivations to demonstrate their validity. Students tend to learn better how to construct derivations when they are engaged in natural demonstrative reasoning. The English-argument exercises supply an appropriate medium for engaging in natural reasoning.

Suggested syllabus for a ten-week quarter

Since I never taught in a quarter system, my suggested syllabus for a ten-week quarter is constructed by combining one of my standard fourteen-week syllabus and a shorter five-week syllabus, which I adopt for more advanced logic courses in order to cover the necessary logic background. My typical five-week syllabus covers only the last two chapters (Predicate Logic and Classical Proof Theory). In order to construct the suggested syllabus below, I envisioned a ten-week course in which the longer syllabus is compressed and the shorter syllabus is expanded to fit ten weeks of classes.

As was suggested above, it is advisable to teach two 75-minute sessions per week instead of three 50-minute sessions weekly. In this suggested syllabus, I am assuming that the instructor aims at covering the entire book. Given the shorter duration of a quarter, I believe that the best approach is to assign substantial amounts of reading to the students and cover only selective topics in class. The book is suitable for self-study. The students can learn a lot on their own, working through the exposition, examples, and solved exercises in the book. As for the exercises, I would leave most of the starred exercises and their solutions to the students and solve in class as many of the non-starred exercises as time permits. Of course, I would begin every lecture by inquiring about any topic or exercise that the students find hard to understand.

Chapter One: Informal Logic. 4 lectures (2 weeks). Spending 2 weeks covering Chapter One is a good investment, since this chapter provides the conceptual foundation of the next three chapters.

Lectures 1–2. Sections 1.1–1.2: taxonomy of arguments, the notion of logical possibility, and the eight logical concepts. I would focus on the definition of an argument, the informal definition of a logical possibility and why it is circular, and the definitions of the eight logical concepts.

Lectures 3–4. Section 1.3: exercises.

Chapter Two: Term Logic (TL). 3 lectures (1.5 weeks). I would spend one lecture only on the theoretical material, mostly covering the TL worldview and the semantics of TL. The students can study the syntax of TL and the definitions of the eight logical concepts on their own. The remaining two lectures will be devoted to solving exercises. It is a good idea to spend a fair amount of time translating written English arguments into TL, as this will prepare the students for carrying out SL and PL translations, which are much more difficult than TL translations.

Chapter Three: Sentence Logic (SL). 3 lectures (1.5 weeks). This is a fast-paced unit. In my judgment, it is important to spend some time discussing the SL worldview, for such discussion gives an intuitive understanding of the semantics of SL. The syntax should be covered in class, since this is the students' first encounter with a recursive grammar. The instructor can save some time by combining the SL worldview and SL semantics into one lecture. The methods of truth tables and of truth analysis can be explained while solving exercises. This is usually my approach when I teach Sentence Logic as part of a condensed five-week logic unit. I would spend only one lecture on the theoretical material and two lectures working on exercises. Again, I would emphasize translation exercises to prepare the students for the important task of translating English into PL and PL into English.

Chapter Four: Predicate Logic (PL). 5 lectures (2.5 weeks). The two important skills that this chapter is meant to teach are translating from and into PL and constructing PL interpretations. The first skill requires a good degree of comfort with PL syntax. I am sure that, with some selectivity, this chapter can be covered in 5 lectures.

Lecture 1. Sections 4.1–4.2: the PL worldview and the syntax of PL. I would invest a significant amount of time in explaining the PL worldview. This will make it easier for the students to understand the semantics of PL. In discussing the syntax of PL, I would focus on nested quantifiers and term construction.

Lectures 2–3. Section 4.3 and parts of Section 4.6: translating PL into English and English into PL and exercises. I would spend these two lectures building translation skills. This is an exercise-driven unit. There is hardly any theoretical material in it. I would work through the examples, the non-starred exercises, and some of the starred exercises. I always conduct translation classes interactively and seldom give a readymade translation. I put the sentence to be translated, whether it is a PL or English sentence, on the blackboard and collect suggested translations from the students and discuss the accuracy of these translations and how to improve them.

Lectures 4–5. Sections 4.4–4.5 and parts of section 4.6: the semantics of PL, logical concepts in PL, and exercises. In this unit, the students learn how to construct PL interpretations. I would leave the definitions of the logical concepts to the students. The best way to understand the notion of a PL interpretation is to work through examples and exercises. I would postpone introducing the formal truth conditions until the end of the unit. The emphasis should be on interpreting PL sentences by giving their natural readings using the semantical assignments of the various interpretations. Some time should be spent on the difference between substitutional and objectual quantification. Examples of PL sets whose models are all infinite are of particular philosophical significance. I usually make a point of discussing at least one example of such sets. The book has an extensive discussion of one example.

Chapter Five: Classical Proof Theory. 5 lectures (2.5 weeks). In a fourteen-week semester I spend about 7 lectures covering the proof theory of SL and PL. A 2-lecture reduction for this unit means fewer exercises and examples. This could have serious implications for the students' ability to build the necessary skills for constructing PL derivations. When I teach this unit as part of a five-week logic session, I focus on long and complex derivations rather than on shorter and simpler ones. The students seem to learn more from constructing complex derivations. I begin with short and simple derivations in order to illustrate the most common inference rules, but I quickly move to long and complex derivations with several subderivations. I also find that building such derivations engages the students more. I would spend only one lecture on the theoretical material and four lectures on the construction of SL and PL derivations and formalizing arguments.

V

As I said at the outset of this introduction, this book is the fruit of more than sixteen years of teaching introductory and advanced logic courses and thinking about the philosophy of logic. The book began in 1995 as a concise collection of notes. I kept adding new paragraphs and exercises to these notes over the years. But the concise version remained concise for many years. I always nurtured the hope that one day I would find the time to unpack that concise version into a self-contained book on symbolic logic. This opportunity came in June 2004 after a very difficult time. In 2001 I fell severely ill with a malady that completely incapacitated me for more than two years. By spring 2004, though I wasn't able to return to teaching, I was feeling well enough to start thinking about logic again. The writing of this book commenced in June 2004. I wrote nonstop and with fervor until the end of June 2005, when I stopped to complete a paper that I had started before I fell ill. After completing that paper, I became occupied with applying for positions to return to academia after the break necessitated by my illness. I came to Lehigh University in fall 2006, and resumed writing this book in spring 2007. Of course, I worked sporadically on it since I had teaching and other duties, as well as another book project. A draft of the book was completed in fall 2011. I revised that draft and added new material to it during the winter of 2012. Thus in total I spent almost six years writing this book.

I am very grateful to Mariam Yaqub for copyediting a mature draft before it was submitted to the publisher; Mariam studied the book carefully and offered feedback, corrections, and suggestions that greatly improved its quality. I am also very grateful to Professor Robert Martin, who copyedited the book formally for Broadview Press and made numerous pedagogical suggestions that significantly enhanced many parts of the expository discussion. I would like to express my gratitude to Ranah Yaqub, who spent many hours verifying all the page references of the Index entries. Thanks are due as well to Stephen Latta and Alex Sager of Broadview Press for their enthusiastic support for the project of this book. Mr. Latta made sure that the various stages of writing, editing, and producing the book progressed efficiently and smoothly. I am thankful also to Tara Lowes of Broadview Press for her meticulous supervision of the production of the book. I would like to acknowledge the contributions of the hundreds of students I taught logic to over the years. They spent many weeks learning the material, asked penetrating questions during class, and wrote comments and made suggestions in their student evaluations. The quality of the coverage in this book owes greatly to the contributions of the students who learned logic from earlier versions of this material. Finally, my deepest gratitude is to my wife, Connie, for her optimistic support and loving care during the difficult time of my illness and for her patience and encouragement during the six-year writing process.

Chapter One

Informal Logic

1.1 Taxonomy of Arguments

1.1.1 **Logic** is the study of reasoning. Since reasoning is carried out by means of arguments, logic may be defined as the science of argument. An argument in this context is not a quarrel; it is the structural unit of any edifice of sentences, statements, propositions, or beliefs that are held together by reason. Precisely speaking, an **argument** is a nonempty collection of **declarative sentences**: one of these sentences is the conclusion of the argument and the others are its premises. An argument may be represented schematically as follows:

$$\frac{\Gamma}{X} \quad \text{or} \quad \Gamma/X$$

where Γ is the set of the **premises**, X is the **conclusion**, and the line, '—' or '/', represents the inferential relation, that is, the **logical link**, between Γ and X.

1.1.1:C COMMENTARY ON 1.1.1

1.1.1:C1 Logic is the "science" of argument not as biology is the science of life (logic is not a natural science), but in the sense that it is an analytic and systematic study. A declarative sentence may be defined as a sentence such that it is sensible to consider it true or false, or to inquire whether it is true or false. The sentences 'Johannes loved Clara' and 'Earth is a bright star' are declarative because it is sensible to inquire whether or not they are true. On the other hand, interrogative and imperative sentences, such as 'What is your name?' and 'Don't hold my hand', are not declarative because it is meaningless to ask whether they are true or false.

This is a *semantical* definition of 'declarative sentence.' The word 'semantical' when attached to a concept indicates that the concept has to do with the *meanings* of certain expressions. Defining declarative sentence in terms of truth and falsity is semantical because the notion of truth depends on the meaning of the sentence, that is, it depends on the statement or assertion that the sentence makes. Later in this book, we will define three *formal* languages: the language of Term Logic (Chapter Two), the language of Sentence Logic (Chapter Three), and the language of Predicate Logic (Chapter Four). These languages are formal because their declarative sentences are defined not semantically but *syntactically* without invoking any semantical concepts. The word 'syntax' refers to the set of rules that allow us to form *grammatical* sentences from a given vocabulary. The syntax of formal languages relies solely on the *forms* or the *structures* of the linguistic expressions. The form or structure of a linguistic expression is the manner in which words of various types are arranged. In every formal language, the class of the declarative sentences is characterized by means of the forms of certain linguistic expressions.

1.1.1:C2 The requirement that the collection be nonempty is made redundant by specifying that one of its members is the conclusion of the argument. Saying that the remaining members (if any) are the premises indicates that we are requiring an argument to have only one conclusion. Our definition of an argument does, however, permit the set of the premises to be empty, finite, or infinite. We explain below why these conditions do not make our definition too permissive or too restrictive.

There are arguments with no premises. We will later encounter many of them. But it suffices to say here that every time we are told that a certain claim is self-evident we are being presented with an argument without premises. For many centuries people believed that the axioms of Euclidean geometry, such as the axiom that there is exactly one line that passes through two points, were true without proof—they were, people believed, self-evident. It is not hard to think of a sentence whose truth is so obvious that most of us would say that it is self-evident. For instance, consider the sentence 'the President of France is a president'; if one hears this sentence and asks us to prove it, we most likely would think that he does not understand its meaning, we most likely would say to him that no proof is needed for this sentence, if you understand the words in this sentence, you would know for sure that it is true; in other words, we believe that this sentence is self-evident. Now you might think that these are not very interesting, or even not very good, arguments. They are, nevertheless, arguments. We did not define interesting or good arguments, but only arguments. We want our definition to include as many arguments as possible; it is not a good practice to rule out some arguments by definition—we might miss out on something.

Are there arguments with infinitely many premises? It is harder to imagine this case, but the answer is affirmative. Consider the following infinite list of sentences, which collectively offer an English translation of the Arabic word '*thalathah*':

Thalathah is a positive integer.
Thalathah is not one.
Thalathah is not two.
Thalathah is not four.
Thalathah is not five.
⋮

Thalathah can only be three if the list above contains, for every positive integer n that is not three, a sentence asserting that *thalathah* is not n. This list may be considered the set of premises of an argument whose conclusion is '*Thalathah* is three'. This argument seems to be a perfectly good argument. Someone who does not know what *thalathah* means in English can learn its English translation on the basis of this argument.

Let us now consider the restriction our definition places on an argument, namely, that an argument must have one conclusion. Surely there are cases in which several conclusions follow from the same set of premises. For example, all the theorems of geometry are ultimately derived from the same set of axioms. These axioms are the premises whose conclusions are the theorems of geometry. Our definition views an argument with several conclusions as several arguments each of which has only one conclusion. Two different arguments may share the same set of premises and have different conclusions.

1.1.1:C3 There is, however, one genuinely permissive aspect of our definition. Any set consisting of premises and a conclusion counts as an argument; our definition does not require that there should be any **relevance** between the premises and the conclusion. Write any list of premises, follow it with a conclusion, and you've got an argument. Here is an example of one such "crazy" argument.

Humans have arms
Birds have wings
———————————
Snow is white

So, why do we permit this kind of set to count as an argument? Because it is quite difficult to give a satisfactory definition of the notion of relevance for arguments in general. At any rate, the situation is not as bad as it might appear. Most of the offending arguments will be classified, once we introduce the concept of validity, as "bad" arguments. Our troubles will not end there, however. As we will see, there remain odd consequences to our choice of leaving relevance out of the definition of a valid argument.

1.1.2 We distinguish two types of logical link: conclusive and probabilistic.

1.1.2a The link of an argument is **conclusive** if and only if it is *logically impossible* that the premises are all true and the conclusion is false.

1.1.2b The link of an argument is **probabilistic** if and only if it is *logically possible* that all the premises are true and the conclusion is false, and the probability of the conclusion being true given the truth of the premises is greater than the probability of the conclusion being true on its own.

1.1.2:C COMMENTARY ON 1.1.2

1.1.2:C1 Here is our first encounter with the notion of logical possibility. We will spend much of this book studying proposed characterizations of this notion. At this early stage we note two things. First, to say that something is logically possible is not to imply that its truth or its existence is probable. Second, to say that something is logically possible is to imply that if we suppose that it is true or that it exists, we will not arrive at a contradiction—that is, a sentence of the form 'Z and not-Z' where 'Z' is replaced by some declarative sentence. For instance, it is logically possible that there is a flying horse. There are no flying horses in our world, but supposing that such a horse exists does not lead to a contradiction. On the other hand, it is logically impossible that four is an odd number; for if it were true, it would follow that four is not even; but since four is divisible by two, it is even after all; hence, four is an even number and four is not an even number, which is a contradiction.

1.1.2:C2 A consequence of our discussion above is that a conclusive link is **truth-preserving**: if the premises are true, the conclusion must be true as well. In other words, the truth of the premises *guarantees* the truth of the conclusion. This is so, because if the link is conclusive then it is logically impossible for the premises to be all true and the conclusion false; that is, supposing that the premises are true and the conclusion is false leads to a contradiction.

Before giving examples of arguments whose links are conclusive, we need to note that the conclusiveness of a link implies a *conditional*, a relation between the truth of the premises and the truth of the conclusion. It is possible that the premises and conclusion of an argument whose link is conclusive are all true; that the premises are false and the conclusion is true; that the premises and conclusion are all false; and that some of the premises are true and some are false, while the conclusion is true, or the conclusion is false. What is not possible, if the link is conclusive, is that the premises are all true and the conclusion is false. Here are a few examples of arguments with conclusive links.

(a) True premises and true conclusion:

All whales are mammals.
All mammals are animals.

All whales are animals.

All men are mortal.
Socrates is a man.

Socrates is mortal.

George Washington was the first President of the United States.
George Washington was born in Virginia.

The first President of the United States was born in Virginia.

(b) False premises and true conclusion:

All Lehigh Students who take Symbolic Logic take Theory of Knowledge.
All Lehigh students who take Theory of Knowledge study Sentence Logic.

All Lehigh students who take Symbolic Logic study Sentence Logic.

Mozart and Verdi were compatriots.
Mozart was a citizen of Italy only.

Verdi was a citizen of Italy.

Thomas Jefferson was the first President of the United States.
Thomas Jefferson was the military leader of the American revolution.

The first President of the United States was the military leader of the American revolution.

(c) False premises and false conclusion:

All mothers are biological mothers.
All biological mothers are protective of their offspring.

All mothers are protective of their offspring.

Mozart and Bizet were compatriots.
Mozart was a citizen of Italy only.

Bizet was a citizen of Italy.

Barack Obama graduated from Yale Law School.
No United States President graduated from Yale Law School.

Barack Obama was not a United States President.

(d) Some true and some false premises and true or false conclusion:

All American Presidents before 2008 were men.
All men are white.

All American Presidents before 2008 were white.

All American Presidents before 2008 were men.
All men are Greek.

All American Presidents before 2008 were Greek.

1.1.2:C3 A probabilistic link is not conclusive. If the link of an argument is probabilistic, then supposing that the premises are all true and the conclusion is false does not lead to a contradiction. This follows from saying that it is logically possible that all the premises are true and the conclusion is false. Recall that being logically possible does not imply being probable. It might be extremely improbable that the premises of a certain argument are all true and the conclusion is false, and yet the case is logically possible. Consider the following three arguments.

Peter was in a car accident.
Peter did not have internal bleeding before the accident.
Peter had internal bleeding immediately after the accident.

Peter's internal bleeding was caused by the car accident.

In a random sample of freshmen at a certain college, 60% took a math course in their first semester.

60% of the freshmen at that college took a math course in their first semester.

John and Mary usually take similar courses.
John had an exam this morning.

Mary also had an exam this morning.

All of these arguments have probabilistic links of various strengths. But in every case it is logically possible that the premises are all true and the conclusion is false. In the first argument, given the truth of the premises, it would be highly improbable that Peter's internal bleeding was caused by something other than the car accident; nevertheless, it is logically possible. The second argument is a standard statistical argument. In this case it is more plausible for the premise to be true and the conclusion false. After all, the sample might not be representative of the whole freshman class; it is a random sample but it might be too small to reflect accurately the composition of the whole population (i.e., the freshman class). The link of the third argument is weak. It is, in fact, quite feasible for the premises to be true and the conclusion false. Just because John and Mary *usually* take *similar* classes, it does not seem likely that they often have exams at the same time.

1.1.3 Since conclusiveness is not a matter of degree (a link is either conclusive or inconclusive), arguments are divided into two *mutually exclusive* and *collectively exhaustive* classes: arguments whose links are conclusive and arguments whose links are inconclusive. In other words, every argument belongs to one of these classes (they are collectively exhaustive) and no argument belongs to both of them (they are mutually exclusive). An argument whose link is conclusive is said to be **deductively valid**. If its link fails to be conclusive, it is called **deductively invalid**. Therefore, every argument is either deductively valid or deductively invalid but not both.

1.1.3:C Commentary on 1.1.3

The link of an argument is inconclusive if it is not conclusive, that is, it is logically possible for all the premises to be true and the conclusion false. We previously called such a link probabilistic if the following probability condition is satisfied: the probability of the conclusion being true given the truth of the premises is greater than the probability of the conclusion being true on its own. It is a matter of presentation. If an argument is presented with the intention that it is deductively valid but its link fails to be conclusive, we refer to it as deductively invalid and to its link as inconclusive. But if an argument is presented with the intention that its link is probabilistic, we usually do not describe it as deductively invalid, even though, precisely speaking, it is deductively invalid if its link is indeed probabilistic.

Since all of the arguments listed in 1.1.2:C2 have conclusive links, they are deductively valid. The arguments listed in 1.1.2:C3 have probabilistic links, so they are, technically speaking, deductively invalid. Here are three more examples of deductively invalid arguments.

All whales are mammals.
Some mammals fly.

Some whales fly.

If you study hard, you will pass the exam.
You passed the exam.

You studied hard.

Thomas Jefferson was not the first President of the United States.
Thomas Jefferson was not born in 1732.

The first President of the United States was born in 1732.

1.1.4 An argument is **deductively sound** if and only if it is deductively valid and all of its premises are true. Deductive soundness, like deductive validity, is not a matter of degree: an argument is either deductively sound or **deductively unsound**. To see why this is so, note that for an argument to be deductively sound it must satisfy two conditions: (1) it must be deductively valid and (2) all of its premises must be true. Each of these properties, deductive validity and truth, is *bivalent*, that is, they do not permit shades and degrees: an argument is either deductively valid or deductively invalid, and a declarative sentence is either true or false.[1] Hence, deductive soundness is a bivalent property as well.

1.1.4:C COMMENTARY ON 1.1.4

The important thing to note about deductively sound arguments is that their conclusions are true. To see this, let Γ/X be any deductively sound argument, where Γ is its set of premises and X is its conclusion. Since Γ/X is deductively sound, it is deductively valid and all the members of Γ are true. To say that Γ/X is deductively valid is to say that it is logically impossible for Γ's members to be all true and X false. This equivalently means that if the members of Γ are true, then X must be true as well. Since, according to the second condition of deductive soundness, the members of Γ are true, X is also true. The diagram below depicts this reasoning.

---

1 For a different point of view see 1.2.2:C1, where examples of declarative sentences that might be non-bivalent are discussed.

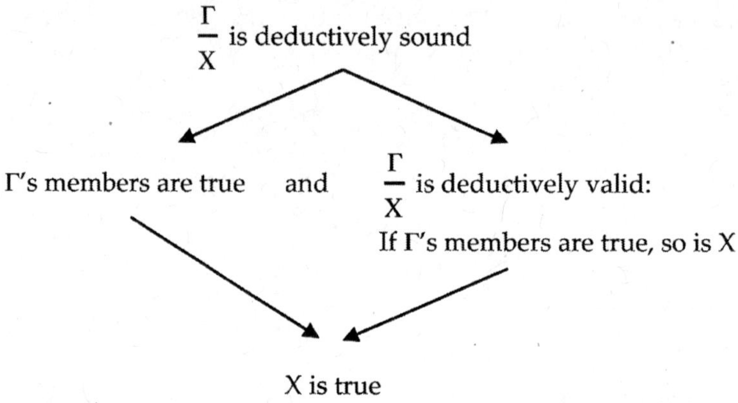

The arguments listed in 1.1.2:C2(a) are deductively sound—they are deductively valid and their premises are true.

1.1.5 **Deductive logic** is the branch of logical theory whose primary component is the study of deductively valid arguments. The following are some of the central questions of deductive logic: What is the nature of deductive validity? What are the natures of other related notions such as logical truth, logical equivalence, and consistency? How do we test for deductive validity and other logical features? Questions about the soundness of deductively valid arguments almost always fall outside the scope of logic, because they are ultimately about the truth and falsity of claims that concern any number of fields.

1.1.5:C COMMENTARY ON 1.1.5

Questions about the truth and falsity of the premises of the third arguments in 1.1.2:C2(a)–(c) belong to the field of United States history. They are not questions of logical theory. Here are two more examples.

√2 is not the ratio of any non-zero integers.
Every rational number that is not zero is the ratio of two non-zero integers.

√2 is not a rational number.

Mozart composed the opera *Don Giovanni* in 1786.
Mozart composed the opera *The Marriage of Figaro* in 1787.

Mozart composed *Don Giovanni* before *The Marriage of Figaro*.

Each of these arguments is deductively valid. To determine whether they are deductively sound or not, we need to know some mathematics and some music history. As it happens, the first argument is deductively sound and the second is not (the two dates are transposed). The main observation here is that the determination of the truth or falsity of these premises does not belong to logical theory but to other fields (in these examples to mathematics and music history).

1.1.6 Sometimes we use the terms 'valid', 'invalid', 'sound', 'unsound', 'validity', and 'soundness' without the modifiers 'deductively' and 'deductive'. An argument is usually described *only* as **deductive** if its link is intended, suggested, or claimed to be conclusive.

1.1.6:C Commentary on 1.1.6

The diagram below describes the taxonomy of deductive arguments. We use the terminology introduced above.

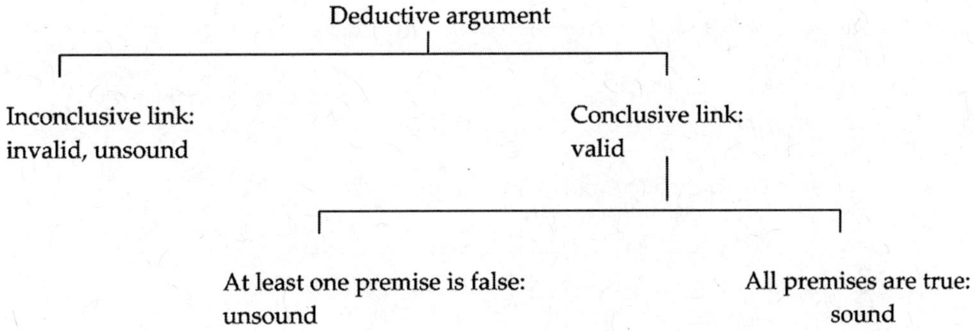

1.1.7 The strength of a probabilistic link, unlike a conclusive link, is a matter of degree. Indeed, it may assume any value between zero and one, excluding zero and one. This feature of **probabilistic arguments** makes it quite difficult and hardly useful to categorize such arguments according to the strength of their links. Instead, probabilistic arguments are classified according to certain general features of the relation between the premises and conclusion. We distinguish two major types of probabilistic argument: abductive and inductive.

1.1.7a An **abductive argument** is *diagnostic*: its conclusion offers an explanation for some or all of the facts reported in the premises. Most often this explanation is causal, that is, it reports causal fact(s) that are responsible for some or all of the effect(s) described in the premises. An abductive inference is commonly referred to as an *inference to the best explanation.*

1.1.7b An **inductive argument** is a probabilistic argument that is not abductive.

1.1.7:C Commentary on 1.1.7

The first argument mentioned in 1.1.2:C3 is abductive, the second and third are inductive.[1] Here are four more examples: the first two are abductive and the last two are inductive.

1 It is possible to think of the second argument mentioned in 1.1.2:C3 as abductive. The conclusion of the argument may be considered as explaining its premise; that is, the fact that 60% of freshmen took a math course in their first semester explains why 60% of the random sample of freshmen took a math course in their first semester.

The dog did not bark the night of the crime.
The dog usually barks at all strangers.
The dog does not bark at the members of the household (family and staff).
All members of the household, except the butler, had alibis during the night of the crime.

The butler did it.

A severe hailstorm struck Charles City, Iowa.
The new cars on the lots of the auto dealerships in Charles City were not damaged before the storm.
Many new cars on the lots of the auto dealerships in Charles City were found to be damaged after the storm.

The hailstorm caused the damage to those new cars.

Governments often lie to their people during time of war.
Our country is engaged in a war now.

Some of the things our government is telling us are lies.

Every student we met so far today had a nametag.
We met many students today.

Every student we will meet today will have a nametag.

Note that we are not claiming that the links of these arguments are probabilistically strong—some of them are and some of them are not. Nevertheless, they are probabilistic links. The conclusions of the first two arguments are causal explanations of the effects described by some of the premises. Hence, these arguments are abductive. It is clear that the last two arguments are not abductive. They are, in fact, *predictive*: their conclusions are predictions based on their premises. Since they are probabilistic arguments and not abductive, they are, according to our definition, inductive. It is important to note that the word 'inductive' is often used to refer to all probabilistic arguments. One often reads in logic books that an argument is either deductive or inductive. We use 'inductive' here in a more restricted sense.

1.1.8 **Probabilistic logic** is the branch of logical theory that is primarily concerned with the study of probabilistic arguments. Its central questions are the following: What is the nature of probabilistic inference? What are the different types of probabilistic arguments? How is the strength of a probabilistic link to be evaluated? Probabilistic logic is often referred to as **inductive logic**.

1.1.9 An argument is said to be **cogent** if it makes a convincing case for its conclusion. Thus its premises must be *plausible* (i.e., believable) and its link must be *sufficiently strong*. How much inferential strength is sufficient depends on the nature and context of the argument. The link of a cogent *deductive* argument must be conclusive. A *probabilistic* argument with a fairly (but not very) strong link and premises that are believed to be true on the basis of strong evidence might not be considered cogent if it is presented to convict a person accused of murder, where the standard for the strength of the link is that the evidence makes the conclusion believable *beyond reasonable doubt*; whereas it might be accepted as cogent in a civil case, where the standard is weaker: the evidence makes the conclusion more probable than its opposite. We give the following general definitions.

1.1.9a A **cogent deductive argument** is a deductively valid argument whose premises are plausible. Since plausibility is a matter of degree, the cogency of a deductively valid argument is a matter of degree as well: how cogent such an argument is depends on how plausible its premises are.

1.1.9b A **cogent probabilistic argument** is one whose link is probabilistically strong and whose premises are plausible. Hence the cogency of any such argument depends on two parameters: the probabilistic strength of its link and the degrees of plausibility of its premises. Of course, both of these factors are matters of degree.

1.1.9:C Commentary on 1.1.9

Is the cogency of a deductive argument the same as its soundness? The answer is negative. There are deductively sound arguments that are not cogent and there are cogent deductive arguments that are unsound. This divergence between the cogency and soundness of a deductive argument lies in the difference between plausibility and truth (recall that both a cogent deductive argument and a deductively sound argument are deductively valid). A sentence is true or false independent of our belief about its truth status. So we might be presented with a sentence against which we have much evidence and for which we have none. This sentence would be quite implausible and it would be rational for us to reject it. However, the sentence might be, unbeknownst to us, true. We can also imagine the converse of this situation: a plausible sentence that is, unbeknownst to us, false.

As an example, suppose that a mad genius proposed in the mid-nineteenth century that space and time are one continuum, that space is curved by the force of gravitational fields, and that space is not Euclidean. He offered no evidence for his claims except the following argument in support of his second claim.

Space curves around a massive star as a result of its gravitational field.
If space is curved around a star and a beam of light passes near the star, it will bend at a certain angle.
The sun is a massive star.

If a light beam passes near the sun, it will bend at a certain angle.

Suppose further that in the mid-nineteenth century there was nothing available that would allow the scientists to evaluate the plausibility of the first two premises of this argument or to test its conclusion. The scientists then believed that Newtonian Gravitational Theory (NGT) was true. They had a large body of evidence supporting this theory. NGT implies that space is separate from time, that space is flat and not affected by any gravitational field, and that it is Euclidean. This contradicts the mad genius's hypothesis. It would be irrational for the scientists in the mid-nineteenth century to abandon NGT for a totally unsupported hypothesis and an argument that lacked any evidence. The argument is valid but its premises were implausible at that time; hence the argument was then non-cogent.

The context changed in the twentieth century. The General Theory of Relativity (GTR), which accords with the mad genius's hypothesis and his argument, was confirmed by many decisive tests including the observation and measurement of the angle of the bending of light passing near the sun. The hypothesis became plausible and the argument cogent. Of course, further discoveries might change the status of the hypothesis and argument. Let us suppose that GTR is more than a well confirmed theory, but is, in fact, true. This supposition implies that the mad genius's hypothesis is true and that his argument is sound. Here, we have an example of a deductively sound argument that was non-cogent in the mid-nineteenth century and became cogent in the twentieth century.

Cogency (and plausibility), therefore, is context-sensitive because it depends on the availability and type of evidence. If the link of an argument is probabilistically strong or is deductively valid, then the cogency of the argument depends on the plausibility of its premises: the more plausible its premises, the more cogent the argument, while the less plausible the premises, the less cogent the argument. Whether a premise is plausible depends on whether we have evidence for it, that is, a good reason to believe it, and the degree of its plausibility depends on the strength of this evidence. Deductive soundness, on the other hand, is not context-sensitive. If an argument is unsound, no amount of evidence can make it sound, and conversely, if it is sound, then there is no context in which it will become unsound.

1.2 Classical Deductive Logic and the Notion of Logical Possibility

1.2.1 By a **logical possibility** we roughly mean any consistent story. In other words, it is a story that describes a real or fictitious situation.[1] Thus the story must not contain an explicit or implicit contradiction, that is, a sentence of the form 'Z and not-Z', where Z is some declarative sentence. The plausibility of the story is irrelevant: if it is consistent (no matter how implausible it may be), then it is a logical possibility.

1.2.1:C COMMENTARY ON 1.2.1

1.2.1:C1 We previously defined deductive validity in terms of the conclusiveness of the link, and we defined conclusiveness in terms of logical possibility. Later we will define deductive validity and other logical concepts, such as logical truth and consistency, directly in terms of logical possibility. It seems that we have already given an adequate definition of logical possibility: it is a story that does not contain explicitly or implicitly a sentence of the form 'Z and not-Z'. Is our task accomplished? The answer is "No"; in fact, we have a long way to go. We will spend most of this book studying three proposals for characterizing the notion of logical possibility. The problem with the definition we gave in 1.2.1 lies with the word 'implicitly'. If the contradiction is explicit, it can be identified in the story: we simply review the story and look for a sentence of the form 'Z and not-Z'. But what does it mean to say that the story contains an *implicit* contradiction? To answer this question let us consider an example.

---

1 This (rough) definition presupposes that real and fictitious situations contain no contradictions.

Here is a story that is based on an example originally described by Bertrand Russell (1872–1970). Imagine a small town called Russell Town, in which lives a man who is the barber of the town. This barber satisfies two conditions: (1) he shaves *all* the men living in Russell Town who do not shave themselves, and (2) he shaves *only* the men living in Russell Town who do not shave themselves. The first condition means that every man living in Russell Town who does not shave himself is shaved by the barber. The second condition means that every person whom the barber shaves is a man living in Russell Town who does not shave himself. Now, does this barber shave himself or not? (Note that there is no third option.) Suppose that he shaves himself; it follows that he is shaved by the barber of the town (he *is* the barber). Therefore, the second condition applies to him: since he is shaved by the barber, he must be a man living in Russell Town who does not shave himself. Thus we obtain a contradiction: the barber shaves himself and he does not shave himself. Let us now suppose the second case, namely that the barber does not shave himself. Since the barber is a man living in Russell Town who does not shave himself, he, according to the first condition, must be shaved by the barber of the town. But he is the barber; this means that he shaves himself after all. So in this case too we obtain a contradiction: the barber shaves himself and he does not shave himself. Since there are only these two cases (the barber either shaves himself or he does not), our story leads to a contradiction.

The way our story led to a contradiction was by deductive reasoning. We took the story as our set of premises and we reasoned deductively from it to a contradiction. We presented, in other words, a deductively valid argument whose premises are the statements of the story and whose conclusion is the contradiction at which we arrived. Here is the argument.

In Russell Town lives a man who is the barber of the town.
This barber shaves all the men living in Russell Town who do not shave themselves.
This barber shaves only the men living in Russell Town who do not shave themselves.

This barber shaves himself and he does not shave himself.

We have now an answer to our original question. A story contains an implicit, but no explicit, contradiction if and only if there is a deductively valid argument whose premises are the sentences of the story and its conclusion is a sentence of the form 'Z and not-Z' and none of the sentences of the story has this form. This definition is problematic; it makes the definition of logical possibility, which we gave in 1.2.1, circular. The circle runs as follows: we defined logical possibility as a consistent story; we defined a consistent story as a story that does not contain an explicit or implicit contradiction; we defined a story that contains an implicit contradiction in terms of deductive validity; we defined deductive validity in 1.1.3 in terms of the conclusiveness of the link; and in 1.1.2a we defined a conclusive link in terms of logical possibility. Thus the definition we gave in 1.2.1 ultimately defines logical possibility in terms of logical possibility. This is why 1.2.1 may be considered as offering a rough explanation of the notion of logical possibility but not a precise definition of it. We will have to work much harder in order to propose definitions of this notion.

1.2.1:C2 There is a very important point to note. Almost always logical possibilities are thought of in relation to some set of natural-language sentences. We are typically given a set Σ of natural-language declarative sentences, and we are interested in **logical possibilities that**

are relevant to Σ. Roughly speaking, a logical possibility p is relevant to Σ if and only if *every* sentence in Σ makes a (true or false) assertion about constituents of p (such as individuals, properties, relations, and states of affairs). Logical possibilities that are relevant to Σ are also referred to as **logical possibilities for** Σ. So, for example, consider this set of sentences: {Pigs fly, Baltimore is in Maryland}. Relevant to this set are a large number of possibilities: each one (perhaps) contains various individuals including pigs, Baltimore, and Maryland, and the property of flying, and the relation of being-located-in; in addition, each possibility specifies a wide variety of states of affairs including that pigs fly (in some possibilities), that pigs don't fly (in other possibilities), and that Baltimore is or isn't in Maryland. Later we shall discuss logical possibilities in much more detail.

In all the logical possibilities that are relevant to Σ, the *language of* Σ is held fixed. In other words, the meanings of the linguistic expressions that occur in the sentences in Σ are not allowed to vary from one logical possibility for Σ to another. For example, if the term 'bachelor' occurs in one of the sentences in Σ and means unmarried adult male, then in every logical possibility for Σ 'bachelor' means unmarried adult male; it cannot mean unmarried adult male in one logical possibility for Σ and, say, unmarried adult female in another logical possibility for Σ. We can express this point compactly in the following slogan: *the language of* Σ *is invariant across all the logical possibilities for* Σ. This issue is more complex and philosophically involved than what this slogan conveys. We will discuss it with some detail in Subsection 4.1:C4 after we gain some experience working with logical possibilities and symbolic representations of them.

1.2.2 We use the expression '**truth value**' to mean *true* or *false*, which we denote as 'T' and 'F', respectively. To say that the truth value of X is true (or false) is simply to say that X is true (or false). We say that a sentence has a **classical truth value** in some logical possibility if and only if the sentence is either true or false, but not both, in that logical possibility. A **bivalent sentence** is a sentence that has a classical truth value in every logical possibility that is relevant to this sentence. A **bivalent language** is a language of which every declarative sentence is bivalent. The sentences and languages with which we deal throughout this book are assumed to be bivalent.

1.2.2:C COMMENTARY ON 1.2.2

1.2.2:C1 Are there non-bivalent languages? This question reduces to the following question: Are there declarative sentences that are not bivalent? Let us focus on the English language. Whether English contains non-bivalent declarative sentences or not is debatable. There are those who insist that every meaningful declarative sentence is bivalent and there are others who argue that there are non-bivalent meaningful declarative sentences. The following sentences might be offered as examples of declarative sentences that are not bivalent.

(a) Pat gave up smoking.
(b) The present king of France is bald.
(c) The third sentence in this list is false.

Consider first the sentence (a). If we say that it is true, then we are asserting that Pat gave up smoking, that is, she used to smoke and then she stopped smoking. If we say the sentence is false, then we are asserting that Pat did not give up smoking. In this case it seems that we are asserting that she used to smoke and continues to smoke. Now suppose that in the actual world Pat never smoked. It seems intuitive to say that in our world the sentence 'Pat gave up smoking' is neither true nor false. This sentence presupposes, one might argue, that Pat has smoked in her life: if the sentence is true, the presupposition is true and Pat stopped smoking, and if the sentence is false, the presupposition is also true and Pat continues to smoke. If this analysis is correct, then in every logical possibility for the sentence 'Pat gave up smoking' in which the presupposition is false (i.e., Pat has never smoked in her life) the sentence has no classical truth value.[1]

Now consider the second sentence, (b). This sentence, too, seems to have a presupposition, namely, that France at present has a king. If the sentence is true, then the present king of France is bald; that is, France at present has a king and this king is bald. If it is false, then the present king of France is not bald, which seems to assert that France at present has a king and this king is not bald. It follows that if this sentence is true or false, its presupposition—that France at present has a king—must be true. Assuming this analysis is correct, we conclude that in every logical possibility for the sentence 'The present king of France is bald' in which the presupposition is false, that is, France presently has no king, the sentence has no classical truth value.

The last sentence, (c), is a version of the famous liar paradox. The sentence (c) asserts that the third sentence in the list is false. Since (c) is itself the third sentence in the list, (c) says of itself that *it* is false. If (c) is bivalent, then it is either true or false in every logical possibility, including our actual world, that is relevant to it. But in our world (c) refers to itself. If (c) is true, then what it asserts must be the case. Since it asserts of itself that it is false, it must be false. So if (c) is true, it is false. Suppose now that (c) is false. Since this is what (c) asserts (namely, that

1 Assume that a sentence S has a presupposition P. By definition of presupposition, P must be true in order for S to be true and for it to be false. Assume that S is a bivalent sentence. Let q be any logical possibility that is relevant to S. Since S is bivalent, S is either true or false in q. Given that P is a presupposition of S, it follows that if S is true in q, P must be true in q; and if S is false in q, P must be true in q as well. We have the following conclusion: if S is a bivalent sentence and P is a presupposition of S, then P must be true in every logical possibility that is relevant to S. According to the explanation of the notion of relevant logical possibility in 1.2.1:C2, if S makes an assertion about some of the constituents of a logical possibility q, P must be true in q. But a typical sentence that has a presupposition does not necessarily meet such a restrictive condition. For instance, if the sentence 'Pat gave up smoking' has the presupposition that Pat smoked in her life, it is surely not true that Pat must have smoked in her life in order for the sentence 'Pat gave up smoking' to make an assertion about the constituents of our world. We may assume that the presupposition is false in our world; yet we clearly understand what this sentence is saying: it is attributing a certain property—namely, the property "giving up smoking"—to some specific person named 'Pat'. There is no question that Pat and the property of giving up smoking are constituents of our world; this property even has a nonempty extension in the actual world. Hence our world is a logical possibility that is relevant to the sentence 'Pat gave up smoking', even though its presupposition is actually false. The only available option is that this sentence is not bivalent. The general conclusion of this reasoning is that if a sentence has a presupposition that is false in at least one logical possibility that is relevant to this sentence, then it is not bivalent. In fact, we can prove even a stronger conclusion regarding bivalent sentences that have presuppositions. The proof is not difficult but it requires concepts and principles that we do not yet have. However, the conclusion we established here is adequate for showing that in a typical case the presence of a presupposition for a sentence entails that the sentence is not bivalent.

(c) is false), the sentence (c) is true after all. Thus if (c) is false, it is true. It seems that we arrived at the following conclusion: (c) cannot be true and cannot be false. This entails that (c) is not bivalent.[1]

If our reasoning above is correct, then the sentences (a)–(c) are not bivalent. This shows that English is not a bivalent language. Since these sentences can be translated into any natural language, we may conclude that no natural language is bivalent. This poses a difficulty for the claim we made in 1.2.2. We said that the sentences and languages with which we work in this book are assumed to be bivalent. In fact, the *formal* languages we introduce in Chapters Two–Four are designed to be bivalent. But at the moment we are discussing *natural* language. If natural languages are not bivalent, then the language with which we work in this chapter is not bivalent. To reconcile this conclusion with the claim made in 1.2.2, we assume that we deal with the portion of English that is bivalent. In other words, rather than considering the set of all declarative English sentences, we only consider the declarative English sentences that are bivalent. This allows us to maintain our claim in 1.2.2 without insisting that English is a bivalent language.

1.2.2:C2 There is a relatively thorny issue that we need to address. Throughout this book, whether we are dealing with logical possibilities or symbolic representations of them, we always assume that these possibilities are relevant to the given sets of natural-language or formal-language declarative sentences. For example, we defined a bivalent sentence as a sentence that has a classical truth value in every logical possibility *for this sentence* (i.e., *that is relevant to this sentence*); and later we will define a logically true sentence as a sentence that is true in every logical possibility *for* this sentence. But what about the logical possibilities that are not relevant to a sentence X? Does X have a meaning there? Does it have a truth value in such possibilities? And if it does, what is this truth value?

By definition, a logical possibility p is not relevant to a declarative sentence X if and only if the assertion made by X is not about *any* of the constituents of p. Let us illustrate with a simple example. Say X is the sentence 'All ravens are black', and p is the logical possibility whose constituents are six specific people and the relation of "being married to." This relation holds between some of these six individuals. p contains no other constituents. So there are no properties in p, including the properties "being a raven" and "being black." Notice that we did not say that the properties "being a raven" and "being black" have *empty extensions* in p, where an extension of a property is the set that consists of all the things that have this property. Many philosophers believe that saying that a property A has an empty extension in a logical possibility q entails that A is a constituent of q but that nothing in q has A. Other philosophers believe that a property cannot have an empty extension: "properties" whose extensions are empty do not exist. In this book we will encounter both views. The system of Term Logic

1 There are many solutions to the liar paradox. Some of them suppose that a liar sentence (such as (c)) is neither true nor false, others suppose that a liar sentence is both true and false, yet others preserve bivalence for the liar sentences but change the way the concept of truth applies to declarative sentences. (There are other options as well.) Discussing any of these solutions is beyond the scope of this book. For a partial and brief survey of approaches to the liar paradox, see R.L. Kirkham, *Theories of Truth* (MIT Press, 1992, ch. 9); for an approach that preserves bivalence and permits the language to contain liar sentences, see A.M. Yaqub, *The Liar Speaks the Truth: A Defense of the Revision Theory of Truth* (Oxford University Press, 1993).

(Chapter Two) presupposes the second view, and the system of Predicate Logic (Chapter Four) presupposes the first view. In order to satisfy all views, we assume that no property exists in the logical possibility p: p's constituents are only the six people and the relation of marriage, which does not have an empty extension in p. The sentence X is not about anything in p.

Some philosophers say that such a sentence has no truth value, classical or otherwise, in p. They might argue that sentences are not the primary bearers of truth values. Rather, the propositions, or statements, or assertions that are made by sentences are the primary bearers of truth values. Sentences are said to have certain truth values only because the assertions they make have these truth values. If X is not about *anything* in p, then X makes no assertion in p: X is an empty vessel; a vacant linguistic entity. Hence X has no truth value in p. If one believes that there are only two truth values—true and false—then this position entails that X is neither true nor false in p. If one believes that there is a third truth value—say "undetermined"—then X is neither true, nor false, nor undetermined in p. Other philosophers say that all sentences that are not about anything but have the linguistic structures of declarative sentences are simply false. For them, "being false" is a property of a declarative sentence that is not true. Since X appears to be a declarative sentence and is not about anything in p, it must be false there. There are other positions, which we will not discuss. Our point is simply to illustrate the philosophical problem of irrelevant logical possibilities (or, "situations").

We will be agnostic about this issue in this book. We will define all our semantical and logical concepts in terms of relevant logical possibilities or representations of them. What happens in logical possibilities that are irrelevant to the sentences with which we are concerned is of no consequence to our conclusions. But this position is a little delicate. We must handle our semantical and logical concepts and their applications to various sets of sentences with adequate care in order to make sure that we remain within the spheres of relevant logical possibilities. It might be thought that the whole issue could be avoided if we restrict our logical possibilities to those that contain huge numbers and varieties of constituents about which *all* the declarative sentences in our ordinary language make assertions. This, in fact, is a much more difficult position to deal with. It would complicate the notion of logical possibility in a manner that makes it resistant to rigorous theorization. Also, almost always we are only concerned with very small portions of the language. It would be perfectly all right to assume that we are working with *partially interpreted languages*, that is, languages of which only specific parts have semantical contents. This is true whether the language is natural or formal. In this book, we will restrict all our concepts, semantical and logical, to the linguistic portions in which the given sentences are composed, and the spheres of relevant logical possibilities to those possibilities about whose constituents the given sentences make assertions; and we shall remain agnostic about the semantical status of the given sentences in logical possibilities that are irrelevant to them.

1.2.3 The concepts of **deductive validity** and **logical consequence** are equivalent; they are two sides of the same coin. An argument Γ/X is deductively valid if and only if its conclusion, X, is a logical consequence of its set of premises, Γ. We introduce the following notation and terminology.

1.2.3a "$\Gamma \models X$" is read "X is a logical consequence of Γ," or "X logically follows from Γ," or "Γ logically implies X."

1.2.3b "$\Gamma \not\models X$" is read "X is not a logical consequence of Γ," or "X does not logically follow from Γ," or "Γ does not logically imply X."

1.2.4 The following two statements are equivalent ways of defining the notions of deductive validity and logical consequence.

1.2.4a An argument Γ/X is **deductively valid** (or $\Gamma \vDash X$) if and only if in every logical possibility for Γ/X in which the members of Γ are all true X is true as well.

1.2.4b An argument Γ/X is **deductively valid** (or $\Gamma \vDash X$) if and only if there is no logical possibility for Γ/X in which the members of Γ are all true and X is false.

1.2.4:C COMMENTARY ON 1.2.4

We previously gave several examples of deductively valid arguments (see 1.1.2:C2 and 1.1.5:C). Here are two more examples.

All women have mothers.
Johanna is a woman.

Johanna has a mother.

If someone wins, each member of the team receives $100.
Ranah won.

Each member of the team received $100.

1.2.5 An argument Γ/X is **deductively invalid** (or $\Gamma \nvDash X$) if and only if there is a logical possibility for Γ/X in which the members of Γ are all true and X is false.

1.2.5:C COMMENTARY ON 1.2.5

In 1.1.3:C we gave examples of deductively invalid arguments. Here are two additional examples.

Some women have children.
Imtithal is a woman.

Imtithal has children.

If everyone wins, each member of the team receives $100.
Ranah won.

Each member of the team received $100.

1.2.6 The following are equivalent definitions of the notion of logical truth.

1.2.6a A sentence is **logically true** if and only if it is true in every logical possibility for it.

1.2.6b A sentence is **logically true** if and only if there is no logical possibility for it in which it is false.

1.2.6:C COMMENTARY ON 1.2.6

Here are five examples of logically true sentences.

(a) Zero is identical with zero.
(b) If the evening star is Venus and the morning star is Venus, then the evening star is the morning star.
(c) If every blond woman is smart and Connie is a blond woman, then Connie is smart.
(d) No Euclidean square has an obtuse angle.
(e) All bachelors are unmarried.

It is important to note the technical sense of 'logically true' here. Occasionally people say that something is logically true to mean that it is reasonable to believe in its truth. This is not the sense in which 'logically true' is used in this book.

1.2.7 The following are equivalent definitions of the notion of logical falsehood.

1.2.7a A sentence is **logically false** if and only if it is false in every logical possibility for it.
1.2.7b A sentence is **logically false** if and only if there is no logical possibility for it in which it is true.

1.2.7:C COMMENTARY ON 1.2.7

Observe that the negation of a logically true sentence is logically false and the negation of a logically false sentence is logically true. The five sentences below are logically false. They are obtained by negating the sentences (a)–(e) listed in 1.2.6:C.

(a) Zero is not identical with zero.
(b) The evening star is Venus and the morning star is Venus, but the evening star is not the morning star.
(c) Every blond woman is smart and Connie is a blond woman, but Connie is not smart.
(d) There is a Euclidean square that has an obtuse angle.
(e) Some bachelors are married.

As is the case with 'logically true', 'logically false', as used in this book, does not mean that it is reasonable to believe in its falsity.

1.2.8 A sentence is **contingent** if and only if it is true in some logical possibilities for it and false in other logical possibilities for it.

1.2.8:C Commentary on 1.2.8

Almost every sentence is contingent. Note that a contingent sentence is neither logically true nor logically false and that the negation of a contingent sentence is contingent as well. Here are four examples of contingent sentences.

- (a) Mariam attends Arizona State University.
- (b) All ravens are black.
- (c) There are flying horses.
- (d) France is a monarchy.

According to definitions 1.2.6–1.2.8, the classes of logically true, logically false, and contingent sentences form mutually exclusive and collectively exhaustive classes. In other words, *every* declarative sentence must belong to *one and only one* of these three classes.

1.2.9 The statements below are equivalent definitions of the relation of logical equivalence.

- 1.2.9a Two sentences are **logically equivalent** if and only if in every logical possibility for them they have identical truth values.
- 1.2.9b Two sentences are **logically equivalent** if and only if there is no logical possibility for them in which they have different truth values.

1.2.9:C Commentary on 1.2.9

The typical case of logical equivalence is that of contingent sentences. If two contingent sentences are logically equivalent, then they are true together in some relevant logical possibilities and false together in the rest of the logical possibilities that are relevant to them. This strong correlation between their truth values (they have the same truth value in every relevant logical possibility) is an indication that in some sense these two sentences have the same content; they describe the same situation, possibly, from different perspectives. The sentences 'Mary is taller than John' and 'John is shorter than Mary' are equivalent; they describe the same fact from two different perspectives. Here are two pairs of logically equivalent sentences.

- (a) 'Thomas is neither impatient nor intolerant' and 'Thomas is patient and tolerant'.
- (b) 'It is not true that every American citizen can vote' and 'There are American citizens who cannot vote'.

There is a counterintuitive consequence of the definitions 1.2.6, 1.2.7, and 1.2.9: all logically true sentences are logically equivalent to each other and all logically false sentences are logically equivalent to each other. To see this, recall that logically true sentences are true in every logical possibility for them; hence, their truth values are identical in every logical possibility that is relevant to them. This implies, by definition 1.2.9a, that all logically true sentences are logically equivalent to each other. Similarly, all logically false sentences have identical truth values (they are false) in every logical possibility for them; therefore, they are logically equivalent to each other. This is counterintuitive because two logically true sentences can have vastly different contents. Consider any two of the logically true sentences listed in 1.2.6:C. For instance, consider (a), which says that zero is identical with zero, and (d), which says that no Euclidean square has an obtuse angle. Two logically false sentences also might (and often do) have very different contents. So the original idea that logically equivalent sentences, in some sense, have the same content does not apply in the case of logically true sentences and of logically false sentences.

1.2.10 A set of sentences is **consistent** if and only if there is a logical possibility for it in which every member of it is true.

1.2.10:C COMMENTARY ON 1.2.10

A set of sentences is consistent if it contains no logically false sentences and there is no conflict between its members, such as 'Mary is taller than John' and 'John is taller than Mary'. Below are two consistent sets of sentences. (A set may be represented by enclosing its members between two braces.)

(a) {Mary is taller than John, Some horses have wings, 2+2 = 4}.
(b) {The evening star is Venus, France is a monarchy, No mammal flies}.

1.2.11 The statements below are equivalent definitions of the notion of inconsistency.

1.2.11a A set of sentences is **inconsistent** if and only if in every logical possibility for it at least one of its members is false.
1.2.11b A set of sentences is **inconsistent** if and only if there is no logical possibility for it in which all its members are true.

1.2.11:C COMMENTARY ON 1.2.11

A set is inconsistent if it contains at least one logically false sentence or if there is a conflict between its members. Here are four examples of inconsistent sets of sentences.

(a) {Connie is smart, Thomas is tolerant, Zero is not identical with zero}.
(b) {Mary is taller than John, John is taller than Sarah, Sarah is taller than Mary}.
(c) {If Fred didn't arrive on Monday or Tuesday, he won't come, Fred didn't arrive on Monday, Fred didn't arrive on Tuesday, but he will come}.
(d) {No pigs can fly, Porky is a flying pig}.

1.2.12 As is clear from definitions 1.2.4–1.2.11, the notion of **relevant logical possibility** is of fundamental importance in the study of the concepts of classical deductive logic: most of these concepts are defined in terms of logical possibility.[1] Thus one of the central tasks of the clas-

---

1 Not all deductive logical concepts are defined in terms of logical possibility. As we will see in Chapter Five, the concept of demonstrative proof, for example, is not defined in terms of logical possibility.

sical theory of deductive logic may be reduced to giving a rigorous characterization of the notion of relevant logical possibility.

1.2.12:C COMMENTARY ON 1.2.12

The classical theory of deductive logic has two central tasks: one is semantical, that is, it has to do with meaning, and the other is formal, that is, it has to do with the forms of sentences. The central semantical task is to give a rigorous characterization of the concept of relevant logical possibility and a precise definition of the notion "true-in-a-logical-possibility" (given bivalence, no separate definition will be needed for the notion "false-in-a-logical-possibility"). A quick survey of the concepts defined in 1.2.4–1.2.11 shows that these concepts are *reduced* to the notions of relevant logical possibility and "true-in-a-logical-possibility." The central formal task is to give a rigorous characterization of the concept of demonstrative proof and a precise definition of the notion of a declarative sentence.

In the following three chapters we deal with the semantical task and with the second part of the formal task, and in the last chapter we deal with the first part of the formal task. In Chapters Two–Four we introduce three logical systems. They are Term Logic (TL), Sentence Logic (SL), and Predicate Logic (PL). Each one of these systems consists of a syntax and a semantics. The syntax defines a type of formal language and the set of its declarative sentences, and the semantics offers characterizations of the notions of relevant logical possibility and of "true-in-a-logical-possibility." The notion of logical possibility is characterized in TL as the concept of a TL diagram, in SL as the concept of SL truth valuation, and in PL as the concept of PL interpretation. In Chapter Five we discuss the proof theory of PL and SL, giving a characterization of the concept of demonstrative proof and describing the relation between the proof theory of PL and SL on one hand and their semantics on the other.

1.2.13 A logical possibility presented to establish a *negative* claim is usually described as a **counterexample**. For instance, a logical possibility for an argument Γ/X in which the members of Γ are all true and X is false is a counterexample to the argument Γ/X, because such a logical possibility establishes the negative claim that the argument is *not* valid. On the other hand, a logical possibility that establishes a *positive* claim is simply referred to as an **example**. For instance, the same logical possibility that is a counterexample to the argument Γ/X may be presented as an example establishing that the set Γ is consistent. We should emphasize, however, that an example just by itself does not show that an argument is valid.

1.3 Exercises

Notes: All answers must be adequately justified. The solutions to the starred exercises are appended to this chapter.

1.3.1 Determine which of the statements below are true and which are false.

- 1.3.1a* No deductively invalid argument is cogent.
- 1.3.1b All deductively valid arguments establish the truth of their conclusions conclusively.
- 1.3.1c* There are many probabilistic arguments that are vastly more cogent than many deductively valid arguments.
- 1.3.1d It is possible for a deductively sound argument to have a false conclusion.
- 1.3.1e* There are circumstances when we are given a deductively sound argument and yet it is rational for us not to accept the conclusion of the argument.
- 1.3.1f* Many (if not most) of the "deductions" described in Sherlock Holmes mysteries are really abductive arguments.
- 1.3.1g* Mathematical proofs (such as a proof of a theorem in geometry) are *intended* to establish that certain arguments are deductively valid.
- 1.3.1h It is possible for a deductively valid argument whose conclusion is false to have all true premises.
- 1.3.1i Every argument whose premises and conclusion are all true is deductively valid.
- 1.3.1j* The conclusion of a deductively valid argument whose premises are all false must be false as well.
- 1.3.1k* A deductively valid argument cannot be made invalid by adding new premises to it.
- 1.3.1l* A deductively invalid argument cannot be made valid by removing one or more of its premises.
- 1.3.1m No deductively valid argument can have a logically false conclusion.
- 1.3.1n An unsound argument cannot be valid.
- 1.3.1o A consistent set of sentences cannot contain a logically false sentence.
- 1.3.1p An inconsistent set of sentences might contain a logically true sentence.
- 1.3.1q There are inconsistent sets whose members are all contingent sentences.
- 1.3.1r* It is possible for two sentences to be logically equivalent, yet their negations are not logically equivalent.
- 1.3.1s The set of premises of a deductively invalid argument must be consistent.
- 1.3.1t* A deductively invalid argument cannot be made valid by adding new premises to it.
- 1.3.1u It is possible for a deductively valid argument whose premises are all contingent to have a logically false conclusion.

1.3.2 Explain why all logically true sentences are logically equivalent to each other, and all logically false sentences are logically equivalent to each other. Can we make a similar claim about contingent sentences? Why?

1.3.3* We use the symbol '≅' to denote the relation of logical equivalence. Show that for all sentences X and Y, if X ≅ Y, then the set {X, not-Y} is inconsistent.

1.3.4* Explain why the assumption of bivalence is needed in 1.2.4, 1.2.6, 1.2.7, and 1.2.11.

1.3.5 Suppose that P1 and P2 are the premises of a deductively valid argument whose conclusion is C. Let Σ be the set consisting of P1, P2, and not-C. Could Σ be consistent?

1.3.6* In order to establish the theorem below we need to make an important assumption, which, given our approach, is indispensable for establishing many theorems of classical informal logic.[1] We will call this assumption the *Conservative Expansion Principle.*

Conservative Expansion Principle: If p is a logical possibility that is relevant to a set Σ of declarative sentences, and Γ is a set of declarative sentences to which p is not relevant, then it is always *possible* to add new constituents to p such that the *expanded* logical possibility p* is relevant to Γ (and, of course, to Σ) and the truth values of the sentences in Σ are *conserved* in p*, that is, each member of Σ has the same truth value in p and p*.

We say 'possible' because there are cases in which the added constituents alter the truth value of a sentence X. It is not difficult to think of such cases. Let X be the sentence 'Some cats are not brown', and p be a logical possibility that contains cats and the property "being brown." Assume that in p, all cats are brown. So X is false in p. Further assume that p does not contain a cat named 'Rumie' nor the property "being orange." Let Z be the sentence 'Rumie the cat is orange'. We add Rumie the cat and the property "being orange" to p. Now we have an expanded logical possibility p* that has one more cat and one more property. Suppose that the new cat, Rumie, is orange. Hence in p*, X and Z are true. The inclusion of these new constituents altered the truth value of X. In other words, p* is not a conservative expansion of p with respect to X. However, we could have added Rumie the cat and the property "being orange" and made Rumie brown. In this case, the truth value of X would not change, and p* would be a conservative expansion of p with respect to X.

Prove that if all the sentences in a set are logically true, then every logical consequence of the set is logically true as well.

1.3.7 Show that if a sentence X is a logical consequence of some set Σ of sentences, then any sentence that is logically equivalent to X is also a logical consequence of Σ.

1.3.8 Is it possible for a logically true sentence and a contingent sentence to be logically equivalent?

1.3.9* What can you determine about the set of the premises of a deductively valid argument whose conclusion is logically false?

---

1 There are other approaches to classical logic that make no use of this assumption. This assumption is provable when translated into the symbolic systems of classical logic. Unfortunately, the informal version of this assumption cannot be proved. We will have to accept it as true on three grounds: (1) there seems to be no counterexample to it, (2) the principle holds in all the symbolic systems of classical logic, and (3) it is needed to establish many important theorems of classical logic.

1.3.10 Assume that the sentence X is a logical consequence of a *consistent* set Γ.

- 1.3.10a Is it possible for X to be logically false?
- 1.3.10b Is it possible for not-X to be a logical consequence of Γ?
- 1.3.10c If Y is another logical consequence of Γ, must the set {X, Y} be consistent?

1.3.11* Let C be a logical consequence of {P1, P2, P3} and of {P1, P2, not-P3}. Show that C is a logical consequence of {P1, P2}.

1.3.12* A passage (or an essay) that defends or criticizes some position by means of an argument is an **argumentative passage** (or an **argumentative essay**). To **schematize** an argument of a passage is to present its premises and conclusion without the repetition, explanation, commentary, or any other discussion found in the passage. Schematization is an act of interpretation. We have to decide which sentence in the passage is the conclusion, which sentences are the premises, and which parts of the passage are complementary discussions. Also, when schematizing an argument, we frequently paraphrase some or all of its premises and conclusion. An interpretation of an argumentative text must be guided by a principle called the *Principle of Charity*. It is the chief principle for interpreting such texts. It may be stated as follows.

The Principle of Charity: In interpreting an argumentative passage attribute to the passage the strongest possible argument that is supported by textual evidence.

Observe that the principle has two parts: the first instructs us to be generous in interpreting an argumentative passage and the second constrains our interpretation within the bounds of textual evidence. These two parts balance each other. The net result is a charitable interpretation of the passage within the confines of the text.

The excerpt below is an argumentative passage, in which the Scottish philosopher David Hume (1711–76) defends the position that the rules of morality are not derived from reason alone. It is taken from *A Treatise of Human Nature*. Schematize the argument and determine its type.

> Since morals have an influence on the actions and affections, it follows that they cannot be derived from reason; and that because reason alone, as we have already proved, can never have any such influence. Morals excite passions and produce or prevent actions. Reason of itself is utterly impotent in this particular. The rules of morality, therefore, are not conclusions of our reason.
>
> No one, I believe, will deny the justness of this inference; nor is there any other means of evading it than by denying that principle on which it is founded. As long as it is allowed that reason has no influence on our passions and actions, it is in vain to pretend that morality is discovered only by a deduction of reason.

1.3.13 In *A Treatise of Human Nature* David Hume rehearsed, almost immediately after the passage cited in the preceding exercise, another argument, already presented, for the claim that reason alone can have no influence on our passions and actions. A simplified version of that argument is in the passage below. Schematize the argument and determine its type.

> Whatever can be produced or prevented by reason alone is capable of being discovered to be true or false. It is clear, however, that nothing is discovered to be true or false unless

it agrees or disagrees with real relations of ideas or with existing matters of fact. Since our passions and actions are not susceptible to such agreement or disagreement, they cannot be produced or prevented by reason alone.

1.3.14 Schematize the argument in the passage below. Is your schematization deductively valid or probabilistic? Cast your schematization the other way by adding or removing a single *relevant* premise; you may not change any of the original premises.

> Most astronomers and cosmologists believe that the universe had a definite beginning. Several theories, the most successful of which is the Big Bang Theory, were proposed to explain the origin and estimate the age of the universe. Recently Dr. Brauer from the Swiss Institute of Astrophysics indicated in a speech addressed to the Association of European Astronomers that new evidence suggests that the universe is much younger than what was previously believed. Dr. Brauer said that recent studies on the chemical composition of celestial bodies have established that the universe currently is almost entirely made of hydrogen, hence it must be concluded that the universe cannot be as old as was previously believed. He further explained that if the universe were indeed as old as had been thought, then there would have been enough time for almost all of the hydrogen that originally existed in the universe to be converted to helium, because hydrogen, as is well known, is steadily and irreversibly converted to helium throughout the universe. (The story is fiction.)

1.3.15* Schematize the argument in the following passage and determine its type.

> My friend Salim believes that his mind and his brain are two different things. He argues as follows. It is clear that my mind is capable of having beliefs. My brain, on the other hand, is simply a machine (albeit a very complex biological machine). Now, it is obvious that only things that are capable of having beliefs can believe in God. Since nothing can be religious unless it can believe in God, and since nothing that is simply a machine can be religious, my mind isn't the same thing as my brain.

1.3.16 The passage below contains two arguments. Schematize each one of them and determine its type

> In 1845 two astronomers John Adams and Urbain Leverrier independently described the characteristics of a hypothetical planet. They predicted that the seventh planet, Uranus, which was discovered in 1781 by William Herschel, was not the last planet in our solar system, as was widely believed at the time. They conjectured the existence of their hypothetical planet. Their reasoning was as follows. If Uranus is the last planet, and if Newtonian physics is correct, then the theoretically calculated orbit of Uranus describes the motion of the planet correctly. But the theoretically calculated orbit of Uranus was shown to differ significantly from the observational orbit, that is, the orbit derived from observations. Since Newtonian physics was then universally accepted, and since it was clear that the observational orbit of Uranus was the correct one, they concluded that Uranus could not be the last planet. Adams and Leverrier, furthermore, calculated mathematically the

characteristics of a hypothetical planet whose gravitational force would account for the perturbations observed in the motion of Uranus. They argued that if such a planet were assumed to exist beyond Uranus, then the theoretical results could be made to agree with the empirical data about the orbit of Uranus. Thus they suggested that this hypothetical planet actually exists. In 1846 a planet with those characteristics was discovered—Neptune, the eighth planet.

1.3.17 The German mathematician, logician, and philosopher Gottlob Frege (1848–1925) believed that the meanings of linguistic expressions cannot be identified with human ideas. The passage below contains a simplified and incomplete version of his original argument. Schematize the argument of the passage and determine its type.

> If the meaning of an expression, say, 'The ancient philosopher who was sentenced to death', is an idea in one's mind, then the expression has as many meanings as there are ideas associated with it. But an expression, such as 'The ancient philosopher who was sentenced to death', has the same meaning for all those who are capable of understanding it. Now, if an expression has as many meanings as there are ideas associated with it, and yet it has the same meaning for so many people, then either their associated ideas are identical with each other or these ideas are very similar to each other. Of course, they are not identical (each one of them can exist independently of the others). On the other hand, if they are very similar, then there must exist a clear way of making relevant comparisons between them. With some reflection, it can be shown that no such way can be found. Therefore, the meaning of an expression cannot be identified with any associated idea.

Solutions to the Starred Exercises

SOLUTIONS TO 1.3.1

1.3.1a False. Probabilistic arguments are, technically speaking, deductively invalid (see 1.1.2b, 1.1.2:C3, and 1.1.3:C); however, many of them are very cogent. Consider, for example, the following probabilistic argument.

> Heavy smoking and lack of exercise are strongly correlated with heart disease later in life.
> Bob has smoked heavily for the past forty years and never exercised.
>
> ---
>
> Bob is at a higher risk for developing heart disease.

Assuming that we have strong evidence to believe that the second premise is true, we can see that this argument is cogent: it has plausible premises and a probabilistically strong link. Its link, however, is not conclusive. We can devise a relevant logical possibility in which the premises are true and the conclusion is false. Imagine that there is a rare gene, as yet undiscovered, that makes parts of the human body produce a certain chemical that reverses the bad effects of heavy smoking and lack of exercise. Suppose that Bob has this gene and so unbeknownst to him and to his doctor, he is not at a higher risk for developing heart disease. This logical possibility, of course, is quite improbable but, as we previously said, a logical possibility need not be probable at all; its only condition is consistency.

1.3.1c True. Many deductively valid arguments have implausible premises; hence they are not cogent. On the other hand, many probabilistic arguments have plausible premises and probabilistically strong links; hence they are cogent. (See 1.1.9 and 1.1.9:C.)

1.3.1e True. If we do not know that the argument is deductively sound—say, we know that it is deductively valid but we do not know if its premises are true—and the premises of the argument seem to us, on the basis of the available evidence, implausible, then it is rational for us not to accept the conclusion. For an example, see 1.1.9:C.

1.3.1f True. Most frequently in these mysteries the inference is from observed effects to their most likely cause—an inference to the best explanation (see 1.1.7a and 1.1.7:C). Let us consider an example. In *A Case of Identity* Sherlock Holmes was visited by a woman who wanted his services for a certain case. After welcoming her and observing her for a while, he asked her: "Do you not find ... that with your short sight it is a little trying to do so much typewriting?" The woman naturally was very surprised to hear his question. After she left, Holmes explained to Dr. Watson how he arrived at the conclusion that she was shortsighted and a typist. He said:

> As you observe, this woman had plush upon her sleeves, which is a most useful material for showing traces. The double line a little above the wrist, where the typewritist presses against the table, was beautifully defined. The sewing-machine, of the hand type, leaves a similar mark, but only on the left arm, and on the side of it farthest from the thumb, instead of being right across the broadest part, as this was. I then glanced at her face, and observing the dint of a pince-nez at either side of her nose, I ventured a remark upon short sight and typewriting, which seemed to surprise her.

Holmes's conclusion causally explains the observed effects: the double line on the plush of the sleeves and the dint at each side of the nose. The remark about the sewing machine is meant to show that the typist-hypothesis is the most likely causal explanation for the double line on the sleeves. This is clearly an abductive argument.

1.3.1g True. Mathematical reasoning is deductive. A mathematical proof is *intended* to establish that a certain mathematical argument is deductively valid. A mathematical proof is either correct or incorrect. Every step of a correct proof must be a valid inference. If a proof contains a step that is not a valid inference, then the proof is incorrect. A **valid inference** is an inference that is truth-preserving. In other words, a valid inference can be represented as a deductively valid argument: an inference from Γ to X is valid if and only if the argument Γ/X is deductively valid. A valid inference is said to be **truth-preserving** because if a conclusion is validly inferred from a set of premises, then in every relevant logical possibility in which the premises are true the conclusion is true as well. In other words, the truth of the premises is "preserved" through the inference.

As an example of a mathematical proof, consider the following short proof for the claim that it is not possible to divide by zero. This proof is a sequence of inferences that establishes the deductive validity of the argument below. The proof is of a type that is called *indirect proof*. In an indirect proof, we assume the opposite of the desired conclusion and then show through a series of valid inferences that this assumption leads to a contradiction. This demonstration allows us to reject the assumption, and hence conclude the opposite of the assumption, which is the desired conclusion. The assumption of an indirect proof is called the *Reductio Assumption*. It is so called because the technical name for an indirect proof is *Reductio Ad Absurdum* (Latin for "reduction to the absurd").

Mathematical Argument

P1	For any number n, n × 0 = 0.
P2	For all numbers n, m, and k, if n = k and m = k, then n = m.
P3	For all numbers n, m, k, if it is possible to divide by k and n × k = m × k, then n = m.
P4	2 ≠ 4
C	It is not possible to divide by 0.

Proof

1. Reductio Assumption: it is possible to divide by 0.
2. From P1: 2 × 0 = 0 and 4 × 0 = 0.
3. From P2 and 2: 2 × 0 = 4 × 0.
4. From P3, 1, and 3: 2 = 4.
5. From P4 and 4: 2 = 4 and 2 ≠ 4, which is a contradiction.
6. From 1 through 5: since the Reductio Assumption leads to a contradiction, it must be false.
7. From 6: it is not possible to divide by 0.

1.3.1j False. For examples see 1.1.2:C2(b).

1.3.1k True. Let Γ/X be a deductively valid argument. Suppose that Σ is the set of sentences resulting from adding new premises to Γ. Thus Γ is a subset of Σ, that is, every member of Γ is also a member of Σ. We want to show that the argument Σ/X is deductively valid. We will show that in every logical possibility for Σ/X in which the members of Σ are true, X is true as well (see 1.2.4a). So let p be a logical possibility for Σ/X in which the members of Σ are true. Since Γ is a subset of Σ, the members of Γ are also true in p. But Γ/X is deductively valid; it follows, by Definition 1.2.4a, that X is true in p. Hence in every logical possibility for Σ/X in which the members of Σ are true X is true as well. Therefore Σ/X is deductively valid.

There is an odd consequence to our answer above. Assume that Γ/X is a deductively valid argument and that Γ is consistent. Now suppose we add a premise to Γ that makes the new set Σ inconsistent—say, we add the negation of one of the members of Γ. According to the answer we gave above, Σ/X is deductively valid. In fact, an argument with an inconsistent set of premises is always valid, no matter what conclusion it might have. In the language of logical consequence, if Σ is inconsistent, then Σ ⊨ X for every sentence X. To see this, suppose that Σ is inconsistent and X is any sentence. Let us ask: Is there a logical possibility for Σ/X in which all the members of Σ are true and X is false? The answer is, of course, "No"; Σ is inconsistent, which means that there is no logical possibility for Σ in which its members are all true (see

1.2.11b). By Definition 1.2.4b, Σ/X is deductively valid. By similar reasoning, we can show that if a sentence X is logically true, then $\Gamma \vDash X$, where Γ is any set of declarative sentences: if there is a logical possibility for Γ/X in which all the members of Γ are true and X is *false*, then there is a logical possibility for X in which it is false; but, of course, there is no such logical possibility because X is logically true (see 1.2.6b); hence there is no relevant logical possibility in which all the members of Γ are true and X is false; it follows that Γ/X is deductively valid. This odd consequence is an outcome of our choice not to require that the premises must be relevant to the conclusion of an argument as a necessary condition for its being deductively valid. This type of validity, i.e., when the premises are not relevant to the conclusion, is called "vacuous validity." If an argument is vacuously valid, it makes little sense to say that the conclusion logically follows from the premises. But we adopted the expression 'logically follows' as a possible reading for the symbol '$\vDash$' because the standard case is when the set of premises is consistent and the conclusion is contingent. In the standard case, there is always relevance between at least some of the premises and the conclusion. There is a branch of logical theory called *Relevance Logic* (also known as *Relevant Logic*), which specializes in studying the various logical systems that offer characterizations of the notion of relevance. Unfortunately, even a brief introduction to the most basic system of Relevance Logic is beyond the scope of this book.[1]

1.3.1l True. Let Γ/X be a deductively invalid argument. Let Σ be the set of sentences resulting from removing one or more premises from Γ. We will show that the argument Σ/X is deductively invalid. Since Γ/X is deductively invalid, there is a logical possibility p for Γ/X in which Γ's members are true and X is false. Since Σ is a subset of Γ (i.e., every member of Σ is also a member of Γ), the members of Σ are all true in p. Thus there is a relevant logical possibility in which the members of Σ are all true and X is false. Σ/X, therefore, is deductively invalid (see 1.2.5).

1.3.1r False. If two sentences are logically equivalent, their negations must be logically equivalent as well. To see this, let X and Y be any logically equivalent sentences. We will show that their negations have identical truth values in every relevant logical possibility (see 1.2.9a). Let p be any logical possibility for not-X and not-Y (i.e., for the *negations* of X and Y). It is clear that p is relevant to X and Y (if not-X and not-Y make assertions about some or all of the constituents of p, then so do X and Y). Since X and Y are logically equivalent, they have identical truth values in p. Thus they are either both true or both false in p. If they are true in p, their negations are both false in p; if they are false in p, their negations are both true in p. It follows that in every logical possibility for not-X and not-Y, they have identical truth values.

1.3.1t False. Consider the following argument.

All of Professor Lund's Introduction to Logic students this term are smart.
Sarah is taking a class with Professor Lund this term.

Sarah is smart.

---

1 An introduction to Relevance Logic (and to many other systems of non-classical logic) is found in *An Introduction to Non-Classical Logic*, second edition, by Graham Priest (Cambridge University Press, 2008). A book that is devoted to Relevance Logic is *Relevant Logic: A Philosophical Interpretation*, by Edwin D. Mares (Cambridge University Press, 2004).

It is possible that Professor Lund is teaching Introduction to Logic and Introduction to Philosophy this term, Sarah is in Introduction to Philosophy, all the students in Introduction to Logic are smart, and Sarah is not smart. Hence there is a logical possibility in which the premises are true and the conclusion is false; by 1.2.5, the argument is deductively invalid. However, if we add a third premise asserting that Professor Lund is teaching only one class this term, the resulting argument would be deductively valid (see 1.2.4).

SOLUTION TO 1.3.3

Let X and Y be any logically equivalent sentences. We want to show that the set {X, not-Y} is inconsistent. We will show that in every logical possibility for the set {X, not-Y}, at least one member of this set is false (see 1.2.11a). Let p be any logical possibility for {X, not-Y}. It is obvious that p is relevant to X and Y. Since X ≅ Y, either X and Y are both true or both false in p (see 1.2.9a). Suppose that they are true in p; in this case not-Y is false in p. Now suppose that they are false in p; so X is false in p. Hence whether X and Y are true or false in p, at least one member of the set {X, not-Y} is false in p. Since p is arbitrary, we may conclude that in *every* logical possibility for the set {X, not-Y} at least one member of {X, not-Y} is false.

SOLUTION TO 1.3.4

We will consider the cases of 1.2.4, 1.2.6, and 1.2.11. Similar reasoning can be given for 1.2.7.

(a) 1.2.4 offers two equivalent definitions for the notion of deductive validity. 1.2.4a asserts that an argument is valid if and only if in every logical possibility for the argument in which the premises are all true the conclusion is true as well. 1.2.4b asserts that an argument is valid if and only if there is no logical possibility for the argument in which the premises are all true and the conclusion is false. Bivalence is needed to ensure that 1.2.4a and 1.2.4b render the same verdict in every case. If bivalence is not presupposed, the two definitions need not be equivalent. To see this, let Γ be a set of bivalent sentences and X a non-bivalent sentence. Suppose that X is not false in any logical possibility for the argument Γ/X, and that there is a logical possibility for Γ/X in which all the members of Γ are true but X fails to have a classical truth value—it has a third value "undetermined." This entails that not every relevant logical possibility in which the members of Γ are all true makes X true as well. Hence, according to 1.2.4a, Γ/X is not deductively valid. However, since we assumed that there is no logical possibility for the argument Γ/X in which X is false, there can be no counterexample to the argument Γ/X. Recall that a counterexample to an argument is a logical possibility in which all the premises are true and the conclusion is *false*. Therefore, according to definition 1.2.4b, Γ/X is deductively valid. The two definitions produce different verdicts about the logical status of the argument Γ/X.

(b) In 1.2.6 we said that 1.2.6a and 1.2.6b are equivalent. 1.2.6a says that a sentence is logically true if and only if it is true in every logical possibility for it. 1.2.6b says that a sentence is logically true if and only if there is no logical possibility for it in which it is false. Suppose that X is a non-bivalent sentence. By 1.2.2, there is a logical possibility that is relevant to X and in which X has no classical truth value—say, it simply fails to have any truth value or it has a third value "undetermined." Suppose that X is true in some relevant logical possibilities and in the rest of them it has no classical truth value or it has the third value "undetermined"; hence it is never false in any relevant logical possibility. 1.2.6a and 1.2.6b produce different verdicts for X: according to 1.2.6a, X is not logically true because it is not true in *every* relevant logical possibility, and according to 1.2.6b it is logically true because there *is no* relevant logical possibility in which it is false. This divergence shows that 1.2.6a and 1.2.6b need not be equivalent if sentences are permitted to be non-bivalent.

(c) 1.2.11 presents two equivalent definitions of the notion of inconsistency. This equivalence of definitions presupposes bivalence. To see this, assume that Γ is a set of non-bivalent sentences that is inconsistent according to definition 1.2.11b. Thus there is no logical possibility for Γ in which Γ's members are all true. Now suppose that there is a logical possibility for Γ in which every member of Γ fails to have a classical truth value—each member has a third value "undetermined." This means that there is a relevant logical possibility in which no member of Γ is false; according to 1.2.11a, Γ is *not* inconsistent. Therefore if bivalence is not presupposed, 1.2.11a and 1.2.11b need not be equivalent.

Solution to 1.3.6

We can express the statement of this theorem compactly by asserting that only logically true sentences logically follow from logically true sentences. This is an important principle, which is universally true in all systems of classical logic as well as in numerous systems of non-classical logic. We invoke the Conservative Expansion Principle in order to prove this principle. So assume that Γ is a set of logically true sentences and that X is a declarative sentence that logically follows from Γ. We want to show that X is logically true, that is, X is true in every logical possibility for X (see 1.2.6a). Let p be any logical possibility for X. Assume that p is relevant to Γ; hence every member of Γ is true in p because the sentences in Γ are logically true. Given that X is a logical consequence of Γ, it follows that X is true in p as well (see 1.2.4a). Now assume the p is not relevant to Γ. According to the Conservative Expansion Principle, there is a logical possibility p* such that p* is relevant to Γ and X, and X has the same truth value in p and p*. Since the members of Γ are logically true sentences, they are all true in p*. But X logically follows from Γ; hence X is true in p*; this entails that it is also true in p. We conclude that X is true in every logical possibility that is relevant to it. This establishes that X is logically true.

Solution to 1.3.9

The set of premises is inconsistent. We will use an indirect proof to justify our answer; that is, we will assume that the set of the premises is consistent and then derive a contradiction from this assumption. Recall that the assumption of an indirect proof is called the "Reductio Assumption." So let Σ/X be a deductively valid argument whose conclusion, X, is logically false. Assume that Σ is consistent; this is the Reductio Assumption. By 1.2.10, there is a logical possibility p for Σ in which all the members of Σ are true. If p is not relevant to X, we invoke the Conservative Expansion Principle to expand p into a logical possibility p* that is relevant to X and Σ and in which the members of Σ are all true. Since Σ/X is deductively valid, X is true in p* (see 1.2.4a). Given that X is logically false, it follows that X is false in p* (see 1.2.7a). We have, thus, the following contradiction: X is both true and false in p*. We conclude that the Reductio Assumption that Σ is consistent must be false.

SOLUTION TO 1.3.11

Assume that q is a logical possibility for the argument {P1, P2}/C in which P1 and P2 are true. If q is not relevant to P3, by the Conservative Expansion Principle, we expand q into a logical possibility q* that is relevant to P1, P2, C, and P3, and in which P1 and P2 are true and the truth value of C is the same as in q . Either P3 is true in q* or not-P3 is true in q*. Assume P3 is true in q*. Since C is a logical consequence of {P1, P2, P3}, C is also true in q*; hence C is true in q. On the other hand, if not-P3 is true in q*, then since C is a logical consequence of {P1, P2, not-P3}, C is true in q*, which means that it is true in q. We conclude that in every relevant logical possibility in which P1 and P2 are true C is true as well. Therefore, C is a logical consequence of {P1, P2}.

SOLUTION TO 1.3.12

Since the Hume passage is the first passage we will schematize in this book, we will study it with exceptional care. It is not a coincidence that our first passage comes from Hume's writing. Hume is held by many as a master of argumentative style. We stand to learn a few things by closely studying a piece of his writing.

The passage has two parts. In the first part Hume presents the argument of the passage; he then repeats the argument paraphrasing its premises and conclusion. In the second part (the second paragraph) Hume offers an analysis of the argument implying that it is deductively valid. Let us take a closer look.

At the outset of the passage Hume states the first premise of the argument flagging it with 'Since', a standard premise indicator (a premise flag). He follows it with the conclusion of the argument, again, flagging it with a conclusion flag: 'it follows'.[1] He then states the second premise of the argument preceded by another premise flag 'because'. This structure is most likely intentional. It is designed to grab the attention of the reader by creating an "inferential shock": the inference is from an extremely plausible premise (some might say "obviously true") to a surprising conclusion. The first premise reports a claim that is hardly controversial. It asserts that morals influence our actions (behavior) and our affections (emotions). No one can seriously deny that moral convictions have such a role in our lives. One look at the heated passions and devoted actions on both sides of the abortion debate should make the plausibility of this premise obvious. The conclusion, on the other hand, is surprising. It flies in the face of a long and wide tradition in philosophy that bases morality on a foundation of reason. According to this tradition, the rules of morality are derived from reason, and from reason alone. So the inferential shock is this: how could such a loaded conclusion be inferred from such an innocuous and obvious premise? The answer is given by the 'because' clause: 'and that because reason alone, as we have already proved, can never have any such influence'. This is the central premise of the argument and it carries most of the inferential burden. Unlike the first premise, the second premise is far from obvious and requires a defense. This is why Hume said, "as we have already proved," referring the reader to an earlier argument and claiming

1 Hume does not need this flag. A 'Since' sentence has two clauses: the one that follows 'Since' immediately and the one that comes after the comma. The first clause is typically a premise and the second is a conclusion. Hume employs the conclusion indicator 'it follows' to stress the inference that this surprising conclusion does, indeed, follow from this obvious premise.

that this argument constitutes a proof for the second premise. What that argument is and how cogent it is need not concern us here. The important thing to note is that since the first premise is highly plausible and, as we will see later, the link is conclusive, the cogency of Hume's argument rests on the plausibility of the second premise.

After stating the argument of the passage, Hume then repeats it in the usual order: first premise, second premise, and conclusion, paraphrasing the premises and conclusion. This time he flags only the conclusion with 'therefore', a typical conclusion indicator.

In the second paragraph of the passage Hume tells us that the inference is clearly correct and that the only way to deny the conclusion is by denying the second premise ('that principle on which it is founded'); as long as the second premise is accepted, the conclusion must be accepted as well. Throughout this discussion, Hume assumes that the first premise is not open to challenge—everyone would accept it as true. Hume's remarks that denying the conclusion leads to denying the second premise and that accepting the second premise leads to accepting the conclusion show that he believes that the argument is deductively valid. Deductively valid arguments have this characteristic: if we accept the premises, we must accept the conclusion, and if we reject the conclusion, we must reject one or more of the premises, because in a deductively valid argument the truth of the premises guarantees the truth of the conclusion.

Now we give a schematization of the argument.

P1	The rules of morality have an influence on our actions and affections.
P2	Nothing that is derived from reason alone has such influence.
C	The rules of morality are not derived from reason alone.

The paraphrasing of P1 and C are justified by textual evidence. For example, the word 'alone' in our paraphrasing of the conclusion is supported by the word 'only' in Hume's third paraphrasing of his conclusion. Our paraphrasing of P2 requires some explanation. In the passage, Hume states the second premise three times. In each one of them he asserts that reason *itself* cannot have an influence on the actions and affections. We paraphrased P2 as saying that what is *derived from reason* alone cannot have such influence. Stating the second premise as Hume does might create a doubt about the argument's deductive validity. One might claim that while reason itself cannot influence our actions and affections, some of the things derived from reason can. We claim that this is not possible: if reason alone cannot influence our actions and affections, then nothing that is derived from reason alone can have such influence. It would be convenient to say that our claim above is guided by the Principle of Charity; unfortunately, though, it is not supported by any textual evidence found in the passage. Our justification for the claim above is based on historical grounds: the way reason is understood by Hume and his contemporaries. Reason is the faculty of the human mind that generates inferences. To say that reason can or cannot influence our actions and affections is to say that what is derived from reason can or cannot have such influence. If we add this understanding of reason to our available evidence, we can claim that the Principle of Charity instructs us to paraphrase P2 in the way we did, as this paraphrasing removes any doubts about the deductive validity of the argument. To be precise, our *schematization* of the argument of the passage is deductively valid. Since we argued that our schematization is a permissible interpretation of the argument, we may conclude that the argument itself is deductively valid.

SOLUTION TO 1.3.15

We will schematize the argument in the third-person mode (one might wish to use the first-person mode, substituting 'My' for 'Salim's').

P1	Salim's mind is capable of having beliefs.
P2	Salim's brain is simply a machine.
P3	Only things that are capable of having beliefs can believe in God.
P4	Nothing can be religious unless it can believe in God.
P5	Nothing that is simply a machine can be religious.
C	Salim's mind isn't the same thing as his brain.

Is this schematization deductively valid or not? Notice that the answer is not immediately obvious. This example shows the inadequacy of our logical intuition when confronted with more complex arguments than the ones we have considered so far. We need better tools than our logical intuitions to deal with complex arguments. The symbolic systems we will study in the remaining chapters of this book will offer us precise tools to handle a large assortment of deductive arguments.

In spite of its appearance, this schematization is deductively invalid. We need first to paraphrase P3 and P4 to make clear the logical relations reported in them. P3 is of the form 'Only X's can be Y'. This means that all those that can be Y are X's. This can be made obvious by considering a familiar example: only American citizens can vote (call this sentence S1). This is a true sentence, while the sentence that all American citizens can vote is false; for instance, American citizens under the age of eighteen cannot vote. So the former does not mean the later. The following sentence, however, is true: all those who can vote are American citizens (call this sentence S2). With some reflection we can see that S1 and S2 have the same content. So P3 can be paraphrased as asserting that all things that can believe in God are capable of having beliefs.

P4 is of the form 'Nothing can be X unless it can be Y'. This means that all those that can be X can be Y. Again, we consider an intuitively clear example: no one can play Beethoven's *Moonlight Sonata* unless he or she can play the piano (call this sentence S3). Let S4 be the sentence 'All those who can play the piano can play Beethoven's *Moonlight Sonata*' and S5 be the sentence 'All those who can play Beethoven's *Moonlight Sonata* can play the piano'. It is clear that S3 and S5 are true and S4 is false. S3, therefore, cannot mean S4. On the other hand, S3 and S5 have the same content. P4, thus, can be paraphrased as the sentence asserting that all things that can be religious can believe in God.

We now present our schematization of the argument of the passage with P3 and P4 paraphrased as above.

P1	Salim's mind is capable of having beliefs.
P2	Salim's brain is simply a machine.
P3	All things that can believe in God are capable of having beliefs.
P4	All things that can be religious can believe in God.
P5	Nothing that is simply a machine can be religious.
C	Salim's mind isn't the same thing as his brain.

There are several types of logical possibilities that make the premises true and the conclusion false. The diagram below represents one such type of logical possibility, which suffices to demonstrate that the argument above is deductively invalid (see 1.2.5).

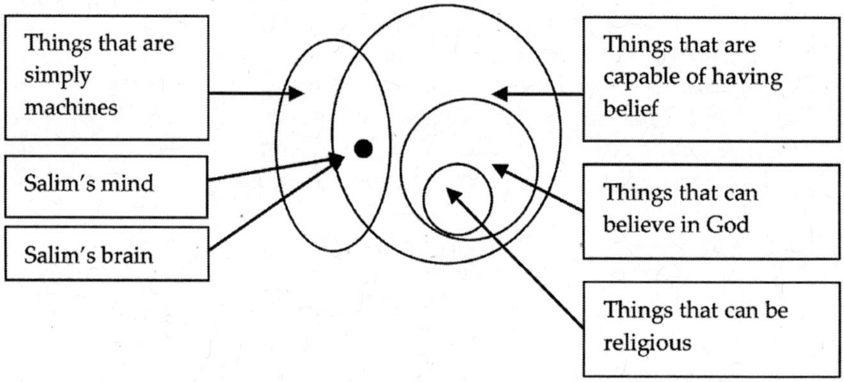

Chapter Two

Term Logic (TL)

2.1 The TL Worldview

Term Logic (TL) is a modern version of Aristotelian Syllogistic Logic. Like every logical system we discuss in this book, TL is based on a certain worldview. A worldview specifies the **basic ingredients of reality**. Every **combination** of what a worldview counts as the basic ingredients of reality constitutes a **logical possibility** according to that worldview. The **TL worldview** specifies **individuals and properties** as the basic ingredients of reality. An individual **instantiates** a property if and only if it has this property. The **extension** of a property in some logical possibility is the set consisting of all the individuals that instantiate the property in that logical possibility. In the TL worldview, every property that is admitted into a logical possibility must have a **nonempty extension**, that is, it must be instantiated by at least one individual in that logical possibility. For example, in order for our actual world to count as a logical possibility, according to the TL worldview, all the properties that have empty extensions, such as the property "being a person taller than twenty feet," must be removed from our world. A consequence of this feature of the TL worldview is that a property, such as the property "being an odd number divisible by two," that *cannot* be instantiated by any individual is not permitted into any logical possibility. The word 'individual' in the philosophical sense designates any single object, such as a human being, a tree, a machine, a country, an idea, or a number. What sorts of individuals are allowed depends on one's philosophical convictions. For instance, if someone believes that mathematical objects can never exist, then she would not include numbers in her list of individuals. In the TL worldview, a logical possibility may consist solely of individuals or may consist of properties and individuals. It cannot consist solely of properties, since every property must be instantiated by at least one individual. Therefore, if there are properties, there are individuals. However, it is common to specify only properties, without specifying any individuals; in such a case the existence of individuals that instantiate these properties is implicitly presupposed. Once a logical possibility is specified, that is, once individuals and extensions of properties are specified, certain relations obtain[1] between them: relations between properties and properties, between individuals and

1 The verb 'obtain' here is used in a peculiar sense that is common in philosophy and logic. The word is ordinarily employed as a transitive verb meaning "to gain" or "to attain." For instance, we say that the scientist obtained certain results from this study or that I obtained many interesting ideas from my discussion with Jones. We usually use the word in logic and philosophy as an intransitive verb meaning "to occur," "to take place," "to be established," "to hold," or "to emerge." We will frequently say things like 'a relation obtains between A and B' (it holds or takes place between A and B) and 'a state of affairs obtains' (it occurs or takes place). This usage may sound odd at first, but the reader will become accustomed to it.

properties, and between individuals and individuals. The TL worldview imposes two restrictions on the language that is used to make assertions about the constituents of a logical possibility: every singular term must refer to an individual in that logical possibility,[1] and every constituent of that logical possibility, whether it is an individual or a property, must be named.[2]

2.1:C Commentary on 2.1

2.1:C1 Before discussing the relations that obtain between properties and individuals, we need to present a very brief introduction to the basic concepts of set theory. We will employ these concepts frequently in this book. A set is a collection of objects. If a set S contains an object c, we also say that c belongs to S or that c is a member of S (symbolically, $c \in S$). The symbolic expression $c \notin S$ means that c does not belong to (is not a member of) S. Two sets are identical if and only if they contain exactly the same objects. There is an empty set, $\varnothing$, which, according to the preceding identity condition, is unique. A set that contains only one object c is called "the singleton of c" and is written as {c}.[3] A set that consists of the objects c_1, c_2, …, and c_n is written as $\{c_1, c_2, \ldots, c_n\}$. Hence a set may be described by listing its members. It may also be described by specifying a property that is exclusively shared by its members. For instance, the property of being a positive even number describes (uniquely) the set whose members are all the positive even numbers. This set may be written as {2, 4, 6, 8, 10, …} or as {n: n is a positive even number}; the last notation is read "The set of all n such that n is a positive even number."

A set S is a subset of a set R (symbolically, $S \subseteq R$) if and only if every member of S is also a member of R; hence every set is a subset of itself, and the empty set is a subset of every set. The expression $S \nsubseteq A$ means that S is not a subset of A, that is, S has a member that is not in A. A set S is a proper subset of a set R (symbolically, $S \subset R$) if and only if S is a subset of R and S and R are not identical, i.e., R contains all the members of S and it contains at least one object that is not a member of S. The expression $S \not\subset R$ means that S is not a proper subset of R, that is, either S is not a subset of R or it is identical with R. It is clear that no set is a proper subset of itself. The intersection of two sets S and R (symbolically, $S \cap R$) is the set consisting of the objects common to both S and R, that is, $c \in S \cap R$ if and only if $c \in S$ and $c \in R$. If the intersection of S and R is empty (i.e., $S \cap R = \varnothing$), we say that they are disjoint. The union of two sets S and R (symbolically, $S \cup R$) is the set consisting of the members of S and the members of R, that is, $c \in S \cup R$ if and only if $c \in S$ or $c \in R$ or both.

2.1:C2 There is a *maximum* of five possible relations between a property and a property, two between an individual and a property, and two between an individual and an individual. Let P and Q be two (not necessarily distinct) properties and d and e be two (not necessarily distinct) individuals. The list and diagrams below describe these nine relations.

1 A singular term might be a proper name, such as 'Socrates', or a definite description, such as 'The great Greek philosopher who was Plato's teacher and who was convicted of corrupting the youth and sentenced to death by poison'. We will discuss singular terms in 2.3.2.

2 These restrictions will be made clear in 2.3.2 and 2.4.1.

3 In standard set theory c is not identical with its singleton.

(a) P and Q are **coextensive:** P and Q have the same extension.

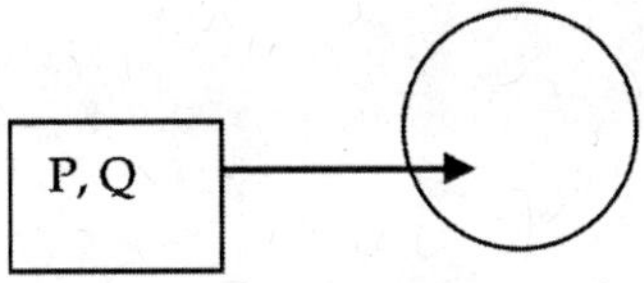

(b) P **strictly includes** Q: the extension of Q is a proper subset of the extension of P.[1]

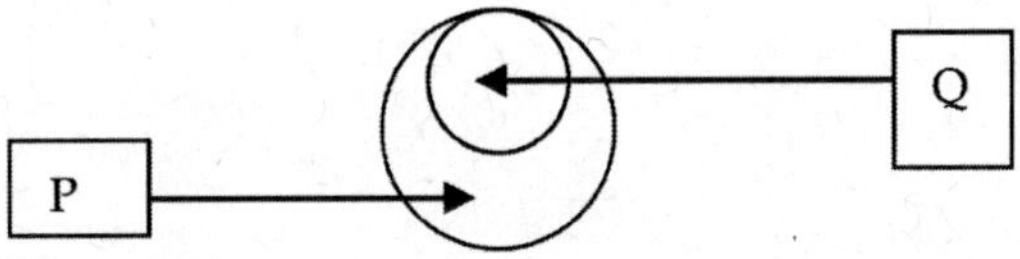

(c) Q **strictly includes** P: the extension of P is a proper subset of Q.

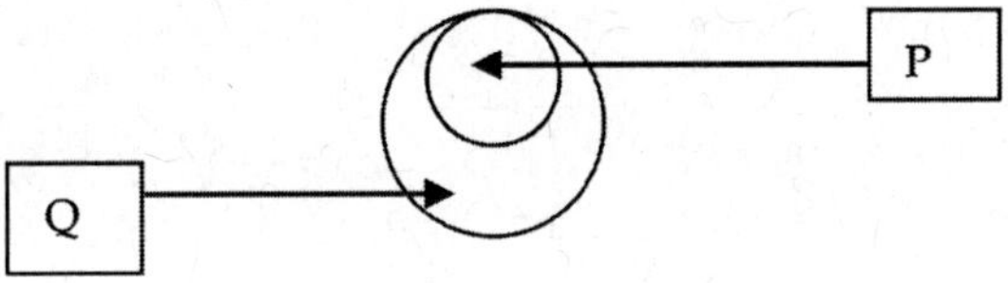

(d) P and Q **strictly overlap**: the extensions of P and Q are not disjoint and neither one is a subset of the other.[2]

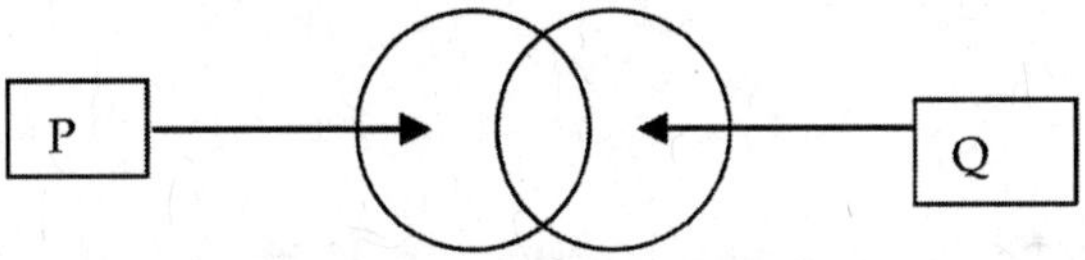

(e) P and Q are **disjoint**: the extensions of P and Q are disjoint.[3]

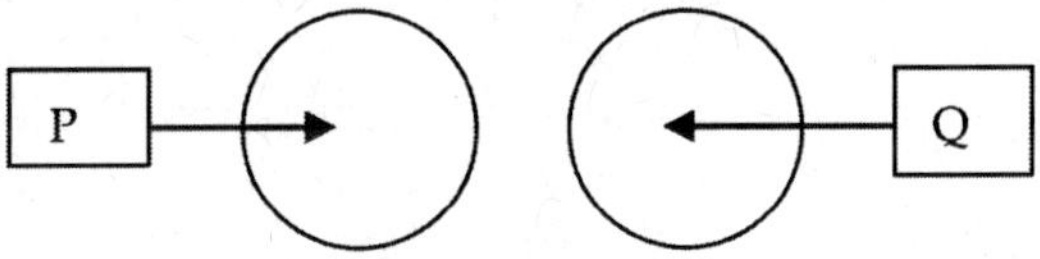

---

1 We say that P *includes* Q if and only if P strictly includes Q or P and Q are coextensive. In other words, P includes Q if and only if the extension of Q is a subset of the extension of P. Hence every property includes itself.

2 We say that P and Q *overlap* if and only if they are not disjoint, that is, they either strictly overlap or one of them includes the other.

3 Some authors describe such properties as *mutually exclusive*. We will avoid this language. The expression 'mutually exclusive properties' is sometimes used to indicate a relation between properties that is significantly more demanding than simply being disjoint. In this usage, two properties are mutually exclusive if and only if no individual that instantiates one of them *can* instantiate the other. In other words, two properties are mutually exclusive if and only if their extensions are disjoint in *every* logical possibility. The properties "being a whale" and "having feathers" (*continued*)

(f) e **instantiates** (i.e., has) the property P: e belongs to the extension of P.

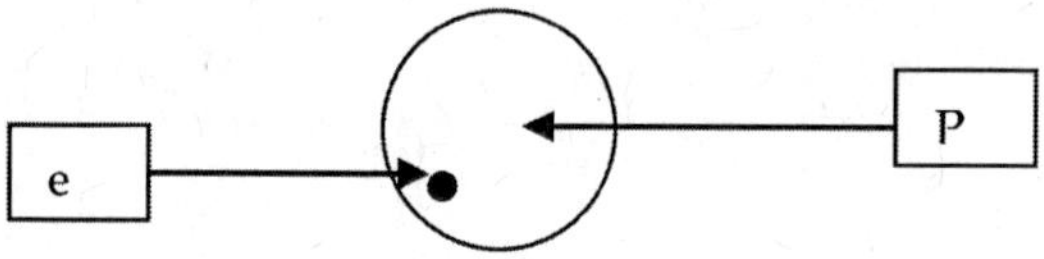

(g) e **lacks** (i.e., does not have) the property P: e does not belong to the extension of P.

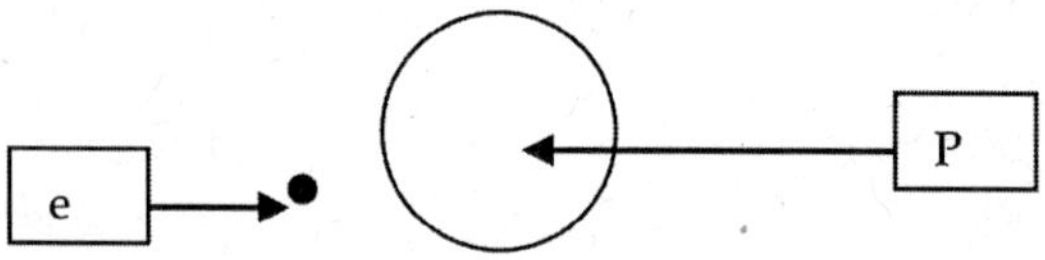

(h) d is **identical with** e: d and e are the same individual.

(i) d is **not identical with** e: d and e are two distinct individuals.

2.1:C3 In the TL worldview, *every* combination of individuals and properties constitutes a logical possibility. This feature of the TL worldview is too permissive, and it will be a source of difficulties for TL. Intuitively, there should be some restrictions on what counts as a *possible* combination of basic ingredients of reality. There are interdependencies between certain properties. For example, the extension of the property "being a Frenchman" is the intersection of the extensions of the properties "being French" and "being a man."[1] In other words, an individual has the property "being a Frenchman" if and only if it has the properties "being French" and "being a man." There is no logical possibility in which there is an individual who is both French and a man but who is not a Frenchman. Hence a specification of extensions for these three properties in which the extension of the property "being a Frenchman" is not the intersection of the extensions of the properties "being French" and "being a man" should not count as a possible combination of basic ingredients of reality. The ordinary conditions of being a Frenchman require that all Frenchmen are French. So the sentence 'All Frenchmen are French' is logically true because it is true in every logical possibility that is *relevant to* this sentence. In

are disjoint in our world because no whale has feathers. However, they are not mutually exclusive because it is *possible* for some whales to have feathers, that is, there is a logical possibility in which some whales have feathers. On the other hand, the properties "being an odd number" and "being an even number" are mutually exclusive because it is *impossible* for a number to be both odd and even, that is, there is no logical possibility in which an odd number is also even.

1 We're assuming that 'Frenchman' is not understood in its old sense—namely, as any French person.

1.2.1:C2, we stated that the meanings of the expressions that are used in a sentence (or a collection of sentences) are not allowed to vary across the logical possibilities that are relevant to this sentence (or to this collection of sentences). So a logical possibility in which the term 'Frenchmen' means, for instance, "men who are Francophone," and not "men who are citizens of France," is not a logical possibility that is relevant to the sentence 'All Frenchmen are French'.

The TL worldview does not allow for any interdependencies between properties. This feature of the TL worldview is reflected in the semantics of TL. In TL, properties are represented as circles. These circles can exhibit any of the relations described in 2.1:C2(a)–(e) above without restrictions. In other words, there are no *necessary* relations between the circles of TL. Thus if the properties "being a Frenchman" and "being French" are represented by two circles, there is no requirement that the first circle be included in the second circle. As will be obvious later, this fact entails that a TL translation of the sentence 'All Frenchmen are French' would fail to be logically true in TL. We will discuss these issues with much elaboration in 2.5.12. At this stage, we only note the fact that ordinary logical possibilities place some restrictions on what combinations of individuals and properties are permissible and what are not, while the TL worldview and TL semantics do not impose any such restrictions.

2.2 The Syntax of TL

2.2.1 The **basic vocabulary** of TL is quite simple. It consists of three categories.

2.2.1a **General terms**, which are the following uppercase italic letters: *A*, *B*, *C*, ..., *U*, *V*, *W*; with numeric subscripts if needed. We use the boldfaced letters '**X**', '**Y**', and '**Z**' as metalinguistic variables ranging over general terms. (An explanation of metalinguistic variables follows in 2.2.1:C2.)

2.2.1b **Singular terms**, which are the following lowercase italic letters: *a*, *b*, *c*, ..., *u*, *v*, *w*; with numeric subscripts if needed. We use the boldfaced letters '**x**', '**y**', and '**z**' as metalinguistic variables ranging over singular terms.

2.2.1c Six **logical words**: 'all', 'some', 'is', 'are', 'no', 'not'

2.2.1:C Commentary on 2.2.1

2.2.1:C1 The worldview upon which a logical system is based guides its syntax and semantics. The basic vocabulary of each of the logical systems described in this book consists of expressions that correspond to basic ingredients of reality and of logical constants (or logical words, or logical symbols), which are used to form more complex expressions from the basic ones. (The basic vocabulary of a logical system may also contain punctuation marks, such as parentheses.) Since the TL worldview specifies individuals and properties as the basic ingredients of reality, the syntax of TL contains expressions that correspond to these two types of ingredients. The singular terms correspond to individuals and the general terms correspond to properties.[1] The log-

1 More precisely, TL general terms correspond to extensions of properties. We will make this point clear in 2.5.12:C2.

ical words allow us to describe the relations that may obtain between these ingredients of reality. Here are a few examples. If *W* corresponds to the property of being a whale and *M* to the property of being a mammal, then the fact that the latter property strictly includes the former can be described by saying that all *W* are *M* and that some *M* are not *W*. If *U* corresponds to the property of being a graduate of Yale University and *b* to Bill Clinton, then the fact that Bill Clinton is a graduate of Yale University can be described by saying that *b* is *U*. If *f* corresponds to the first President of the United States and *g* to George Washington, then the fact that George Washington and the first President of the United States are the same individual could be expressed by saying that *f* is *g*.

2.2.1:C2 In the study of language it is important to distinguish between two languages: The *object language* and the *metalanguage*. The object language is the language under study and the metalanguage is the language in which the object language is discussed. In our particular case the object language is the language of TL and the metalanguage is English augmented with appropriate symbols. When a singular or general term is mentioned in the metalanguage, it is customary to enclose the letter between two single-quotation marks, ‘ ’. However, in order to simplify our notation, almost always we will not follow this usage where there is no cause for misunderstanding. When we need to say something about all or many general terms or singular terms, we use metalinguistic variables (i.e., variables in the metalanguage) to range over general terms or over singular terms. We designate the boldfaced letters **X**, **Y**, **Z**, and **x**, **y**, **z** to be the metalinguistic variables ranging over general and singular terms, respectively.

2.2.2 The **TL sentences** are the expressions generated from the basic vocabulary of TL according to the eight **formation rules** listed below. If **X** and **Y** are any general terms of TL and **x** and **z** are any singular terms of TL, then every expression that has one of the following forms is a TL sentence.

2.2.2a	all **X** are **Y**	2.2.2e	**z** is **Y**
2.2.2b	no **X** is **Y**	2.2.2f	**z** is not **Y**
2.2.2c	some **X** are **Y**	2.2.2g	**x** is **z**
2.2.2d	some **X** are not **Y**	2.2.2h	**x** is not **z**

2.2.2:C COMMENTARY ON 2.2.2

In order to illustrate the applications of these formation rules let us consider the following example. We will write down all the TL sentences that can be generated from this set of basic vocabulary {*E*, *F*, *c*, *d*, *g*}.[1]

(a) all **X** are **Y** (4 sentences):[2]
all *E* are *E*, all *F* are *F*, all *E* are *F*, all *F* are *E*.

---

1 Technically speaking, this is incomplete; a basic vocabulary must also include the logical constants (i.e., the six logical words). But since the logical constants are part of every basic vocabulary, it is sufficient to list only the general and singular terms.

2 Each of the sentence forms 2.2.2a–2.2.2d generates n^2 sentences where n is the number of general terms: there are n possible substitutions for **X** and n for **Y**.

(b) no **X** is **Y** (4 sentences):
no E is E, no F is F, no E is F, no F is E.

(c) some **X** are **Y** (4 sentences):
some *E* are *E*, some *F* are *F*, some *E* are *F*, some *F* are *E*.

(d) some **X** are not **Y** (4 sentences):
some *E* are not *E*, some *F* are not *F*, some *E* are not *F*, some *F* are not *E*.

(e) **z** is **Y** (6 sentences):[1]
c is *E*, *c* is *F*, *d* is *E*, *d* is *F*, *g* is *E*, *g* is *F*.

(f) **z** is not **Y** (6 sentences):
c is not *E*, *c* is not *F*, *d* is not *E*, *d* is not *F*, *g* is not *E*, *g* is not *F*.

(g) **x** is **z** (9 sentences):[2]
c is *c*, *c* is *d*, *c* is *g*, *d* is *c*, *d* is *d*, *d* is *g*, *g* is *c*, *g* is *d*, *g* is *g*.

(h) **x** is not **z** (9 sentences):
c is not *c*, *c* is not *d*, *c* is not *g*, *d* is not *c*, *d* is not *d*, *d* is not *g*, *g* is not *c*, *g* is not *d*, *g* is not *g*.

2.2.3 The type of syntax described above may be called **non-recursive generative grammar**, because the collection consisting of *all* the TL sentences can be *generated* (produced) by the eight formation rules listed above when applied to the basic vocabulary of TL. None of these rules is **iterative** (i.e., **recursive**), in the sense that none can be applied more than once in the construction of any sentence. This last feature guarantees that only a finite number of TL sentences can be generated from a finite list of basic vocabulary.[3]

2.2.3:C COMMENTARY ON 2.2.3

The example given in 2.2.2:C illustrates how the formation rules of TL are not iterative and shows that only finitely many TL sentences can be generated from a finite set of basic vocabulary. Natural languages, on the other hand, contain iterative rules that generate infinitely many sentences from a finite list of vocabulary. For example, the clause 'who met a tall man' may be appended to the sentence 'I met a tall man' any number of times. Thus if there is no preset limit on the length of sentences, this operation may be iterated indefinitely generating

---

1 If n is the number of general terms and m the number of singular terms, each of the rules 2.2.2e and 2.2.2f generates m×n sentences, since there are m possible substitutions for **z** and n for **Y**.

2 Each of the rules 2.2.2g and 2.2.2h generates m^2 sentences, where m is the number of singular terms, since there are m possible substitutions for **x** and m for **z**.

3 This condition is sufficient but not necessary. An iterative rule that can be applied only a finite number of times in the construction of any sentence cannot generate infinitely many sentences from a finite set of vocabulary. The universal-quantifier and the existential-quantifier rules of Predicate Logic are of this sort (see 4.2.7:C).

infinitely many sentences from the finite set of vocabulary {'I', 'met', 'a', 'tall', 'man', 'who'}. Here are the first few iterations. The number to the left of the sentence represents the number of times the operation is applied.

[0] I met a tall man.
[1] I met a tall man who met a tall man.
[2] I met a tall man who met a tall man who met a tall man.
[3] I met a tall man who met a tall man who met a tall man who met a tall man.

2.3 Translating English into TL

2.3.1 A **general term** of some natural language L is a word or a phrase that does or could stand for more than one individual without being ambiguous. More precisely, an L general term names a property, whether this property has an empty extension or not. The words 'rabbit', 'woman', and 'happy', and the phrases 'student at Harvard University', 'living president of the United States', and 'flying pig' are examples of general terms. The **extension** of an L general term is the extension of the property that this term names. An individual **satisfies** a general term if and only if it instantiates the property that this term names. An English general term *whose extension is nonempty* is translated as a TL general term.

2.3.1:C COMMENTARY ON 2.3.1

The condition that only general terms whose extensions are nonempty can be translated as TL general terms is in accordance with the TL worldview. We said in 2.1 that this worldview does not admit properties whose extensions are empty. We intuitively think that there are such properties. For instance, the property "being a living president of the United States who is taller than seven feet" has an empty extension. This means that in our world no individual satisfies the general term 'a living president of the United States who is taller than 7 feet', that is, this term has an empty extension in our world. We will call a general term whose extension is empty an **empty general term.**[1]

We will see later that the semantics of TL are based on a certain type of diagram called a "TL diagram." In such a diagram, a TL general term is assigned to a circle. It is natural to think of this circle as standing for the **referent** of the general term in that TL diagram (these circles are

1 Although some philosophers call such terms "non-referring general terms," we will try to avoid this expression when dealing with *natural-language* general terms. The reason for this is that we want to give consideration to the position that permits empty general terms to have referents. Holders of this position argue that the **referent** of a term is typically the object the term names. Since empty terms name properties whose extensions are empty, these terms actually *refer* to those properties; and hence empty general terms are referring terms whose referents have empty extensions. A more standard position is to consider any individual that satisfies a general term as a referent of that term. Thus a general term has as many referents as there are individuals in its extension. This implies that an empty general term has no referents; and hence it can correctly be described as a non-referring term. We will remain neutral regarding whether empty general terms may be considered referring terms or not. When dealing with natural language general terms, we will refrain from describing empty general terms as referring or as non-referring terms.

meant to correspond to extensions of properties). If a TL general term were allowed to name no circle, it would be a genuine case of a general term that names no object. Since every TL general term must refer to a circle in every TL diagram that is relevant to that term, and since these circles are assumed to be nonempty, we can assert that TL does not permit a general term to be empty or non-referring.

It is debatable whether a natural-language general term can fail to name a property or, at least, the extension of a property. Some philosophers think that certain *disjunctive* "general terms" do not name anything, including properties. An example might be the disjunctive expression 'a student at Lehigh University or a marine mammal of the order Cetacea'. If this expression names a single property, it would have to be the property "being a student at Lehigh University *or* being a marine mammal of the order Cetacea." But it seems doubtful that our world contains such a single property. Roughly speaking, properties are thought of as *states* or *types* of individuals. It seems doubtful that there is a single state or a single type of individual that may be described as being a student at Lehigh University *or* being a marine mammal of the order Cetacea. What this expression seems to be naming is a disjunction of two properties: "being a student at Lehigh University" and "being a marine mammal of the order Cetacea." If this is correct and if this expression is indeed a general term, then we have a case of an English general term that actually names no property.

An alternative position here is to allow properties corresponding to all general terms; thus, to insist that "being a student at Lehigh University *or* being a marine mammal of the order Cetacea" is a property after all. One attraction of this position is that the distinction between properties that are ultimately, in reality, disjunctive and those that are not can seem artificial, a matter of merely how our language orders things.

We would like to maintain philosophical neutrality as much as possible, but sometimes there is no escape from making a philosophical commitment. This is one of these times. We either allow general terms to name no properties or we do not. We defined a general term, in 2.3.1, as an expression that names a property, whether this property's extension is empty or not. Hence we committed ourselves to the position that expressions that name no properties are not genuine general terms. This position seems to cohere well with the TL worldview. As stated in 2.1, the TL worldview specifies individuals and properties as the only basic ingredients of reality. It is natural, therefore, to think of terms as being of two types: terms that name individuals and terms that name properties. We call the first type "singular terms" and the second type "general terms".

A general term of some natural language L, by its nature, does or could stand for more than one individual. This feature does not make a general term ambiguous. The phrase 'living president of the United States' stands, at the time of writing this paragraph, for four individuals. The phrase is not ambiguous because it stands for more than one individual—representing multiple individuals is part of its semantical function. A general term, however, can be **ambiguous** if it names two different properties. An example is the word 'kiwi': kiwi is a type of bird and a type of fruit. It is ambiguous not because it stands for more than one individual but because it names two different properties. TL does not allow general terms to be ambiguous. Therefore, an ambiguous general term must be disambiguated before it can be translated into TL. Of course, in this case, the original term would be translated as more than one TL general term.

2.3.2 A **singular term** of some natural language L is a word or a phrase that stands for exactly one individual and could not stand for more than one individual without being ambiguous. The words 'Canada', 'Plato', and 'Microsoft', and the phrases 'The President of the United States' and 'The smallest positive even number' are examples of singular terms. The individual that a singular term names is called the **referent** of the singular term. An English singular term is translated as a TL singular term.

2.3.2:C COMMENTARY ON 2.3.2

If a singular term stands for more than one individual, it is **ambiguous**. For example, the singular term 'George Bush' is ambiguous because it names at least two people. Our definition does not permit such terms to be counted as singular terms. The restriction is reasonable. We use singular terms primarily to refer to specific individuals. If a singular term has two referents, we wouldn't know which referent is mentioned when the term is used. The restriction, also, is not severe. If a singular term is ambiguous, it can always be disambiguated by adding words to it to produce new singular terms each of which stands for exactly one of the referents of the original term. For instance, 'President George Bush' may be disambiguated by adding the middle name 'Herbert' or 'Walker' to it (assuming, of course, that each of the new singular terms refers to exactly one individual). Just about every proper name of a person in fact is the name of more than one person. We can still count most uses of personal proper names as unambiguous, however, because in the context of their utterance, it is usually clear to speaker and hearer (or reader) that only one individual is meant. We said in 2.3.1:C that TL does not permit general terms to be ambiguous. Similarly, TL does not permit singular terms to be ambiguous.

Our definition of singular terms requires that they have referents. In other words, non-referring "terms" are not considered, according to our definition, genuine singular terms. In note 2, p. 80, we will briefly discuss a philosophical position that has a similar implication about singular terms. Ordinarily, however, we use many non-referring terms as singular terms. For example, the expressions 'Santa Claus', 'The present king of France', and 'The largest even number' are all non-referring, but we ordinarily think of them as singular terms. If we believe that there are genuine singular terms that have no referents, then we must relax the conditions of definition 2.3.2 to allow for non-referring singular terms. In this case, we must restrict the English singular terms that can be translated as TL singular terms to those that have referents. TL does not permit a singular term to be non-referring.

2.3.3 English **universal quantifiers** such as 'all', 'every', 'each', and 'any' are translated into TL as 'all'. English **existential quantifiers** such as 'there is', 'there exists', 'at least one', 'one or more', 'several', and 'some' (understood as indicating the existence of at least one individual) are translated as 'some'. 'Some' is usually used in English to mean "there are some but not all." In other words, 'some are X' is usually understood to mean that some are X and some are not X. This 'some' is not the 'some' of TL. Another slight mismatch with ordinary English is that 'some' is usually used to mean "more than one," but in TL, one (or more) is "some."

2.3.4 The following list shows how **English idioms** are translated into TL.[1]

2.3.4a English idioms: All **X**'s are **Y**. **X**'s are **Y**'s. Every (each, any) **X** is **Y**. Whatever (whoever) is **X** is **Y**. Those that (who) are **X** are **Y**. If something (someone) is **X**, then it (he, she) is **Y**. Only **Y**'s are **X**'s. None but **Y**'s are **X**'s. Nothing (no one) is **X** unless it (he, she) is **Y**. A thing (person) is not **X** unless it (he, she) is **Y**. A thing (person) is **X** only if it (he, she) is **Y**.
TL translation: all **X** are **Y**

2.3.4b English idioms: No **X** is **Y**. **X**'s are not **Y**'s. Every (each, any) **X** is non-**Y**. Whatever (whoever) is **X** is not **Y**. Those that (who) are **X** are not **Y**. If something (someone) is **X**, then it (he, she) is not **Y**. Nothing (no one) that is **X** is **Y**. There is no **X** that (who) is **Y**. There are no **X**'s that (who) are **Y**.
TL translation: no **X** is **Y**

2.3.4c English idioms: Some **X**'s are **Y**. At least some **X**'s are **Y**. At least one **X** is **Y**. There is (exists) an **X** that is **Y**. There are (exist) **X**'s that are **Y**.
TL translation: some **X** are **Y**

2.3.4d English idioms: Some **X**'s are not **Y**. At least some **X**'s are not **Y**. At least one **X** is not **Y**. There is (exists) an **X** that is not **Y**. There are (exist) **X**'s that are not **Y**. Not all **X**'s are **Y**'s. Not every (each) **X** is **Y**.
TL translation: some **X** are not **Y**

2.3.4:C Commentary on 2.3.4

These idioms are all clear except, perhaps, for the last five idioms listed in 2.3.4a. The best way to show that these idioms mean that all **X**'s are **Y** is to give intuitively clear examples. The sentences 'Only American citizens are entitled to receive Social Security benefits' and 'None but American citizens are entitled to receive Social Security benefits' have the same content as the sentence 'All those who are entitled to receive Social Security benefits are American citizens'. It should be apparent that none of these sentences means that all American citizens are entitled to receive Social Security benefits. Also, each of the sentences 'Nothing is bright unless it is visible', 'A thing is not bright unless it is visible', and 'A thing is bright only if it is visible' means that all things that are bright are visible. It is clear that they do not mean that all visible things are bright.

1 As mentioned in 2.2.1a, the boldfaced letters '**X**' and '**Y**' are metalinguistic variables that range over TL general terms. Our usage here is hybrid: when used in TL sentence forms, they substitute for TL general terms; and when used in English idioms, they substitute for English general terms whose TL translations are the corresponding TL general terms. We employ this hybrid usage only in 2.3.4.

2.4 The Semantics of TL

2.4.1 The semantics of TL sentences is given by TL diagrams. A **TL diagram** D *for* a set Γ of TL sentences, that is, a TL diagram that is *relevant to* Γ, is a collection of circles, dots, or both such that

2.4.1a every general term that occurs in some member of Γ refers to a circle in D, and every singular term that occurs in some member of Γ refers to a dot in D;

2.4.1b every circle in D is the referent of some general term; and every dot in D is the referent of some singular term;

2.4.1c no two circles in D are the referent of the same general term, and no two dots in D are the referent of the same singular term; and

2.4.1d no circle in D is the referent of a singular term and no dot in D is the referent of a general term.

We shall refer to the circle in D that is the referent of a general term **X** as the **X**-circle in D, and to the dot in D that is the referent of a singular term **z** as the **z**-dot in D. We refer to the circles and dots that constitute a TL diagram D as "the constituents of D." While every general or singular term that occurs in a member of Γ must refer to a constituent of D, a general or a singular term that refers to a constituent of D need not be a term that occurs in a member of Γ, that is, some constituents of D may be the referents of general and singular terms that do not occur in any member of Γ. D is said to be a TL diagram for a TL sentence Y if and only if it is a TL diagram for a set containing Y.[1]

2.4.1:C COMMENTARY ON 2.4.1

It is clear from the definition of a TL diagram that *non-referring terms* and *ambiguous terms*, whether general or singular, are not allowed in TL. The condition that every term that occurs in a member of Γ must refer to a circle or a dot in D prevents terms from being non-referring; and the condition that no two circles or two dots can be the referent of the same general term or the same singular term, respectively, prevents terms from being ambiguous. Furthermore, the condition that every circle and dot must be the referent of some term ensures that every constituent of D has a name but permits constituents to have multiple names. In other words, there are no *unnamed objects* in any TL diagram, but there might be objects that have more than one name.

1 We are assuming the set-theoretic principle that for every object α, there is a set that is the *singleton* of α, that is, the set whose only member is α. We denote the singleton of α as $\{\alpha\}$. Thus for every TL sentence X, there is always a set that contains X—namely the singleton of X.

Let us make these points clear by means of an example. Take Γ to be the following set: {some *B* are not *E*, all *B* are *F*, *a* is *F*, *a* is not *c*}. The four diagrams below are TL diagrams *for* Γ.

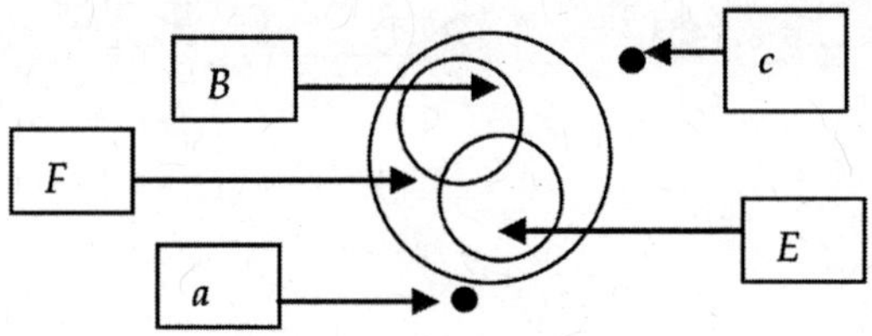

D1

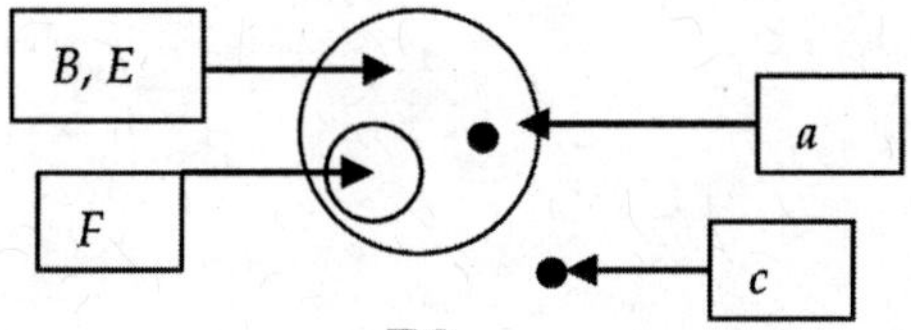

D2

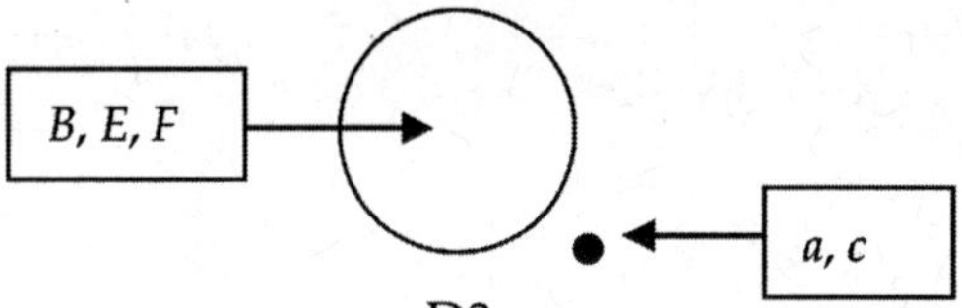

D3

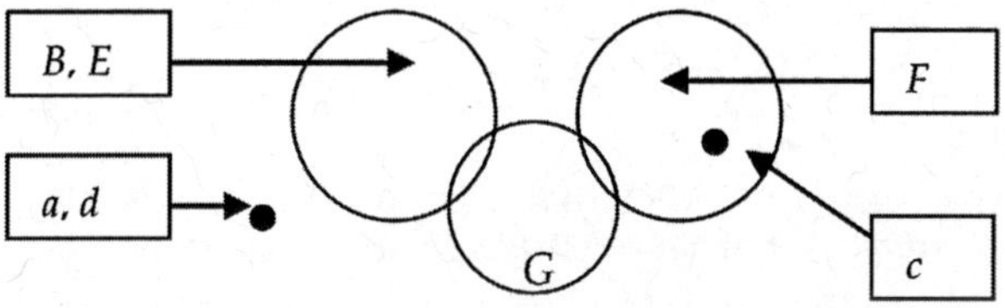

D4

In D1 every general term occurring in a member of Γ refers to exactly one circle and every singular term refers to exactly one dot; hence there are no non-referring or ambiguous terms. Every circle and dot in D1 is the referent of exactly one term; hence there are no unnamed objects and no object has multiple names. In D2, D3, and D4, there are also no non-referring terms, ambiguous terms, nor unnamed objects, but there are objects with multiple names. In D2 there is only one circle that is the referent of the general terms *E* and *B*; in D3, one circle is the referent of all the general terms and one dot is the referent of all the singular terms; and in D4, one circle is the referent of *E* and *B* and one dot is the referent of *a* and *d*. Thus in D2 and D4 a circle has two names, in D3 the circle has three names and the dot has two names, and in D4 a dot has two names. D4 is different from the other diagrams. The *a*-dot is also named *d*, which does not occur in any member of Γ, and there is a circle named in *G*, which also does not occur in any member of Γ. Our definition permits diagrams such as D4 to count as TL diagrams for Γ.

The three diagrams below are *not* TL diagrams for the set Γ.

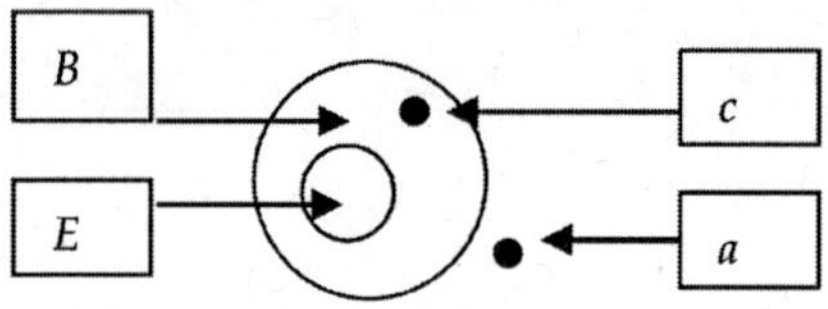

D5

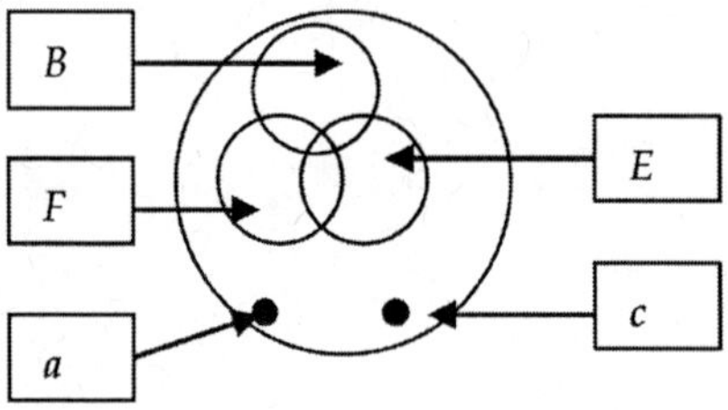

D6

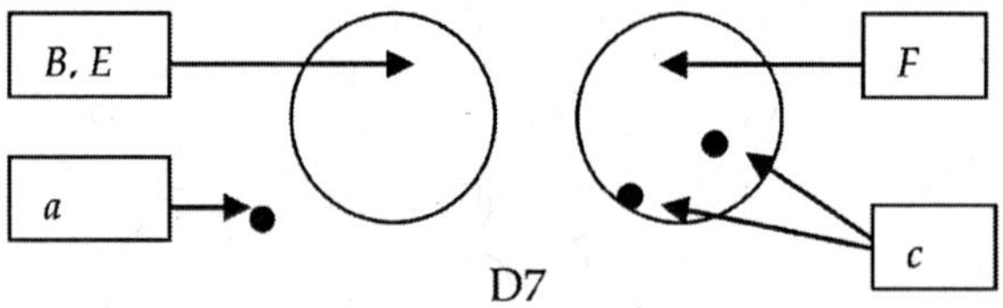

D7

D5 leaves the general term *F* without a referent. In D6 there is a circle that has no name. D7 assigns two referents to the singular term *c*.

We did not define a TL diagram in itself but a TL diagram *for* a set of TL sentences, that is, a TL diagram that is *relevant to* a set of TL sentences. Thus the notion of a TL diagram is always relative to a given set of TL sentences. As we will see later, the notion of a TL diagram will be proposed as a characterization of the notion of logical possibility. Since we are interested only in logical possibilities that are relevant to sets of declarative sentences, TL diagrams are defined only in relation to sets of TL sentences. However, occasionally we may speak of a TL diagram without explicit reference to a set of TL sentences when the context makes it clear to which set of TL sentences the TL diagram is relevant.

Note carefully that a TL diagram relevant to a set of sentences need not represent the truth of any of the sentences in that set. Thus, for example, there are two TL diagrams relevant to the set of (only one) sentence {Obama is the 44th president}. One diagram contains only one dot, labeled both *o* and *p* (where *o* stands for Obama and *p* stands for the 44th president); and this represents the truth of that sentence; another contains two dots, one labeled *o*, the other *p*; this represents the falsity of that sentence (saying that Obama is not the 44th president).

2.4.2 Two TL diagrams D1 and D2 for a set Γ of TL sentences are **similar with respect to** Γ if and only if (1) for every singular term **z** that occurs in a member of Γ, the **z**-dot in D1 corresponds to the **z**-dot in D2, and for every general term **X** that occurs in a member of Γ, the **X**-circle in D1 corresponds to the **X**-circle in D2; and (2) the relations between the circles and dots in D1 that are the referents of the terms that occur in the members of Γ *mirror precisely* the relations between the corresponding circles and dots in D2. TL diagrams that are similar with respect to Γ are also referred to as **similar Γ diagrams**.

2.4.2:C COMMENTARY ON 2.4.2

Let us consider an example of similar TL diagrams. Take Γ to be the set {some *A* are *B*, all *E* are *B*, *c* is *E*, *c* is not *A*}. The following two TL diagrams are similar Γ diagrams.

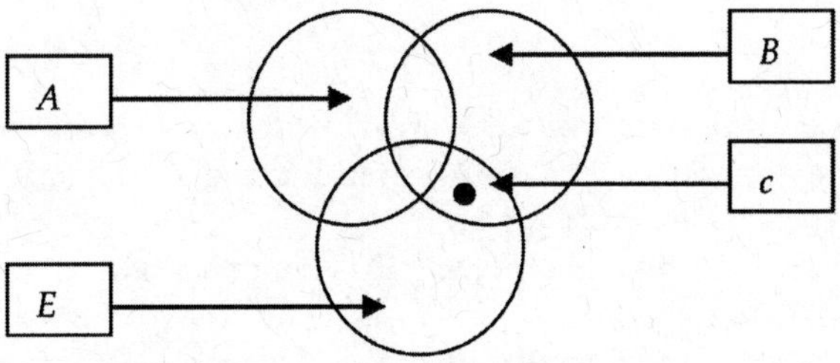

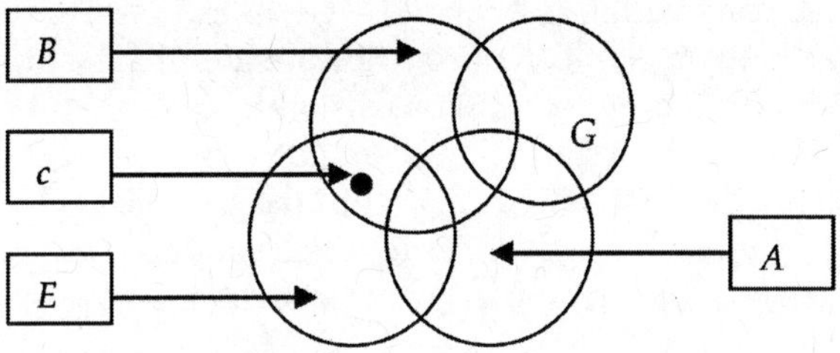

It is clear that the *A*-circle, the *B*-circle, the *E*-circle, and the *c*-dot in the first diagram have the same relations to each other as their counterparts in the second diagram. Recall that a TL diagram for Γ may contain circles and dots that represent general and singular terms that do not occur in Γ. For each TL diagram, there is an infinite number of other non-identical but similar TL diagrams.

On the other hand, the TL diagrams below are not similar Γ diagrams.

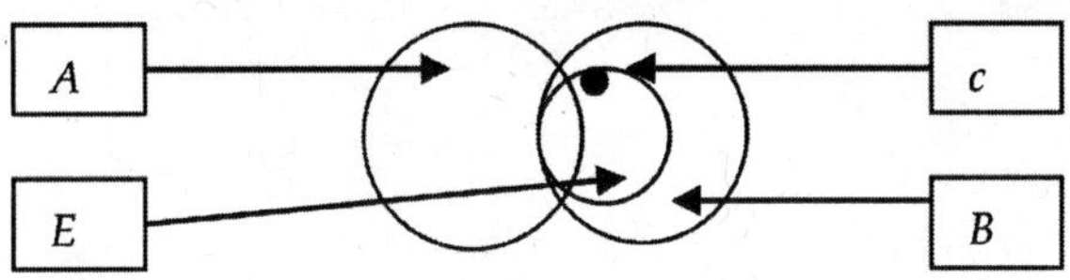

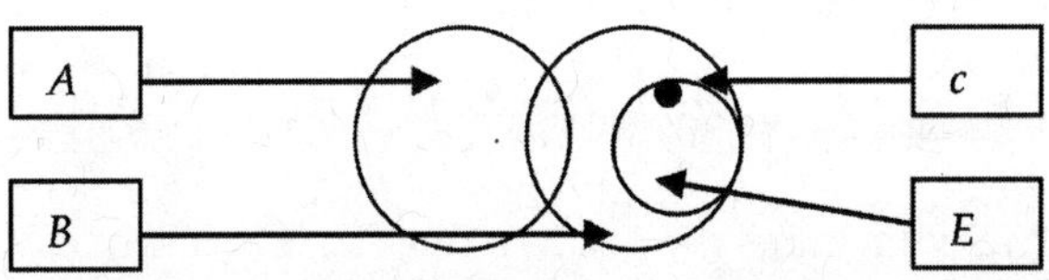

The relation between the *A*-circle and the *E*-circle in the first diagram differs from their relation in the second diagram: they strictly overlap in the first and they are disjoint in the second.

2.4.3 The **truth conditions** of a TL sentence are the conditions under which the sentence is true on a relevant TL diagram. Said differently, a sentence is true on a TL diagram D if and only if D satisfies its truth conditions, and it is false on D if and only if D is relevant to the sentence and does not satisfy its truth conditions. The following list describes the truth conditions of every TL sentence on any TL diagram D for this sentence. In stating the conditions below, we will employ the same terminology we introduced in 2.1:C2 to describe the possible relations between properties. It is natural to use the same terminology since the diagrams presented there explain the meanings of these expressions when used to describe the possible relations between circles. Let **X** and **Y** be any general terms of TL and **x** and **z** be any singular terms of TL.

2.4.3a A TL sentence of the form 'all **X** are **Y**' is true on D if and only if the **Y**-circle includes the **X**-circle, that is, the **X**-circle and the **Y**-circle are identical or the **Y**-circle strictly includes the **X**-circle.

2.4.3b A TL sentence of the form 'no **X** is **Y**' is true on D if and only if the **X**-circle and the **Y**-circle are disjoint.

2.4.3c A TL sentence of the form 'some **X** are **Y**' is true on D if and only if the **X**-circle and the **Y**-circle overlap, that is, the **X**-circle and the **Y**-circle strictly overlap or one of them includes the other (or they are identical)—or simply, if and only if the **X**-circle and the **Y**-circle are not disjoint.

2.4.3d A TL sentence of the form 'some **X** are not **Y**' is true on D if and only if the **Y**-circle does not include the **X**-circle, that is, the **X**-circle and the **Y**-circle strictly overlap, the **X**-circle strictly includes the **Y**-circle, or they are disjoint.

2.4.3e A TL sentence of the form '**z** is **Y**' is true on D if and only if the **z**-dot is inside the **Y**-circle.

2.4.3f A TL sentence of the form '**z** is not **Y**' is true on D if and only if the **z**-dot is outside the **Y**-circle.

2.4.3g A TL sentence of the form '**x** is **z**' is true on D if and only if the **x**-dot and the **z**-dot are identical, that is, **x** and **z** refer to the same dot in D.

2.4.3h A TL sentence of the form '**x** is not **z**' is true on D if and only if the **x**-dot and the **z**-dot are not identical, that is, **x** and **z** refer to two distinct dots in D.

If every sentence in a set Γ is true on D, we say that D **satisfies** Γ or that D is a **model** of Γ.

2.4.3:C COMMENTARY ON 2.4.3

(a) There are two types of diagrams that make a TL sentence of the form 'all **X** are **Y**' true.

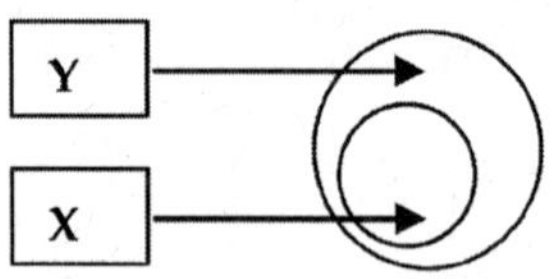

The **Y**-circle strictly includes the **X**-circle

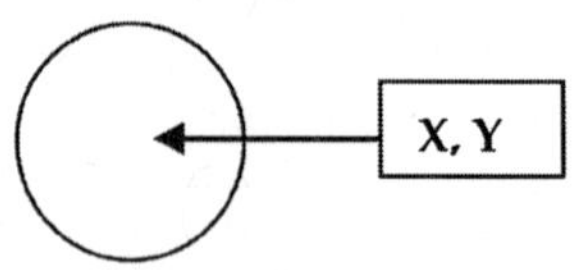

The **Y**-circle and **X**-circle are identical

(b) There is only one type of diagram that makes a TL sentence of the form 'no **X** is **Y**' true.

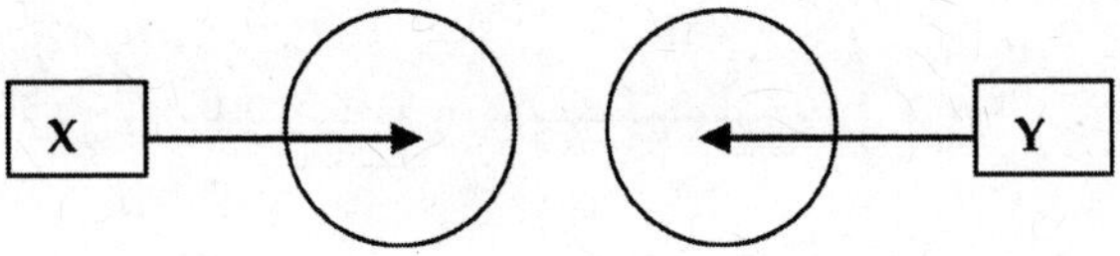

The **X**-circle and the **Y**-circle are disjoint

(c) There are four types of diagrams that make a TL sentence of the form 'some **X** are **Y**' true.

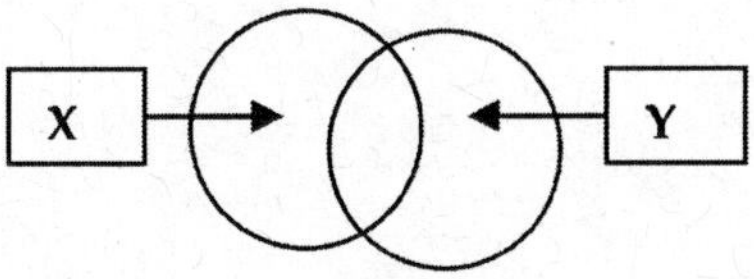

The **X**-circle and the **Y**-circle strictly overlap

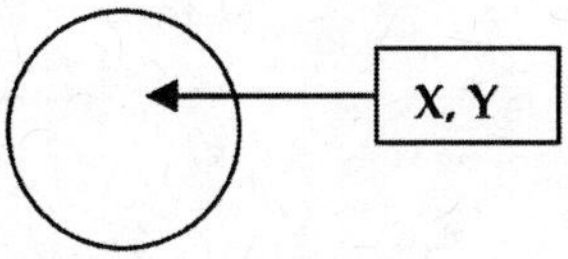

The **X**-circle and the **Y**-circle are identical

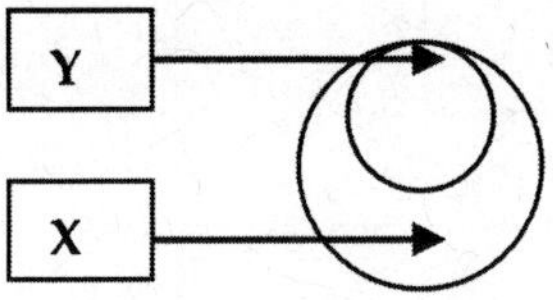

The **X**-circle strictly includes the **Y**-circle

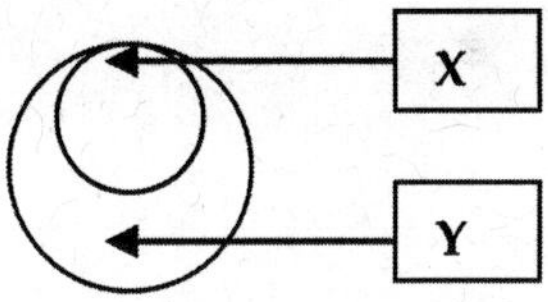

The **Y**-circle strictly includes the **X**-circle

(d) There are three types of diagrams on which a TL sentence of the form 'some **X** are not **Y**' is true.

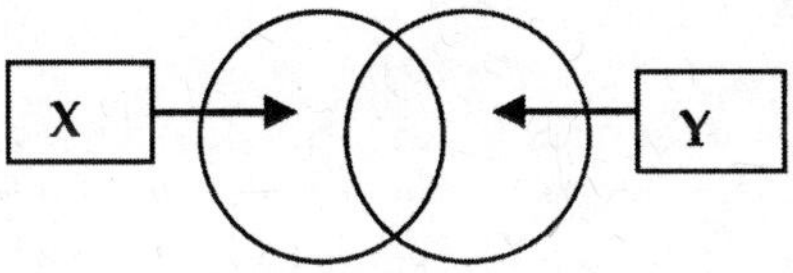

The **X**-circle and the **Y**-circle strictly overlap

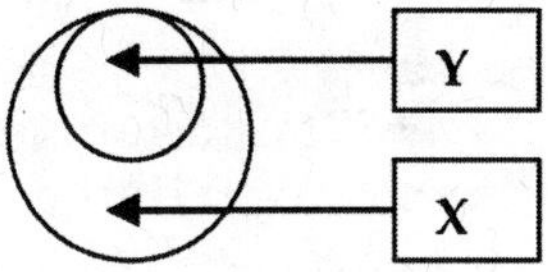

The **X**-circle strictly includes the **Y**-circle

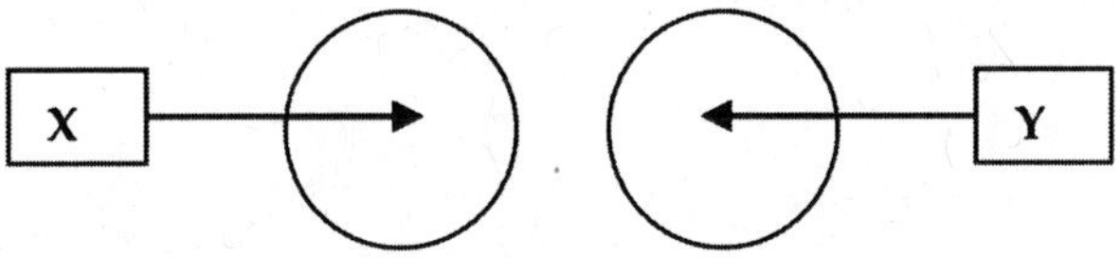

The **X**-circle and the **Y**-circle are disjoint

(e) There is one type of diagram on which a TL sentence of the form '**z** is **Y**' is true.

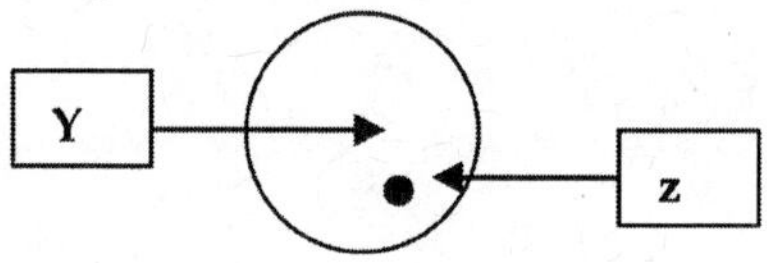

The **z**-dot is inside the **Y**-circle

(f) There is one type of diagram on which a TL sentence of the form '**z** is not **Y**' is true.

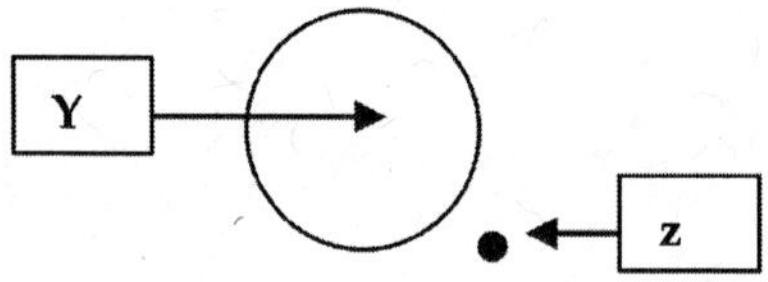

The **z**-dot is outside the **Y**-circle

(g) There is one type of diagram on which a TL sentence of the form '**x** is **z**' true.

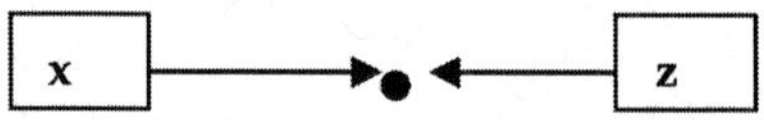

The **x**-dot and the **z**-dot are identical

(h) There is one type of diagram that makes a TL sentence of the form '**x** is not **z**' true.

The **x**-dot and the **z**-dot are distinct

2.4.4 For every TL sentence X in a set Γ of TL sentences, if X is true on a TL diagram D for Γ, then it is true on every TL diagram that is **similar** to D with respect to Γ; and if it is false on D, then it is false on every TL diagram that is similar to D with respect to Γ. Hence, under certain conditions, similar Γ diagrams behave as one diagram regarding the semantical and logical status of Γ.

2.4.4:C COMMENTARY ON 2.4.

We frequently use the expression 'up to similarity' in this chapter. In order to explain this expression, we introduce the notion of a **representative class of** TL **diagrams for** Γ. Let Γ be any nonempty set of TL sentences. Let Diag(Γ) be a collection of TL diagrams such that:

(a) every member of Diag(Γ) is a TL diagram for Γ,
(b) no two members of Diag(Γ) are similar to each other with respect to Γ, and
(c) if D is any TL diagram for Γ, then there is a TL diagram D* in Diag(Γ) that is similar to D with respect to Γ (D and D* could be identical).

Diag(Γ) here is a set of the maximum number of non-similar TL diagrams that are relevant to Γ.

There are infinitely many collections that satisfy (a)–(c) above. The reason is that for each TL diagram in one collection, there are infinitely many different but similar TL diagrams any one of which might replace it in another Diag(Γ) collection. There is, however, a one-to-one correspondence between any two such collections: the corresponding TL diagrams are similar to each other with respect to Γ. We call each set that satisfies the conditions (a)–(c), that is, every Diag(Γ), a "representative class of TL diagrams for Γ." Under certain conditions, which are determined by the context, a representative class of TL diagrams for Γ is sufficient for determining the semantical and logical status of Γ and of any of its members.

Now we are ready to explain the expression 'up to similarity'. When we say, for instance, that these TL diagrams are, *up to similarity*, all the TL diagrams for Γ, we mean that these TL diagrams are *all* the diagrams in some representative class of TL diagrams for Γ, that is, they are all the members of some Diag(Γ). It should be clear that if Γ is finite, every Diag(Γ) is also finite. Using the terminology we introduced, we say in this case that there is, up to similarity, a finite number of diagrams for Γ. Of course, the diagrams that are, up to similarity, all the TL diagrams for some set Γ of TL sentences are, by definition, non-similar to each other with respect to Γ.

2.5 Logical Concepts in TL

2.5.1 A TL **argument** is defined, similar to 1.1.1, as a nonempty collection of TL sentences: one of these sentences is the conclusion of the argument and the others (if any) are its premises.

2.5.2 The intuitive notion of a **logical possibility** that is relevant to a set of declarative sentences is characterized in TL as the well-defined notion of a **TL diagram** that is relevant to a set of TL sentences. We have already given a precise definition of what it means for a TL sentence to be true or false on a TL diagram, and we have fully identified the TL expressions that form grammatical TL sentences. (The TL sentences are meant to represent English declarative sentences.) Thus the **logical concepts** introduced in 1.2.4–1.2.11 can now be defined in TL by simply replacing the phrase 'logical possibility' with 'TL diagram' and making the necessary adjustments. The definitions of these concepts in TL are listed in 2.5.3–2.5.10.

2.5.3a A TL argument Γ/X is **deductively valid** (or Γ ⊨ X) if and only if on every TL diagram for Γ/X on which the members of Γ are all true **X** is true as well.

2.5.3b A TL argument Γ/X is **deductively valid** (or Γ ⊨ X) if and only if there is no TL diagram on which the members of Γ are all true and **X** is false.

2.5.3:C COMMENTARY ON 2.5.3

The following TL argument is deductively valid.

e is *A*
e is *B*
____
some *A* are *B*

We can show that it is deductively valid by considering, up to similarity, all the TL diagrams for the argument and showing that either (a) on every TL diagram on which the premises are true the conclusion is true as well, or (b) on every TL diagram on which the conclusion is false at least one of the premises is false. There is, however, a simpler way. Rather than considering *all* the TL diagrams for the argument we consider *only* those TL diagrams for the argument that (a) make the premises true or (b) make the conclusion false. If we follow the first option, then we need to show that the conclusion is true on the TL diagrams that make the premises true; and if we choose the second option, we need to show that at least one of the premises is false on every TL diagram that makes the conclusion false.

Since this is our first example, we will consider, up to similarity, *all* the TL diagrams for the argument and indicate the truth values of the sentences on them.

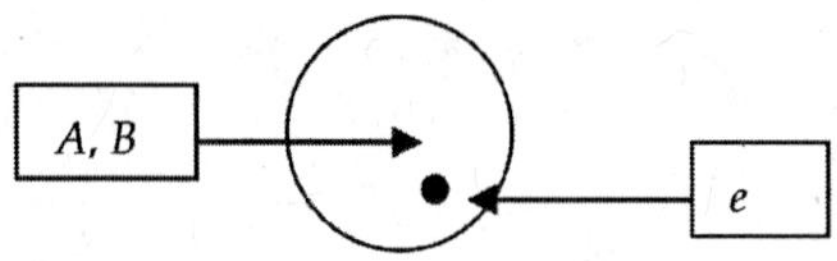

D1: *e* is *A* (true), *e* is *B* (true), some *A* are *B* (true)

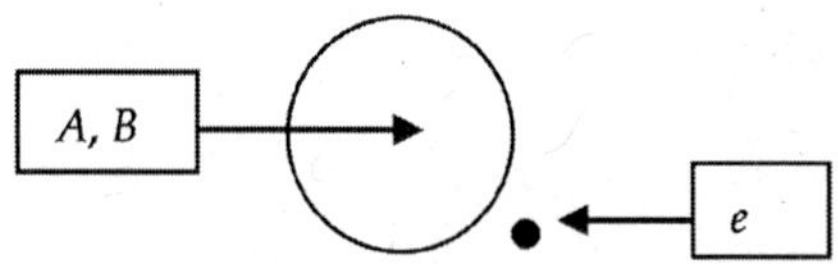

D2: *e* is *A* (false), *e* is *B* (false), some *A* are *B* (true)

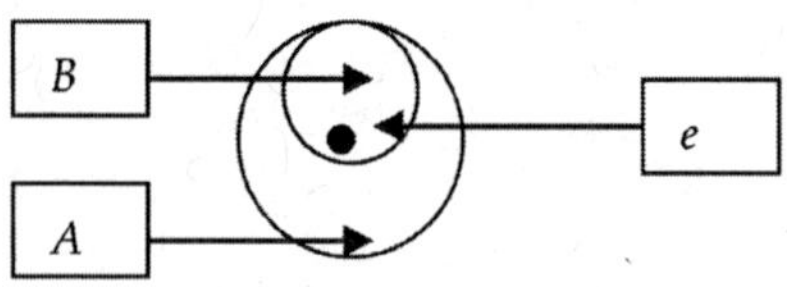

D3: *e* is *A* (true), *e* is *B* (true), some *A* are *B* (true)

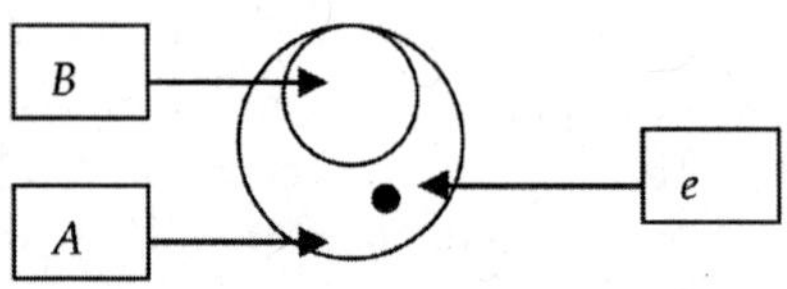

D4: *e* is *A* (true), *e* is *B* (false), some *A* are *B* (true)

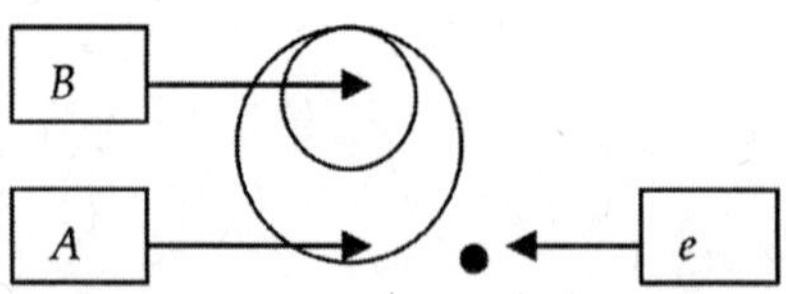

D5: *e* is *A* (false), *e* is *B* (false), some *A* are *B* (true)

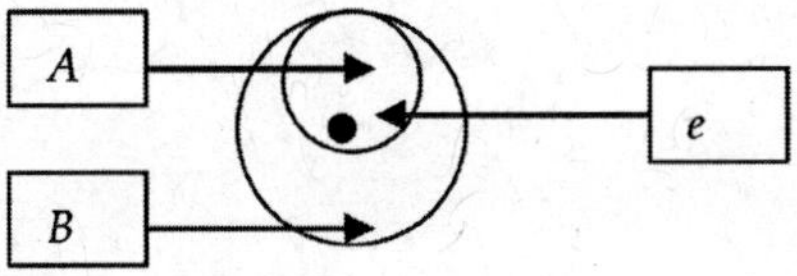

D6: *e* is *A* (true), *e* is *B* (true), some *A* are *B* (true)

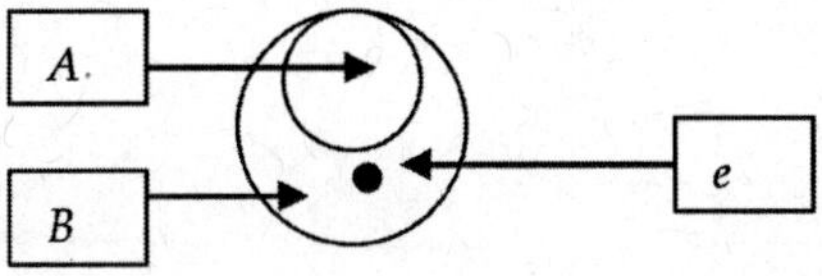

D7: *e* is *A* (false), *e* is *B* (true), some *A* are *B* (true)

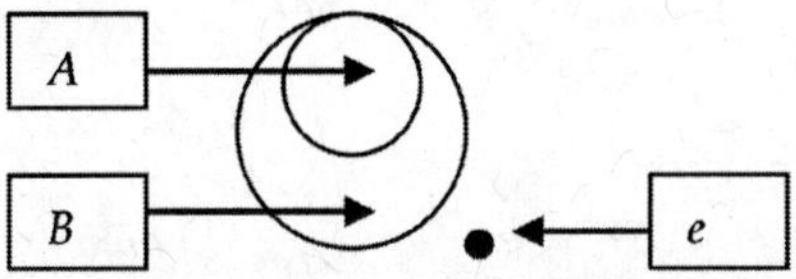

D8: *e* is *A* (false), *e* is *B* (false), some *A* are *B* (true)

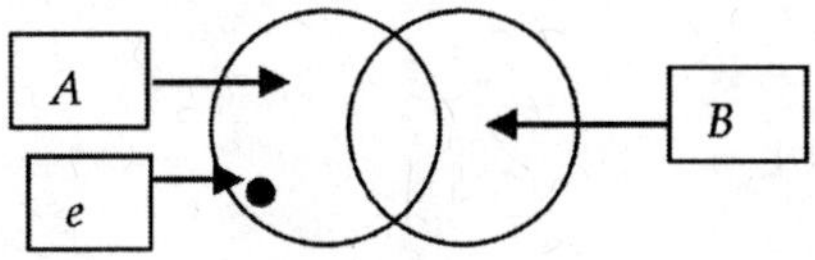

D9: *e* is *A* (true), *e* is *B* (false), some *A* are *B* (true)

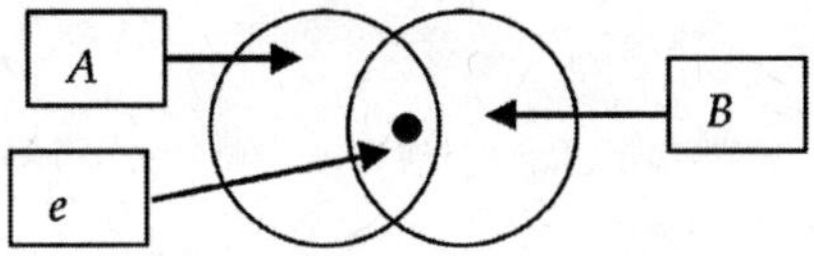

D10: *e* is *A* (true), *e* is *B* (true), some *A* are *B* (true)

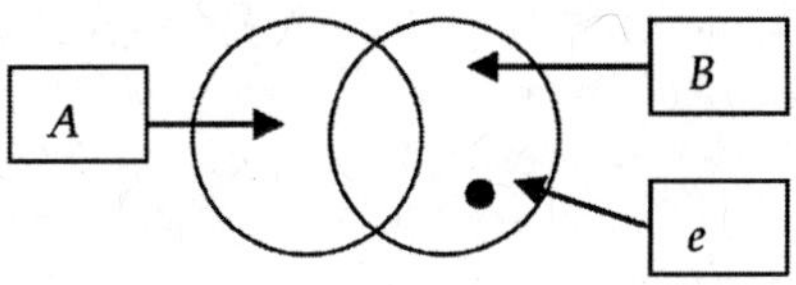

D11: *e* is *A* (false), *e* is *B* (true), some *A* are *B* (true)

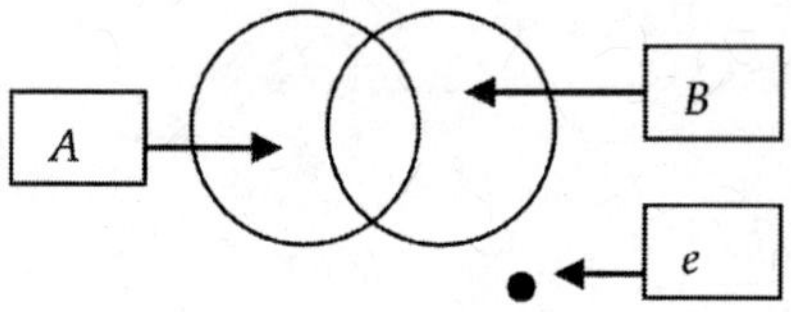

D12: *e* is *A* (false), *e* is *B* (false), some *A* are *B* (true)

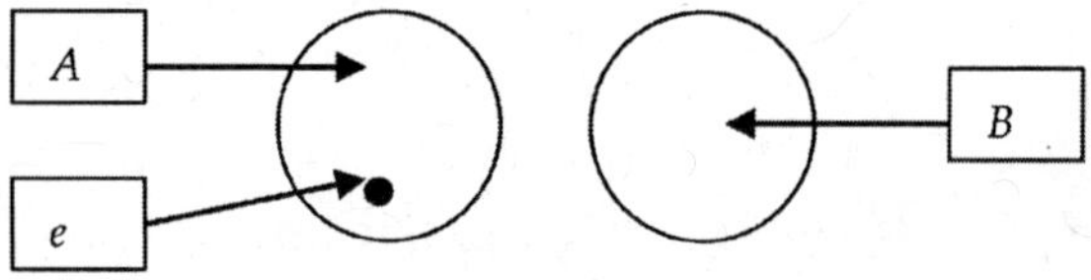

D13: *e* is *A* (true), *e* is *B* (false), some *A* are *B* (false)

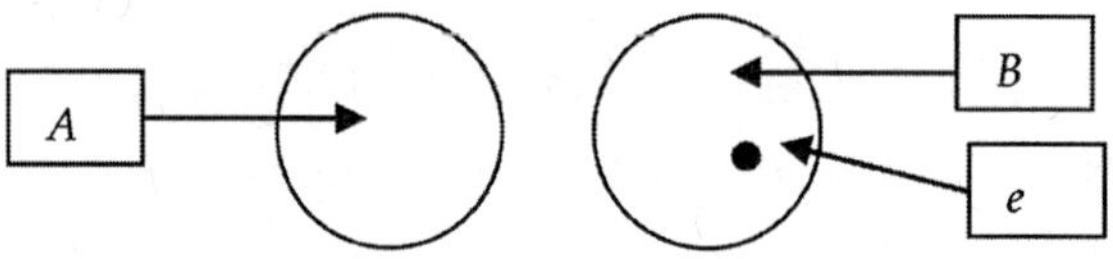

D14: *e* is *A* (false), *e* is *B* (true), some *A* are *B* (false)

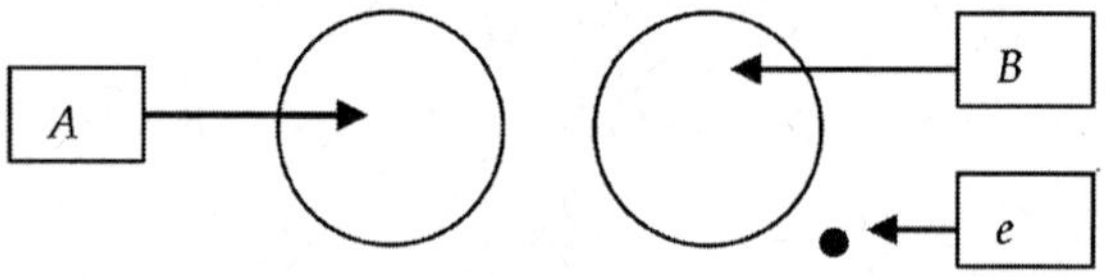

D15: *e* is *A* (false), *e* is *B* (false), some *A* are *B* (false)

The TL diagrams above are all the members of a certain representative class of TL diagrams for the argument. None of them makes the premises true and the conclusion false. Therefore, the TL argument is deductively valid.

It is clear that we could have made matters much easier for ourselves had we followed a strategy which looks only at diagrams in which (a), above, is the case; or one which looks only at diagrams in which (b), above, is the case. The first strategy—considering all the relevant TL diagrams on which the premises are true—requires that we focus on four diagrams, D1, D3, D6, and D10, and verify that on each one of them the conclusion is true as well. The second strategy—considering all the relevant TL diagrams on which the conclusion is false—requires that we focus only on D13, D14, and D15, and confirm that on each one of these diagrams at least one premise is false.

2.5.4 A TL argument Γ/X is **deductively invalid** (or $\Gamma \nvDash X$) if and only if there is a TL diagram on which all the members of Γ are true and X is false.

2.5.4:C COMMENTARY 2.5.4

The TL argument below is deductively invalid. There is a relevant TL diagram on which the premises are true and the conclusion is false.

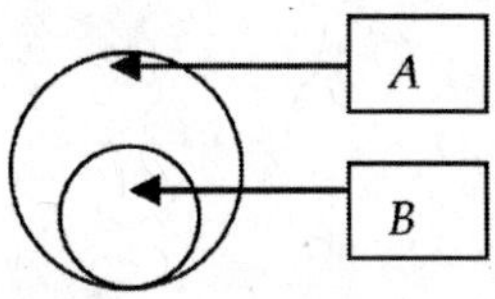

some *A* are *B*
some *A* are not *B*

---

some *B* are not *A*

some *A* are *B* (true), some *A* are not *B* (true), some *B* are not *A* (false)

2.5.5a A TL sentence is **logically true** if and only if it is true on every TL diagram for this sentence.

2.5.5b A TL sentence is **logically true** if and only if there is no TL diagram on which it is false.

2.5.5:C COMMENTARY ON 2.5.5

The TL sentence 'some *K* are *K*' is logically true. There is, up to similarity, only one TL diagram for this sentence and it makes the sentence true.

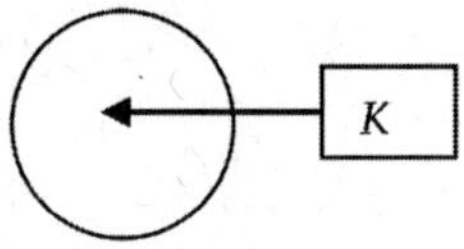

some *K* are *K* (true)

2.5.6a A TL sentence is **logically false** if and only if it is false on every TL diagram for this sentence.

2.5.6b A TL sentence is **logically false** if and only if there is no TL diagram on which it is true.

2.5.6:C COMMENTARY ON 2.5.6

The sentence '*e* is not *e*' is logically false. The sentence has, up to similarity, only one TL diagram and it makes the sentence false.

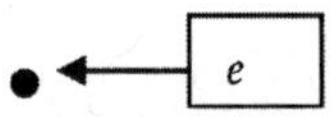

e is not *e* (false)

2.5.7 A TL sentence is **contingent** if and only if it is true on some TL diagrams and false on other TL diagrams.

2.5.7:C COMMENTARY ON 2.5.7

The TL sentence 'some *E* are not *K*' is contingent: it is true on D1 and false on D2.

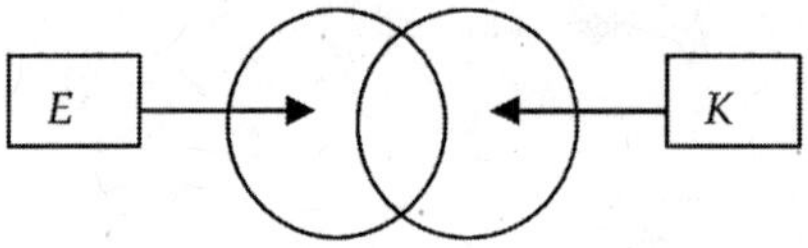

D1: some *E* are not *K* (true)

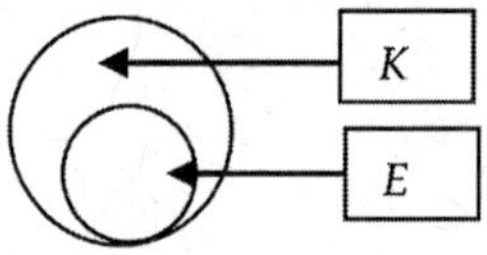

D2: some *E* are not *K* (false)

2.5.8a Two TL sentences are **logically equivalent** if and only if on every TL diagram for these sentences they have identical truth values.

2.5.8b Two TL sentences are **logically equivalent** if and only if there is no TL diagram on which they have different truth values.

2.5.8:C COMMENTARY ON 2.5.8

The TL sentences 'some *E* are *K*' and 'some *K* are *E*' are logically equivalent. There are, up to similarity, five TL diagrams for these sentences. On each of these diagrams the sentences have identical truth values.

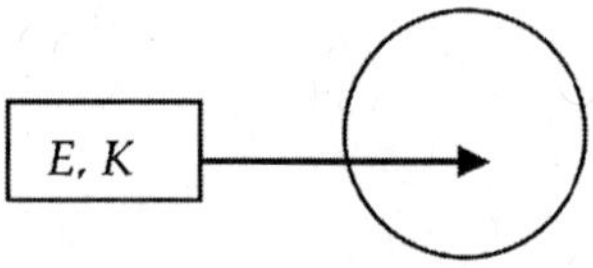

D1: some *E* are *K* (true), some *K* are *E* (true)

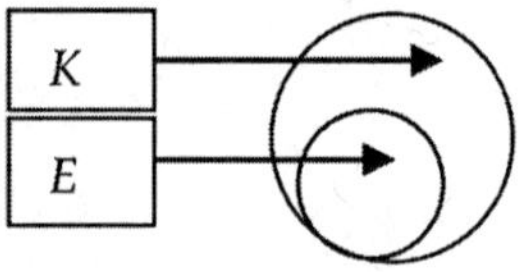

D2: some *E* are *K* (true), some *K* are *E* (true)

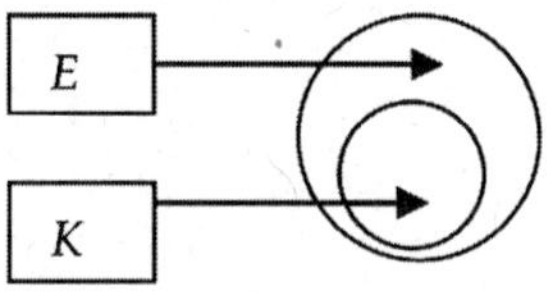

D3: some *E* are *K* (true), some *K* are *E* (true)

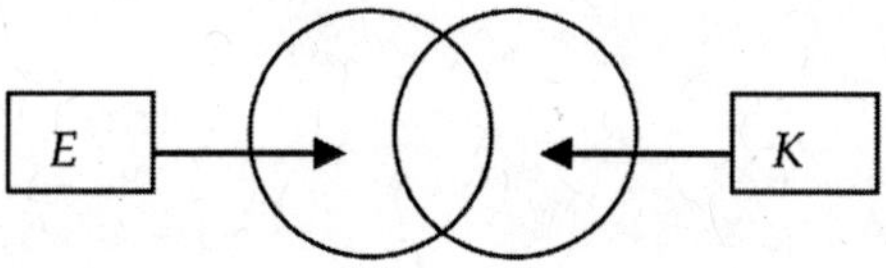

D4: some *E* are *K* (true), some *K* are *E* (true)

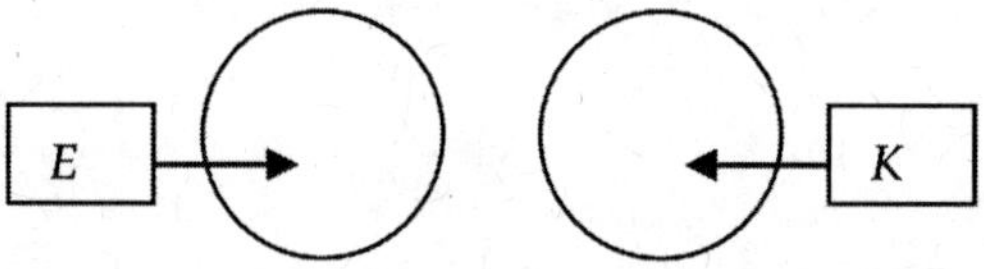

D5: some *E* are *K* (false), some *K* are *E* (false)

2.5.9 A set of TL sentences is **consistent** if and only if there is a TL diagram on which every member of the set is true, that is, there is a TL diagram that satisfies the set (or simply, if and only if the set has a TL model).

2.5.9:C COMMENTARY ON 2.5.9

Consider the following set of TL sentences: {*a* is not *b*, *a* is *M*, *b* is not *N*, all *N* are *M*}. The TL diagram below makes every member of the set true; hence the set is consistent.

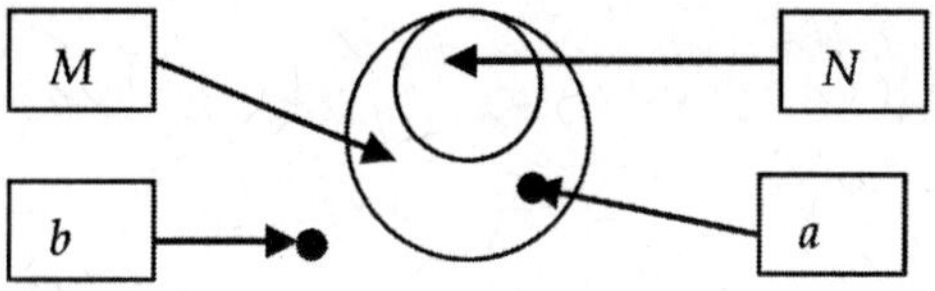

a is not *b* (true), *a* is *M* (true), *b* is not *N* (true), all *N* are *M* (true)

2.5.10a A set of TL sentences is **inconsistent** if and only if on every TL diagram for this set at least one member of the set is false.

2.5.10b A set of TL sentences is **inconsistent** if and only if there is no TL diagram on which all the members of the set are true, that is, there is no TL diagram that satisfies the set (or simply, if and only if the set has no TL model).

2.5.10:C COMMENTARY ON 2.5.10

Let us show that the following set is inconsistent: {*a* is *K*, all *K* are *N*, *b* is not *N*, *b* is *a*}. We do not have to consider *all* the non-similar TL diagrams for this set. It is sufficient to consider all the relevant non-similar diagrams that make as many as possible of the members of the set true and verify that on each of these diagrams at least one of the remaining members of the set is false. There are, up to similarity, two TL diagrams for this set on which the first three members of the set are true. Each one of these diagrams makes the last member, '*b* is *a*', false. The set, therefore, is inconsistent.

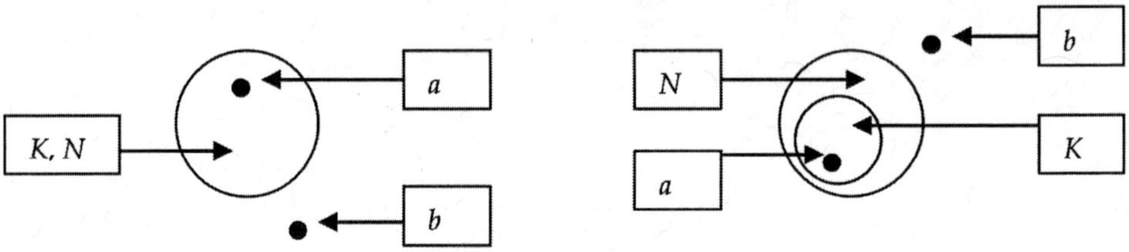

a is *K* (true), all *K* are *N* (true), *b* is not *N* (true), *b* is *a* (false)

2.5.11 We said in 2.4.4:C that there is, up to similarity, a finite number of TL diagrams for any given *finite* set of TL sentences. We also said that under certain conditions, which are determined by the context, an appropriate representative class of TL diagrams for a given set is sufficient for determining the semantical and logical status of the set and of any of its members. There is nothing creative about determining these conditions. For instance, if we are interested in deciding whether a TL argument Γ/X is valid or not, we should not consider a representative class of TL diagrams for Γ that are irrelevant to X. Rather, we should choose a representative class of TL diagrams for the argument Γ/X. Using the notation we introduced in 2.4.4:C, we say that any Diag(Γ/X) will be sufficient for determining the logical status of the argument Γ/X. On the other hand, if we are interested in finding whether Γ is consistent or not, any Diag(Γ) will be adequate for reaching the correct decision. So the process of choosing an appropriate Diag(Σ), where Σ is any finite set of TL sentences, can be specified as a deterministic mechanical procedure. Moreover, there are deterministic mechanical procedures for drawing these diagrams and determining the truth values of the members of Σ on each one of them. Thus *in principle* we can determine by surveying a finite list of relevant TL diagrams whether any of the logical concepts defined above is applicable or not in any given case if the set of TL sentences under consideration is finite. For this reason we say that these **logical concepts** are **decidable** in TL.

2.5.11:C COMMENTARY ON 2.5.11

A **concept is decidable** if and only if there is an **effective decision procedure** for determining whether something is subsumed under the concept or not. A decision procedure is effective if and only if it can, in principle, be followed mechanically (i.e., without any creative aspects) and after a finite number of deterministic steps leads to the correct answer, whether it is "Yes" or "No." Not every effective procedure is a decision procedure. An **effective procedure** is simply a deterministic mechanical procedure that after finitely many steps produces a certain result. There are effective procedures that produce the answer "Yes" when and only when the correct answer is "Yes," but that might not produce any answer when the correct answer is "No"; we refer to such procedures as 'effective Yes-procedures'. Also, there are effective procedures that produce the answer "No" when and only when the correct answer is "No," but that might not produce any answer when the correct answer is "Yes"; such procedures are referred to as 'effective No-procedures'. According to our definition here, an effective *decision* procedure must be able to produce *both* answers. A concept that has only an effective Yes-procedure is called a **semidecidable concept**. When we study Predicate Logic, we will encounter semidecidable concepts.

For any given finite set of TL sentences there are, up to similarity, finitely many TL diagrams. The process of describing these TL diagrams and determining the truth values of the given TL sentences on each of these diagrams is a mechanical process. This process provides us with an effective decision procedure for determining whether or not any of the eight logical concepts, defined in 2.5.3–2.5.10, applies in any given case, if the set of TL sentences under consideration is finite. We describe this fact by saying that these logical concepts are decidable in TL.

Let us illustrate this procedure by means of an example. Consider the following (finite) set of TL sentences: Γ = {*a* is *b*, *a* is *N*, *b* is *N*, all *N* are *N*, *b* is *a*, *a* is not *a*}. There are, up to similarity, six TL diagrams for Γ. Here is a list of these diagrams together with the truth values of the members of Γ on each diagram.

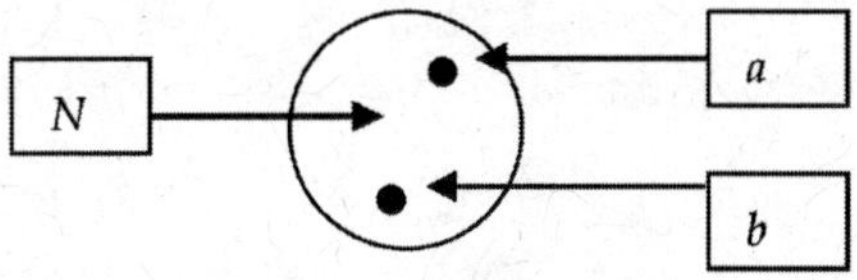

D1: *a* is *b* (false), *a* is *N* (true), *b* is *N* (true),
all *N* are *N* (true), *b* is *a* (false), *a* is not *a* (false)

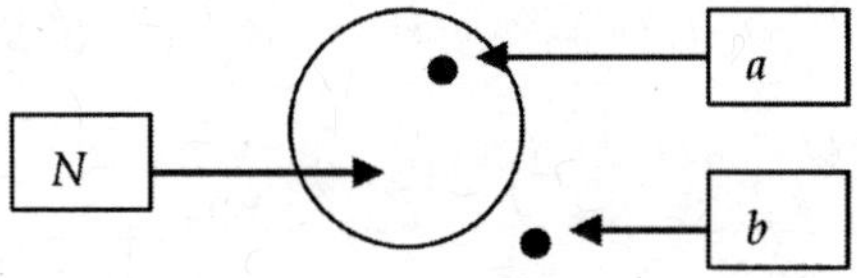

D2: *a* is *b* (false), *a* is *N* (true), *b* is *N* (false),
all *N* are *N* (true), *b* is *a* (false), *a* is not *a* (false)

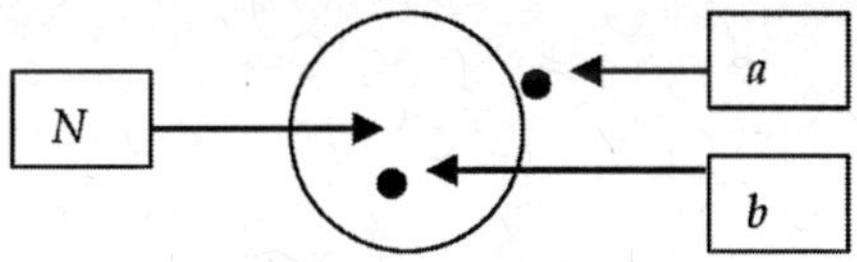

D3: *a* is *b* (false), *a* is *N* (false), *b* is *N* (true),
all *N* are *N* (true), *b* is *a* (false), *a* is not *a* (false)

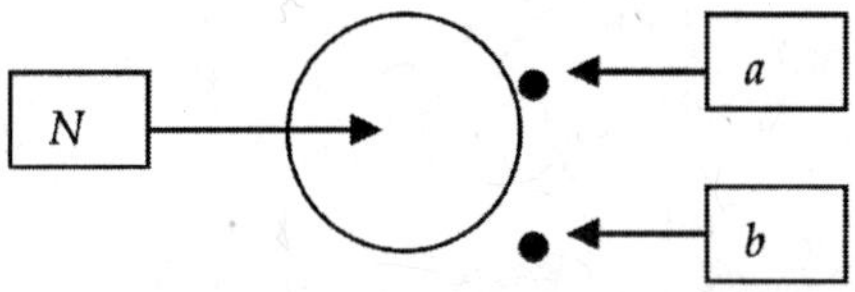

D4: *a* is *b* (false), *a* is *N* (false), *b* is *N* (false),
all *N* are *N* (true), *b* is *a* (false), *a* is not *a* (false)

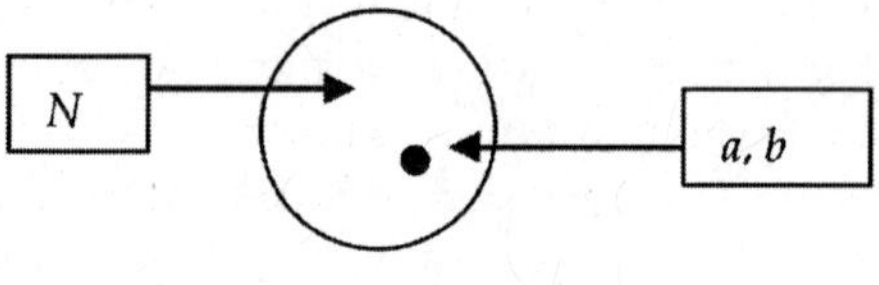

D5: *a* is *b* (true), *a* is *N* (true), *b* is *N* (true), all *N* are *N* (true), *b* is *a* (true), *a* is not *a* (false)

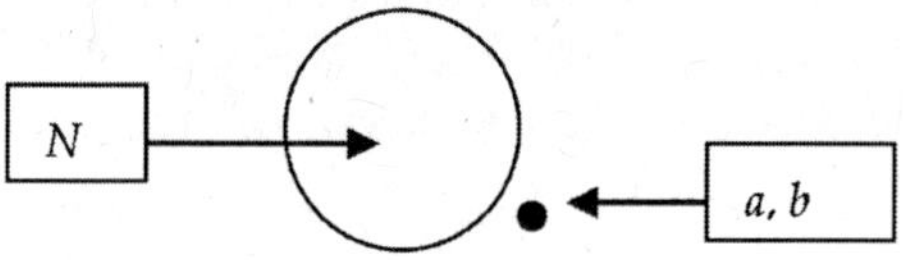

D6: *a* is *b* (true), *a* is *N* (false), *b* is *N* (false), all *N* are *N* (true), *b* is *a* (true), *a* is not *a* (false)

Now we are ready to answer any question about the applicability of the logical concepts defined in 2.5.3–2.5.10 to Γ, or members of Γ. Below are some of these questions and their answers. (It should be clear that D1–D6 include representative classes of TL diagrams for all the subsets of Γ.)

(a) Is Γ consistent? No. None of D1–D6 makes all the members of Γ true. (See 2.5.9.)

(b) Are '*a* is *b*' and '*b* is *a*' logically equivalent? Yes. On every TL diagram for these sentences, they have identical truth values. (See 2.5.8a.)

(c) Are '*a* is *N*' and '*b* is *N*' logically equivalent? No. D2 makes the first sentence true and the second false. (See 2.5.8b.)

(d) Is '*b* is *N*' a logical consequence of the set {'*a* is *b*', '*a* is *N*'}? Yes. On every relevant TL diagram (D5) on which the members of the set are true, '*b* is *N*' is true as well. (See 2.5.3a.)

(e) Is '*a* is *N*' logically true? No, it is contingent. '*a* is *N*' is true on D1 and false on D3. (See 2.5.5b and 2.5.7.)

(f) Is 'all *N* are *N*' contingent? No, it is logically true. 'all *N* are *N*' is true on all the TL diagrams for it. (See 2.5.7 and 2.5.5a.)

(g) Is '*a* is not *a*' contingent? No, it is logically false. The sentence is false on every TL diagram for it. (See 2.5.7 and 2.5.6a.)

(h) Is the argument {'*a* is *N*', '*b* is *N*'}/'*a* is *b*' deductively invalid? Yes. On D1 the premises of the argument are true and the conclusion is false. (See 2.5.4.)

(i) Is the set Δ whose members are all the members of Γ except for the sentence '*a* is not *a*' consistent? Yes. D5 makes every member of Δ true. (See 2.5.9.)

We said that these logical concepts are decidable in TL. They are decidable because their applicability can be determined by the effective decision procedure illustrated above. However, they are decidable *in principle* because this procedure is not practical if Γ is large (add the sentence 'some *N* are *K*' to Γ and see how many (non-similar) TL diagrams the new set has).

2.5.12 **The TL worldview is incomplete**. It correctly identifies certain types of individuals and certain types of properties as basic ingredients of reality, but there are other types of individuals and properties and other basic ingredients of reality that the TL worldview does not recognize. For example, there are relations, such as the relation "being a sister of"; fictitious objects, such as Sherlock Holmes; and properties with empty extensions, such as the property "being a living US President who is seven feet tall." Perhaps it is not fair to fault the TL worldview for excluding fictitious objects, as many philosophers do not believe that there are such objects. But properties with empty extensions as well as relations seem to be part of the landscape of reality. Because the TL worldview is incomplete, there are TL diagrams that fail to represent any relevant logical possibilities and there are, *arguably*, relevant logical possibilities that are not represented by any TL diagrams. This conclusion has an important implication regarding the use of TL tools to test for the deductive validity or invalidity of natural-language arguments. The presence of TL diagrams that do not correspond to relevant logical possibilities and of relevant logical possibilities that do not correspond to TL diagrams implies that there are natural-language arguments that cannot be translated *faithfully* into TL. A TL argument is a **faithful translation** of a natural-language argument if and only if the interdependencies that exist between the terms of the TL argument are *precisely those* interdependencies that exist between the terms of the natural-language argument. In order for the deductive validity or invalidity of a TL argument to be indicative of the logical status of the natural-language argument of which the TL argument is a translation, the TL argument must be a faithful translation of the natural-language argument. Determining whether a TL argument faithfully translates a natural-language argument is not an exact science: the faithfulness of a translation cannot be determined by precise methods. But in most cases we should be able to decide whether a TL translation distorts the original argument or is a correct formulation of it.

2.5.12:C COMMENTARY ON 2.5.12

2.5.12:C1 First, the conclusion stated above needs to be made more precise. In a certain sense, every TL diagram represents a logical possibility if there is no restriction on the range of logical possibilities considered. But if we restrict the range of logical possibilities to those that are *relevant* to the situation at hand, then, as stated above, there are cases in which some TL diagrams fail to represent any of the relevant logical possibilities, and there *seem* to be cases in which some relevant logical possibilities are not represented by any TL diagrams. We need to explain the notion of a relevant logical possibility and the notion of a TL diagram that represents a logical possibility.

The first notion was already introduced in 1.2.1:C2. We revisit it here. Let L be some natural language, Σ a set of declarative sentences of L, and p any logical possibility. Roughly speaking, p is said to be a logical possibility *for* Σ, or that is *relevant to* Σ, if and only if every sentence in Σ makes a (true or false) assertion about some or all of the constituents of p. Recall that the L expressions that occur in the members of Σ must have the same meanings across all the logical possibilities for Σ. Furthermore, we always assume that the L expressions that correspond to the logical words of TL have the same meaning in all the logical possibilities that are relevant to any set of declarative sentences of L.

We now define the relation of representation. In order to define this relation, we need to introduce some important notation. As above, Σ is a set of declarative sentences of a natural language L. We assume that a specific translation key is given. A translation key for Σ is a list that associates the L general and singular terms that occur in the members of Σ with TL general and singular terms.

(a) Σ^{TL} is the set of the TL sentences that are the translations of the sentences in Σ according to the given translation key.
(b) Voc(Σ^{TL}) is the set that consists of the logical words of TL and of the TL basic vocabulary that occurs in the members of Σ^{TL}, that is, the TL basic vocabulary of which the members of Σ^{TL} are composed.
(c) Voc(Σ) is the set that consists of L expressions that correspond to all the logical words of TL, and of the L expressions that are translated into the general and singular terms in Voc(Σ^{TL}).
(d) The TL sentence X^{TL} that is composed of Voc(Σ^{TL}) is a translation of the L sentence X that is composed of Voc(Σ).

It is important to note that we are not claiming that every sentence composed of Voc(Σ) must have, on the basis of the given translation key, a TL translation composed of Voc(Σ^{TL}). Natural languages are significantly richer and more flexible than any formal language. This implies that there might be cases of Σ and Σ^{TL} such that some of the sentences that are composed of Voc(Σ) cannot be translated into any TL sentences composed of Voc(Σ^{TL}) if the translation key is held fixed. In fact, it is not hard to think of an example of such a case. Say the TL general terms *D*, *K*, and *G*, translate the English general terms 'doors', 'keys', and 'objects that are green' respectively. According to this translation key, we can translate many English sentences into TL. For instance 'All doors are green' is translated as 'all *D* are *G*', 'No key is a door' as 'no *K* is *D*', 'Some green objects are keys' as 'some *G* are *K*', and 'Some doors are not green objects' as 'some *D* are not *G*'. However, according to this translation key, there is no TL translation of the sentence 'Every door key is green'. 'Door key' is a compound general term, and TL has no means for constructing compound terms out of their simpler components. (We could, of course, introduce an additional general term *C* to translate 'doorkey'.) We are ready to define the relation of representation.

Representation: A TL diagram D for Σ^{TL} **represents** a logical possibility p for Σ if and only if for every TL sentence **Z** that is composed of Voc(Σ^{TL}) and for every L sentence X that is composed of Voc(Σ), if **Z** is a translation of X (i.e., **Z** is X^{TL}), then **Z** is true (or false) on D if and only if X is true (or false) in p.

We can now state the conclusion of 2.5.12 more precisely. There are cases of Σ and Σ^{TL}, such that some of the TL diagrams for Σ^{TL} fail to represent any of the logical possibilities for Σ, and there *seem* to be cases of Σ and Σ^{TL}, such that some of the logical possibilities for Σ are not represented by any TL diagrams for Σ^{TL}.

2.5.12:C2 In this subsection we make a few important observations about the definition, stated above, of the relation of representation. First, a TL diagram represents a logical possibility always with respect to some set of L sentences and some set of TL translations of these L sentences. So, precisely speaking, we should say that D represents p with respect to Σ-Σ^{TL}. However, we will try to avoid this cumbersome terminology. If the sets Σ and Σ^{TL} are given within the context, we will simply speak of D as representing p. If the context does not clearly specify these sets, we will say that a TL diagram for Σ^{TL} represents a logical possibility for Σ.

Second, the definition implies that a certain *structural correspondence* holds between constituents of D and constituents of p. Observe that we did not say that there is a structural correspondence between *the* constituents of D and *the* constituents of p. In most cases such a correspondence is partial: it holds only between some constituents of D and some constituents of p. Furthermore, the structural similarity between D and p is typically imperfect: D and p could have some structural similarities and many structural dissimilarities. The structural correspondence between D and p is typically restricted to the relations between constituents that can be described by TL sentences composed of Voc(Σ^{TL}) and L sentences composed of Voc(Σ) that the TL sentences translate.

To illustrate this point, we consider an example. Suppose that a TL diagram D for Σ^{TL} represents a logical possibility p for Σ. Further suppose that *t* and *r* are two singular terms and *A*, *J*, *B*, *G*, and *S* are general terms in Voc(Σ^{TL}). The structural relations between the *t*-dot, *r*-dot, *A*-circle, *J*-circle, *B*-circle, *G*-circle, and *S*-circle are depicted in the following diagram of D.

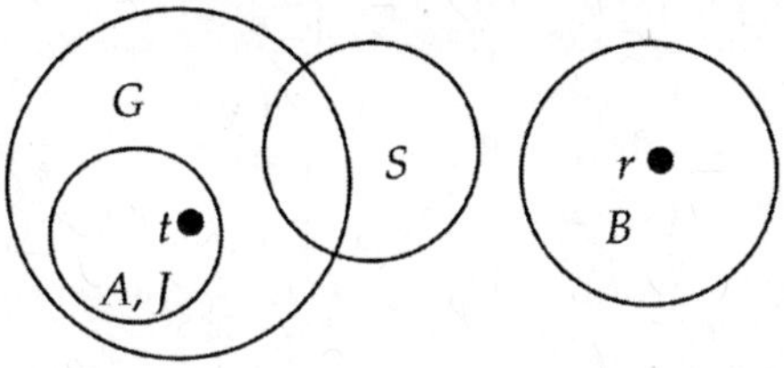

Since *t*, *r*, *A*, *J*, *B*, *G*, and *S* are terms in Voc(Σ^{TL}), there must be two singular terms and four general terms in Voc(Σ) that the TL terms translate. Say, *t* translates 'The fruit that fell on Newton's head', *r* translates 'The fruit that Johnny ate yesterday', *A* translates 'Apples', *J* translates 'Juicy', *B* translates 'Bananas', *G* translates 'Green', and *S* translates 'Sour'. The definition of representation requires that all the TL sentences that are true (or false) on D translate sentences that are true (or false) in p. Thus we have the following structural relations in p.

1. The extensions of 'Apple' and 'Juicy' are identical in p.
   Reason: Since the TL sentences 'all *A* are *J*' and 'all *J* are *A*' are true on D, and since they are translation of the English sentences 'All apples are juicy' and 'All juicy things are apples', the English sentences are also true in p.
2. The extension of 'Green' strictly includes the extension of 'Apple'.
   Reason: The English sentences 'All apples are green' and 'Some green things are not apples' are true in p because their TL translations 'all *A* are *G*' and 'some *G* are not *A*' are true on D.
3. The extension of 'Green' strictly overlaps with the extension of 'Sour'.
   Reason: The English sentences 'Some green things are sour'. 'Some green things are not sour', and 'Some sour things are not green' are true in p, since their TL translations 'some *G* are *S*', 'some *G* are not *S*', and 'some *S* are not *G*' are true on D.
4. The extensions of 'Apple' and of 'Sour' are disjoint.

Reason: The English sentence 'No apple is sour' is true in p since its TL translation 'no *A* is *S*' is true on D.

5. The extensions of 'Banana' and 'Sour' are disjoint, and the extensions of 'Banana' and 'Green' are also disjoint.
   Reason: The English sentences 'No banana is sour' and 'No banana is green' are true in p because their TL translations 'no *E* is *S*' and 'no *E* is *G*' are true in p.
6. The referents of 'The fruit that fell on Newton's head' and 'The fruit that Johnny ate yesterday' are not identical.
   Reason: Since the TL sentence '*t* is not *r*' is true on D and it is a translation of the English sentence 'The fruit that fell on Newton's head is not the fruit that Johnny ate yesterday', the English sentence is true in p.[1]
7. The referent of 'The fruit that fell on Newton's head' belongs to the extension of 'Apples'.
   Reason: The English sentence 'The fruit that fell on Newton's head was an apple' is true in p because its TL translation '*t* is *A*' is true on D.
8. The referent of 'The fruit that Johnny ate yesterday' belongs to the extension of 'Bananas'.
   Reason: The English sentence 'The fruit that Johnny ate yesterday was a banana' is true in p since its TL translation '*r* is *B*' is true on D.

Thus the *t*-dot and *r*-dot in D correspond to the referents of 'The fruit that fell on Newton's head' and 'The fruit that Johnny ate yesterday' in p, respectively, and the *A*-circle, *J*-circle, *B*-circle, *G*-circle, and *S*-circle in D correspond to the extensions of 'Apples', 'Juicy', 'Bananas', 'Green', and 'Sour' in p, respectively; and the structural relations between these two dots and five circles in D precisely mirror the structural relations between the corresponding individuals and extensions in p.

The third observation we make about the definition of representation is that the preceding discussion discloses a certain feature of TL. TL is an **extensional system**: the circles of a TL diagram represent extensions of properties and not the properties themselves. Coextensive properties, that is, properties that have the same extension, are represented in a TL diagram by a single circle. We should emphasize that the TL worldview need not be extensional: properties need not be reduced to their extensions. Consider a hypothetical example. Say there is a logical possibility in which the Lehigh students who are enrolled in Calculus I at time t are, by sheer coincidence, exactly the same students who are enrolled in Symbolic Logic at t. Call this logical possibility "q." Hence in q, the properties "being a Lehigh student who is enrolled in Calculus I at t" and "being a Lehigh student who is enrolled in Symbolic Logic at t" are coextensive. Any TL diagram D that represents q must represent these two properties as a single circle. To be precise, what D represents is not these two "different" properties but their common extension. However, the logical possibility q is independent of its TL representation: what goes on in D need not determine what goes on in q. If D can represent only the extensions of these properties, q is under no requirement to treat these two properties as one property. A TL diagram does not have the wherewithal to capture the fact that two coextensive properties might be independent of each other. The TL worldview, on the other hand, does not have to suffer from this limitation. The basic ingredients of reality in this worldview are individuals

1 This is not an independent fact. It is entailed by the rest of the facts mentioned in this list.

and properties. Extensions could be treated as non-basic ingredients of reality: their existence depends on the existence of properties and individuals.

There are many philosophers who reduce properties to their extensions. Those philosophers, therefore, believe that if two properties in some logical possibility happen to have the same extension, then they must be the same property (in that logical possibility). We will not debate this position here. We will remain neutral about whether the TL worldview is extensional or not. We simply note that, although the semantics of TL is extensional, the TL worldview need not be extensional. This position does not rule out the possibility that there might be very good reasons for attributing extensionality to the TL worldview or for denying that it is an extensional worldview. We will simply refrain from making a philosophical commitment either way. Our only commitment is that the extensionality of the semantics of TL does not necessitate that the foundation of TL be also extensional.

Although TL semantics is extensional, and hence circles represent extensions of properties and not properties, we will allow ourselves the latitude of speaking of a circle in a TL diagram as representing a property in some logical possibility. So long as no misunderstanding is expected, we will "pretend" that TL circles represent properties. In situations where this pretense is likely to cause some misunderstanding, we will refrain from speaking in this manner; we will make it explicit that the circles of TL diagrams represent only extensions of properties.

2.5.12:C3 Before we discuss the truth status of the conclusion stated at the end of 2.5.12:C1, we need to give an example that illustrates the notions and notation we introduced in the preceding subsections. Consider the following set Σ of English declarative sentences and their TL translations. We will always follow a two-step procedure in translating sets of English sentences into a symbolic language. We first give a translation key of English expressions into the basic vocabulary of the symbolic language, and then we give translations of the English sentences using the translation key.

The set Σ

S1	No penguin is a mammal.
S2	All whales are mammals.
S3	Some mammals can swim.
S4	Lassie is a mammal.

Translation Key

P: Penguins
M: Mammals
W: Whales
S: Individuals that can swim.
l: Lassie

The Set Σ^{TL}

$S1^{TL}$	no *P* is *M*
$S2^{TL}$	all *W* are *M*
$S3^{TL}$	some *M* are *S*
$S4^{TL}$	*l* is *M*

$Voc(\Sigma^{TL})$ = {*P*, *M*, *W*, *S*, *l*, all, some, no, not, is, are}
Voc(Σ) = {penguins, mammals, whales, individuals that can swim, Lassie, all, some, no, not, is, are}

Examples of **X** and $\mathbf{X}^{TL}$

S5	Lassie is a whale.	$S5^{TL}$	*l* is *W*
S6	All whales can swim.	$S6^{TL}$	all *W* are *S*
S7	No penguin can swim.	$S7^{TL}$	no *P* is *S*
S8	Some mammals are not whales.	$S8^{TL}$	some *M* are not *W*
S9	Some individuals that can swim are not mammals.	$S9^{TL}$	some *S* are not *M*

Note that we allow a certain degree of flexibility in constructing sentences that are composed of Voc(Σ). For instance, 'No penguin can swim' uses the singular form 'penguin', which is not in Voc(Σ), and 'Lassie is a whale' uses the indefinite article 'a' and the singular form 'whale', none of which is in Voc(Σ). We will try to stay as close as possible to the precise forms of the expressions in Voc(Σ), but we will also allow ourselves a certain degree of freedom in order to avoid writing sentences that may sound less natural, ambiguous, or even infelicitous, such as 'No penguin is individuals that can swim'. After all, we are dealing with a natural language and not a formal language. Formal languages are far more rigid than natural languages.

D below is a TL diagram for Σ^{TL}.

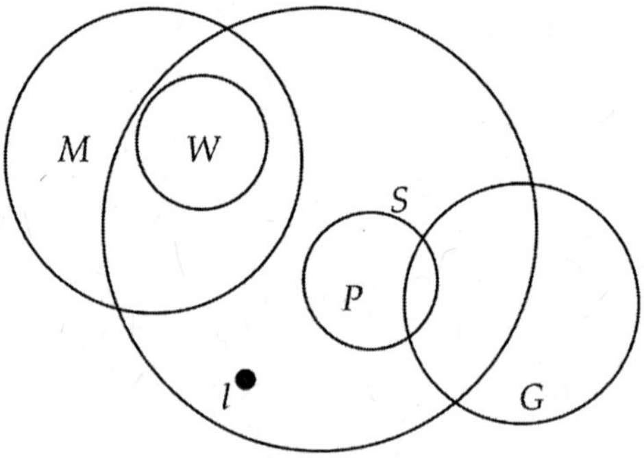

D: no *P* is *M* ($S1^{TL}$, true); all *W* are *M* ($S2^{TL}$, true); some *M* are *S* ($S3^{TL}$, true); *l* is *M* ($S4^{TL}$, false)
l is *W* ($S5^{TL}$, false); all *W* are *S* ($S6^{TL}$, true); no *P* is *S* ($S7^{TL}$, false);
some *M* are not *W* ($S8^{TL}$, true); some *S* are not *M* ($S9^{TL}$, true)

We describe a logical possibility p for Σ that is represented by D. p consists of *only* the following individuals and properties.

A shark called 'Lassie'
A cat called 'Simsim'
The property "being a penguin"
The property "being a whale"
The property "being a mammal"
The property "being able to swim"
The property "being domesticated"

Finally, we describe the relations that hold between the constituents of p. Lassie the shark is neither a mammal nor a penguin, but it can swim. Simsim the cat is a mammal that cannot swim, and it is the only domesticated individual in p. (So, in p, the extension of the property "being domesticated" is the singleton of Simsim.) All whales are mammals and they are capable of swimming. Penguins can swim but are not mammals. There are mammals that are not whales but can swim. All these properties are assumed to be *bivalent* and have nonempty extensions. In 1.2.2 we introduced the notion of a bivalent sentence. Bivalence can be extended to properties: a property Q is bivalent if and only if every individual either instantiates or lacks Q, but it cannot instantiate and lack Q at the same time. It is clear that p is a logical possibility *for* Σ since every member of Σ makes an assertion about constituents of p, and as far as we can tell there is no explicit or implicit contradiction in the description we gave of p.

In order to argue that D and p satisfy the condition stated in the definitions of representation (recall that this condition, which we will call the **Representation Condition**, states that for every X^{TL}, X^{TL} is true (false) on D if and only if X is true (false) in p), we note the following correspondence relations between the constituents of D and p. These relations are entailed by the translation key and the descriptions of D and p.

1. The *l*-dot corresponds to a shark whose name is 'Lassie'.
2. The *P*-circle corresponds to the property "being a penguin."
3. The *W*-circle corresponds to the property "being a whale."
4. The *M*-circle corresponds to the property "being a mammal."
5. The *S*-circle corresponds to the property "being able to swim."
6. The relations between the constituents of D that are the referents of the general and singular terms in Voc(Σ^{TL}) mirror precisely the relations between the corresponding constituents of p.

It is possible to give a demonstrative proof that the Representation Condition holds for D and p. After all, there are only finitely many TL sentences that can be composed of Voc(Σ^{TL}); we can compare the truth values that these sentences have on D with the truth values, in p, of the English sentences they translate, and verify that all corresponding sentences have identical truth values. This, however, is a tedious proof. Hence we will be content with an intuitive argument for the claim that the Representation Condition holds for D and q. Let X be any sentence composed of Voc(Σ) that has a TL translation X^{TL} composed of Voc(Σ^{TL}). X^{TL} makes an assertion about constituents of D that are the referents of general and singular terms in Voc(Σ^{TL}). Since these constituents of D correspond to constituents of p and X^{TL} is a translation of X, it is plausible to assume that X makes a similar assertion about the corresponding constituents of p. In general, the truth value of a sentence depends on the assertion it makes and the relations that hold between the constituents that are mentioned in the sentence: if the sentence describes these relations correctly, it is true; and if it describes them incorrectly, it is false.[1] Hence the truth values of X^{TL} and X depend on the assertions they make and the relations that hold between the constituents that are mentioned in these sentences. Given that X^{TL} and X make "similar" assertions about constituents that correspond to each other, and given that the relations between the constituents of D that are the referents of the general and singular terms in Voc(Σ^{TL}) mirror precisely the relations between the corresponding constituents of p, it is reasonable to expect that X^{TL} and X have the same truth value. The conclusion of this argument is that D and p satisfy the Representation Condition. Hence D represents p.

2.5.12:C4 We will refer to a TL diagram that represents a relevant logical possibility as **genuine** and to one that represents no relevant logical possibility as **superfluous**. There are cases in which the set of genuine TL diagrams is a *proper subset* of the set of all relevant TL diagrams. To see an example of a superfluous TL diagram, consider the following argument.

P All the Chess Club members are students in Symbolic Logic.

C All biological mothers of the Chess Club members are biological mothers of students in Symbolic Logic.

Suppose that part of the meaning of 'biological mother' is that everyone has exactly one biological mother.[2] According to the definition of relevant logical possibility, this meaning must be held fixed in all the logical possibilities that are relevant to this argument. There are many logical possibilities for this argument. These logical possibilities, however, may be divided

---

1 The notion of truth described in this paragraph is a version of the *correspondence conception* of truth. A common formulation of this conception is that a sentence (or a proposition) is true if and only if it corresponds to the facts. The correspondence conception of truth is the most popular conception among philosophers. By choosing a version of this conception as our operative notion of truth in this argument, we are not making a philosophical commitment to it. The argument could be articulated using a different conception of truth. However, the correspondence conception seems most natural for this argument. In Chapter Three, we will see that the semantics of Sentence Logic is committed to the correspondence conception of truth.

2 The reason for this restrictive meaning is to guard against certain interpretations of 'biological mother' that would allow someone to have more than one biological mother. If this is possible, then one could offer different interpretations of the general terms 'biological mother of a Chess Club member' and 'biological mother of a student in Symbolic Logic'; say the first term means "the youngest biological mother" and the second "the oldest biological mother." According to this reading, it is possible for someone who has several biological mothers to be a member of the Chess Club and a student in Symbolic Logic while his biological mother as a member of the Chess Club is not his biological mother as a student of Symbolic Logic. Such a possibility would allow for the premise to be true and the conclusion false. This entails that the argument would be deductively invalid. It is worth noting that if 'being a biological mother of' is interpreted as designating a relation (and not a property) between two individuals, then this argument would be deductively valid without further restrictions on the meaning of 'biological mother'. We demonstrate this fact in Chapter Five, when we present a translation of a structurally similar argument (Exercises 5.5.7c). If one believes that the restriction we placed on the meaning of the term 'biological mother' is unwarranted, then one is free to add a second premise to the argument asserting that everyone has one and only one biological mother. The resulting argument is still invalid in TL.

into three types: logical possibilities in which the premise and conclusion are true, logical possibilities in which the premise and conclusion are false, and logical possibilities in which the premise is false and the conclusion is true. The reader can easily construct examples of the first two types. Here is an example of the third type. Imagine that there are only three members in the Chess Club—Jody, Amar, and James—and that Amar and James are students in Symbolic Logic but Jody is not. Thus P1 is false in this logical possibility. Imagine further that Jody and Jill have the same biological mother, and that Jill is a student in Symbolic Logic. So Jody's biological mother is the biological mother of a student in Symbolic Logic. The conclusion, hence, is true in this logical possibility.

Let us now translate this English argument into TL.

Translation Key

C: Members of the Chess Club
S: Students in Symbolic Logic
M: Biological mothers of Chess Club members
N: Biological mothers of students in Symbolic Logic

TL Argument

S1 all *C* are *S*

S2 all *M* are *N*

Unlike the English argument, the TL argument presents no logical relation between the premise and the conclusion. The premise and conclusion of the English argument are connected by the phrase 'biological mother of', which stands for the *relation* of (biological) motherhood. TL cannot capture this connection because the basic vocabulary of TL contains no symbols for relations. TL basic vocabulary is a reflection of its worldview, which does not recognize relations as basic ingredients of reality. Hence there is an interdependency between the premise and the conclusion of the English argument, which is not captured by the TL translation. According to the definition of 'faithful translation', this TL argument is not a faithful translation of the original English argument. Since Predicate Logic (Chapters Four and Five) has the means to express relations, it is an adequate system for capturing the logical relations that exist between the various parts of the English arguments.

There are many TL diagrams for the TL argument. We will focus on four of them.

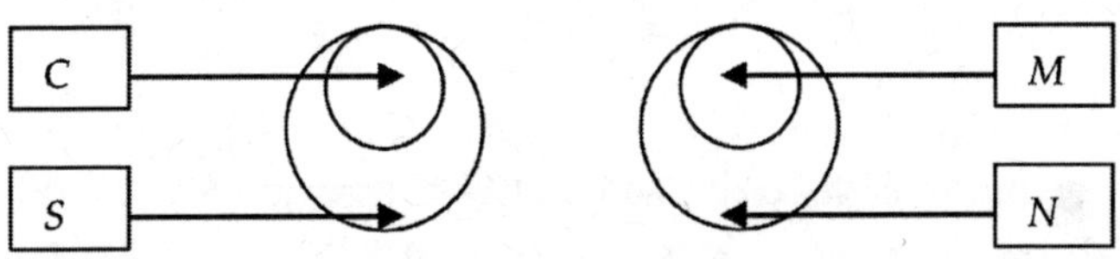

D1: all C are *S* (true), all *M* are *N* (true)

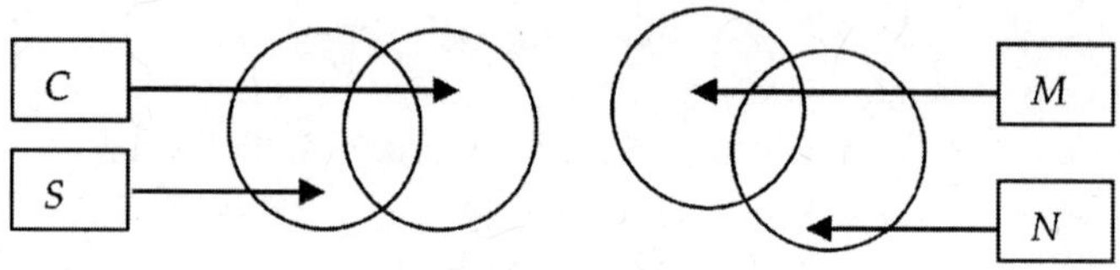

D2: all *C* are *S* (false), all *M* are *N* (false)

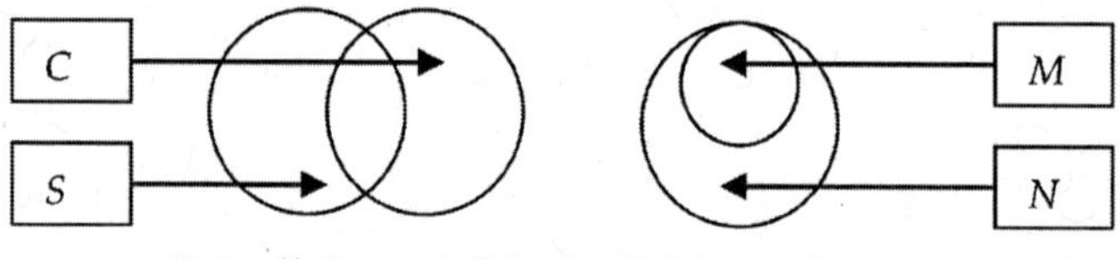

D3: all *C* are *S* (false), all *M* are *N* (true)

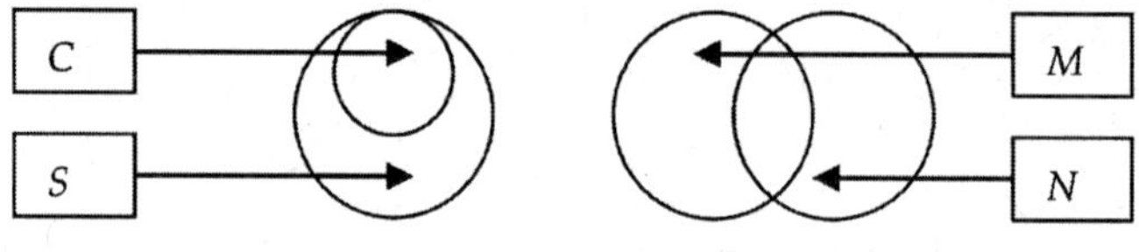

D4: all *C* are *S* (true), all *M* are *N* (false)

Each of D1, D2, and D3 represents a logical possibility of one of the three types mentioned previously. These TL diagrams, therefore, are genuine. The last TL diagram, D4, is superfluous: it does not represent any logical possibility for the English argument. Using the translation key to describe in English the situation depicted by D4, we obtain the following story: it is a situation in which every member of the Chess Club is a student in Symbolic Logic, yet some of the biological mothers of members of the Chess Club are not the biological mothers of any students in Symbolic Logic. This story does not represent a logical possibility, that is, the situation the story describes is an impossible situation. The story contains an implicit contradiction: the first clause entails, given the meaning of 'biological mother', that all the biological mothers of the Chess Club members are the biological mothers of students in Symbolic Logic, and the second clause entails that not all the biological mothers of the Chess Club members are the biological mothers of students in Symbolic Logic. Thus this story entails a statement of the form 'X and not-X'.

The English argument is clearly deductively valid but its TL translation is deductively invalid. This demonstrates our claim that if a TL argument that is a translation of a natural-language argument does not capture correctly the interdependencies between the terms of the original argument, then the logical status of the TL argument might not be indicative of the logical status of the natural-language argument. It is important to note that the shortcoming of the TL translation is not due to a bad translating job; rather, it is due to the limitation of the expressive power of TL, which is a manifestation of the limitation of its worldview.

It is not hard to generate other examples of deductively valid English arguments whose TL translations are invalid. TL is also not equipped to deal with compound general terms and compound singular terms. Here are two such arguments.

Everyone who oversleeps and takes the bus to school will be late for class.
Mariam was not late for class.
Mariam took the bus to school.

Mariam did not oversleep.

All state presidents have proper manners.

The President of France has proper manners.

Below are the TL translations of these arguments.

Translation Key

A: Those who oversleep and take the bus to school
T: Those who take the bus to school
S: Those who oversleep
L: Those who are late for class
m: Mariam
P: State presidents
M: Those who have proper manners
f: The President of France

First TL argument

all *A* are *L*
m is not *L*
m is *T*

---

m is not *S*

The second TL argument

all *P* are *M*

---

f is *M*

These TL arguments are deductively invalid. TL cannot capture the interdependency between the term 'those who oversleep and take the bus to school' and the terms 'those who oversleep' and 'those who take the bus to school' because TL has no means of representing compound general terms, such as terms that are conjunctions of other terms. The second TL argument does not capture the interdependency between the singular term 'the President of France' and the general term 'state presidents'. Therefore, the TL translations of these English arguments are not faithful. In fact, none of the TL diagrams that are counterexamples to the TL translations of these English arguments represents a relevant logical possibility, that is, none of them is genuine.

The second TL argument can be improved with the addition of a meaning-postulate. **Meaning-postulates** are premises in the symbolic language that capture certain aspects of the meanings of the natural-language expressions. The meaning-postulate the second argument needs is '*f* is *P*'. It is clear that the following TL argument is deductively valid: all *P* are *M*; *f* is *P*; therefore, *f* is *M*. The first TL argument, on the other hand, cannot be improved even with the addition of meaning-postulates. The expanded TL argument below is still deductively invalid.

all *A* are *L*
all *A* are *T*
all *A* are *S*
m is not *L*
m is *T*

---

m is not *S*

The TL diagrams below are counterexamples to, respectively, the first TL argument, the expanded version of the first TL argument, and the second TL argument.

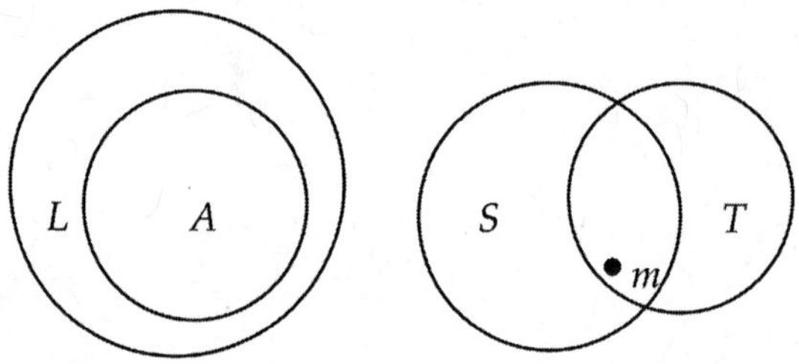

D1: all *A* are *L* (true); *m* is not *L* (true); *m* is *T* (true); *m* is not *S* (false)

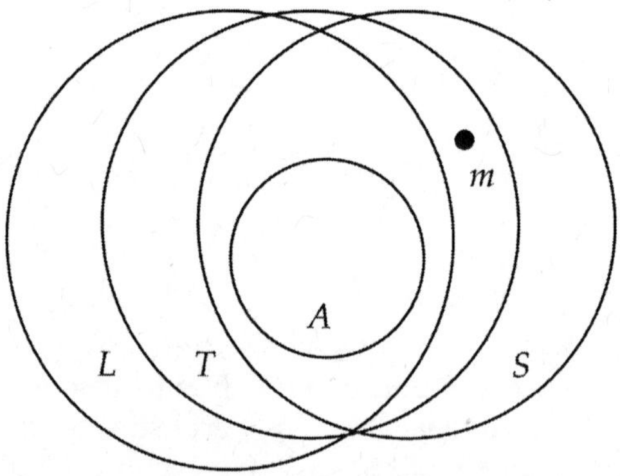

D2: all *A* are *L* (true); all *A* are *T* (true); all *A* are *S* (true); *m* is not *L* (true); *m* is *T* (true); *m* is not *S* (false)

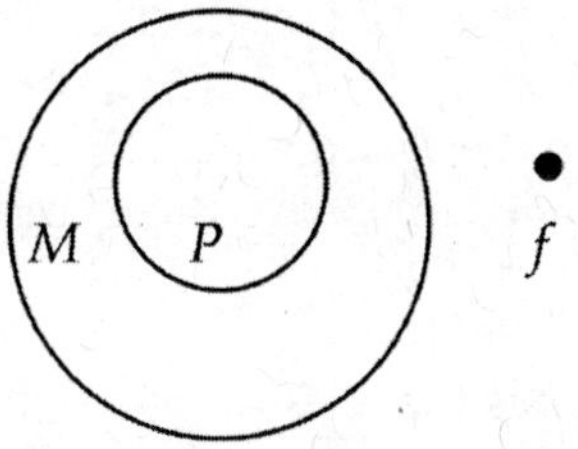

D3: all *P* are *M* (true); *f* is *M* (false)

The reason that D2 does not represent a relevant logical possibility is that D2 allows for the intersection of *S* and *T* to be larger than *A*. If we translate this into English we obtain an impossible situation, in which there are people who oversleep and who take the bus to school (because they are in the intersection of *S* and *T*) but they either do not oversleep or do not take the bus to school (because they are outside *A*). In order for TL to capture the compound general term 'those who oversleep and take the bus to school', it must be able to describe a situation in which the intersection of *S* and *T* does not exceed *A*, that is, it must be able to describe the relations between *S*, *T*, and *A* that are depicted by the following diagram.

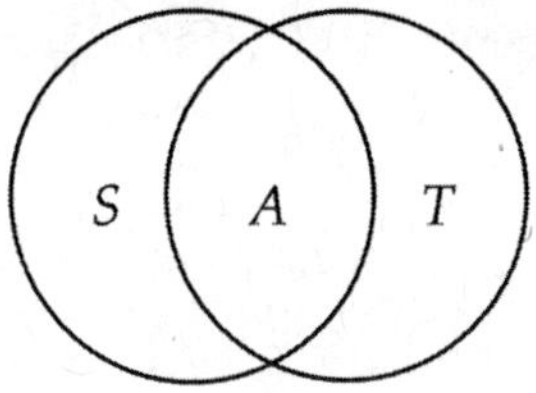

An interesting fact about TL is that no collection of TL sentences can describe precisely these relations between *S*, *T*, and *A*. However, if one takes the definition of a TL diagram in terms of *circles* and dots literally, then the diagram above is not really a TL diagram since the intersection of the *S*-circle and *T*-circle is not a circle but a simple closed curve.

2.5.12:C5 The four invalid TL arguments presented in 2.5.12:C4 have many counterexamples. But all of them are superfluous. If we describe in English, using the translation keys, the situations depicted by these TL diagrams, we arrive at impossible situations, each of which contains an explicit or implicit contradiction. Thus these TL counterexamples cannot be used to construct counterexamples to the English arguments.

The discussion above suggests that when such a case occurs, that is, when the counterexamples to a TL translation of an English argument are all superfluous, then we have strong evidence that the English argument is deductively valid. We say "strong evidence" instead of "conclusive evidence" because it is conceivable that there might be a counterexample to the English argument that cannot be represented by any TL diagram for the TL argument. On the other hand, if one of the TL counterexamples is genuine, then we have conclusive evidence that the original argument is deductively invalid. For a genuine TL diagram, by definition, represents a relevant logical possibility. Hence if there is a genuine TL diagram that is a counterexample to the TL argument, then there is a logical possibility that it is a counterexample to the original English argument.

It is not hard to demonstrate the truth of the previous assertion. Let Σ/C be an English argument. Using the notation we introduced in 2.5.12:C1, we represent the TL translation of this argument as Σ^{TL}/C^{TL}. Suppose that the TL argument is deductively invalid. Further suppose that one of the TL diagrams that are counterexamples to the TL argument is genuine. Let this TL diagram be D. Hence the members of Σ^{TL} are true and C^{TL} is false on D. Since D is genuine, there is a logical possibility p for the English argument Σ/C that is represented by D. Given the definition of representation, it follows that C is false in p (because C^{TL} is false on D), and that every sentence **X** in Σ is true in p (because $\mathbf{X}^{TL}$ is true on D). Thus p is a counterexample to the English argument, which means that the English argument is also deductively invalid.

We have, therefore, a procedure to follow when the TL translation of an English argument is deductively invalid and there is a reason to think that the TL argument might not be a perfectly faithful translation of the English argument. Here is a step-by-step description of this procedure.

(a) Survey the TL diagrams that are counterexamples to the TL argument. Using the translation key, describe in English, individually or collectively, these counterexamples. We refer to the resulting English descriptions as "stories."
(b) Using informal reasoning, determine whether any of these stories depicts a logical possibility, that is, whether any is a consistent story: it contains no explicit or implicit contradiction.
(c) If there is such a consistent story, that is, if the TL diagram on the basis of which the story is constructed is genuine, then you have a counterexample to the English argument; hence it is deductively invalid.
(d) If there is no consistent story, that is, if the TL diagrams on the basis of which these stories are constructed are all superfluous, then it is very likely that there is no counterexample to the English argument; hence it is very likely that the argument is deductively valid.

The procedure described above is not airtight, because the last step does not give us conclusive reason but only strong probability.

2.5.12:C6 Cases in which English arguments are deductively invalid while their TL translations are valid are usually controversial, for many philosophers find them unconvincing. The examples that have been proposed and of which I know rely either on designating properties with empty extensions (such as the property of being a winged horse) or on presupposing different extensions for "common" terms. It is obvious that if there are logical possibilities in which some properties have empty extensions, then such possibilities cannot be represented by TL diagrams. TL is not equipped to deal with properties whose extensions are empty: the TL worldview presupposes that every property has at least one instantiation.

Consider, for instance, the following English argument.

All winged horses are horses.
All winged horses have wings.

Some horses have wings.

This argument seems to be deductively invalid. Our actual world appears to be a counterexample to this argument. If we allow hypothetical statements to be true, even when some of the properties mentioned in these statements have empty extensions, then the premises of this argument are true in our world, because they describe hypothetical situations in which the properties have the relations mentioned in the premises. However, the conclusion is false in the actual world, since there are no winged horses.

The TL translation of this argument is deductively valid. Here is the translation.

Translation Key

W: Winged horses
H: Horses
N: Individuals that have wings

TL Argument

all *W* are *H*
all *W* are *N*

---

some *H* are *N*

If we draw the (non-similar) relevant TL diagrams on which the conclusion is false, we will see that at least one of the premises must be false. The reason that this TL argument is valid is that no TL general term is allowed to be empty. Hence there are individuals that are *W*; but then, if the premises are true, those individuals are both *H* and *N*, which means that the *H*-circle and the *N*-circle overlap (i.e., some *H* are *N*). If the English argument is indeed deductively invalid, then this is a case of an invalid English argument whose TL translation is valid. The TL argument exhibits an interdependency between *W*, *H*, and *N*—namely, that they share common individuals—that is not in the English argument. This shows that the TL argument is not a faithful translation of the English argument. We have a demonstration, therefore, of the claim that if a TL argument is not a faithful translation of an English argument, then there might be relevant logical possibilities that are not represented by any TL diagrams for the TL argument. In the winged horse example, our world, in which the premises of the English argument are true and the conclusion is false, cannot be represented by any TL diagram for the TL translation of this argument, because the TL argument has no counterexamples.

Many philosophers reject this criticism of TL on the grounds that it is unfair to fault TL for a certain *defensible* philosophical position that its worldview presupposes. The TL worldview excludes properties with empty extensions because, one might argue, TL presupposes that the reality of a property is not above and beyond the reality of the individuals that instantiate this property. So if no such individuals exist, the property has no reality. Therefore anyone who accepts this philosophical position will not be moved by this criticism.

Other philosophers might argue that in any logical possibility in which there are no winged horses—making the conclusion false—the premises are false because they attribute properties (being a horse and having wings) to fictitious objects (winged horses). Those philosophers might affirm that every fictional statement is false.[1] According to this position, a logical possibility in which there are no winged horses is not a counterexample to the English argument, since the premises are false in such a logical possibility,[2] yet a counterexample must make the premises

1 In this book, the adjectives 'fictional' and 'fictitious' are used in two different senses. Roughly speaking, something is fictitious if it is imaginary and not actual, but its existence is possible, that is, it is not an impossible object. Something is fictional if it describes or designates a fictitious situation or fictitious object. Thus fictional things are not fictitious. For instance the names 'Sherlock Holmes' and 'Pegasus' are fictional names because they designate two fictitious objects: Sherlock Holmes and Pegasus. But the names themselves are not fictitious; they are actual linguistic entities. Similarly the sentence 'Moby Dick destroyed Ahab's boat' is a fictional sentence because it describes a fictitious situation, but it is not a fictitious sentence since it is an actual linguistic entity that can be written and uttered.

2 This brings us to the thorny problem of *negative existentials*, which are sentences of the form 'There are no Ps' or 't does not exist', where 'P' is to be replaced with a general term and 't' with a singular term. The first form is usually referred to as 'general negative existential' and the second 'singular negative existential'. For example 'There are no winged horses' is a general negative existential, and 'Pegasus does not exist' is a singular negative existential. The problem in its most general formulation concerns *all* negative existentials. However, its most conspicuous formulation concerns empty general terms and non-referring singular terms. There are a few versions of this problem. We will discuss only one of them. Consider the sentence 'There are no winged horses'. It seems that this sentence is *about* winged horses, which are fictitious objects. If all statements about fictitious objects are false, then the sentence 'There are no winged horses' must be false. This verdict appears to be wrong, for the sentence describes something that is the case—namely, that our world contains no winged horses.

We have several options here, the most common of which are: (1) allow some fictional statements to be true; (2) deny that negative existentials whose terms are empty or non-referring are about fictitious objects, and hence affirm that they are not fictional statements; (3) affirm that all statements that contain empty or non-referring terms, including negative existentials, are neither true nor false, and hence affirm that 'There are no winged horses' and 'Pegasus does not exist' are not true; (4) affirm that all statements that contain empty or non-referring terms, including negative existentials, are false, and hence "bite the bullet" and affirm that 'There are no winged horses' and 'Pegasus does not exist' are false.

All these options seem problematic. If we follow the first option, then we need a criterion for determining when a fictional statement is true and when it is false. For instance, are the sentences 'If there were winged horses, then some horses would be able to fly', 'Moby Dick is a fictitious whale', and 'Little Johnny believes that Santa Clause is a fat, jolly, old man' true or false? If one says that all of these are true, then we want to know why they are true. Why is the sentence 'Moby Dick is a fictitious whale' true and the sentence 'Moby Dick is a whale' false? It is not a satisfying answer to say that the former is true because it says something that is the case and the second says something that is not the case. The two sentences appear to have the same semantical function. Each sentence attributes a certain property to a certain object: the former attributes the property of being a fictitious whale to Moby Dick and the latter attributes the property of being a whale to Moby Dick. If there is no Moby Dick, why is the first attribution the case and the second not the case? And if it is all right to attribute properties to fictitious objects, what makes a fictitious object have a certain property and lack another? These are hard questions, and they call for a theory of nonexistent objects, or for a theory of truth for statements about nonexistent objects.

The second option is the most popular. It is based on a theory first advanced by the British philosopher and logician Bertrand Russell (1872–1970). The theory is a little too involved for an introductory logic book; but if we allow ourselves some latitude and imprecision, we can give a general (though, rough) idea of the theory. Russell's solution is to treat negative existentials that contain empty or non-referring terms as the denial of existential sentences that are about the actual objects of our world and not about nonexistent objects. So to say that there are no winged horses (*continued*)

true. On the other hand, in any logical possibility in which there are winged horses the premises and the conclusion are true because, by definition, winged horses are horses and have wings. Thus such a logical possibility is also not a counterexample to the English argument. This shows

---

is to say that it is not the case that there are objects of our world that are winged horses, where 'winged horse' is a general term whose extension is empty (i.e., an empty general term). So 'There are no winged horses' is not about "winged horses"; rather it is about the actual objects of the world. It simply denies that any object of our world has the property "being a winged horse."

Singular negative existentials are trickier. Still roughly speaking, Russell's theory asserts that non-referring singular terms are really complex general terms that are disguised as singular terms, and that singular negative existentials are really general negative existentials that are disguised as singular sentences. Thus, for example, the *logical form* (i.e., the *real form*) of the sentence 'Moby Dick does not exist' is 'It is not the case that there is an object of our world that is a white sperm whale that satisfies the descriptions found in the classic novel by the American author Herman Melville'. Therefore, the sentence 'Moby Dick does not exist' is not about a nonexistent object called 'Moby Dick', and its logical form is not 'm is not E', where m is the non-referring singular term 'Moby Dick' and E is the general term 'existent'. The sentence is really the denial of an existential sentence that is about the actual objects of the world and a certain novel. According to this solution, both sentences 'There are no winged horses' and 'Moby Dick does not exist' are, as expected, true. On the other hand, an affirmative existential, whether singular or general, that contains an empty or non-referring term must be understood as a general existential sentence also about the actual objects of the world. So the logical form of the sentence 'Moby Dick exists' is 'There is one and only one object of our world that is a white sperm whale that satisfies the descriptions found in Melville's classic novel'. In other words, the sentence affirms the existence of an object that satisfies a certain description. Since no object of our world satisfies this description, the sentence 'Moby Dick exists' is, as expected, false.

Russell's theory faces much-discussed difficulties. We will mention only one of them. Consider the sentence 'The present king of France is bald'. Russell's theory gives an analysis of this sentence similar to its analysis of 'The present king of France exists'. The logical form of the latter is 'There is one and only one object of our world who is a present king of France'. Similarly, the logical form of the former is 'There is one and only one object of our world who is a present king of France and is bald'. This sentence is false if and only if (1) there is no present king of France, or (2) there is more than one present king of France, or (3) there is one and only one present king of France who is not bald. Since the first condition is true, the sentence 'The present king of France is bald' is false. Observe that these three conditions are mutually exclusive, that is, if one of them is true the others are false. Hence to deny the sentence 'The present king of France if bald' is to assert one of these three conditions. This makes the sentence 'The present king of France is not bald' ambiguous. If it asserts the first condition, then it is true, since there is no object of our world that has the property of being a present king of France. If it asserts the second condition, then it is false, since the second condition affirms the existence of more than one present king of France where in actuality there is none. If it asserts the third condition, then it is also false, since the third condition affirms the existence of exactly one present king of France who is not bald, where our world contains no such individual. Thus the sentence 'The present king of France is not bald' is true if the negation is understood as denying the existence of a present king of France, it is false if the negation is understood as denying *only* the uniqueness of this king, and it is also false if the negation is understood as denying the attribution of baldness to the present king of France.

Some philosophers think that this analysis is one of the strengths of the theory, others think that treating a sentence such as 'The present king of France is not bald' as ambiguous is a problem for the theory. There is something counterintuitive about saying that the sentence 'The present king of France is bald' posits the existence of such a king, and hence it is false, while the sentence 'The present king of France is not bald' might be denying the existence of such a king, and hence it might be true. In other words, it is counterintuitive to say that the negation article 'not' in the sentence 'The present king of France is *not* bald' might be denying the existence of the king and not (*continued*)

that the English argument, contrary to appearances, is deductively valid, and hence there is no divergence between the logical status of the English argument and its TL translation.

There are relevant logical possibilities that contain properties whose extensions are empty, and which show more clearly a logical possibility that cannot be represented by any TL diagram. Of course, these examples are still subject to the objection stated above—namely, that TL presupposes a philosophically defensible position that outlaws any property with an empty extension. However, as we will see later, these examples are not easily dismissed by this objection. These are not properties that involve fictitious objects; rather they involve causal features of real objects. The presence of such logical possibilities implies that the philosophical position that TL presupposes might not be as defensible as some philosophers think. We will consider an example of such a logical possibility in Subsection 4.5.13:C6.

A more convincing criticism of TL would be to produce an invalid English argument whose terms have nonempty extensions in all the relevant logical possibilities and show that its TL translation is valid. This would be very telling, because in this case there would be a logical possibility that is a counterexample to the English argument while there is no TL diagram that is a counterexample to the TL argument. This would prove that there are relevant logical possibilities in which no property has an empty extension, and which cannot be represented by any TL diagrams.

Philosophers have given examples of arguments that, they believe, satisfy these conditions. Here is an argument that belongs to the medieval Islamic tradition.

P1 All the pilgrims who die on their way to Mecca will not perform the hajj this year.
P2 All those who will not perform the hajj this year will try to perform the hajj next year.

C All the pilgrims who die on their way to Mecca will try to perform the hajj next year.

---

his baldness. On the other hand, attributing ambiguity to this sentence introduces a notion of *scope*, which has proved to be philosophically significant. In very simple terms, the scope of an existential quantifier in a sentence S is the part of S to which the quantifier applies. So we may interpret the implicit existential quantifier in the sentence 'The present king of France is not bald' as applying to the whole sentence including the negation word 'not' (in this case, we say that the quantifier has a *wide scope*), or as applying only to the affirmative part of the sentence, which excludes the word 'not' (in this case, we say that the quantifier has a *narrow scope*). The wide-scope interpretation may be expressed as 'There is one and only one king of France who is not bald', and the narrow-scope interpretation may be expressed as 'It is not the case that there is one and only one king of France who is bald'. In the first reading the negation word 'not' occurs within the scope of the existential quantifier, and hence the negation is a denial of the king's baldness; and in the second reading the negation expression 'It is not the case that' occurs outside the scope of the existential quantifier, and hence the negation is a denial of the whole existential assertion.

The third and fourth options (see above) imply that the sentences 'The present king of France does not exist' and 'The present king of France is bald' are both not true. Perhaps, the second sentence is indeed not true; perhaps, it is even meaningless, since it seems to presuppose the existence of something that does not exist. But to say that the first sentence is not true is problematic. The first sentence seems to follow from the negative existential 'There is no present king of France', and the latter from 'France presently has no king', which is, on the face of it, clearly true. But if A follows from B, B from C, and C is true, then both A and B must be true as well. The preceding assertion is, in fact, true in many (but not all) logical systems, including some non-classical systems. There are other solutions to the problem of negative existentials, which we will not discuss. Our goal was merely to create an appreciation of this philosophical problem.

In the actual world, the first premise is true and the second premise and the conclusion are false. The falsity of the second premise, however, appears to be a contingent matter; if so, then there is a logical possibility, call it "q," in which the second premise is also true. q is a logical possibility in which the premises are true and the conclusion is false; therefore the English argument is deductively invalid.

The TL translation of this argument, however, is deductively valid.

Translation Key

D: Pilgrims who die on their way to Mecca
H: Those who will not perform the hajj this year
T: Those who will try to perform the hajj next year

TL Argument

S1 all *D* are *H*
S2 all *H* are *T*

S3 all *D* are *T*

It is clear that any TL diagram for the argument that makes S1 and S2 true also makes S3 true. The reader can verify this by drawing the four non-similar relevant TL diagrams on which S1 and S2 are true. This shows that the logical possibility q cannot be represented by any TL diagram for the TL argument above, since q is a counterexample to the English argument and the TL translation of this argument has no counterexamples.

There is a standard objection to this case. The objection runs as follows. In any logical possibility in which there are pilgrims who died on their way to Mecca, some of those who will not perform the hajj this year could not try to perform it next year, because some of those pilgrims who will not perform the hajj this year are pilgrims who died on their way to Mecca and it is clear that dead pilgrims cannot try to perform any hajj in the future. So if q is a logical possibility in which there are pilgrims who died on their way to Mecca, then q is a logical possibility in which P2 must be false. q, therefore, cannot make P1 and P2 true together. This shows that q is not a counterexample to the English argument because a counterexample must make the premises all true. On the other hand, if q is a logical possibility in which P2 is true, then there are no pilgrims who died on their way to Mecca because such pilgrims cannot try to perform the hajj next year. This entails that in q the property "being a pilgrim who died on his or her way to Mecca" has an empty extension, yet we assumed that all the terms of the English argument have nonempty extensions in all the relevant logical possibilities. In this case, q would not be a logical possibility that is relevant to the English argument, and hence it cannot serve as a counterexample to the argument.

There might be another type of logical possibility. If it is logically possible for the dead to be resurrected, we could argue that there are logical possibilities in which both P1 and P2 are true. Imagine a world, *w*, in which all the pilgrims who died on their way to Mecca are resurrected after this year's hajj, and they, like all the others who did not perform the hajj this year, will try to perform the hajj next year. Hence in *w* P1 is true, P2 is true, *and* C is true. So *w* is not a counterexample to the English argument because the conclusion is true in *w*.

According to this analysis, there are two types of *relevant* logical possibilities: one in which the truth of P1 entails the falsity of P2 (the dead cannot be resurrected) and one in which both P1 and P2 are true (the dead can be resurrected). The first type of logical possibility does not furnish a counterexample to the English argument because in it the premises are not both true, and the second type also does not generate a counterexample to the argument because in it the conclusion is true. This demonstrates that there are no relevant logical possibilities in which P1 and P2 are true and C is false. Therefore, contrary to appearances, the English argument is deductively valid. The conclusion of this analysis is that there is no divergence between the logical status of the English argument and the logical status of the TL argument: both are deductively valid.

The only way to make both premises true and the conclusion false while insisting that all the general terms have nonempty extensions, the objection continues, is to introduce different extensions for the middle term. The truth of P1 presupposes a context in which there are pilgrims who died on their way to Mecca, and in which dead pilgrims cannot perform *this* year's hajj. The falsity of the conclusion presupposes a context in which there are pilgrims who died on their way to Mecca, and in which dead pilgrims cannot try to perform *next* year's hajj. Holding both of these contexts fixed, the truth of P2 presupposes a context in which all the individuals who will not perform the hajj this year *and who will stay alive until next year's hajj* will try to perform the hajj then. These contexts make the general term 'those who will not perform the hajj this year' (call it '*T*'), as used in the English argument, ambiguous. In P1 the extension of *T* consists of all the individuals who will not perform the hajj this year, including those who will die before reaching Mecca. In P2 the extension of *T* consists of only those individuals who will not perform the hajj this year and who will remain alive until the next hajj. The extension of *T* in the second premise is a proper subset of its extension in P1.[1] But, as we explained in 2.3.1:C and 2.3.2:C, TL does not allow general or singular terms to be ambiguous. Hence the term *T* must be disambiguated. The English argument below is a reformulation of the original argument after *T* has been disambiguated.

P1 All the pilgrims who die on their way to Mecca will not perform the hajj this year.
P2* All those who will not perform the hajj this year and who will stay alive until next year's hajj will try to perform the hajj next year.

C All the pilgrims who die on their way to Mecca will try to perform the hajj next year.

This argument is invalid. The inference is blocked, because the term 'one who will not perform the hajj this year' is not the same term as 'one who will not perform the hajj this year and will stay alive until next year's hajj'. These terms have different extensions if their extensions are not empty. In other words, the inference is invalid because there is no *common middle term*. Once the term 'one who will not perform the hajj this year' is disambiguated, a faithful TL translation of the English argument would look like this.

1 If we were to make both extensions identical and keep P1 and P2 true, those who died before reaching Mecca would have to be resurrected after this year's hajj in order to try to perform the next hajj. In such a situation C would be true; yet we assumed that C is false.

S1*	all D are H_1
S2*	all H_2 are T
S3	all D are T

where H_1 and H_2 stand, respectively, for the second term of P1 and the first term of P2*. It is clear that the TL argument is deductively invalid. The reader can verify this by drawing a TL diagram on which S1* and S2* are true and S3 is false. In this case as well there is no divergence between the logical status of the English argument and the logical status of the TL argument: both are deductively invalid.

In sum, the standard objection argues that if, on the one hand, we assume that the terms of the argument have nonempty extensions and assume that there is a common middle term, then either the truth of P1 entails the falsity of P2 (the dead pilgrims cannot be resurrected) or the truth of P1 and P2 entails the truth of C (the dead pilgrims are resurrected). In such a case, therefore, there is no counterexample to the English argument, because either the premises cannot both be true or if they are true, the conclusion is true as well. Hence the English argument and its TL translation are deductively valid. On the other hand, if we assume that P1 and P2 are true and C is false, then the middle term would be ambiguous and, once it is disambiguated, there would be no common middle term. In this case, the English argument and its TL translation would be deductively invalid. Therefore in either case there is no divergence between the logical status of the English argument and the logical status of its TL translation. This objection is meant to show that this proposed English argument fails to demonstrate the existence of a relevant logical possibility that cannot be represented by a TL diagram.

2.5.12:C7 If a TL argument is a **faithful translation** of a natural-language argument, we can avoid all the previous difficulties. The logical status of a TL argument that is a faithful translation of a natural-language argument is indicative of the logical status of the original argument: they are either both deductively valid or both deductively invalid. The unfortunate thing is that we cannot give a proof of this claim. But we can argue in general that it seems plausible to assume that the logical status of an argument is determined by the interdependencies that exist between its various parts. A faithful symbolic translation of a natural-language argument exhibits *all and only* the interdependencies that exist between the various parts of the natural-language argument. Therefore, since both arguments have exactly the same interdependencies between their *corresponding* parts, and since the logical status of an argument is determined by these interdependencies, both arguments must have the same logical status. Throughout this book we will assume that this conclusion is correct; and hence we will determine the logical status of English arguments by determining the logical status of the symbolic arguments that are *faithful* translations of the English arguments.

2.6 Exercises

Note: All answers must be adequately justified.

2.6.1* Consider the following set of TL basic vocabulary: {*a*, *J*, *K*}. Let Δ be the set of all the TL sentences that can be generated from this set of basic vocabulary.

2.6.1a How many members does Δ have?
2.6.1b Identify three TL sentences in Δ such that one is logically true, one is logically false, and one is contingent.
2.6.1c Give an example of a pair of logically equivalent TL sentences in Δ and of a pair of sentences in Δ that are not logically equivalent.
2.6.1d Is Δ a consistent set?
2.6.1e What are the diagrams for the following set of TL sentences: {all *J* are *K*, all *K* are *J*}? Determine the truth values of these sentences on each diagram. Is this set consistent?

2.6.2 Determine which of the TL arguments below is deductively valid and which is invalid.

2.6.2a* all *D* are *H*
no *F* is *D*
———————
some *H* are not *F*

2.6.2b* some *A* are *B*
some *B* are not *L*
some *L* are not *A*
———————
some *A* are not *L*

2.6.2c all *E* are *W*
all *T* are *W*
some *T* are not *E*
———————
some *W* are not *T*

2.6.2d *e* is *F*
all *F* are *G*
no *G* is *V*
d is *V*
———————
e is not *d*

2.6.3 Recall that a set of TL sentences is consistent if and only if there is *at least one* TL diagram on which the members of the set are all true (alternatively, there is a TL diagram that satisfies the set; the set has a model). Now, a set of TL sentences is said to be **categorical** if and only if there is, up to similarity, *at most one* model for the set. Hence, every inconsistent set is categorical (it has no model), and, more important, every set that is both consistent and categorical has, up to similarity, *exactly one* model. Are the sets below consistent? Are they categorical?

2.6.3a* {some *Q* are not *R*, some *R* are not *Q*, some *Q* are *R*}.
2.6.3b* {all *K* are *M*, all *M* are *N*, *c* is *J*, no *N* is *J*, *c* is *K*}.
2.6.3c {all *A* are *B*, some *A* are *S*, some *A* are not *S*, some *B* are not *A*, all *S* are *B*}.
2.6.3d {all *H* are *F*, *d* is *F*, *e* is not *H*, *d* is *e*}.

2.6.4 Demonstrate that each of the claims below is true.

2.6.4a* The set {all *G* are *K*, some *G* are *H*, no *K* is *H*} is inconsistent.
2.6.4b* The TL sentence 'some *B* are not *A*' is not a logical consequence of the set {some *A* are not *B*, some *B* are *A*}.
2.6.4c The TL sentences 'no *M* is *N*' and 'no *N* is *M*' are logically equivalent.
2.6.4d The TL sentence 'some *G* are *K*' is a logical consequence of {*s* is *J*, *r* is *K*, all *J* are *G*, *s* is *r*}.

2.6.5 The English sentence 'All people are not vicious dog owners' is ambiguous—it has four different meanings. Translate this sentence into four different TL sentences each of which represents one possible meaning of the English sentence.

2.6.6 Translate into TL the schematizations of the arguments found in the exercises listed below. Determine whether the TL arguments are valid or not. What can you conclude about the original English arguments?

2.6.6a* Exercise 1.3.12
2.6.6b Exercise 1.3.13
2.6.6c Exercise 1.3.15

2.6.7 Schematize the arguments in the following passages, translate them into TL, and test for deductive validity.

2.6.7a* The great composer Beethoven is immortal. This follows from the facts that Beethoven is Indian, that all Indians are Buddhist, and that Buddhists are immortal.

2.6.7b Since Earth is a planet and since at least one planet has water, we must conclude that Earth has water.

2.6.7c* I have come to believe that some fruits have feathers. My reasoning is as follows. We know that a kiwi is a type of bird, and that some fruits are kiwis. Now, since all birds have feathers, some fruits must have feathers as well.

2.6.7d* Only meaningful statements constitute knowledge. A statement is meaningful only if it is empirically verifiable. But it is clear that there is no true generalization that is empirically verifiable, and that some mathematical theorems are true generalizations. Therefore, not every mathematical theorem constitutes knowledge.

2.6.7e Only humans think about moral dilemmas. Whatever thinks about moral dilemmas must have a mind. Therefore, some humans have minds.

2.6.7f It seems to me that we can establish that penguins don't fly without engaging in expensive empirical studies or tedious field observations. For it is totally obvious that some birds don't fly. Penguins are birds. It immediately follows that penguins don't fly.

2.6.8 Use the methods of TL to show that the arguments in the following passages are deductively valid.

2.6.8a* It might be thought that the sentences '$16 = 16$' and '$4^2 = 16$' are expressing exactly the same statement. But I believe that these sentences are expressing two different statements. For the statement that $16 = 16$ is self-evident while the statement that $4^2 = 16$ is a mathematical theorem. Now, since no self-evident statement requires a proof and since every mathematical theorem requires a proof, the two statements are not the same.

2.6.8b Define a thing to be *accessible to our senses* if and only if it or some of its effects can be experienced by one or more of our senses. Now, given this definition, we all could agree that only things that are accessible to our senses can be known by us. But whatever is accessible to our senses is capable of affecting our sense organs. It is quite obvious that no future event has such a capability. Therefore, future events cannot be known by us.

2.6.8c The divine command theory is the thesis that moral principles are due to the *arbitrary* will of God. For instance, someone who believes in this theory would say that there is nothing inherently wrong with murder; murder is wrong simply because God said so. Such a person might argue for this claim as follows. Any sentence that expresses a moral claim (such as 'Cheating is wrong') is contingent. No sentence is contingent unless it is capable of being true. Whatever is capable of being true is possible. The sentence 'Murder is good' expresses a moral claim. Since anything that is possible can be made true by God, the sentence 'Murder is good' can be made true by God.

2.6.8d* David is a universal skeptic. A universal skeptic believes that he or she does not know anything. Those who have such a belief must (at least) know that they are thinking, and whoever knows that he or she is thinking actually knows something. It follows that there are universal skeptics who actually know something.

2.6.8e The doctrine of natural rights asserts that people have certain rights, such as the right to liberty, that are theirs solely because they are humans. Some political philosophers advocate this theory while others think it is mistaken. An argument that is sometimes given against the theory is this. If something is a natural right, then it is an inevitable consequence of human nature. Such consequences, however, cannot be taken away. But every right that people might have can be suspended (under certain circumstances). Nothing can be suspended unless it can be taken away. It follows that people don't have natural rights.

2.6.9 Each of the passages below contains a two-part argument. Schematize each part of each argument, translate it into TL, and test for deductive validity.

2.6.9a* Some philosophers, such as Plato (428–348 BCE) and René Descartes (1596–1650), believe in innatism, which is the thesis that humans have innate ideas, i.e., inborn concepts or knowledge. The following is a simplified version of an argument for innatism that Plato gave in one of his dialogues (the *Meno*). Consider a certain young child who has never been taught geometry before. By asking appropriate questions and drawing a few illustrations, you can make the child arrive at some geometrical knowledge (such as that the diagonal of a square divides the square into two congruent triangles). It is clear, however, that only those who can comprehend geometrical concepts (such as the concept of a straight line and the concept of congruent triangles) are capable of arriving at geometrical knowledge. Therefore, some people who have never been taught geometry can comprehend geometrical concepts. But we can now argue that if a person has never been taught geometry, and yet is capable of comprehending geometrical concepts, then that person must have those concepts innately. It follows that the young child has innate geometrical concepts.

2.6.9b Philosophers and non-philosophers have long supposed that ordinary empirical objects (i.e., objects of our sense experience, such as chairs, cows, and stars) are composed of material substance and qualities. Material substance is the most basic "stuff" to which all material objects are reduced if all (or almost all) their qualities (shape, color, smell, taste, etc.) are removed. The Irish philosopher George Berkeley (1685–1753) advanced in *The Principles of Human Knowledge* (and other works) the provocative view that material substance does not exist: empirical objects are nothing but collections of sensible qualities, which are ideas existing in the mind (hence, this view is a version of idealism). Berkeley's argument for this position is complex. We present here only a stage of the argument in a simplified form. The conclusion of this stage is that it is not rational for humans to believe in the existence of material substance. This conclusion is reached via a two-part argument. In the first part we argue that since material substance, by definition, is not a quality of empirical objects, and since none but qualities of empirical objects can be experienced by our senses, material substance is not an aspect of empirical objects that we can know; and that is because we cannot know an aspect of empirical objects unless it can be experienced by our senses. Given the first part, we now argue that if we cannot know something, then it is not rational for us to believe in its existence; therefore it is not rational for us to believe in the existence of material substance.

2.6.10 Consider the following simple argument. All horses are animals; therefore all heads of horses are heads of animals.[1] Translate the argument into TL. Show that the TL argument is deductively invalid. Explain why the TL counterexamples cannot be used to construct counterexamples to the English argument.

---

1 This argument is historically known as De Morgan's counterexample to Aristotelian Logic.

Solutions to the Starred Exercises

SOLUTIONS TO 2.6.1

2.6.1a 22 members. Each of the sentence forms 2.2.2a–2.2.2d has two metalinguistic variables '**X**' and '**Y**'. Our basic vocabulary contains two general terms '*J*' and '*K*'; they can be substituted for **X** and for **Y**. So each of these forms generates 4 sentences. Forms 2.2.2e–2.2.2h allow one substitution for **x** and **z**, which is the singular term '*a*', and two substitutions for **Y**, which are the general terms '*J*' and '*K*'. Hence each of 2.2.2e and 2.2.2f generates 2 sentences and each of 2.2.2g and 2.2.2h generates one sentence.

2.6.1b 'some *K* are *K*' is logically true, '*a* is not *a*' is logically false, and '*a* is *J*' is contingent. There is, up to similarity, only one TL diagram for the first sentence, and it makes the sentence true (see 2.5.5a).

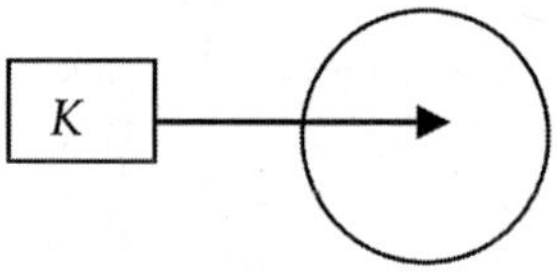

some *K* are *K* (true)

The second sentence also has, up to similarity, only one TL diagram, which makes the sentence false (see 2.5.6a).

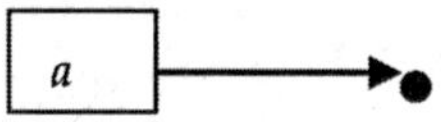

a is not *a* (false)

Below are two TL diagrams. The first makes the sentence '*a* is *J*' true and the second makes it false (see 2.5.7).

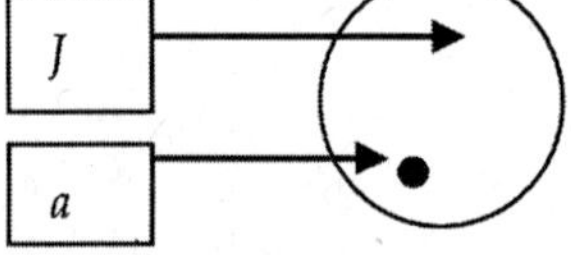

a is *J* (true)

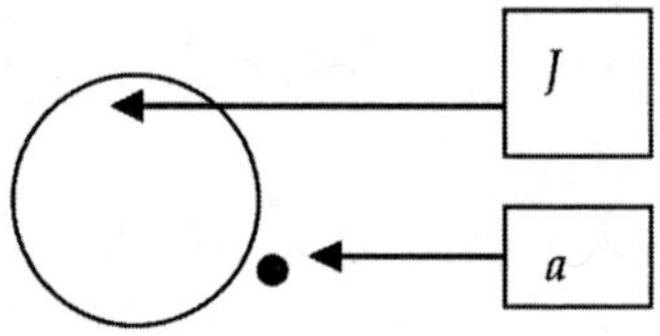

a is *J* (false)

Other logically true sentences are: all *J* are *J*, all *K* are *K*, some *J* are *J*, *a* is *a*

Other logically false sentences are: no *J* is *J*, no *K* is *K*, some *J* are not *J*, some *K* are not *K*

Other contingent sentences are: all *J* are *K*, all *K* are *J*, no *J* is *K*, no *K* is *J*, some *J* are *K*, some *K* are *J*, some *J* are not *K*, some *K* are not *J*, *a* is *K*, *a* is not *J*, *a* is not *K*

2.6.1c The sentences 'some *J* are *K*' and 'some *K* are *J*' are logically equivalent and the sentences 'some *J* are not *K*' and 'some *K* are not *J*' are not. The first two sentences have identical truth values on each of their TL diagrams (see 2.5.8a).

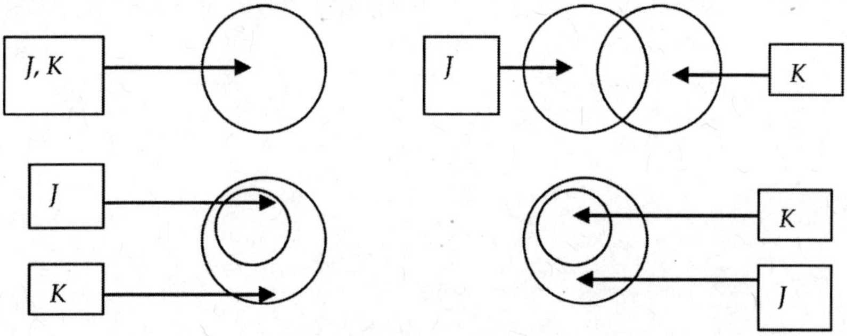

some *J* are *K* (true), some *K* are *J* (true)

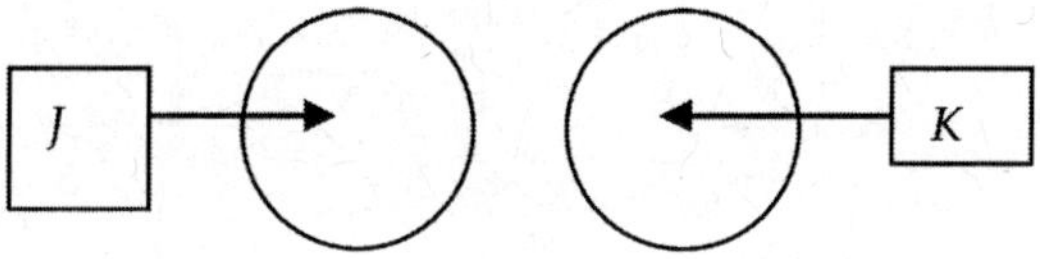

some *J* are *K* (false), some *K* are *J* (false)

The diagram below makes 'some *J* are not *K*' true and 'some *K* are not *J*' false (see 2.5.8b).

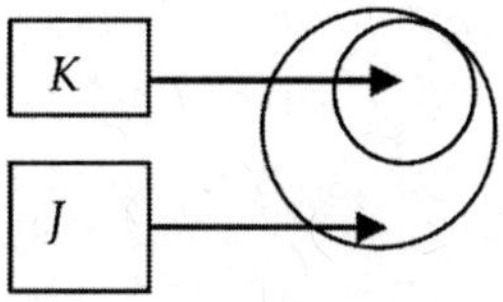

some *J* are not *K* (true), some *K* are not *J* (false)

Another logically equivalent pair is: no *J* is *K*, no *K* is *J*; in addition, each logically true sentence is logically equivalent to every other logically true sentence; and each logically false sentence is logically equivalent to every other logically false sentence. And, of course, every sentence is logically equivalent to itself. All other pairs are not logically equivalent.

2.6.1d No, Δ is inconsistent. Δ contains logically false sentences such as '*a* is not *a*' and 'no *K* is *K*'. There is no TL diagram on which these sentences are true (see 2.5.6b). It follows that there is no TL diagram on which the members of Δ are all true (see 2.5.10b).

2.6.1e The TL diagrams for this set are, up to similarity, the first five diagrams listed in 2.6.1c above. 'all *J* are *K*' is true only on the first and third diagrams, and 'all *K* are *J*' is true only on the first and fourth diagrams. The set is consistent because each of its members is true on the first diagram (see 2.5.9).

SOLUTIONS TO 2.6.2

2.6.2a Deductively valid. Consider a representative class of TL diagrams for the argument in which the premises are true. (See 2.4.4:C for the definition of 'representative class of TL diagrams for Γ'.)

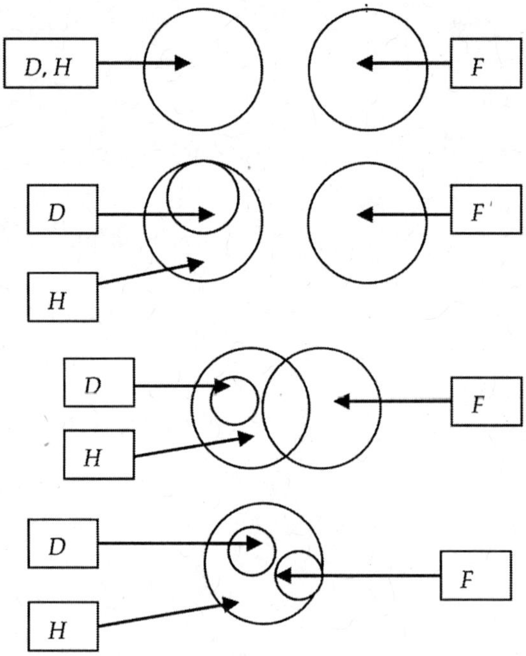

all *D* are *H* (true), no *F* is *D* (true) / some *H* are not *F* (true)

On each of the diagrams above the conclusion, 'some *H* are not *F*', is true as well. Hence, the argument is deductively valid (see 2.5.3a).

2.6.2b Deductively invalid. The following TL diagram makes the premises true and the conclusion false (see 2.5.4).

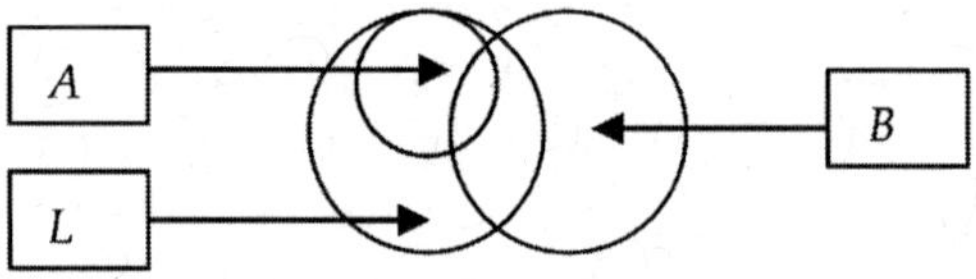

some *A* are *B* (true), some *B* are not *L* (true),
some *L* are not *A* (true) / some *A* are not *L* (false)

SOLUTIONS TO 2.6.3

2.6.3a Consistent and categorical. The diagram below is, up to similarity, the only model of the set.

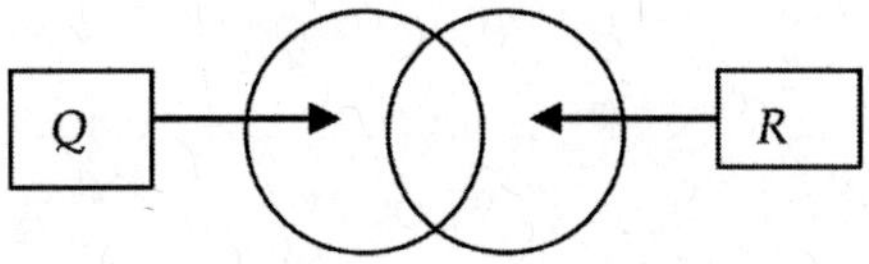

some *Q* are not *R* (true), some *R* are not *Q* (true), some *Q* are *R* (true)

2.6.3b Inconsistent and, therefore, categorical. We will consider a representative class of TL diagrams (D1–D4 below) for the set on which the first four sentences are true, and will confirm that the last sentence, '*c* is *K*', is false on each of these diagrams. Thus there is no TL diagram on which the members of the set are all true (see 2.5.9).

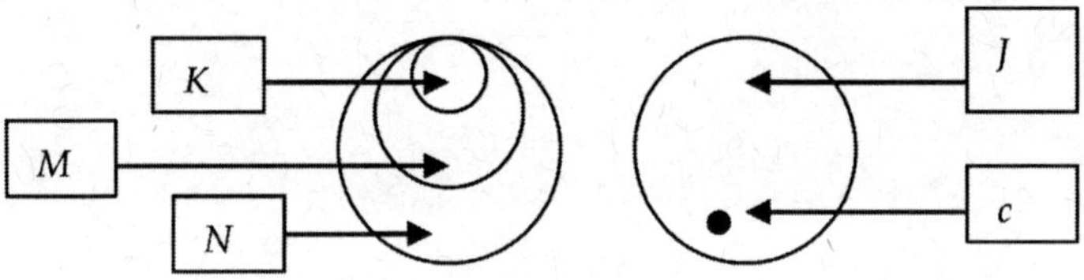

D1: all *K* are *M* (true), all *M* are *N* (true),
c is *J* (true), no *N* is *J* (true), *c* is *K* (false)

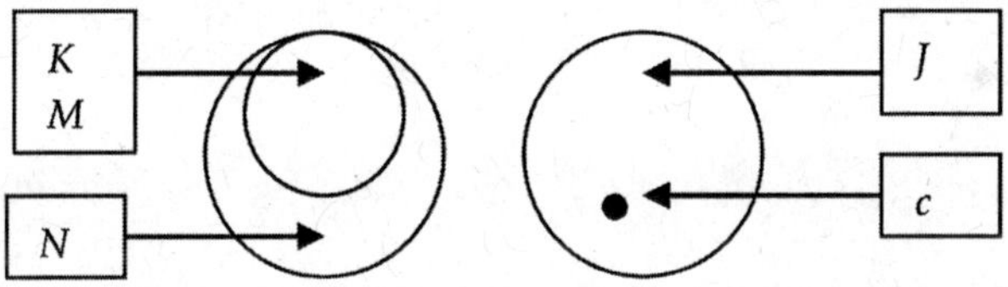

D2: all *K* are *M* (true), all *M* are *N* (true),
c is *J* (true), no *N* is *J* (true), *c* is *K* (false)

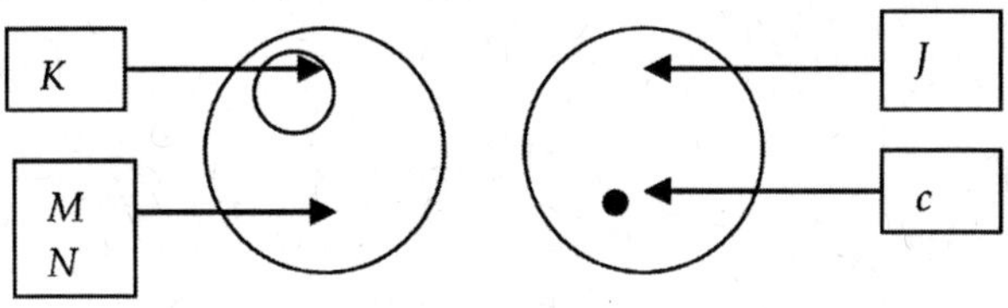

D3: all *K* are *M* (true), all *M* are *N* (true),
c is *J* (true), no *N* is *J* (true), *c* is *K* (false)

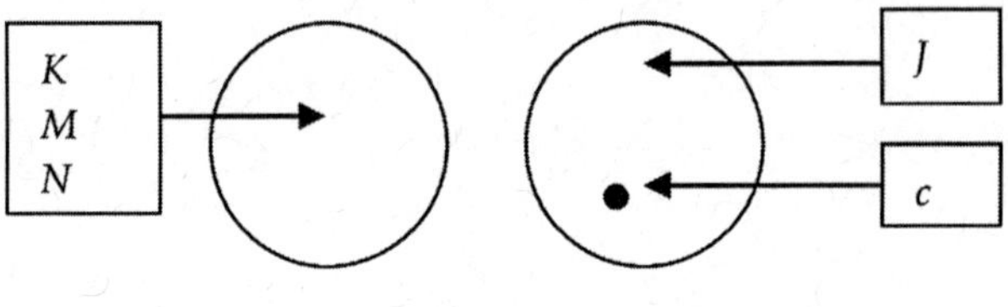

D4: all *K* are *M* (true), all *M* are *N* (true),
c is *J* (true), no *N* is *J* (true), *c* is *K* (false)

D2, D3, and D4 are produced by identifying (i.e., making identical) two or more of the *K*-circle, *M*-circle, and *N*-circle. TL diagrams such as D2–D4 will be referred to as **standard variants** of the original diagram (in this case, D1). In general, if a TL diagram D contains nested circles, then its standard variants are the TL diagrams that are obtained from D by identifying two or more of the nested circles. This is merely a definition. We are not claiming that a diagram and its standard variants *always* agree on the truth values of the sentences for which they are diagrams. This is a useful terminology, which we will invoke in the remainder of this chapter.

Solutions to 2.6.4

2.6.4a The TL diagrams below are, up to similarity, the only diagrams that make the first and third sentences true. On each of these diagrams the second sentence is false. Thus there is no TL diagram on which every member of the set is true (see 2.5.9).

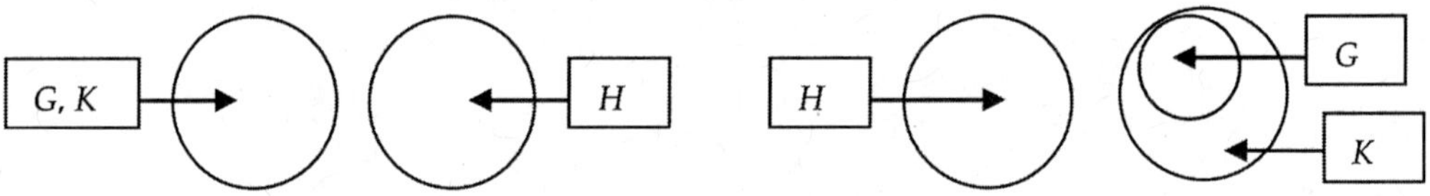

all *G* are *K* (true), some *G* are *H* (false), no *K* is *H* (true)

2.6.4b The following TL diagram makes the members of the set true and 'some *B* are not *A*' false (see 2.5.4).

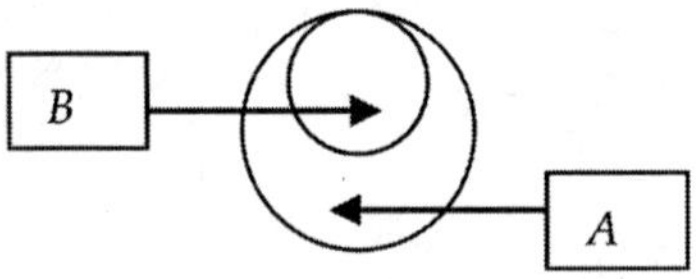

Some *A* are not *B* (true), some *B* are *A* (true) / some *B* are not *A* (false)

SOLUTION TO 2.6.6

2.6.6a

Schematization of the Argument

P1 The rules of morality have influence on our actions and affections.
P2 Nothing that is derived from reason alone has such influence.

C The rules of morality are not derived from reason alone.

Translation Key

M: The rules of morality
I: Things that have influence on our actions and affections
R: Things that are derived from reason alone

TL Argument[1]

S1 all *M* are *I*
S2 no *R* is *I*

S3 no *M* is *R*

The TL argument is deductively valid. As is clear from the diagrams below, every TL diagram that makes the premises true makes the conclusion true as well (see 2.5.3a).

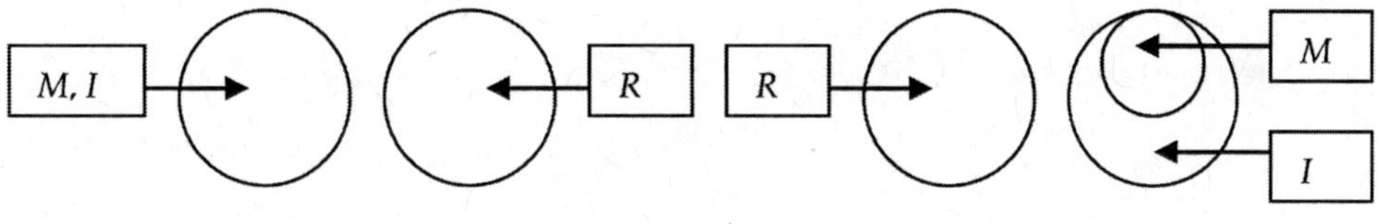

all *M* are *I* (S1; true), no *R* is *I* (S2; true) / no *M* is *R* (S3; true)

The TL argument is a faithful translation of the schematized argument, which is, as explained in the solution to 1.3.12, a correct schematization of the original argument (the argument in the passage). Since the TL argument is deductively valid, the original argument is valid too (see 2.5.12:C7).

1 In translating English arguments into TL, the reader is referred to 2.3.4.

SOLUTIONS TO 2.6.7

2.6.7a

Schematization of the argument

P1 Beethoven is Indian.
P2 All Indians are Buddhist.
P3 Buddhists are immortal.

C Beethoven is immortal.

Translation Key

b: Beethoven
I: Indians
D: Buddhists
M: Immortal

TL Argument

S1 *b* is *I*
S2 all *I* are *D*
S3 all *D* are *M*

S4 *b* is *M*

The TL argument is deductively valid. D1–D4 below are, up to similarity, the only TL diagrams on which the premises are all true. The conclusion, S4, is true on each of these diagrams (see 2.5.3a).

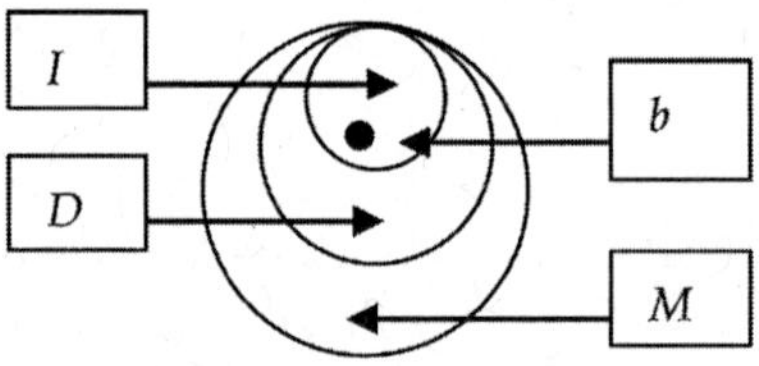

D1: *b* is *I* (S1; true), all *I* are *D* (S2; true), all *D* are *M* (S3; true) / *b* is *M* (S4; true)

D2, D3, and D4 are the standard variants of D1. Since the schematization and TL translation are straightforward, the argument in the passage is also deductively valid.

It should be noted, however, that one should always try to understand what is really meant by every English argument; in this case, given the unlikelihood that anyone seriously advances this argument, it is perhaps futile to figure out what is "really meant." In any case, the sentence 'Buddhists are immortal' is most plausibly interpreted as the claim that Buddhists do not (literally) die; but the sentence 'Beethoven is immortal' suggests a different sense of immortality: that one's achievements live on after one's death. Under these interpretations, 'immortal' does not keep the same sense throughout: it actually has two different meanings, so should correspond to two different circles in a TL diagram. But this would make the argument invalid. Compare exercise 2.6.7c, where this duality of meaning of the same word is more apparent.

2.6.7c

Schematization of the argument

P1 A kiwi is a type of bird.
P2 Some fruits are kiwis.
P3 All birds have feathers.

C Some fruits have feathers.

The conclusion of the argument is stated at the beginning of the passage and then repeated at the end. The argument is schematized from the author's point of view; for the argument begins with 'My reasoning is as follows'. Because of this point of view, 'we know that' is omitted from P1.[1]

Translation key

K_1: Kiwi (the bird)
B: Birds
F: Fruits
K_2: Kiwi (the fruit)
R: Things that have feathers

TL Argument

S1: all K_1 are B
S2: some F are K_2
S3: all B are R

S4: some F are R

1 Knowledge, in the philosophical sense, entails truth. So, for example, to say that Sarah knows that Erin is in love with Theodore entails that it is true that Erin is in love with Theodore.

S1 is a correct translation of P1. In this context 'A is a type of B' means that all A's are B's. The TL argument is deductively invalid. The diagram below makes the premises, S1–S3, true and the conclusion, S4, false (see 2.5.4).

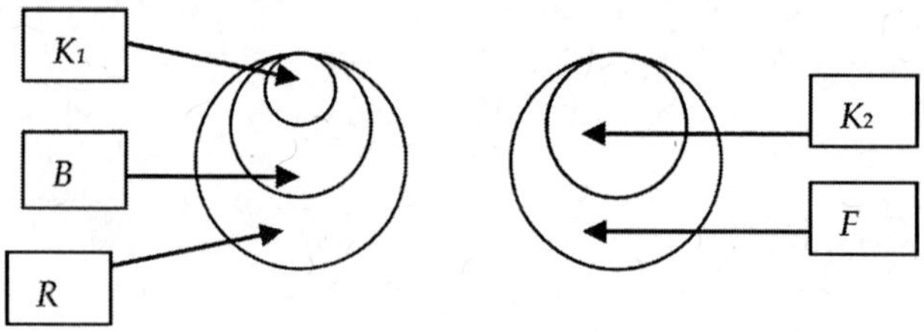

all K_1 are B (S1; true), some F are K_2 (S2; true),
all B are R (S3; true) / some F are R (S4; false)

This TL diagram is genuine. It depicts a logical possibility in which the word 'Kiwi' is an ambiguous general term; it stands for a type of bird and a type of fruit. Let us disambiguate the word by calling the bird "Kiwi1" and the fruit "Kiwi2." Now, this logical possibility tells us that there is no relation between fruits and things that have feathers (so the conclusion is false), and that, indeed, Kiwi2 is a type of fruit, Kiwi1 is a type of bird, and all birds have feathers (so the premises are all true). This logical possibility, therefore, is a counterexample to the schematized argument. We explained above that our schematization is a correct presentation of the argument in the passage. So the original English argument is deductively invalid.

Not every TL diagram that is a counterexample to the TL argument above is genuine. There are superfluous counterexamples to the TL argument; that is, there are TL diagrams that make the premises of the TL argument all true and the conclusion false, yet these TL diagrams do not depict any logical possibilities that are relevant to the English argument. Here is one such diagram.

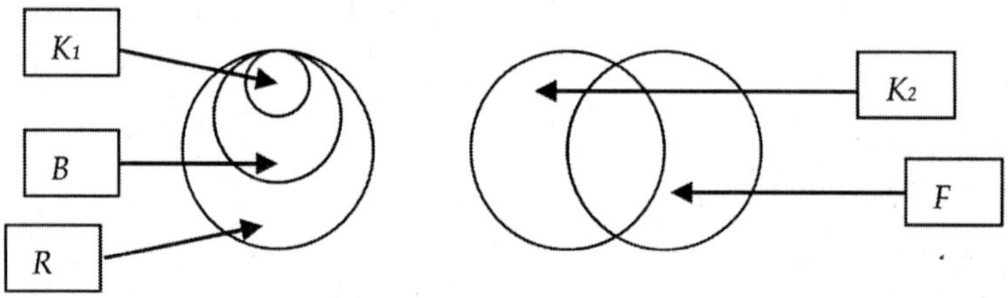

all K_1 are B (S1; true), some F are K_2 (S2; true),
all B are R (S3; true) / some F are R (S4; false)

If we describe this TL diagram using the translation key above, we obtain the following story: there is no relation between fruits and things that have feathers, Kiwi1 is indeed a type of bird and all birds have feathers, but not all Kiwi2's are fruits—some of them are and some of them are not. The last clause is the problem. No logical possibility relevant to the English argument permits some of the fruits called "Kiwi2" to be non-fruits. The way we defined "Kiwi2" as a type of fruit restricts all relevant logical possibilities to those situations in which every Kiwi2 is a fruit.[1] This TL diagram, therefore, is superfluous. This example illustrates the procedure we discussed in 2.5.12:C5. That procedure implies that the mere existence of superfluous TL diagrams does not guarantee that the English argument is deductively valid. The presence of superfluous TL diagrams does not rule out the existence of a genuine one. Our first counterexample to the TL argument is a genuine TL diagram, while the second is superfluous.[2] In order for the English argument to be deductively valid in spite of the deductive invalidity of its TL translation, *all* the counterexamples to the TL argument must be superfluous.

The English argument is an instance of a type of fallacy called "the fallacy of equivocation." An argument is a **fallacy of equivocation** if it contains an ambiguous word or expression so that the argument is deductively valid if the ambiguous word or expression is treated as having a single meaning and is invalid if the word or expression is disambiguated. The schematization above would be deductively valid if the ambiguous word 'kiwi' were considered as having a single meaning, but once the ambiguity is exposed, the argument is clearly invalid. That is why it is a fallacy: it has the appearance of a deductively valid argument but in reality it is invalid.

The fallacy of equivocation makes for good dramatic use. William Shakespeare employs it at the end of *Macbeth* with unmatched dramatic flair. Early in the play an apparition of a bloody child conjured by three witches tells Macbeth:

> Be bloody, bold, and resolute, laugh to scorn
> The power of man, for none of woman born
> Shall harm Macbeth.

So Macbeth replies: "Then live, Macduff: what need I fear of thee?" At the end of the play Macbeth reaffirms his belief that he cannot be harmed by anyone. He says to Macduff while Macduff is about to strike him with his sword:

> Let fall thy blade on vulnerable crests;
> I bear a charmed life, which must not yield
> To one of woman born.

To which Macduff replies:

> Despair thy charm;
> And let the angel whom thou still hast served
> Tell thee, Macduff was from his mother's womb
> Untimely ript.

---

1 It is important to stress that language is held fixed in all the logical possibilities that are relevant to some natural-language argument. In other words, when we consider the various logical possibilities that are relevant to some natural-language argument, the meanings of the expressions that occur in the argument must not be allowed to vary. (See1.2.1:C2.)

2 The presence of superfluous TL diagrams suggests that the TL argument is not a faithful translation of the English argument. There is, in fact, an interdependency between the terms 'Kiwi2' and 'fruit': all Kiwi2 are fruit. This interdependency is not captured in the TL translation of the English argument.

Macbeth's implied reasoning is as follows:

> I shall not be harmed by anyone of woman born.
> Everyone is of woman born.
> ______________________________
> I shall not be harmed by anyone.

Macbeth interpreted 'of woman born' to mean "having a mother." Macduff, on the other hand, interpreted 'of woman born' to mean "naturally born." If one interprets the expression 'of woman born' in both premises as having a single meaning (either Macbeth's or Macduff's reading), the argument is deductively valid but unsound (see 1.1.6:C); for on Macbeth's reading the first premise is false (Macbeth could be harmed by Macduff) and on Macduff's reading the second premise is false (Macduff was not naturally born). But if one interprets 'of woman born' in the first premise to mean "naturally born" and in the second premise to mean "having a mother," then both premises are true but the argument is deductively invalid. This argument is an example of the fallacy of equivocation.

2.6.7d

Schematization of the argument

P1 Only meaningful statements constitute knowledge.
P2 A statement is meaningful only if it is empirically verifiable.
P3 There is no true generalization that is empirically verifiable.
P4 Some mathematical theorems are true generalizations.
____
C Not every mathematical theorem constitutes knowledge.

Translation Key

M: Meaningful statements
K: Things that constitute knowledge
E: Things that are empirically verifiable
G: True generalizations
T: Mathematical theorems

TL Argument

S1 all *K* are *M*
S2 all *M* are *E*
S3 no *G* is *E*
S4 some *T* are *G*
____
S5 some *T* are not *K*

The TL argument is deductively valid. We will establish this by showing that there is no TL diagram on which the premises, S1–S4, are all true and the conclusion, S5, is false (see 2.5.3b). Instead of considering all the relevant TL diagrams on which S1–S4 are true or all the relevant TL diagrams on which S5 is false, we will follow a different strategy in order to reduce the number of TL diagrams we need to consider. We will consider only those TL diagrams for the argument on which S5 (the conclusion) is false and S1–S3 are true and verify that S4 is false on each of these diagrams. There are, up to similarity, only seven relevant TL diagrams that make S5 false and S1–S3 true.

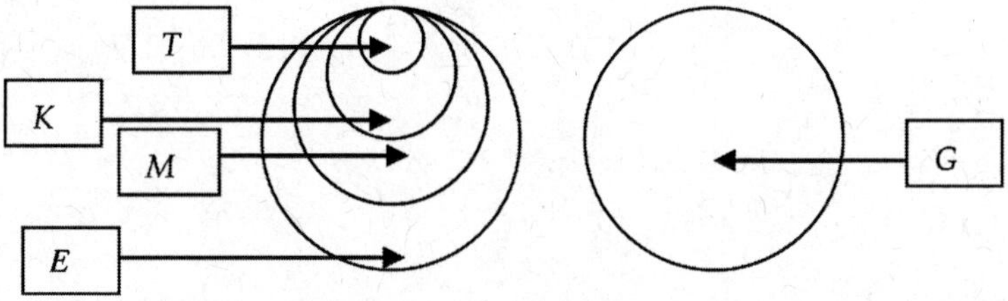

all *K* are *M* (S1; true), all *M* are *E* (S2; true), no *G* is *E* (S3; true),
some *T* are *G* (S4; false) / some *T* are not *K* (S5; false)

The other six diagrams are the standard variants of the TL diagram above.

The schematization is almost a verbatim presentation of the argument in the passage. The phrase 'it is clear that' is one of a host of phrases, such as 'it is obvious that', 'we all could agree that', 'it is apparent that', that typically indicate the author's belief that the following premise (or premises) needs no argument. Given the idioms listed in 2.3.4a, the TL translation is straightforward. Hence, the verdict of deductive validity can be carried over to the original argument.

SOLUTIONS TO 2.6.8

2.6.8a

Schematization of the argument

P1	The statement that 16 = 16 is self-evident.
P2	The statement that $4^2 = 16$ is a mathematical theorem.
P3	No self-evident statement requires a proof.
P4	Every mathematical theorem requires a proof.
	———————————————
C	The statement that 16 = 16 and the statement that $4^2 = 16$ are not the same.

Translation Key

s_1: The statement that 16 = 16
s_2: The statement that $4^2 = 16$
E: Self-evident statements

P: Statements that require proofs
M: Mathematical theorems

We restricted the term '*P*' to statements because it is clear from the context that we are talking about statements and not things in general.

TL Argument

S1 s_1 is *E*
S2 s_2 is *M*
S3 no *E* is *P*
S4 all *M* are *P*

S5 s_1 is not s_2

The TL argument is deductively valid. The two diagrams below are, up to similarity, all the TL diagrams that make the premises true; the conclusion is true on these diagrams (see 2.5.3a).

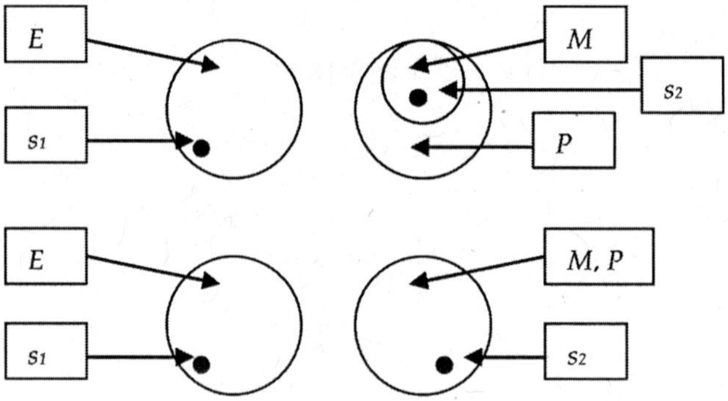

s_1 is *E* (S1; true), s_2 is *M* (S2; true), no *E* is *P* (S3; true), all *M* are *P* (S4; true) / s_1 is not s_2 (S5; true)

It is obvious that the TL translation is faithful and the schematization is correct. Therefore, the argument in the passage is deductively valid as well.

2.6.8d

Schematization of the argument

P1 David is a universal skeptic.
P2 A universal skeptic believes that he or she does not know anything.
P3 Those who believe that they do not know anything must know that they are thinking.
P4 Whoever knows that he or she is thinking actually knows something.

C There are universal skeptics who actually know something.

Translation Key

d: David
U: Universal skeptics
B: Those who believe that they do not know anything
T: Those who know that they are thinking
K: Those who actually know something

TL Argument

S1 *d* is *U*
S2 all *U* are *B*
S3 all *B* are *T*
S4 all *T* are *K*

S5 some *U* are *K*

Translating P2 as S2, perhaps, dilutes the meaning of P2. P2 seems to be a definition. It defines the term 'universal skeptic'. The most likely interpretation of P2 seems to be this: A person is called "a universal skeptic" if and only if this person believes that he or she does not know anything. In other words, P2 implies that the set consisting of those who are universal skeptics is precisely the same set consisting of those who believe that they do not know anything. A faithful translation of P2 into TL would give us two TL sentences 'all *U* are *B*' and 'all *B* are *U*'. Since the former is sufficient for the deductive validity of the argument, we conveniently select it as the sole translation of P2. It would do no harm to include a fifth premise, say S2*, asserting that all *B* are *U*, but it would be logically superfluous.

The TL argument is deductively valid. The following TL diagram and its standard variants are, up to similarity, all the diagrams on which the premises, S1–S4, are true.

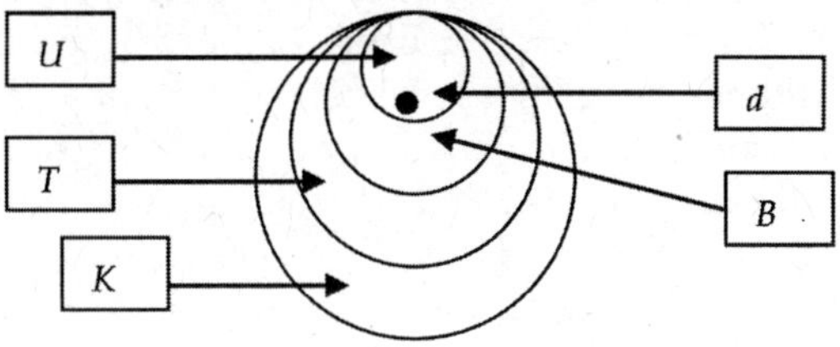

d is *U* (S1; true), all *U* are *B* (S2; true), all *B* are *T* (S3; true), all *T* are *K* (S4; true) / some *U* are *K* (S5; true)

We can easily verify that the conclusion, S5, is true on all these diagrams (see 2.5.3a). Since the TL translation is adequate (strictly speaking, it is not faithful because it omits the implied premise 'all *B* are *U*') and the schematization is correct, we conclude that the argument in the passage is deductively valid.

There are two important observations to make about this argument and its translation. First, a stronger claim follows from S1–S4; namely, that all *U* are *K*. This mirrors the logical relations in the English argument: the claim that *all* universal skeptics actually know something follows from P1–P4. But as stressed previously, we do not make changes in the argument unless we have good reasons to do so. The original argument has the weaker conclusion that *some* universal skeptics actually know something and that is the conclusion of our schematization.

Second, the first premise, S1, is extraneous to the deductive validity of the TL argument. This should be apparent from the diagram above. If the *d*-dot is inside or outside the *U*-circle, or if it doesn't exist at all, the *K*-circle still includes the *U*-circle. The presence and location of the *d*-dot is irrelevant to the logical status of the TL argument.

The question now: Is P1 also extraneous to the deductive validity of the schematized argument? In other words, is P1 superfluous? The answer depends on the way we understand the remaining premises, P2–P4. If we understand these premises as asserting relations between groups of existing (real) people, then P1 is superfluous. This way of understanding P2–P4 is in accord with the TL worldview. Recall that in this view every property has a *nonempty* extension. Thus if the extensions of the general terms occurring in the premises P2–P4 are understood to be nonempty, P1 would be superfluous. Of course, it is not a requirement that the English argument adopt the TL worldview. We could understand these premises (or at least P2 and P3) as asserting *hypothetical* relations between concepts. In this case, there need not be anyone who is a universal skeptic, nor anyone who believes that he or she does not know anything. Indeed, there need not be people at all. This way of understanding the premises P2–P4 is consistent with a *subjunctive* reading of them: if there *were* **X**'s, they *would* be **Y**. The sentence 'All universal skeptics actually know something' would still follow from P2–P4 (P1 would be extraneous to this deduction). This conclusion is understood to mean that if someone is a universal skeptic, then he or she actually knows something; it is irrelevant to the truth of this conclusion whether such a person exists or not.

On the other hand, the sentence 'Some universal skeptics actually know something' would not follow from P2–P4, if these premises were interpreted hypothetically. We take 'Some' to mean "there is at least one." If the hypothetical reading of the premises P2–P4 (or at least P2 and P3) is consistent with there being no universal skeptics, then 'There is at least one universal skeptic who actually knows something' cannot follow from P2–P4. We need P1 in this case. The relevance of P1 is that it ensures that there is at least one universal skeptic (namely, David). If there is such a person, then the general terms that occur in P2–P4 stand for at least one individual. (In a manner of speaking, P1 "anchors" these terms.) Thus if P2–P4 are interpreted hypothetically, the conclusion that some universal skeptics actually know something does not follow from P2–P4 but it does follow from P1–P4. P1, in this case, is not superfluous.

How should we understand P2–P4? There are grounds for reading at least P2 and P3 hypothetically. As we said previously, P2 seems to be a definition. Definitions typically do not entail the existence of the objects to which the definiendum (i.e., the term that is being defined) applies. For example, we can define a *super even number* to be any even number greater than 2 that is *not* the sum of two prime numbers. If the Goldbach conjecture, which says that every even number greater than 2 is the sum of two prime numbers, is true, then there are no super even numbers. The definition, however, is meaningful even though there might be no numbers to which the definiendum applies. This view of definition makes a hypothetical reading of P2 (and hence, of P3) sound reasonable, which means that P1 is relevant to the deductive validity of the English argument.

This implies that if we drop P1 and read P2 and P3 hypothetically, the resulting English argument, {P2, P3, P4}/C, would be deductively invalid. Nevertheless, as we demonstrated previously, the TL argument, {S2, S3, S4}/S5, which is a translation of the new English argument, is deductively valid. It seems that we have a case in which the English argument is deductively invalid while its TL translation is valid. But as we explained in 2.5.12:C6, English arguments in which some of the premises are hypothetical statements that contain general terms with empty extensions are controversial cases, and there are philosophers who argue that such natural-language arguments might be, contrary to appearances, deductively valid.

The same problem surfaces when dealing with non-referring English singular terms, such as 'Sherlock Holmes'. There are interpretations of fictional statements that render the following argument deductively invalid.

Sherlock Holmes lives on Baker Street.
Sherlock Holmes plays the violin.

There is someone who lives on Baker Street and plays the violin.

The TL argument below is a translation of this argument.

s is *B*
s is *V*

---

some *B* are *V*

where '*s*' stands for "Sherlock Holmes," '*B*' for those who live on Baker street, and '*V*' for those who play the violin. As the reader can readily verify, the TL argument is deductively valid.

Some philosophers argue that the reason for this divergence between the two verdicts is that the TL argument is not a faithful translation of the English argument, since the TL argument contains an interdependency between the terms '*B*' and '*V*'—namely that they apply to the same individual—that does not exist in the English argument. If those philosophers are correct, then we have a case in which there is a logical possibility for the English argument that is not represented by any TL diagram for the TL argument. This logical possibility is one in which the premises of the English argument are true and its conclusion is false. No TL diagram for the TL argument represents this logical possibility because there is no TL diagram on which the premises of the TL argument are true and its conclusion is false.

Similar to the case of general terms with empty extensions, there are philosophers who dispute this analysis. They argue that there is really no divergence between the logical status of the TL argument and the logical status of the English argument, because statements that attribute properties to fictitious objects are false. Hence in any logical possibility that contains no real individual called "Sherlock Holmes," whose attributes match those mentioned in the famous mysteries of Sir Arthur Conan Doyle, the premises are false. Such logical possibilities do not constitute counterexamples to the English argument. On the other hand, in any logical possibility in which there is an individual who is called "Sherlock Holmes," who lives on Baker Street, and who plays the violin, the conclusion is true. Thus in this case too there is no counterexample to the English argument. So the English argument, again contrary to appearances, is deductively valid—its logical status is the same as the logical status of the TL argument.

We will see in Chapter 4 that Predicate Logic (PL) allows general terms to have empty extensions. So PL is a more appropriate system for capturing the universal-skeptic argument faithfully. However, PL singular terms, like TL's, cannot be non-referring. This implies that PL, like TL, faces the same difficulty (if there is one) with arguments concerning fictitious objects. There are other logical systems, which are called "free logics," that permit singular terms to have no referents. The study of these logics is beyond the scope of this book.

SOLUTION TO 2.6.9

2.6.9a

Schematization of the first part of the argument

P1 The young child has never been taught geometry before.
P2 The young child is capable of arriving at geometrical knowledge.
P3 Only those who can comprehend geometrical concepts are capable of arriving at geometrical knowledge.

C1 Some people who have never been taught geometry can comprehend geometrical concepts.

The first and second premises in the passage are couched in performative language. We were told to *consider* a young child who has never been taught geometry and that we can *make* the child arrive at some geometrical knowledge, by *performing* certain actions—namely, asking questions and drawing illustrations. This language has roots in Plato's *Meno*. Socrates (Plato's primary interlocutor) performs an "experiment" with a slave boy in the household of Meno. Socrates asks the boy a series of questions and draws a few illustrations in the sand; he helps the boy to reach some geometrical knowledge. The performative language in the passage retains some of the flavor of Plato's argument, but it is extraneous to the logical relations in the argument. P1 and P2 as paraphrased in our schematization extract the relevant information from the premises in the passage. The third premise and the first conclusion show us how to paraphrase the first two premises.

Translation Key

c: The young child
N: Those who have never been taught geometry
K: Those who are capable of arriving at geometrical knowledge
G: Those who can comprehend geometrical concepts

TL Argument

S1 *c* is *N*
S2 *c* is *K*
S3 all *K* are *G*

S4 some *N* are *G*

We will show that the TL argument is deductively valid by showing that there is no TL diagram on which S1–S3 are true and S4 is false (see 2.5.3b). An easy way to show this is to consider the non-similar TL diagrams that make S4 false and S2 and S3 true, and demonstrate that S1 is false on these diagrams. There are, up to similarity, only two such diagrams.

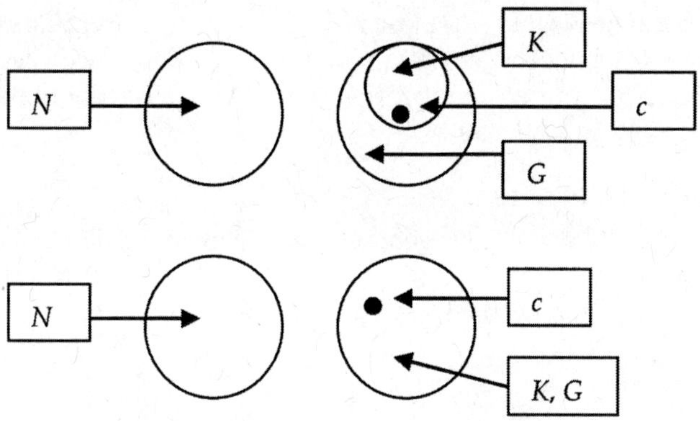

c is *N* (S1; false), *c* is *K* (S2; true),
all *K* are *G* (S3; true) / some *N* are *G* (S4; false)

We explained above that our schematization is an accurate presentation of the first part of the argument in the passage. The TL translation is faithful. Therefore, the first part of the argument is deductively valid.

Schematization of the second part of the argument

P4 If a person has never been taught geometry, and yet is capable of comprehending geometrical concepts, then that person must have those concepts innately.
[P5 The young child is a person who has never been taught geometry, and yet is capable of comprehending geometrical concepts.]

C2 The young child has innate geometrical concepts.

We placed the second premise, P5, in brackets because it is implicit in the passage. P4 speaks of people in general and states that those who have never been taught geometry and yet are capable of comprehending geometrical concepts must have these concepts innately. The second conclusion, C2, asserts that a specific person—namely, the young child—has innate geometrical concepts. Clearly there is a gap between P4 and C2. We need another premise telling us that the young child is one of those people who are described in P4. Such a premise is available to us from the first part. Exercise 2.6.9 states that the passage contains a two-part argument. The second part of a two-part argument typically depends on the first part; most likely the conclusion of the first part enters as a premise in the second part. In the passage above, the second part of the argument begins with the phrase 'But we can now argue'. The natural way to understand this phrase is to take it as saying "given the preceding argument (or, conclusion) we can now argue...". P5 is a stronger version of C1. While C1 speaks of some people, P5 speaks of a specific person. P5 logically follows from P1–P3. P1 asserts that the young child has never been taught geometry and P2 and P3 entail that the young child is capable of comprehending geometrical concepts. The author could have chosen P5, instead of C1, as the conclusion of the first part. As is clear from the TL translation of the first part, the reason we know that some *N* are *G* is because we know that there is at least one individual common to both *N* and *G*. This individual is *c*, the young child. Given this textual evidence, the Principle of Charity directs us to include P5 in our schematization of the second part of the argument.

Translation Key

A: Those who have never been taught geometry and, yet, are capable of comprehending geometrical concepts
I: Those who have innate geometrical concepts
c: The young child

TL Argument

S5 all *A* are *I*
S6 *c* is *A*

S7 *c* is *I*

Of course, the TL argument is deductively valid. Hence the second part is valid as well, since we defended above the correctness of our schematization and it is clear that the TL translation is faithful. The TL diagrams below are, up to similarity, all the diagrams on which S5 and S6 are true; S7 is also true on these diagrams (see 2.5.3a).

all *A* are *I* (S5; true), *c* is *A* (S6; true) / *c* is *I* (S7; true)

There is an interesting observation to make about the first part of the argument. We said above that P5 logically follows from the premises P1–P3. For the convenience of the reader we reproduce this argument below.

P1 The young child has never been taught geometry before.
P2 The young child is capable of arriving at geometrical knowledge.
P3 Only those who can comprehend geometrical concepts are capable of arriving at geometrical knowledge.

P5 The young child is a person who has never been taught geometry, and yet is capable of comprehending geometrical concepts.

Since P5 logically follows from P1–P3, the argument is deductively valid. However, its TL translation is deductively invalid. TL has no means of expressing conjunctions. We would not be able to translate P5 as '*c* is *N* and *c* is *G*' (or '*c* is *N* and *G*'), which is the natural rendering of P5. P5 would have to be translated as '*c* is *A*', where '*A*' is the translation of the English compound general term 'having never been taught geometry and yet capable of comprehending geometrical concepts'. We obtain the following TL argument.

S1 *c* is *N*
S2 *c* is *K*
S3 all *K* are *G*

S8 *c* is *A*

The reader can easily verify that the TL argument is deductively invalid. However, all the TL diagrams that are counterexamples to this TL argument are superfluous. They have to be, since the English argument {P1, P2, P3}/P5 is deductively valid, and hence no relevant logical possibility is a counterexample to it.

This example is interesting because it shows that superfluous TL diagrams could be the result of the expressive limitation of TL and not only of the incompleteness of its worldview. The TL diagrams on which the premises, S1–S3, are true display the dual membership of *c* in the extensions of *N* and *G*. It is the expressive limitation of TL that prevents TL from describing this fact by the compound sentence '*c* is *N* and *c* is *G*' or by the simple sentence '*c* is *NG*', where '*NG*' is a compound term that stands for '*N* and *G*'. Although TL cannot translate the English argument {P1, P2, P3}/P5 faithfully, it can capture a modified argument whose premises are P2 and P3 and whose conclusion is 'The young child can comprehend geometrical concepts' (call this sentence C3). The TL translation of the English argument {P2, P3}/C3 is: *c* is *K*, all *K* are *G*; therefore, *c* is *G*. It is obvious that the new English argument and its TL translation are deductively valid. Now we have two TL sentences: '*c* is *N*' and '*c* is *G*'. These sentences together express the English statement that the young child is a person who has never been taught geometry and yet is capable of comprehending geometrical concepts, which is P5.

Chapter Three

Sentence Logic (SL)

3.1 The SL Worldview

Sentence Logic (SL)[1] is a simple symbolic system that serves as the common foundation of many modern logical systems. In the **SL worldview**, **states of affairs** are the **basic ingredients of reality**. Any collection of states of affairs, with a specification as to which states obtain and which states fail to obtain,[2] constitutes a **logical possibility** according to this worldview. A *fact* is a state of affairs that obtains. For convenience, we shall refer to a state of affairs that fails to obtain as a *negative fact*. Hence in the SL worldview, any combination of facts and negative facts designates a logical possibility. We should stress that the facts and negative facts that constitute a logical possibility are a "pre-selected" subset of the totality of all states of affairs. We do not require logical possibilities to be *maximal* in the sense that every state of affairs must either obtain or not obtain in a logical possibility. Thus, for example, if the logical possibility concerns Jonathan's graduation party and who, among his friends, is attending the party, then whether or not the United States' men's soccer team will win the next World Cup need not be determined in this logical possibility. An important feature of this worldview is that there are no interdependencies between states of affairs, that is, each may obtain or fail to obtain independently of the status of any other state of affairs. The SL worldview presupposes a certain conception of truth that is called **the correspondence conception**. Roughly stated, this conception asserts that a declarative sentence is true if and only if it corresponds to a fact. A more precise formulation of this conception is this: a declarative sentence X is true in a logical possibility q if and only if the state of affairs that X represents (describes) obtains in q; and X is false in q if and only if the state of affairs that X represents fails to obtain in q.

3.1:C Commentary on 3.1

It is not easy to give a precise definition of the notion of a state of affairs. But we can give a rough explanation of this notion. A state of affairs is what a *simple* declarative sentence represents. For instance, the simple declarative sentence 'Santa Fe is the capital of New Mexico' represents the state of affairs that Santa Fe is the capital of New Mexico, that is, the state of affairs consisting of the city of Santa Fe, the state of New Mexico, and the relation "being the capital of," which holds between the city and the state. A simple declarative sentence is, roughly speaking, a declarative sentence that contains no sentential connectives (which connect two sentences together, to make one larger one) such as 'and', 'or', 'although', and 'if-then', and no quantifiers such as 'all', 'every', 'some', and 'there is'.

---

1 Sentence Logic is also called "Propositional Logic" and "Sentential Logic."
2 See page 37, note 1 of Chapter Two for this sense of 'obtain'.

We will illustrate the SL worldview by means of an example. Imagine that there are two cats Tom and Sylvester and a mat. The relevant states of affairs are two.

T Tom is on the mat.
S Sylvester is on the mat.

Depending on which state obtains and which does not, there are four logical possibilities.

q1 Tom and Sylvester are on the mat (T and S).
q2 Tom is on the mat but Sylvester is not (T and not-S).
q3 Tom is not on the mat but Sylvester is (not-T and S).
q4 Neither Tom nor Sylvester is on the mat (not-T and not-S).

The following table describes these possibilities in a compact form.

Logical possibility	The status of T	The status of S
q1: T and S	Obtains (fact)	Obtains (fact)
q2: T and not-S	Obtains (fact)	Does not obtain (negative fact)
q3: not-T and S	Does not obtain (negative fact)	Obtains (fact)
q4: not-T and not-S	Does not obtain (negative fact)	Does not obtain (negative fact)

As we will explain later, in SL logical possibilities are characterized as *truth valuations*. An SL truth valuation is a "linguified" logical possibility; it is a logical possibility at the level of language. Rather than presenting a combination of facts and negative facts, we present a collection of "basic" sentences with an assignment of truth values. The idea is this: if a sentence corresponds to a fact, it receives the truth value T (true), and if it corresponds to a negative fact, it receives the truth value F (false). This is another formulation of the correspondence conception of truth, but now it is applied to SL sentences. An assignment of truth values to "basic" sentences is an SL truth valuation (in 3.4.1 we give a precise definition of an SL truth valuation). The four logical possibilities listed above correspond in SL to four truth valuations.[1] Let the SL sentence *A* translate the English sentence 'Tom is on the mat' and *B* translate 'Sylvester is on the mat'. The symbol '$\neg$' is used in SL as the sign of negation; for example, '$\neg A$' is the SL translation of 'Tom is *not* on the mat'. The table below displays the logical possibilities q1–q4 and their corresponding SL truth valuations.

Logical possibility	Truth valuation	The truth value of *A*	The truth value of *B*
q1: T and S	V1: *A* and *B*	T	T
q2: T and not-S	V2: *A* and $\neg B$	T	F
q3: not-T and S	V3: $\neg A$ and *B*	F	T
q4: not-T and not-S	V4: $\neg A$ and $\neg B$	F	F

1 We should have said that these logical possibilities correspond to four *classes* of truth valuations. We will make this notion clear in 3.4.1:C1.

3.2 The Syntax of SL

3.2.1 As in TL, the **basic vocabulary** of SL is simple and consists of three categories. However, it is based on sentences rather than terms.

3.2.1a **Sentence letters**, which are the following uppercase italic letters: *A*, *B*, *C*, ..., *U*, *V*, *W*; with numeric subscripts if needed.[1]
3.2.1b Five **logical symbols**:[2] ¬, ∧, ∨, →, ↔[3]
3.2.1c **Parentheses**:[4] '(' and ')'

3.2.1:C COMMENTARY ON 3.2.1

It is natural that SL basic vocabulary comprises sentence letters. In fact, as we will see below, SL syntax as a whole is sentence-based. This feature mirrors the SL worldview, which is based on states of affairs. SL sentence letters correspond to states of affairs. The five logical symbols give us the wherewithal to form complex sentences from the sentence letters. The parentheses allow us to display the order in which various syntactical operations are applied.

3.2.2 The **sentences** of SL are either atomic or compound. The **atomic** sentences of SL are the sentence letters. The **compound** sentences of SL are those expressions constructed from the atomic sentences by applying any finite number of times one or more of the five **formation rules** listed below.[5] The letter 'R' stands for "formation rule." So, for instance, R∧ and R→ stand for the formation rules that generate compound sentences from other sentences by combining them with the signs ∧ and →, respectively. Let **X** and **Y** be any SL sentences.

R¬: ¬**X** is an SL sentence.
R∧: (**X**∧**Y**) is an SL sentence.
R∨: (**X**∨**Y**) is an SL sentence.
R→: (**X**→**Y**) is an SL sentence.
R↔: (**X**↔**Y**) is an SL sentence.

3.2.2:C COMMENTARY ON 3.2.2

3.2.2:C1 The logical symbols and parentheses are called "the logical vocabulary of SL," and the sentence letters "extra-logical vocabulary." Consider the following set of SL basic vocabulary: {*M*, *N*}.[6] We can generate an infinite number of compound SL sentences from this small set of

---

1 To be technically precise, every time we mention a symbol of the object language in the metalanguage we should enclose this symbol between single quotes, ' '. However, as in Chapter 2, almost always we will not follow this usage where there is no cause for misunderstanding.
2 These are the logical constants of SL.
3 Another standard list of SL logical symbols is: '~' or '–' instead of '¬', '&' or '•' instead of '∧', '⊃' instead of '→', and '≡' instead of '↔'.
4 The parentheses are SL punctuation marks.
5 Our definition *restricts* the compound sentences of SL to those expressions that can be *generated* from the basic vocabulary by these formation rules. (Observe that these rules are iterative.)
6 Every set of SL basic vocabulary must also contain the logical vocabulary of SL. For convenience we list only the extra-logical vocabulary.

basic vocabulary. The table below displays some of these sentences together with the atomic sentences and the formation rules that are used to construct these compound sentences. The order in which the rules are listed represents their order of application. For example, the second line lists three compound sentences, $\neg N$, $\neg\neg N$, and $\neg\neg\neg N$ (first column); they are formed from the atomic sentence N (second column) by applying the formation rule R$\neg$ once, twice, and thrice, respectively, to N (third column). The fifth line lists the compound sentence $(\neg N \vee M)$, the atomic sentences N and M, from which the sentence is composed, and the formation rules R$\neg$ and R$\vee$, which are used to form the sentence $(\neg N \vee M)$. The order of R$\neg$ and R$\vee$ indicates the order in which these rules are applied to generate $(\neg N \vee M)$. It is not important, at this stage, to know the semantics ("meanings") of the logical symbols. We only need to know how to apply the formation rules in order to generate compound sentences that contain these symbols.

SL sentences	Sentence letters used	Formation rules applied
M, N	M, N	None (atomic sentences)
$\neg N, \neg\neg N, \neg\neg\neg N$	N	R$\neg$, R$\neg$ R$\neg$, R$\neg$ R$\neg$ R$\neg$
$(N \wedge N), (N \vee N), (N \rightarrow N), (N \leftrightarrow N)$	N	R$\wedge$, R$\vee$, R$\rightarrow$, R$\leftrightarrow$
$(N \wedge M), (N \vee M), (N \rightarrow M), (N \leftrightarrow M)$	M, N	R$\wedge$, R$\vee$, R$\rightarrow$, R$\leftrightarrow$
$(\neg N \vee M)$	M, N	R$\neg$ R$\vee$
$((N \wedge M) \rightarrow M)$	M, N	R$\wedge$ R$\rightarrow$
$((\neg((N \vee M) \rightarrow N) \rightarrow M) \leftrightarrow M)$	M, N	R$\vee$ R$\rightarrow$ R$\neg$ R$\rightarrow$ R$\leftrightarrow$
$(\neg((N \leftrightarrow N) \vee (N \rightarrow M)) \rightarrow N)$	M, N	R$\leftrightarrow$ and R$\rightarrow$ (applied first) R$\vee$ R$\neg$ R$\rightarrow$
$(\neg((N \leftrightarrow N) \vee (N \rightarrow M)) \rightarrow \neg N)$	M, N	R$\leftrightarrow$ and R$\rightarrow$, and R$\neg$ (applied first) R$\vee$ R$\neg$ R$\rightarrow$
$(\neg((N \leftrightarrow N) \vee (N \rightarrow M)) \rightarrow (\neg N \wedge M))$	M, N	R$\leftrightarrow$ and R$\rightarrow$, and R$\neg$ (applied first) R$\vee$ and R$\wedge$ (applied second) R$\neg$ R$\rightarrow$
$\neg(((\neg(M \rightarrow \neg M) \rightarrow (M \vee M)) \wedge M) \leftrightarrow M)$	M	R$\neg$ and R$\vee$ (applied first) R$\rightarrow$ R$\neg$ R$\rightarrow$ R$\wedge$ R$\leftrightarrow$ R$\neg$

We can see from this table that SL syntax permits the formation of very long, complex sentences from a finite set of sentence letters. As the last sentence demonstrates, the set of sentence letters could consist of one member only. The last four sentences show that it is cumbersome to list the formation rules applied in linear order; there is a need to branch out. Later we will introduce *construction trees*, which display the "syntactical genealogy" of SL sentences in a manner that allows for branching out.

3.2.2:C2 The five logical symbols are usually referred to as **sentential connectives**. The symbols $\wedge$, $\vee$, $\rightarrow$, and $\leftrightarrow$ are called *binary connectives* because they connect two (not necessarily distinct) sentences to form a new sentence, and $\neg$ is called a *unary connective* because it applies to one sentence. It is a bit strange to call $\neg$ "a connective" since it does not, precisely speaking, connect sentences to form a new sentence. But the usage is traditional (and harmless), and we shall follow it in this book. The **immediate component** of an SL sentence of the form $\neg$**X** is **X** and its **main connective** is $\neg$. The immediate components of SL sentences of the forms (**X**$\wedge$**Y**), (**X**$\vee$**Y**), (**X**$\rightarrow$**Y**), and (**X**$\leftrightarrow$**Y**) are **X** and **Y** and their main connectives are $\wedge$, $\vee$, $\rightarrow$, and $\leftrightarrow$, respectively.

An important observation to make is that the formation rules of the binary connectives invoke the use of parentheses, while the formation rule of the unary connective ¬ does not. The reason is that the application of ¬ engenders no ambiguity, but the applications of ∧, ∨, →, and ↔ would generate ambiguities if parentheses were not employed.[1] The sign ¬ always applies to the sentence that is immediately to the right of the sign. For example, in the SL sentence $(\neg(A \rightarrow B) \wedge D)$ the sign ¬ applies to $(A \rightarrow B)$ because this is the sentence that it is immediately to the right of ¬. In any compound SL sentence that contains ¬ there is exactly one SL sentence that is immediately to the right of (an occurrence of) ¬. On the other hand, we need the parentheses in R∧, R∨, R→, and R↔ in order to determine the scope of the connectives. For instance, the expression $A \rightarrow B \rightarrow D$ is ambiguous; it could be $((A \rightarrow B) \rightarrow D)$ or $(A \rightarrow (B \rightarrow D))$.

3.2.2:C3 Complex SL sentences contain many connectives and parentheses. There are rules that help us determine certain syntactical features of SL sentences. We call these rules **bookkeeping rules**. We will discuss two such rules. The first allows us to determine whether the parentheses in an SL expression are "balanced," that is, whether every left parenthesis corresponds to a unique right parenthesis and every right parenthesis corresponds to a unique left parenthesis. If the parentheses are not balanced, the expression is not a sentence of SL. We call this rule the "Balance Rule." The second rule finds the main connective of an SL sentence. We call it the "Main-Connective Rule."[2]

The Balance Rule: We use this rule for SL expressions that have parentheses. We assign the number one to the leftmost parenthesis. We proceed to the right adding 1 for every left parenthesis and subtracting 1 for every right parenthesis. The parentheses in the SL expression are balanced when and only when the rightmost parenthesis receives the number zero and no parenthesis in the expression receives a negative number.

The Main-Connective Rule: It is important that, when this rule is applied, the SL sentence has not been modified according to any convention (we will introduce a convention later). If the leftmost character is ¬, then the main connective of the SL sentence is ¬. If the leftmost character is '(', we apply the Balance Rule. Since the expression is an SL sentence, its parentheses are balanced. This means that the number assigned to its rightmost parenthesis is 0. We find the parentheses at which we reached number 1 in the process of counting. There should be either one, two, or three parentheses at which we reached number 1. If we reached 1 only once, the main connective of the SL sentence is its binary connective; for, in this case, the sentence contains only one binary connective. If we reached 1 twice, we examine the character that is immediately to the right of the second parenthesis assigned 1. If this character is a binary connective, then it is the main connective of the sentence. If this character is the sentence's right-

1 If the formation rules were different, the need for parentheses might be eliminated. Reverse Polish Notation, for instance, requires no parentheses. According to this notation the connective always comes first. So $(A \rightarrow B)$ becomes $\rightarrow AB$, $\neg(A \rightarrow B)$ becomes $\neg \rightarrow AB$, and $(\neg(A \rightarrow B) \wedge D)$ becomes $\wedge \neg \rightarrow ABD$. While Reverse Polish Notation is suitable for automation, it is not intuitive.

2 We should stress that while the first rule applies to any SL expression, the second applies only to SL sentences.

most parenthesis, then the first binary connective that occurs in the sentence is its main connective. If we reached 1 thrice, the main connective is the binary connective that is immediately to the right of the second parenthesis assigned 1.

Let us discuss a few examples to illustrate the utility and application of these rules. We apply the Balance Rule to the SL expressions listed below. In order to make matters clear, we write the number assigned to a parenthesis by our counting process as a subscript to that parenthesis.

E1: $(_1\neg(_2A\wedge B)_1\leftrightarrow C)_0$

E2: $(_1\neg A\wedge(_2B\leftrightarrow C)_1)_0$

E3: $\neg(_1A\wedge(_2B\leftrightarrow C)_1)_0$

E4: $(_1(_2\neg K\rightarrow(_3L\wedge D)_2)_1\rightarrow(_2\neg(_3N\vee M)_2)_1)_0$

E5: $(_1(_2(_3\neg\neg K\wedge D)_2\vee C)_1\leftrightarrow(_2\neg(_3N\leftrightarrow M)_2\rightarrow G)_1)_0$

E6: $(_1\neg(_2P\vee S)_1\rightarrow Q)_0\vee H)_{-1}\rightarrow(_0(_1C\vee A)_0$

E7: $(_1(_2(_3(_4K\rightarrow H)_3\rightarrow C)_2\vee J)_1\leftrightarrow(_2\neg(_3L\wedge P)_2\vee\neg Q)_1$

E8: $(_1(_2P\vee S)_1\wedge\neg\neg U)_0\wedge(_1(_2D\vee H)_1\wedge\neg(_2B\rightarrow S)_1)_0$

The first three expressions are SL sentences. Of course, their parentheses are balanced. E1 begins with '(' and contains two parentheses that are assigned number 1, the second of which is followed by the binary connective $\leftrightarrow$. Therefore the main connective of E1 is $\leftrightarrow$. E2 also begins with '(' and contains two parentheses that are assigned number 1, but the second parenthesis assigned 1 is followed by E2's rightmost parenthesis. Hence, the main connective of E2 is the first binary connective occurring in the sentence. This connective is $\wedge$. E3 begins with $\neg$; its main connective, therefore, is $\neg$.

The SL expression E4 has balanced parentheses but it is not an SL sentence. $\neg(N\vee M)$ is enclosed between parentheses, yet the rule R$\neg$ does not call for outer parentheses. The parentheses of E6 and E7 are not balanced. E7 violates the first condition of the Balance Rule and E6 violates the second condition. E7's rightmost parenthesis does not receive the number 0 and E6 contains a parenthesis that is assigned a negative number.

E5 and E8 have balanced parentheses. E5 is an SL sentence while E8 is not. In E5 we reach number 1 three times; hence, its main connective is the binary connective that is immediately to the right of the second parenthesis assigned number 1. This connective is the first $\leftrightarrow$ (from the left, of course). In E8 we reach number 1 five times. This is sufficient to show that the expression E8 is not an SL sentence. We said previously that in SL sentences we reach number 1 once, twice, or thrice. In fact, E8 is only missing the outermost parentheses. If we enclose E8 between parentheses, it becomes an SL sentence whose main connective is the second $\wedge$. Here is the resulting sentence.

$$(_1(_2(_3P\vee S)_2\wedge\neg\neg U)_1\wedge(_2(_3D\vee H)_2\wedge\neg(_3B\rightarrow S)_2)_1)_0$$

We will introduce a convention in 3.2.5 that permits us to drop the outermost parentheses. Recall, however, that we required at the outset of our discussion of the second bookkeeping rule that the SL sentence must not be modified by any convention. The rule, in order to deliver the right results, can only be applied to SL sentences in their original syntactical formats.

The Main-Connective Rule is very effective in dealing with complex SL sentences. In most cases, however, when the SL sentences are of relatively simple structures, we do not need to rely on this method. We can locate the main connective by virtualizing how the sentence is built from its basic components. For instance, consider the SL sentence $((\neg K \rightarrow (L \wedge D)) \rightarrow \neg (N \vee M))$. This sentence is built from the basic components K, L, D, N, and M as follows: first, we form $(N \vee M)$ by connecting N and M by $\vee$; second we place $\neg$ to the left of $(N \vee M)$ to get $\neg (N \vee M)$; third, we form $(L \wedge D)$ by connecting L and D by $\wedge$; fourth, we place $\neg$ to the left of K to get $\neg K$; fifth, we form $(\neg K \rightarrow (L \wedge D))$ by connecting $\neg K$ and $(L \wedge D)$ by $\rightarrow$; and finally, we form $((\neg K \rightarrow (L \wedge D)) \rightarrow \neg (N \vee M))$ by connecting $(\neg K \rightarrow (L \wedge D))$ and $\neg (N \vee M)$ by $\rightarrow$. Thus the main connective is the last connective used to form this sentence, that is, the second $\rightarrow$.

3.2.3 We use the following **terminology**. $\neg \mathbf{X}$ is the *negation* of **X**. $(\mathbf{X} \wedge \mathbf{Y})$ is the *conjunction* of **X** and **Y**, and they are the *conjuncts* of $(\mathbf{X} \wedge \mathbf{Y})$. $(\mathbf{X} \vee \mathbf{Y})$ is the *disjunction* of **X** and **Y**, and they are the *disjuncts* of $(\mathbf{X} \vee \mathbf{Y})$. $(\mathbf{X} \rightarrow \mathbf{Y})$ is the (*material*) *conditional* of **X** and **Y**; **X** is the *antecedent* and **Y** is the *consequent* of $(\mathbf{X} \rightarrow \mathbf{Y})$. $(\mathbf{X} \leftrightarrow \mathbf{Y})$ is the (*material*) *biconditional* of **X** and **Y**.

3.2.3:C Commentary on 3.2.3

The main connective of the SL sentence determines whether the sentence is a negation, conjunction, disjunction, conditional, or biconditional. The **X** and **Y** in these sentences are their immediate components and they can be compound sentences as well. For instance, the sentence below is the biconditional of $(\neg (A \wedge B) \vee C)$ and $((C \rightarrow \neg A) \rightarrow D)$. We apply the Main-Connective Rule to show that the main connective is, indeed, $\leftrightarrow$.

$$({}_1({}_2\neg({}_3A \wedge B)_2 \vee C)_1 \leftrightarrow ({}_2({}_3C \rightarrow \neg A)_2 \rightarrow D)_1)_0$$

The sentences $(\neg (A \wedge B) \vee C)$ and $((C \rightarrow \neg A) \rightarrow D)$ are the immediate components of the biconditional.

3.2.4 The **construction tree** of an SL sentence **X** displays the main connectives and the immediate components of **X** and of the sentential components of **X**. The tree terminates with the atomic components of **X**.

3.2.4:C Commentary on 3.2.4

We have already defined what we mean by 'an immediate component of **X**'. If an SL sentence **Y** occurs in **X**, **Y** is said to be a *sentential component* of **X**. If a sentential component of **X** is an atomic sentence, it is called an *atomic component* of **X**. As the case with the Main-Connective Rule, construction trees are drawn for SL sentences that are not modified by any convention.

Below is the construction tree of the biconditional mentioned in 3.2.3:C.

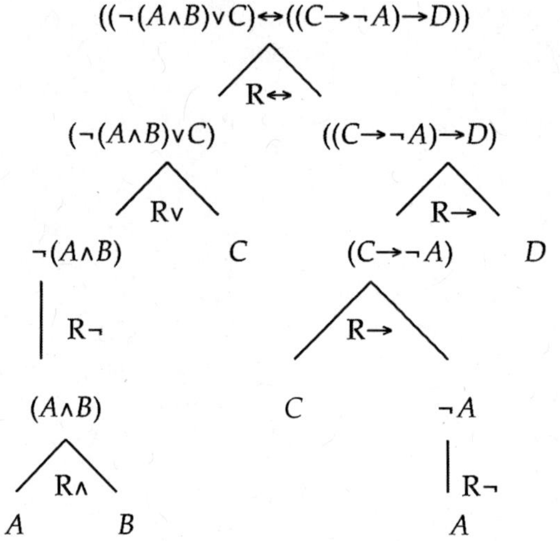

The SL sentences that are listed below a node are all the *proper* sentential components of the sentence at the node. We say 'proper' because every SL sentence **X** is a sentential component of itself, since **X** occurs in **X**. Hence **Y** is a proper sentential component of **X** if **Y** is not **X** and it is a sentential component of **X**. The sentences that are *immediately* below a node are the immediate components of the sentence at the node, and the formation rule displayed immediately below a node is the rule applied to form the sentence at the node from its immediate components. These formation rules introduce the main connectives of the sentences they form. Hence, for example, the formation rule displayed immediately below the biconditional is R↔ and immediately below the disjunction is R∨. The tree terminates with the atomic components of the biconditional. The atomic sentences listed below a node are the atomic components of the sentence at the node.

Construction trees give us a test for "SL-sentencehood": an SL expression is a sentence of SL if and only if it has a *complete* construction tree, that is, all the branches of its construction tree terminate with atomic sentences. To see how this works, we draw the construction tree of the SL expression E4 listed in 3.2.2:C3.

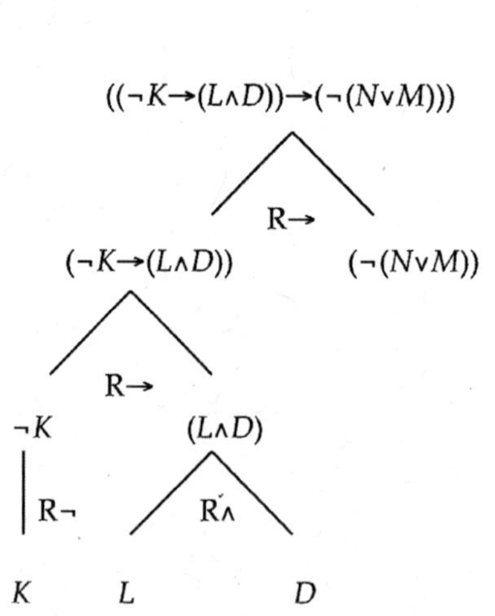

The expression $(\neg(N \vee M))$ cannot be decomposed; it is not an SL sentence. None of the formation rules listed in 3.2.2 can generate this expression. The most likely candidate is R$\neg$ applied to the SL sentence $(N \vee M)$. But such an application generates $\neg(N \vee M)$, not $(\neg(N \vee M))$. R$\neg$ does not introduce outer parentheses. The construction tree of E4 is not complete, that is, not all its branches terminate with atomic sentences.

Sometimes when an SL expression is missing certain parentheses, it has no unique main connective. In this case every possible decomposition of the SL expression does not produce a complete construction tree. Here is an example: $((P \vee \neg S) \rightarrow Q \wedge \neg V)$. There are two possible decompositions of this expression,[1] none of which produces a complete construction tree.

First decomposition:

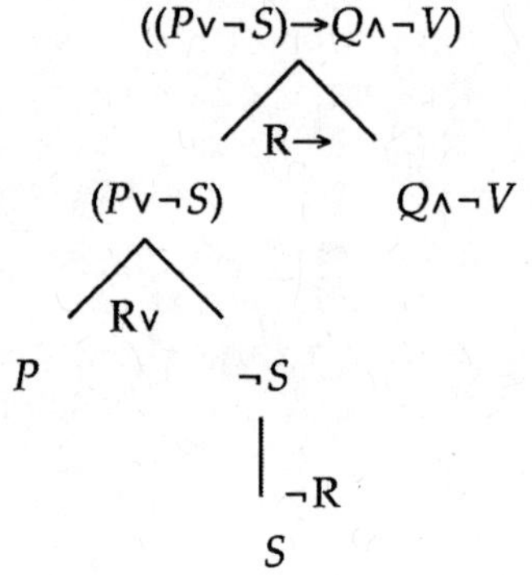

Second decomposition:

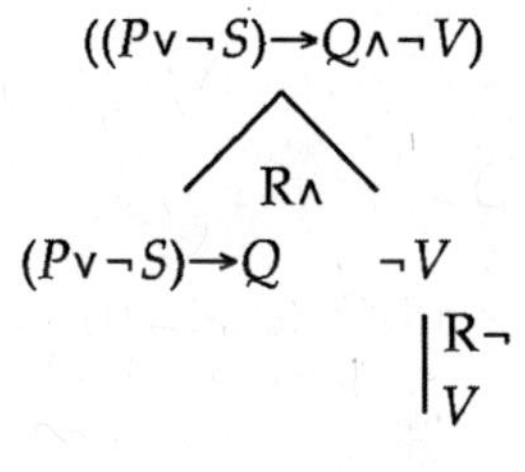

In the first decomposition the expression $Q \wedge \neg V$ is not an SL sentence; it is missing the outermost parentheses, which are part of any binary-connective formation rule. Thus it cannot be decomposed: no formation rule can generate $Q \wedge \neg V$. In the second decomposition the expression $(P \vee \neg S) \rightarrow Q$ is not an SL sentence, for the same reason: it is missing the outermost parentheses.

3.2.5 We will almost always follow the **convention** of dropping the outermost parentheses. For example, we write $A \rightarrow B$ instead of $(A \rightarrow B)$ and $(A \rightarrow B) \wedge C$ instead of $((A \rightarrow B) \wedge C)$.

1 There is no conclusive reason why one should not attempt to decompose the expression as a disjunction. In this case there are three possible decompositions of the expression. The third possibility, of course, also does not generate a complete construction tree.

3.2.6 The type of syntax described in 3.2.2 is usually called **generative recursive grammar**. It is so called because every compound sentence of SL can be *generated* from the basic vocabulary by applying one or more of the five formation rules any finite number of times. Thus these rules can be *iterated* indefinitely (they are **iterative** or **recursive** rules). This feature implies that the number of SL sentences that can be generated from a finite list of basic vocabulary is infinite.

3.2.6:C COMMENTARY ON 3.2.6

We said in 2.2.3 and 2.2.3:C that the formation rules of TL are not iterative and that, therefore, only finitely many TL sentences can be generated from a finite list of TL basic vocabulary. We also explained in 2.2.3:C that natural languages contain iterative operations. We discussed the example of the clause 'who met a tall man', which can be appended any number of times to the sentence 'I met a tall man'. SL is like natural languages: it contains iterative rules. In fact, all of SL formation rules are iterative. We use the word 'recursive' in this context as synonymous with the word 'iterative'. SL formation rules are *indefinitely* iterative, that is, they can be repeated without limit. When we discuss the syntax of Predicate Logic in Chapter Four, we will see examples of iterative rules that cannot be repeated indefinitely.

The number of compound SL sentences that can be generated from a single sentence letter by applying any of the five formation rules is infinite. Consider, for example, the sentence letter '*D*' and the formation rule R¬. R¬ could be applied to '*D*' generating '¬*D*', then applied to '¬*D*' generating '¬¬*D*', and so on. There is no upper limit on the number of times the sign ¬ can be written to the left of '*D*'. The same observation applies to R∧, R∨, R→, and R↔. For instance, the SL sentence '(((((*D*→*D*)→*D*)→*D*)→*D*)→*D*)' results from applying R→ to '*D*' and '*D*', then to '(*D*→*D*)' and '*D*', then to '((*D*→*D*)→*D*)' and '*D*', and so on. In principle, R→ could be iterated in this manner indefinitely, generating infinitely many SL sentences.

3.3 Translating English into SL

3.3.1 A **bivalent "basic" sentence** is an affirmative sentence containing no sentential connectives, such as 'and', 'or', and 'if-then', and asserting a claim that is either true or false but not both. The sentences 'Michael is a logician', 'The cat was on the mat', and 'New Mexico is south of Colorado' are examples of bivalent "basic" sentences. However, this definition allows quite complex sentences, such as 'Little Susie believes that Santa Claus lives at the North Pole', to count as "basic" sentences. Bivalent "basic" sentences are translated into SL as **atomic sentences**. English **negation** and **denial** expressions, such as 'not', 'it is not the case that', and 'it is false that', are translated into SL as ¬. English **conjunction** connectives (such as 'and'), **disjunction** connectives (such as 'or' when used in the **inclusive** sense), **conditional** connectives (such as 'if-then'), and **biconditional** connectives (such as 'if and only if') are translated **truth-functionally** into SL as ∧, ∨, →, and ↔, respectively. We should note that some of these English words are not always used as sentential connectives. For example, the sentence 'Fred ate a ham and cheese sandwich' does not mean that Fred ate a ham sandwich and Fred ate a cheese sandwich. On the other hand, the sentence 'Fred ate ham and cheese' usually means that Fred ate ham and Fred ate cheese. The word 'and' in the first sentence is not functioning as a sentential connective, while in the second it is a sentential connective.

3.3.1:C Commentary on 3.3.1

3.3.1:C1 The connective ∨ corresponds in English to an **inclusive** 'or'. Once we specify the semantics of SL, we will see that every SL sentence of the form **X**∨**Y** is true if and only if **X** is true or **Y** is true or *both.* It is the 'both' clause that gives ∨ its inclusive sense. In English 'or' can be inclusive or exclusive. If 'or' in an English sentence of the form 'A or B' is exclusive, then the sentence is false when its disjuncts are *both* true. Whether a particular usage of 'or' is inclusive or exclusive mostly—but not always—depends on the context of the usage. For example, if it is clear from the context that we can do only one thing this afternoon, then the disjunction 'We may go to the zoo or the movies this afternoon' is exclusive. On the other hand, in a typical context, the disjunction 'You may take a notebook or a pencil from the supply room' is inclusive. But there are contexts that leave the usage of 'or' ambiguous: it can be interpreted either inclusively or exclusively. Legal language usually makes the inclusive use of 'or' clear by employing connectives such as 'and/or' or stating explicitly 'or both'.[1] In this book we will translate *all* English disjunction connectives into the inclusive SL connective ∨ *unless* there is a *logical* reason for translating the English disjunction in the exclusive sense.[2]

3.3.1:C2 We need to explain the expression 'truth-functionally'. Let * be a binary connective in some language. Let P and Q be two declarative sentences of that language. We say that * is used truth-functionally in P*Q if and only if the truth value of P*Q depends solely on the truth values of P and Q; in other words, the truth value of P*Q is a *function* of the truth values of P and Q. If * is always used truth-functionally, we say that * is a truth-functional connective.

All SL connectives are truth-functional. English connectives, on the other hand, may be employed truth-functionally or not. The conditional connective 'if' is one of those connectives. A well-known case is that of the subjunctive conditional. A subjunctive conditional may be roughly defined as a conditional whose antecedent and consequent are false, but the truth of the conditional depends on the causal relations between the states of affairs described by the antecedent and consequent. Consider, for example, the following conditionals:

(a) If kangaroos had no tails, they would topple over.
(b) If kangaroos had no tails, they would be able to fly.

It seems reasonable to assert the truth of (a) and the falsity of (b). The truth values of these conditionals are not determined by the truth values of their component sentences. The antecedent and consequent of each conditional are false (kangaroos have tails, they don't topple over, and they aren't able to fly). So the difference between the truth values of the conditionals must be attributed to factors other than the truth values of their component sentences. The truth of (a) presumably is the result of the causal connection between kangaroos' having tails and their ability to maintain balance: if they lacked tails, they would lose balance and topple over. But there is no causal connection between kangaroos' having tails and their lack of ability to fly:

1 The sentences 'Violators will be subject to fine and/or imprisonment' and 'Violators will be subject to fine or imprisonment or both' are typical examples of legal language.
2 By a 'logical reason' we mean a reason that pertains to the logical relations in the argument.

not having tails would not make kangaroos capable of flying. The use of 'if' in these conditionals, therefore, is not truth-functional.

There is another reason why we restrict the SL translations of English connectives to the truth-functional usage of these connectives. Many English connectives have connotations beyond their truth-functional employment. In other words, even if an English connective is used truth-functionally, there might be aspects to the meaning of the connective that are not captured by its truth-functionality. The conditional connective 'even if' is a case in point. The truth value of the conditional

(c) Even if it rains, I'll go to the party

seems to depend solely on the truth value of its consequent. Arguably, (c) is false if and only if it rains and I don't go to the party or it doesn't rain and I don't go to the party, that is, the 'even if' conditional is false if and only if its consequent is false.[1] But (c) has further connotations beyond its truth-functionality. The use of 'even if' in (c) suggests that typically if it rains I don't go outside but this time, contrary to my typical practice, I am determined to go to the party *even if* it rains.[2] This connotation is not part of the truth-functional usage of 'even if' in (c). When (c) is translated into SL, this connotation is not carried over to the SL sentence. Such non truth-functional connotations cannot be translated into SL. Another example that illustrates the same point is the sentence 'It is snowing but I am going outside'. 'But' is usually translated into SL as the conjunction connective $\wedge$. But the SL sentence is missing something important: the use of 'but' in the English sentence implies that there is a tension between snowing and going outside. Usually when it is snowing, people are inclined to stay indoors. This tension between the meanings of the two sentences 'It is snowing' and 'I am going outside' is not part of the truth-functionality of the English sentence. Its SL translation only captures its truth-functionality—namely, that the compound sentence is true when and only when it is snowing and I am going outside.

---

1 It is reasonable to suppose that 'Even if it rains, I'll go to the party' truth-functionally means that if it rains, I'll go to the party, and if it doesn't rain, I'll go to the party, which is equivalent to saying that if it rains or doesn't rain, I'll go to the party. The antecedent of the last conditional is logically true; hence the truth value of the conditional depends solely on the truth value of the consequent: if the consequent is true, the conditional is true, and if the consequent is false, the conditional is false.

2 All 'even if' conditionals imply that the truth of the antecedent typically leads to the falsity of the consequent. The following pair of conditionals exemplifies this usage.

(d) Even if you don't invite me, I'll come to your party.
(e) Even if you invite me, I'll come to your party.

(d) sounds perfectly sensible while (e) strikes us as odd and in need of further elaboration. (d) displays a typical employment of 'even if'. The truth of the antecedent 'you don't invite me' ordinarily leads to the falsity of the consequent 'I'll come to your party'. Using the language of facts, we say that the fact of your not inviting me to the party ordinarily results in my not going to your party. (d) indicates that in spite of this typical connection between your not inviting me and my not going to the party, in this case, I will go to your party *even if* you do not invite me. The oddity of (e) is due to the fact that being invited to a party is usually a reason for going to the party; so it is not clear why someone needs to affirm that she will go to the party *even if* she is invited.

3.3.2 The following list shows how **English idioms** are translated into SL.[1]

3.3.2a English idioms: **X** and **Y**. Both **X** and **Y**. **X** as well as **Y**. Although **X**, **Y**. **X** yet **Y**. **X** but **Y**.
SL translation: **X∧Y**

3.3.2b English idiom: Neither **X** nor **Y**.
SL translation: **¬X∧¬Y**

3.3.2c English idioms: **X** or **Y** (in the inclusive sense). **X** or **Y** or both. Either **X** or **Y** (in the inclusive sense). Either **X** or **Y** or both. **X** and/or **Y**.
SL translation: **X∨Y**

3.3.2d English idioms: **X** or **Y** (in the exclusive sense). **X** or **Y** but not both. Either **X** or **Y** (in the exclusive sense). Either **X** or **Y** but not both.
SL translation: **(X∨Y)∧¬(X∧Y)**

3.3.2e English idioms: If **X**, then **Y**. If **X**, **Y**. **Y** if **X**. **Y** when **X**. **X** only if **Y**. **X** only when **Y**. **X** is sufficient (or a sufficient condition) for **Y**. **Y** is necessary (or a necessary condition) for **X**.
SL translation: **X→Y**

3.3.2f English idioms: **X** unless **Y**. Unless **Y**, **X**.
SL translation: **¬Y→X**

3.3.2g English idioms: **X** if and only if **Y**. **X** when and only when **Y**. **X** just in case **Y**. **X** is sufficient and necessary (or a sufficient and necessary condition) for **Y**.
SL translation: **X↔Y**

3.4 The Semantics of SL

3.4.1 The semantics of SL sentences is given by truth valuations. V is an SL **truth valuation** *for* a set Γ of SL sentences, that is, V is *relevant to* Γ, if and only if there is a collection Σ of atomic SL sentences such that

3.4.1a every atomic component of a member of Γ is in Σ;[2] and
3.4.1b V assigns *exactly one* truth value, T (true) or F (false), to every atomic sentence in Σ.

Σ may contain atomic sentences that are not atomic components of any sentence in Γ. V also assigns, according to the truth conditions listed in 3.4.2, truth values to all the compound SL sentences that are generated from the atomic sentences in Σ; and hence since the members of Γ are composed of atomic sentences in Σ, V assigns truth values to all the members of Γ. V is said to be a truth valuation for an SL sentence if it is a truth valuation for a set containing that sentence.

---

1 The boldfaced letters '**X**' and '**Y**' are metalinguistic variables that range over SL sentences. Our usage here is hybrid: when used in SL sentence forms, they substitute for SL sentences; and when used in English idioms, they substitute for English sentences whose SL translations are the corresponding SL sentences. We employ this hybrid usage only in 3.3.2.

2 As explained in 3.2.4:C, our definition of sentential components makes every SL sentence a sentential component of itself. Hence, if **X** is an atomic sentence, its sole atomic component is **X**.

Two SL truth valuations V and V* for Γ are called **equivalent with respect to** Γ just in case[1] for every atomic component **X** of a member of Γ, V and V* assign the same truth value to **X**.

3.4.1:C COMMENTARY ON 3.4.1

3.4.1:C1 We will give an example of an SL truth valuation. Let Γ be the following set of SL sentences.

$\{\neg\neg\neg A, B \rightarrow \neg\neg\neg A, (D \vee B) \rightarrow A, B, (E \wedge A) \rightarrow \neg(D \vee A)\}$

$\neg\neg\neg A$ has one atomic component A. $B \rightarrow \neg\neg\neg A$ has two atomic components: A and B. $(D \vee B) \rightarrow A$ has three atomic components: A, B, and D. B has only one atomic component, namely itself. $(E \wedge A) \rightarrow \neg(D \vee A)$ has three atomic components: A, D, and E. Therefore the atomic components of the members of Γ are A, B, D, and E. According to our definition, any assignment of truth values to these atomic components is an SL truth valuation for Γ. Here are three of these truth valuations.

V1 assigns T to *A*, F to *B*, T to *D*, and T to *E*.

V2 assigns T to *A*, F to *B*, F to *D*, and T to *E*.

V3 assigns F to *A*, T to *B*, T to *D*, and F to *E*.

However, there are many other truth valuations for Γ. Indeed, there are infinitely many truth valuations for Γ. A truth valuation for Γ may assign truth values to atomic sentences that do not occur in the members of Γ. Since there are infinitely many atomic sentences (i.e., sentence letters) in SL, there are infinitely many SL truth valuations that assign truth values to the atomic components of the members of Γ and to additional atomic sentences. Consider, for examples, the following three truth valuations for Γ.

V1* assigns T to *A*, F to *B*, T to *D*, T to *E*, F to *G*, and F to *H*.

V2* assigns T to *A*, F to *B*, F to *D*, T to *E*, F to *G*, T to *K*, and T to *M*.

V3* assigns F to *A*, T to *B*, T to *D*, F to *E*, T to *J*, F to *C*, F to *M*, and F to *N*.

It is clear that we can expand this list indefinitely. Every time we add a new atomic sentence or change a truth-value assignment, we introduce a new truth valuation for Γ. SL vocabulary allows the use of numerical subscripts for sentence letters; this ensures that the collection of all atomic SL sentences is infinite.

However, there is something peculiar about the two lists of SL truth valuations above: every truth valuation and its "starred" counterpart assign the same truth values to *A*, *B*, *D*, and *E*, which are the atomic components of the members of Γ. According to our definition above, each pair of V1 and V1*, V2 and V2*, and V3 and V3* is a pair of equivalent truth valuations

1 The expression 'just in case' is a logical expression that means "if and only if."

with respect to Γ. Let V be any truth valuation for Γ. The set of *all* the truth valuations that are equivalent to V with respect to Γ is called *an equivalence class with respect* to Γ. These equivalence classes have a very important property: they are *mutually exclusive* and *collectively exhaustive* of the truth valuations for Γ. Being mutually exclusive means that any two equivalence classes are disjoint, that is, they do not share any members (if they share at least one member, then they are identical); and being collectively exhaustive of the truth valuation for Γ means that every truth valuation for Γ must belong to one of these classes. In other words, every truth valuation for Γ must belong to *one and only one* equivalence class with respect to Γ.

The reader might have observed that if Γ is a finite set of SL sentences, then the number of equivalence classes with respect to Γ is finite. As we will see in 3.4.2, the truth value of a compound SL sentence is determined by the truth values of its atomic components and the types of the sentential connectives that occur in the sentence. This implies that any two equivalent truth valuations with respect to Γ assign the same truth values to the members of Γ. In other words, all the truth valuations in an equivalence class with respect to Γ agree on the truth values they assign to the sentences in Γ. Because of this feature of equivalence classes with respect to Γ, every member of such class can serve as a *representative* of the whole class. If there are n equivalence classes with respect to Γ, then there are n different ways of distributing truth values to the atomic components of the members of Γ. Thus if we are interested in the various ways of assigning truth values to the members of Γ, it is sufficient to consider only n truth valuations for Γ each of which is a representative of one of the equivalence classes with respect to Γ. We describe this situation by saying that there are, *up to equivalence*, n different truth valuations for Γ, or that there are n nonequivalent truth valuations for Γ.

The members of the set Γ above have four atomic components, each of which may be assigned one of two truth values, T or F. Hence, there is a total of 2×2×2×2 (= 2^4) different possible ways of assigning truth values to these atomic components. In other words, there is, up to equivalence, a total of 16 different SL truth valuations for Γ. The table below displays these truth valuations in a systematic fashion.

V	*A*	*B*	*D*	*E*
V1	T	T	T	T
V2	T	T	T	F
V3	T	T	F	T
V4	T	T	F	F
V5	T	F	T	T
V6	T	F	T	F
V7	T	F	F	T
V8	T	F	F	F
V9	F	T	T	T
V10	F	T	T	F
V11	F	T	F	T
V12	F	T	F	F
V13	F	F	T	T
V14	F	F	T	F
V15	F	F	F	T
V16	F	F	F	F

The column headed by V displays the names of the 16 truth valuations, V1–V16. Each of the columns headed by *A*, *B*, *D*, and *E* displays the truth values assigned to this atomic sentence by the various truth valuations listed in the V-column. The assignments that an SL truth valuation makes are displayed in the row headed by that truth valuation. For instance, V9 assigns F to *A*, T to *B*, T to *D*, and T to *E*.

3.4.1:C2 In general, if there are n atomic sentences (n is greater than 0), the total number of nonequivalent SL truth valuations for these atomic sentences is 2^n. Each atomic sentence can be assigned one of *two* truth values; hence there are 2×2×...×2 (n times) different ways of assigning truth values to those n atomic sentences. There are different procedures for listing these truth valuations systematically. We will agree on one procedure and we will follow it throughout this chapter. Here is our procedure.

(a) The first column always displays the names of the 2^n truth valuations; they are V1–V2^n.
(b) After the first column, we list n columns; each column is headed by an atomic sentence. The columns are ordered according to the alphabetical order of the atomic sentences (recall that atomic sentences are sentence letters). We call these columns "the atomic-sentence columns."
(c) We divide the *first* atomic-sentence column into 2^1 segments. We fill the cells of the first segment with 'T's and of the second segment with 'F's. In other words, the top half of this column will have 'T's, and the bottom half will have 'F's.
(d) We divide the *second* atomic-sentence column into 2^2 segments. We fill the cells of the first segment with 'T's, of the second with 'F's, of the third with 'T's, and of the fourth with 'F's.
(e) In general, we divide the kth atomic-sentence column into 2^k segments. We fill the cells of each segment with either 'T's or 'F's, beginning with 'T's, and alternating the 'T's and 'F's for consecutive segments. Because we will always have an even number of segments, the cells of the last segment are always filled with 'F's.
(f) If we follow this procedure correctly, the last, nth, atomic-sentence column is divided into 2^n segments, each of which consists of one cell only. These cells display truth values ordered as follows: TFTFTF...TF.

SL truth valuations assign truth values to the compound sentences as well. However, unlike the *arbitrary* assignment of truth values to the atomic sentences, the assignment of truth values to the compound sentences is fully determined by the truth values of their atomic components and the truth conditions of the connectives that occur in these compound sentences. Once we specify the truth conditions of the SL connectives, we will revisit the example discussed in 3.4.1:C1 and determine the truth values of the members of Γ.

3.4.2 The **truth conditions** of an SL sentence **X** are the conditions that determine the truth value of **X** on any given truth valuation for **X**. Said differently, **X** is true on an SL truth valuation V if and only if V satisfies the truth conditions of **X**, and **X** is false on V if and only if V is relevant to **X** and V does not satisfy the truth conditions of **X**. If V assigns T to **X** (i.e., **X** is true on V), we write 'V(**X**) = T', and if V assigns F to **X** (i.e., **X** is false on V), we write 'V(**X**) = F'. Below is a complete list of those truth conditions. Let **X** and **Y** be any SL sentences and V be any truth valuation that assigns truth values to **X** and **Y**.

V(¬**X**) = T	if and only if	V(**X**) = F.
V(**X**∧**Y**) = T	if and only if	V(**X**) = T and V(**Y**) = T.
V(**X**∨**Y**) = T	if and only if	V(**X**) = T or V(**Y**) = T or both.
V(**X**→**Y**) = T	if and only if	V(**X**) = F or V(**Y**) = T or both.
V(**X**↔**Y**) = T	if and only if	V(**X**) = V(**Y**).

If every sentence in a set Γ is true on V, we say that V **satisfies** Γ or that V **is a model of** Γ.

3.4.2:C COMMENTARY ON 3.4.2

3.4.2:C1 There are two important things to note about the truth conditions listed above. First, **X** and **Y** stand for any SL sentences that receive truth values on V. **X** and **Y** could be atomic or compound. Second, these 'if and only if' statements specify the conditions under which an SL sentence is true and the conditions under which the SL sentence is false. The language of SL, like every language with which we deal in this book, is bivalent. Thus on any given truth valuation for some SL sentence, the sentence is either true or false but not both. The truth conditions above are expressed as biconditionals. This means that if the assertion made by the right-hand side of one of these biconditionals obtains, then the sentence mentioned in the left-hand side of the biconditional is true on V; if it fails to obtain and V is relevant to the sentence, then the sentence is false on V. For example, the truth condition of **X**∧**Y** is that **X** and **Y** are both true on V. If this condition obtains, i.e., if **X** and **Y** are true on V, then **X**∧**Y** is true on V as well; but if this condition fails to obtain, i.e., if **X** or **Y** is false on V (in the inclusive sense), then **X**∧**Y** is false on V.

It is customary to describe these truth conditions by means of truth tables. Each truth table shows the possible assignments of truth values to **X** and **Y** and the corresponding truth values of the compound sentence. We have five SL connectives, so we construct five tables.

X	¬X
T	F
F	T

Truth table for ¬

X	Y	X∧Y
T	T	T
T	F	F
F	T	F
F	F	F

Truth table for ∧

X	Y	X∨Y
T	T	T
T	F	T
F	T	T
F	F	F

Truth table for ∨

X	Y	X→Y
T	T	T
T	F	F
F	T	T
F	F	T

Truth table for →

X	Y	X↔Y
T	T	T
T	F	F
F	T	F
F	F	T

Truth table for ↔

There are two ways to think about the truth conditions (tables) of the SL connectives. According to the first way, these conditions require no justification; the truth condition of each connective defines its meaning. The meaning of the connective, in turn, justifies the English-SL translations we listed in 3.3.1. For example, we are justified in translating truth-functionally an English conjunction into an SL conjunction because the meaning given to SL conjunctions by their truth conditions makes them the best SL candidates to capture truth-functionally English conjunctions. The truth condition of an SL conjunction asserts that **X**∧**Y** is true on any given V if and only if both **X** and **Y** are true on V. This truth condition gives meaning to the symbol ∧. Now, we know that ∧ connects two SL sentences to produce an SL sentence that is true when and only when its immediate components are true. This makes the syntactical and semantical role of ∧ in SL similar to 'and' in English when 'and' is employed truth-functionally. 'And' also connects two sentences to produce a conjunction that is true when and only when its conjuncts are true. We are justified, therefore, to call **X**∧**Y** "a conjunction" and to translate 'and', or any other English connective that has the same truth-functional role as 'and', into ∧, and vice versa.

The second way reverses the direction of justification. According to the second way, we are justified in specifying these truth conditions for the SL connectives because these connectives are intended to capture the truth-functional employments of their English counterparts. We shall adopt the second approach here and explain the justification behind the truth tables (conditions) above based on the intended meaning of these connectives.

The truth tables for ¬, ∧, and ∨ are clear. We want ¬ to represent negation in SL. English negation is commonly understood to deny the truth value of the sentence it negates. Since we deal with only bivalent sentences, not-**X** is true if **X** is false and not-**X** is false if **X** is true. This explains why the truth table for ¬ shows F for ¬**X** when **X** is T, and T for ¬**X** when **X** is F. **X**∧**Y** is supposed to be a conjunction; we want it to capture the meaning of '**X** and **Y**', where 'and' is used truth-functionally. We seem to be willing to commit ourselves to the truth of '**X** and **Y**' when and only when we are ready to commit ourselves to the truth of **X** and to the truth of **Y**. If we believe that **X** is false or **Y** is false or both, we should deny that '**X** and **Y**' is true. Because ∧ is meant to capture 'and' truth-functionally and because 'and' has the semantical role just described, the truth table for ∧ is fully justified: **X**∧**Y** is true if both **X** and **Y** are true and false if either or both are false. ∨ is intended to capture the meaning of 'or' when used inclusively. Thus we say that '**X** or **Y**' is true if **X** is true or **Y** is true or both and '**X** or **Y**' is false if both **X** and **Y** are false. This explains the choices made in the truth table for ∨.

The truth table for → requires some explanation. But in order to understand the choices we made in this truth table, we must first discuss the truth table for ↔. ↔ is intended to capture the English biconditional connective 'if and only if'. An English biconditional of the form '**X** if and only if **Y**' *usually* asserts that **X** and **Y** are either both true or both false—that is, they have identical truth values. If **X** and **Y** have different truth values, the biconditional '**X** if and only if **Y**' is false. Since we want ↔ and 'if and only if' to have similar semantical roles, the truth table for ↔ is, therefore, justified. → is meant to capture the English 'if-then' when used truth-functionally. The first, second, and fourth rows of the truth table seem to have some intuitive appeal. If I say to my daughter, "If you wash the car, I'll give you $20," you will not accuse me of making a false promise if she washes the car and I give her $20 or if she doesn't wash the car and I don't give her $20. This explains the assignment of T to **X**→**Y** in the first and fourth rows of the truth table for →. But if she washes the car and I don't give her $20, I am guilty of making a false promise. This explains the assignment of F to **X**→**Y** in the second row of the truth table for →. The third row seems arbitrary. It is, however, an outcome of three factors: (1)

X→Y is bivalent, (2) → is truth-functional, and (3) → is weaker in commitment than ↔. We intuitively expect my promise 'I'll give you $20 if you wash the car' to be less committed than 'I'll give you $20 if and only if you wash the car'. The bivalence of **X→Y** entails that we have only two options for any relevant truth valuation V: **X→Y** is true on V or **X→Y** is false on V. The truth-functionality of → entails that all truth valuations that assign F to **X** and T to **Y** must assign the *same* truth value to **X→Y**. If we assign F to **X→Y** when **X** is false and **Y** is true, the truth table for → would be identical with the truth table for ↔. → would translate, in this case, 'if and only if'. But we want → to translate the weaker 'if-then'; hence, we have no choice but to assign T to **X→Y** when **X** is false and **Y** is true. In other words, we see the difference between 'I'll give you $20 if you wash the car' and 'I'll give you $20 if and only if you wash the car' as residing solely in the case when I give my daughter $20 in spite of her not washing the car. The less committed promise of 'if-then' is true in this case and the more committed promise of 'if and only if' is false in this case. In all other cases they have identical truth values.[1] A truth-functional conditional that has the truth conditions of → is called a **material conditional** (or **material implication**).

The best way to memorize these truth conditions is to memorize their **singular cases**. An assignment of truth value to a compound sentence is singular if it does not result in a disjunction of truth-value assignments to its immediate components. The singular case for a conjunction is the assignment of T: **X∧Y** is true if and only if both **X** and **Y** are true; it is false otherwise. The singular case for a disjunction is the assignment of F: **X∨Y** is false if and only if both **X** and **Y** are false; it is true otherwise. The singular case for a (material) conditional is the assignment of F: **X→Y** is false if and only if **X** is true and **Y** is false; it is true otherwise. A biconditional has no singular case. But it is not hard to remember the truth conditions of a biconditional: **X↔Y** is true if and only if **X** and **Y** have the same truth value; it is false otherwise.

3.4.2:C2 We now revisit the example discussed in 3.4.1:C1 and determine the truth values of the compound sentences in Γ. We will construct one truth table for all the sentences in Γ. We determine the truth value of a compound sentence using a bottom-up procedure, beginning with the simplest sentential components and ending with the original compound sentence. For instance, let V be the truth valuation such that V(*A*) = T, V(*B*) = T, V(*D*) = F, and V(*E*) = F. To determine the truth value of $(E \wedge A) \rightarrow \neg(D \vee A)$ on V we employ the following bottom-up procedure.

(a) The simplest sentential components of $(E \wedge A) \rightarrow \neg(D \vee A)$ are the atomic components *A*, *D*, and *E*. Their truth values on V are given above.
(b) The second simplest sentential components of the sentence are $E \wedge A$ and $D \vee A$. According to the truth tables for ∧ and ∨, $V(E \wedge A) = F$ and $V(D \vee A) = T$.
(c) At the next stage we deal with the more complex sentential component $\neg(D \vee A)$. Given the truth table for ¬ and that $V(D \vee A) = T$, $V(\neg(D \vee A)) = F$.
(e) Now we are ready to determine the truth value of the original sentence on V. $V(E \wedge A) =$ F and $V(\neg(D \vee A)) = F$; by the truth table for →, we obtain that $V((E \wedge A) \rightarrow \neg(D \vee A)) = T$.

1 This analysis shows, I hope, that any philosophically involved justification of the truth conditions of **X→Y** is wrongheaded. Later we will have reasons to question the adequacy of the material conditional '→' as a representation of the English indicative conditional.

When we construct the truth table for a compound SL sentence **X**, we follow the procedure described in 3.4.1:C2 for listing the SL truth valuations for **X** and we write the truth values of the sentential components of **X** under the main connectives of those components. For example, the truth value, on a given truth valuation, of ¬(*D*∨*A*) is written under the sign ¬ and of (*E*∧*A*)→¬(*D*∨*A*) under the sign →. The main connective and the truth values of the sentence **X** are boldfaced.

V	*A*	*B*	*D*	*E*	**¬**	¬	¬	*A*	*B*	**→**	¬	¬	*A*	(*D*	∨	*B*)	**→**	*A*	*B*	(*E*	∧	*A*)	**→**	¬	(*D*	∨	*A*)
V1	T	T	T	T	**F**	T	F	T	T	**T**	T	F	T	T	T	T	**T**	T	T	T	T	T	**F**	F	T	T	T
V2	T	T	T	F	**F**	T	F	T	T	**T**	T	F	T	T	T	T	**T**	T	T	F	F	T	**T**	F	T	T	T
V3	T	T	F	T	**F**	T	F	T	T	**T**	T	F	T	F	T	T	**T**	T	T	T	T	T	**F**	F	F	T	T
V4	T	T	F	F	**F**	T	F	T	T	**T**	T	F	T	F	T	T	**T**	T	T	F	F	T	**T**	F	F	T	T
V5	T	F	T	T	**F**	T	F	T	F	**T**	T	F	T	T	T	F	**T**	T	F	T	T	T	**F**	F	T	T	T
V6	T	F	T	F	**F**	T	F	T	F	**T**	T	F	T	T	T	F	**T**	T	F	F	F	T	**T**	F	T	T	T
V7	T	F	F	T	**F**	T	F	T	F	**T**	T	F	T	F	F	F	**T**	T	F	T	T	T	**F**	F	F	T	T
V8	T	F	F	F	**F**	T	F	T	F	**T**	T	F	T	F	F	F	**T**	T	F	F	F	T	**T**	F	F	T	T
V9	F	T	T	T	**T**	F	T	F	T	**F**	F	T	F	T	T	T	**F**	F	T	T	F	F	**T**	F	T	T	F
V10	F	T	T	F	**T**	F	T	F	T	**F**	F	T	F	T	T	T	**F**	F	T	F	F	F	**T**	F	T	T	F
V11	F	T	F	T	**T**	F	T	F	T	**F**	F	T	F	F	T	T	**F**	F	T	T	F	F	**T**	T	F	F	F
V12	F	T	F	F	**T**	F	T	F	T	**F**	F	T	F	F	T	T	**F**	F	T	F	F	F	**T**	T	F	F	F
V13	F	F	T	T	**T**	F	T	F	F	**T**	F	T	F	T	T	F	**F**	F	F	T	F	F	**T**	F	T	T	F
V14	F	F	T	F	**T**	F	T	F	F	**T**	F	T	F	T	T	F	**F**	F	F	F	F	F	**T**	F	T	T	F
V15	F	F	F	T	**T**	F	T	F	F	**T**	F	T	F	F	F	F	**T**	F	F	T	F	F	**T**	T	F	F	F
V16	F	F	F	F	**T**	F	T	F	F	**T**	F	T	F	F	F	F	**T**	F	F	F	F	F	**T**	T	F	F	F

Note that we did not define the notion of truth valuation independently of any set of SL sentences. Our truth valuations are always relative to sets of SL sentences. V1–V16 in the table above are, up to equivalence, all the *relevant* truth valuations for Γ. We note that none of the truth valuations V1–V16 *satisfies* Γ; that is, none of them makes its members, ¬¬¬*A*, *B*→¬¬*A*, (*D*∨*B*)→*A*, *B*, and (*E*∧*A*)→¬(*D*∨*A*), all true.

3.4.3 The **truth table** for an SL sentence **X** displays the truth values of **X** and of its sentential components on *all* SL truth valuations for **X**. Given our discussion in 3.4.1:C1, it is sufficient to consider *only* one representative of every equivalence class with respect to **X**. Since an SL sentence has finitely many atomic components, its truth table will display only finitely many truth valuations. A truth table for a finite set Γ of SL sentences is a "combined" truth table for all the sentences in Γ. A combined truth table for several SL sentences displays, up to equivalence, all the truth valuations for these sentences, their truth values, and the truth values of their sentential components on these truth valuations. The truth table given in 3.4.2:C2 above is an example of a combined truth table for several SL sentences.

3.4.3:C COMMENTARY ON 3.4.3

We have already encountered a truth table in 3.4.2:C2. We said there that a truth table is a bottom-up approach to determining the truth values of a compound sentence **X**. We begin by listing all the relevant SL truth valuations (we follow the procedure discussed in 3.4.1:C2). We, thus, list the truth values assigned to the atomic components of **X** by all the relevant SL truth valuations. We then determine, using the truth conditions mentioned in 3.4.2, the truth values of the simplest compound sentential components of **X**, then of the second simplest compound sentential components, and so on until we reach **X**.

Sometimes we construct the truth table not for a single SL sentence but for a collection of SL sentences. The truth table we constructed in 3.4.2:C2 is of this sort. Whether we construct a table for a single sentence or a collection of sentences depends on the question we are trying to answer by means of the truth table. We will discuss in 3.5.11 how the method of constructing truth tables provides us with an effective decision procedure for answering all sorts of questions about the logical status of an SL sentence or a collection of SL sentences. We will not be able to demonstrate this procedure until we have defined logical concepts in SL, but we can, at least, illustrate the main idea.

Consider the following set of SL sentences:

$$\Sigma = \{(H \rightarrow K) \vee \neg J,\ J \wedge L,\ \neg(L \leftrightarrow H)\}$$

Suppose we would like to know whether Σ is *satisfiable*, that is, whether there is an SL truth valuation on which every member of Σ is true. The truth table we constructed for the set Γ in 3.4.2:C2 showed that Γ is not satisfiable. None of the truth valuations V1–V16 satisfies Γ. To answer our question about Σ we construct a combined truth table for its members. Note that the members of Σ have four atomic components—*H*, *J*, *K*, and *L*—so the truth table for Σ has 2^4 (16) rows.

V	*H*	*J*	*K*	*L*	(H	→	K)	**∨**	¬	*J*	*J*	**∧**	*L*	**¬**	(*L*	↔	*H*)
V1	T	T	T	T	T	T	T	**T**	F	T	T	**T**	T	**F**	T	T	T
V2	T	T	T	F	T	T	T	**T**	F	T	T	**F**	F	**T**	F	F	T
V3	T	T	F	T	T	F	F	**F**	F	T	T	**T**	T	**F**	T	T	T
V4	T	T	F	F	T	F	F	**F**	F	T	T	**F**	F	**T**	F	F	T
V5	T	F	T	T	T	T	T	**T**	T	F	F	**F**	T	**F**	T	T	T
V6	T	F	T	F	T	T	T	**T**	T	F	F	**F**	F	**T**	F	F	T
V7	T	F	F	T	T	F	F	**T**	T	F	F	**F**	T	**F**	T	T	T
V8	T	F	F	F	T	F	F	**T**	T	F	F	**F**	F	**T**	F	F	T
V9	F	T	T	T	F	T	T	**T**	F	T	T	**T**	T	**T**	T	F	F
V10	F	T	T	F	F	T	T	**T**	F	T	T	**F**	F	**F**	F	T	F
V11	F	T	F	T	F	T	F	**T**	F	T	T	**T**	T	**T**	T	F	F
V12	F	T	F	F	F	T	F	**T**	F	T	T	**F**	F	**F**	F	T	F
V13	F	F	T	T	F	T	T	**T**	T	F	F	**F**	T	**T**	T	F	F
V14	F	F	T	F	F	T	T	**T**	T	F	F	**F**	F	**F**	F	T	F
V15	F	F	F	T	F	T	F	**T**	T	F	F	**F**	T	**T**	T	F	F
V16	F	F	F	F	F	T	F	**T**	T	F	F	**F**	F	**F**	F	T	F

Examining the (main) truth values in the compound-sentence columns, we discover that V9 and V11 satisfy Σ, therefore Σ is satisfiable.

3.4.4 A **truth analysis** for a set Γ of SL sentences is a top-down analysis that begins by assigning initial truth values to the members of Γ, and then either **confirming** the initial assignment by **recovering** an SL truth valuation that assigns the initial truth values to the members of Γ, or **refuting** the initial assignment by obtaining a contradiction, which occurs when a sentential component of a member of Γ is assigned both T and F.

3.4.4:C COMMENTARY ON 3.4.4

3.4.4:C1 The method of constructing truth tables is an effective procedure for answering any question about the applicability of the logical concepts (defined in 3.5) to a set of SL sentences or to its members. But it is not practical. The number of rows in a truth table increases exponentially with the number of atomic components. If we have 6 atomic components, the truth table consists of 2^6 (that is, 64) rows. But, as we noticed in the table for Σ constructed in 3.4.3:C, the question about the satisfiability of Σ was answered by the existence of V9 or V11. We only needed to find one truth valuation that satisfies Σ in order to answer the question affirmatively; the existence of the other fifteen truth valuations was irrelevant.

Fortunately there is a method that allows us to search for an SL truth valuation such as V9 or V11 without listing all the truth valuations for Σ. This method is called "truth analysis." It is a top-down method that begins with a truth value assignment to a compound sentence **X** and ends either with an assignment of truth values to the atomic components of **X** or with a contradiction.

Let us illustrate the procedure with a simple example. We take **X** to be the SL sentence $(E \wedge A) \rightarrow \neg(D \vee A)$. Say we are interested in finding an SL truth valuation that assigns F to **X**; we only want to find one such truth valuation, if it exists. We begin by assigning F to **X**. We will always write the truth value of a compound sentence as a superscript to its main connective and of an atomic sentence as a superscript to the sentence itself. So far we have made one truth-value assignment.

$$(E \wedge A) \rightarrow^{F} \neg(D \vee A)$$

We know from the truth conditions of the conditional that the conditional is false when and only when its antecedent is true and its consequent is false. So we have this:

$$(E \wedge^{T} A) \rightarrow^{F} \neg^{F}(D \vee A)$$

Given the truth conditions of the conjunction and negation, we obtain the following analysis:

$$(E^{T} \wedge^{T} A^{T}) \rightarrow^{F} \neg^{F}(D \vee^{T} A)$$

'*A*' in the conjunction receives the truth value T, so we copy this truth value to all the occurrences of '*A*' in **X**; there is only one other occurrence, which is in the disjunction.

$$(E^{T} \wedge^{T} A^{T}) \rightarrow^{F} \neg^{F}(D \vee^{T} A^{T})$$

So far, our analysis proceeded in a deterministic manner. Beside the initial choice of F, we had no choice in making the truth-value assignments we made. But now we are presented with a choice. The truth condition of the disjunction produces two assignments of truth values to 'D'. They are:

$(D^{T} \vee^{T} A^{T})$ and $(D^{F} \vee^{T} A^{T})$

Anyone of these assignments is sufficient for our purposes here. Each of these assignments together with the assignments of T to A and E defines an SL truth valuation on which the sentence $(E \wedge A) \rightarrow \neg(D \vee A)$ (**X**) is false. We intended to find one truth valuation that makes **X** false; we found two. In fact, these are all the nonequivalent SL truth valuations that make **X** false. The first truth valuation, V1, assigns T to A, D, and E and the second, V2, assigns T to A and E and F to D. We say that our analysis *confirmed* the initial assignment by *recovering* the truth valuations V1 and V2.

3.4.4:C2 The truth table for **X** has 8 rows. Constructing the truth table for **X** entails listing 8 different truth valuations with all their truth-value assignments to the sentential components of **X**. The truth-analysis method produced in a few steps the desired truth valuation. But this method could also "branch out" into many options. However, the five rules we list below show how to apply the truth-analysis method in general *and* how to minimize the number of options we have to consider.

(a) We begin by making an initial assignment of truth values to the given SL sentences. The truth analysis will either **confirm** the initial truth-value assignment by **recovering** an SL truth valuation or **refute** the initial truth-value assignment by reaching a contradiction for every available option.
(b) Since no matter which sequence of sentences or sentential components we follow in our truth analysis the *final conclusion* (decision) is invariant,[1] we should always analyze the truth values of the sentences or sentential components that give us the fewest options.
(c) If a truth value of an atomic component is "discovered," we should copy this truth value to all the occurrences of this atomic sentence. We refer to a truth value of an atomic component as an **atomic value**. So the rule says: once an atomic value is obtained, it should be assigned to all the occurrences of the atomic sentence.
(d) We should halt the analysis once a truth valuation is recovered.
(e) If we reach a contradiction, we should analyze another available option (if there is any). If all available options lead to contradictions, then the initial truth-value assignment has been refuted. That is, it has been shown to be impossible.

1 By 'the final conclusion' we do not mean the immediate outcome we reach by following a certain path. Such conclusions are not invariant. One path might lead to a contradiction, another to a truth valuation. What is invariant is the final conclusion (decision) we reach by applying the truth-analysis method. The final conclusion could be one of two: either the initial truth-value assignment is refuted (all available options lead to contradictions) or the initial truth-value assignment is confirmed (a truth valuation is recovered). It should be made clear that in the last case different analyses may recover different truth valuations. What is invariant, in this case, is that there is a truth valuation that delivers the initial truth-value assignment.

We illustrate these rules by means of two examples. We take **X** as before to be the SL sentence $(E\wedge A)\to\neg(D\vee A)$. But this time we want to find (if there is any) an SL truth valuation that makes **X** true. So, we begin by assigning T to the conditional (the first rule of truth analysis).

$$(E\wedge A)\to^{T}\neg(D\vee A)$$

We are faced with three options: the antecedent and consequent true, the antecedent false and the consequent true, and the antecedent and consequent false. Based on the truth conditions of the conjunction, negation, and disjunction, the first option yields the smallest number of choices. So according to the second rule of truth analysis, we should begin with this option. Hence we assign T to the antecedent and consequent.

$$(E\wedge^{T}A)\to^{T}\neg^{T}(D\vee A)$$

By analyzing the conjunction, we get

$$(E^{T}\wedge^{T}A^{T})\to^{T}\neg^{T}(D\vee A)$$

Following the third rule of truth analysis, we copy the truth value of A to its occurrence in the disjunction.

$$(E^{T}\wedge^{T}A^{T})\to^{T}\neg^{T}(D\vee A^{T})$$

Now we analyze the negation and the disjunction.

$$(E^{T}\wedge^{T}A^{T})\to^{T}\neg^{T}(D^{F}\vee^{F}A^{<T,\,F>})$$

The analysis yields a contradiction. A receives both T and F (we write '<T, F>' as a superscript to A to indicate the contradiction). We say that the option of assigning T to the antecedent and consequent is **closed**. According to the fifth rule of truth analysis, we consider another available option. So, this time we assign F to the consequent and T to the antecedent; this option leads to fewer choices than the third option, which assigns F to the antecedent and consequent.

$$(E\wedge^{F}A)\to^{T}\neg^{T}(D\vee A)$$

We follow the path that yields the fewest choices. So we analyze the negation and the disjunction and copy the truth value of A to its occurrence in the conjunction.

$$(E\wedge^{F}A^{F})\to^{T}\neg^{T}(D^{F}\vee^{F}A^{F})$$

At this stage we have two options: we may either assign T or F to E. Either option recovers a truth valuation that delivers the initial truth-value assignment—namely **X** is true. We randomly select the first option: E is true. Here is the recovered truth valuation: $V(A) = F$, $V(D) =$

F, and V(E) = T. The fourth rule of truth analysis directs us to halt the analysis here. We fulfilled our goal: the SL truth valuation V makes **X** true. There are other truth valuations that also make **X** true. We can recover all of them if we follow all the available options. But one truth valuation is sufficient, so we stop here.

This is our second example. The truth table we constructed in 3.4.2:C2 shows that the set Γ (whose members are $\neg\neg\neg A$, $B\rightarrow\neg\neg A$, $(D\vee B)\rightarrow A$, B, and $(E\wedge A)\rightarrow\neg(D\vee A)$) is unsatisfiable, i.e., no truth valuation is a model of Γ. Our goal in this case is not to find an SL truth valuation that fulfills a certain condition but to demonstrate that no truth valuation fulfills a certain condition. The truth table in 3.4.2:C2 displays, up to equivalence, all the SL truth valuations for Γ, allowing us to survey all those truth valuations and confirm that none of them satisfies Γ. The truth analysis method is also suitable for this purpose and here too is typically faster than the truth-table method. Given the rules (especially the fifth rule) of truth analysis listed above, this method in this case works as follows. We begin by assigning T to all the members of Γ. We proceed by analyzing the truth values of the members of Γ and of their sentential components, invoking the second rule to reduce the number of options we have to consider. We must then show that *every* available option leads to a contradiction, that is, an assignment of T *and* F to a member of Γ or to one of its sentential components. In other words, we must show that every available option is *closed*. Such an analysis establishes that the initial assignment of T to all the members of Γ is impossible, that is, there is no truth valuation that makes the members of Γ all true.

Thus we begin by assigning T to every member of Γ; this is the assignment we want to refute.

$$\neg^{T}\neg\neg A \qquad B\rightarrow^{T}\neg\neg A \qquad (D\vee B)\rightarrow^{T}A \qquad B^{T} \qquad (E\wedge A)\rightarrow^{T}\neg(D\vee A)$$

We copy the truth value of B to all its occurrences in the sentences above and we analyze *all* the sentences that do not lead to branching out (i.e., they do not give us options)

$$\neg^{T}\neg^{F}\neg^{T}A^{<T,\,F>} \qquad B^{T}\rightarrow^{T}\neg^{T}\neg^{F}A^{T} \qquad (D\vee^{T}B^{T})\rightarrow^{T}A^{T} \qquad B^{T} \qquad (E^{F}\wedge^{F}A^{T})\rightarrow^{T}\neg^{F}(D\vee^{T}A^{T})$$

The analysis above produces a contradiction. (The contradiction could be attached to any of the occurrences of A.) Note that we didn't have to consider the truth value of D. We obtain a contradiction regardless of the truth value of D: whatever the truth value of D might be, we land on a contradiction. The analysis in effect branches out at this stage yielding two options: in one of them D is true, in the other it is false. The analysis above shows that both options are closed. The initial truth-value assignment is refuted and the set Γ, therefore, is unsatisfiable. The presence of a contradiction demonstrates that there is no SL truth valuation that makes the members of Γ all true. It should be emphasized, however, that there is no unique way of obtaining a contradiction. If an initial truth-value assignment is refuted, several contradictions may be reached by (slightly) different analyses.

Truth analysis is our method of choice. We will employ it in the commentaries and solutions in this chapter unless a question explicitly calls for the use of the truth-table method.

3.5 Logical Concepts in SL

3.5.1 As in 1.1.1, an SL **argument** is defined as a nonempty collection of SL sentences: one of these sentences is the conclusion of the argument and the others are its premises.

3.5.2 The intuitive notion of a **logical possibility** that is relevant to a set of declarative sentences is characterized in SL as the notion of an SL **truth valuation** that is relevant to a set of SL sentences. We defined what it means for an SL sentence to be true or false on a relevant truth valuation, and we identified the SL expressions that form grammatical SL sentences. Thus the **logical concepts** introduced in 1.2.4–1.2.11 can now be given precise definitions in SL. We simply replace the expression 'logical possibility' with 'truth valuation' in those definitions and make the necessary adjustments. The resulting statements are listed in 3.5.3–3.5.10.

3.5.2:C COMMENTARY ON 3.5.2

We said in 3.1:C that an SL truth valuation is a "linguified" logical possibility. According to SL worldview a logical possibility is a collection of states of affairs with a specification as to which states obtain and which fail to obtain. An SL truth valuation is a collection of atomic sentences with a specification as to which sentences are true and which are false. The intuitive idea is that true atomic sentences represent states of affairs that obtain (i.e., facts) and false atomic sentences represent states of affairs that fail to obtain (i.e., negative facts). An SL truth valuation, thus, is a "linguified" analogue of a logical possibility. SL, however, takes a conceptually economical approach to the notion of logical possibility. Instead of defining a logical possibility as a collection of facts and negative facts and a corresponding truth valuation as a collection of true and false atomic sentences, SL characterizes logical possibilities as truth valuations. Intuitively, we think of SL truth valuations as representing logical possibilities. In 3.5.12, we will address the issue of how successful SL is in representing logical possibilities.

3.5.3a An SL argument Γ/**X** is **deductively valid** (or $\Gamma \vDash$ **X**) if and only if on every truth valuation for Γ/**X** on which all the members of Γ are true **X** is true as well.

3.5.3b An SL argument Γ/**X** is **deductively valid** (or $\Gamma \vDash$ **X**) if and only if there is no truth valuation on which the members of Γ are all true and **X** is false.

3.5.3:C COMMENTARY ON 3.5.3

We use the truth-analysis method to show that $\neg J$ is a logical consequence of Γ, which is the following set: $\{J \leftrightarrow H, (K \wedge H) \rightarrow S, \neg(S \vee \neg K)\}$. To distinguish between the premises and the conclusion, we list the members of Γ first, then a slash '/', then $\neg J$. Our goal is to show that there is no truth valuation on which the members of Γ are all true and $\neg J$ is false. In other words, we want to show that the following assignment of truth values is impossible.

$$J \leftrightarrow^{T} H \qquad (K \wedge H) \rightarrow^{T} S \qquad \neg^{T}(S \vee \neg K) \qquad / \qquad \neg^{F} J$$

In analyzing the truth values above, we follow the rules of truth analysis. Hence, we analyze all the sentences and sentential components that do not lead to branching out. If we "discover" an atomic value, we copy it to all the occurrences of the atomic sentence.

$J^{T}\leftrightarrow^{T}H^{T} \qquad (K^{T}\wedge^{T}H^{T})\rightarrow^{<T,F>}S^{F} \qquad \neg^{T}(S^{F}\vee^{F}\neg^{F}K^{T}) \quad / \quad \neg^{F}J^{T}$

We reach a contradiction. The conditional $(K\wedge H)\rightarrow S$ receives T and F. As mentioned previously, different analyses may produce different contradictions. Our analysis led deterministically, that is, without branching out, to a contradiction: the initial assignment is refuted. This demonstrates that it is impossible to make the premises true and the conclusion false. $\Gamma/\neg J$ is, therefore, deductively valid.

3.5.4 An SL argument **Γ/X** is **deductively invalid** (or $\boldsymbol{\Gamma \nvDash X}$) if and only if there is an SL truth valuation on which the members of **Γ** are all true and **X** is false.

3.5.4:C COMMENTARY ON 3.5.4

The following SL argument is deductively invalid.

$(A\vee\neg D)\wedge(D\rightarrow C)$
$\neg C\leftrightarrow E$
$\neg(K\vee\neg E)$

$A\wedge K$

In this example our goal is to recover a relevant truth valuation that makes the premises true and the conclusion false. We begin by assigning T to the premises and F to the conclusion.

$(A\vee\neg D)\wedge^{T}(D\rightarrow C) \qquad \neg C\leftrightarrow^{T}E \qquad \neg^{T}(K\vee\neg E) \quad / \quad A\wedge^{F}K$

We analyze the sentences and sentential components that do not lead to branching out and we copy every atomic value to all the occurrences of the atomic sentence.

$(A\vee^{T}\neg^{T}D^{F})\wedge^{T}(D^{F}\rightarrow^{T}C^{F}) \qquad \neg^{T}C^{F}\leftrightarrow^{T}E^{T} \qquad \neg^{T}(K^{F}\vee^{F}\neg^{F}E^{T}) \quad / \quad A\wedge^{F}K^{F}$

'A' presents us with a choice. It could be true or false. Either case, we recover a truth valuation. Here is one of them: V(A) = T, V(C) = F, V(D) = F, V(E) = T, and V(K) = F. V is an SL truth valuation on which the premises of the argument are all true and the conclusion is false. Hence the SL argument is deductively invalid.

3.5.5a An SL sentence is **logically true** if and only if it is true on every SL truth valuation for that sentence.

3.5.5b An SL sentence is **logically true** if and only if there is no SL truth valuation on which it is false.

3.5.5:C COMMENTARY ON 3.5.5

Consider the SL sentence $((M\rightarrow\neg N)\wedge P)\rightarrow\neg((N\wedge M)\vee\neg P)$. This is a logically true sentence. We prove this using truth analysis. Since our purpose is to demonstrate that there is no truth valuation on which the sentence is false, we begin by assigning F to the sentence and conclude by reaching a contradiction for every option we might encounter.

$$((M\rightarrow\neg N)\wedge P)\rightarrow^{F}\neg((N\wedge M)\vee\neg P)$$

We proceed by analyzing the conditional, invoking the rules of truth analysis discussed in 3.4.4:C2.

$$((M^{T}\rightarrow^{<T,\,F>}\neg^{F}N^{T})\wedge^{T}P^{T})\rightarrow^{F}\neg^{F}((N^{T}\wedge^{T}M^{T})\vee^{T}\neg^{F}P^{T})$$

The conditional $M\rightarrow\neg N$ receives contradictory truth values. Thus we have refuted the initial truth-value assignment. The sentence cannot be made false; it is logically true.

3.5.6a An SL sentence is **logically false** if and only if it is false on every SL truth valuation for that sentence.

3.5.6b An SL sentence is **logically false** if and only if there is no SL truth valuation on which it is true.

3.5.6:C COMMENTARY ON 3.5.6

The SL sentence $((A\vee R)\leftrightarrow J)\wedge(J\wedge\neg(\neg R\rightarrow A))$ is logically false. We initially assign T to it and then refute this assignment. Here is the analysis.

$$((A^{F}\vee^{<T,\,F>}R^{F})\leftrightarrow^{T}J^{T})\wedge^{T}(J^{T}\wedge^{T}\neg^{T}(\neg^{T}R^{F}\rightarrow^{F}A^{F}))$$

The disjunction receives contradictory truth values. This shows that it is impossible to make the original sentence true; it is logically false.

3.5.7 An SL sentence is **contingent** if and only if it is true on some SL truth valuations and false on other truth valuations.

3.5.7:C COMMENTARY ON 3.5.7

Showing a sentence to be contingent requires two different truth analyses. One is aimed at recovering a truth valuation on which the sentence is true and the other at recovering a truth valuation on which the sentence is false. Let **X** be the SL sentences $((A\rightarrow C)\rightarrow(B\vee D))\vee(\neg(D\vee E)\wedge(G\wedge\neg A))$. We will show that **X** is contingent. This sentence is fairly complex; we might want to use the Main-Connective Rule to locate its main connective. But in order to use this rule we have to bring back the outermost parentheses. Below is the resulting sentence.

$$(_{1}(_{2}(_{3}A\rightarrow C)_{2}\rightarrow(_{3}B\vee D)_{2})_{1}\vee(_{2}\neg(_{3}D\vee E)_{2}\wedge(_{3}G\wedge\neg A)_{2})_{1})_{0}$$

The main connective, therefore, is the second $\vee$.

We begin our truth analysis by assigning T to **X**. Our goal is to recover a truth valuation from this assignment.

$$((A\to C)\to(B\lor D))\lor^{T}(\neg(D\lor E)\land(G\land\neg A))$$

According to the truth conditions of the disjunction, we have three options: both disjuncts are true, the first true and the second false, and the first false and the second true. There is a helpful aspect to recovering truth valuations: we don't need to consider all options. Thus the prudent thing to do is to consider the option that minimizes the number of choices we have to make. If the path we follow leads to a truth valuation, we halt our analysis. If we reach a contradiction, we try another option. This helpful feature is not available to us when our goal is to refute the initial assignment. In such a case, we must show that all available options lead to contradictions. Returning to **X**, we choose the third option (the first disjunct false and the second true). This option leads to fewer choices. We analyze as far as we can go without branching out.

$$((A^{F}\to^{T}C)\to^{F}(B^{F}\lor^{F}D^{F}))\lor^{T}(\neg^{T}(D^{F}\lor^{F}E^{F})\land^{T}(G^{T}\land^{T}\neg^{T}A^{F}))$$

'*C*' presents us with a choice; it could be true or false. Both options are equally good. Neither leads to a contradiction. So we arbitrarily choose '*C*' to be true. The recovered truth valuation is this: V1(*A*) = F, V1(*B*) = F, V1(*C*) = T, V1(*D*) = F, V1(*E*) = F, and V1(*G*) = T. **X** is true on V1.

Now we assign F to **X** and analyze as far as we can go without branching out.

$$((A\to^{T}C)\to^{F}(B^{F}\lor^{F}D^{F}))\lor^{F}(\neg(D^{F}\lor E)\land^{F}(G\land\neg A))$$

We have several options here. They are not totally independent of each other. We choose any set of options that generates no contradiction. The following analysis will do.

$$((A^{T}\to^{T}C^{T})\to^{F}(B^{F}\lor^{F}D^{F}))\lor^{F}(\neg^{T}(D^{F}\lor^{F}E^{F})\land^{F}(G^{T}\land^{F}\neg^{F}A^{T}))$$

We recover the following truth valuation: V2(*A*) = T, V2(*B*) = F, V2(*C*) = T, V2(*D*) = F, V2(*E*) = F, and V2(*G*) = T. It is clear that **X** is false on V2.

3.5.8a Two SL sentences are **logically equivalent** if and only if on every SL truth valuation for these sentences they have identical truth values.

3.5.8b Two SL sentences are **logically equivalent** if and only if there is no SL truth valuation on which they have different truth values.

3.5.8:C COMMENTARY ON 3.5.8

Like contingency, logical equivalence requires two different truth analyses. But unlike contingency, the goal of these analyses is not to recover truth valuations but to refute the initial truth-value assignments. These truth analyses demonstrate that it is impossible for the two sentences to have different truth values. So we suppose that the first is true and the second is false, and we refute this supposition; then we suppose that the first is false and the second is true, and we refute this supposition too.

We will use the outlined procedure to show that the sentences below are logically equivalent. We call the first sentence **X** and the second **Y**.

$$(J \vee \neg K) \rightarrow ((K \leftrightarrow M) \vee (B \rightarrow M)) \qquad ((\neg K \leftrightarrow M) \wedge (\neg M \wedge B)) \rightarrow (K \wedge \neg J)$$

We begin by assigning T to **X** and F to **Y** and analyze these truth values as far as we can go without branching out.

$$(J^{T} \vee^{T} \neg^{F} K^{T}) \rightarrow^{T} ((K^{T} \leftrightarrow^{F} M^{F}) \vee^{<T,\,F>} (B^{T} \rightarrow^{F} M^{F})) \quad ((\neg^{F} K^{T} \leftrightarrow^{T} M^{F}) \wedge^{T} (\neg^{T} M^{F} \wedge^{T} B^{T})) \rightarrow^{F} (K^{T} \wedge^{F} \neg^{F} J^{T})$$

Our analysis did not branch out at all and it generated a contradiction. This shows that there is no truth valuation on which **X** is true and **Y** is false. That was half the story. We now need to show that there is no truth valuation on which **X** is false and **Y** is true. So we assign F to **X** and T to **Y**. Our goal is to refute this assignment as well.

$$(J^{T} \vee^{T} \neg^{F} K^{T}) \rightarrow^{F} ((K^{T} \leftrightarrow^{F} M^{F}) \vee^{F} (B^{T} \rightarrow^{F} M^{F})) \quad ((\neg^{F} K^{T} \leftrightarrow^{T} M^{F}) \wedge^{T} (\neg^{T} M^{F} \wedge^{T} B^{T})) \rightarrow^{T} (K^{T} \wedge^{<T,\,F>} \neg^{F} J^{T})$$

Again, we did not branch out and we reached a contradiction. This refutes the initial assignment and establishes our goal.

3.5.9 A set of SL sentences is **consistent** if and only if there is an SL truth valuation on which every member of the set is true, that is, there is a truth valuation that satisfies the set (or simply, if and only if the set has an SL model).

3.5.9:C COMMENTARY ON 3.5.9

We show that the set Γ below is consistent.

$$\{(K \rightarrow \neg B) \vee D,\ \neg D \wedge J,\ L \wedge (J \leftrightarrow A),\ L \wedge \neg (R \vee Q)\}$$

We assign T to every member of Γ and then proceed to recover a truth valuation.

$$(K \rightarrow^{T} \neg B) \vee^{T} D^{F} \qquad \neg^{T} D^{F} \wedge^{T} J^{T} \qquad L^{T} \wedge^{T} (J^{T} \leftrightarrow^{T} A^{T}) \qquad L^{T} \wedge^{T} \neg^{T} (R^{F} \vee^{F} Q^{F})$$

We have three options for $K \rightarrow^{T} \neg B$. Since *K* and *B* do not occur in the other sentences, each option recovers a truth valuation. We arbitrarily select one of these options: *K* and ¬*B* are true (*B* is false). The truth valuation we recovered is this: V(*A*) = T, V(*B*) = F, V(*D*) = F, V(*J*) = T, V(*K*) = T, V(*L*) = T, V(*Q*) = F, and V(*R*) = F. All the members of Γ are true on V; therefore the set is consistent.

3.5.10a A set of SL sentences is **inconsistent** if and only if on every SL truth valuation for that set at least one member of the set is false.

3.5.10b A set of SL sentences is **inconsistent** if and only if there is no SL truth valuation on which the members of the set are all true, that is, there is no truth valuation that satisfies the set (or simply, if and only if the set has no SL model).

3.5.10:C Commentary on 3.5.10

We consider the following set Σ of SL sentences.

$$\{A \rightarrow (S \vee E),\ \neg H,\ H \leftrightarrow (B \vee Q),\ (B \rightarrow Q) \rightarrow A,\ (E \wedge S) \rightarrow H,\ B \vee (S \leftrightarrow E)\}$$

This is an inconsistent set. We assign T to all its members and then refute this assignment by showing that all options lead to contradictions. (Observe that this set has 6 atomic components; a truth table for this set consists of 64 rows.)

$$A^{T} \rightarrow^{T} (S \vee^{T} E) \quad \neg^{T} H^{F} \quad H^{F} \leftrightarrow^{T} (B^{F} \vee^{F} Q^{F}) \quad (B^{F} \rightarrow^{T} Q^{F}) \rightarrow^{T} A^{T} \quad (E \wedge^{F} S) \rightarrow^{T} H^{F} \quad B^{F} \vee^{T} (S \leftrightarrow^{T} E)$$

At this point the analysis branches out. $S \vee^{T} E$ generates three options for S and E (TT, TF, and FT). $E \wedge^{F} S$ generates three options for E and S (FF, TF, and FT). $S \leftrightarrow^{T} E$ generates two options for S and E (TT and FF). Either choice we follow will produce the same *final* conclusion—namely, the assignment of T to every member of Σ is impossible. Of course, the prudent thing to do is to consider only $S \leftrightarrow^{T} E$ because it generates the fewest options. But if someone made a tactical error and adopted $S \vee^{T} E$ or $E \wedge^{F} S$, she will still reach the same final conclusion. Let us intentionally be imprudent and consider all possible options for *E and S*, TT, FF, TF, and FT, in order to demonstrate our point.

Take first E to be true and S false, or E false and S true. Either option makes $S \leftrightarrow E$ false, which contradicts its assignment of T. Now take E and S to be true. On this supposition $E \wedge S$ turns out to be true, contrary to its assignment of F. Finally, let E and S be false; $S \vee E$ now is false, which also contradicts its giving assignment of T. Below are the four options analyzed and the contradictions to which they lead.

(a) $A^{T} \rightarrow^{T} (S^{F} \vee^{T} E^{T}) \quad \neg^{T} H^{F} \quad H^{F} \leftrightarrow^{T} (B^{F} \vee^{F} Q^{F}) \quad (B^{F} \rightarrow^{T} Q^{F}) \rightarrow^{T} A^{T} \quad (E^{T} \wedge^{F} S^{F}) \rightarrow^{T} H^{F} \quad B^{F} \vee^{T} (S^{F} \leftrightarrow^{\langle T, F\rangle} E^{T})$

(b) $A^{T} \rightarrow^{T} (S^{T} \vee^{T} E^{F}) \quad \neg^{T} H^{F} \quad H^{F} \leftrightarrow^{T} (B^{F} \vee^{F} Q^{F}) \quad (B^{F} \rightarrow^{T} Q^{F}) \rightarrow^{T} A^{T} \quad (E^{F} \wedge^{F} S^{T}) \rightarrow^{T} H^{F} \quad B^{F} \vee^{T} (S^{T} \leftrightarrow^{\langle T, F\rangle} E^{F})$

(c) $A^{T} \rightarrow^{T} (S^{T} \vee^{T} E^{T}) \quad \neg^{T} H^{F} \quad H^{F} \leftrightarrow^{T} (B^{F} \vee^{F} Q^{F}) \quad (B^{F} \rightarrow^{T} Q^{F}) \rightarrow^{T} A^{T} \quad (E^{T} \wedge^{\langle T, F\rangle} S^{T}) \rightarrow^{T} H^{F} \quad B^{F} \vee^{T} (S^{T} \leftrightarrow^{T} E^{T})$

(d) $A^{T} \rightarrow^{T} (S^{F} \vee^{\langle T, F\rangle} E^{F}) \quad \neg^{T} H^{F} \quad H^{F} \leftrightarrow^{T} (B^{F} \vee^{F} Q^{F}) \quad (B^{F} \rightarrow^{T} Q^{F}) \rightarrow^{T} A^{T} \quad (E^{F} \wedge^{F} S^{F}) \rightarrow^{T} H^{F} \quad B^{F} \vee^{T} (S^{F} \leftrightarrow^{T} E^{F})$

No matter which option we follow, we reach a contradiction. The initial truth-value assignment is refuted. Hence Σ is inconsistent.

3.5.11 It is clear that there is, up to equivalence, a finite number of SL truth valuations for any *finite* set of SL sentences. The method of truth tables provides us with a mechanical procedure for listing these truth valuations and determining the truth values of the members of the set on every one of them. Hence we have an effective decision procedure for determining the applicability of each of the logical concepts defined above, if the given set of SL sentences is finite. We simply survey the rows of the truth table for that set, searching for truth valuations that determine whether a certain concept applies or not. Therefore these **logical concepts** are **decidable** in SL, as there is a mechanical process to check whether or not they apply in any given case.

3.5.11:C Commentary on 3.5.11

In 2.5.11:C we gave general definitions of the notions of decidable concept and effective decision procedure. In order to make this chapter self-contained, we will reproduce these definitions here. An **effective decision procedure** is a procedure that can, in principle, be followed mechanically (i.e., without any creative aspects) and after a finite number of deterministic steps leads to the correct answer, whether it is "Yes" or "No." Not every effective procedure is a decision procedure. An **effective procedure** is simply a deterministic mechanical procedure that after finitely many steps produces a certain result. There are effective procedures that produce the answer "Yes" when and only when the correct answer is "Yes," but that might not produce any answer when the correct answer is "No"; we refer to such procedures as 'effective Yes-procedures'. Also, there are effective procedures that produce the answer "No" when and only when the correct answer is "No," but that might not produce any answer when the correct answer is "Yes"; such procedures are referred to as 'effective No-procedures'. An effective *decision* procedure must be able to produce *both* answers. A **concept is decidable** if and only if there is an effective decision procedure for determining whether something is subsumed under the concept or not. A concept that has only an effective Yes-procedure is called a **semidecidable concept**.

The truth-table method is an effective decision procedure for determining the applicability of any of the logical concepts defined in 3.5.3–3.5.10. Let us illustrate this procedure by means of an example. Take Δ to be the following set of SL sentences:

{$D \lor E$, $D \rightarrow (J \leftrightarrow K)$, $\neg E \lor L$, $\neg J$, $\neg(\neg L \land K)$, $K \rightarrow (J \rightarrow K)$, $L \land \neg(D \rightarrow L)$}

We construct the truth table for Δ (it consists of 32 rows). Because we cannot fit all the sentences in one table, we will divide Δ's truth table into two tables.

V	D	E	J	K	L	D	∨	E	D	→	(J	↔	K)	¬	E	∨	L	¬	J	¬	(¬	L	∧	K)
V1	T	T	T	T	T	T	**T**	T	T	**T**	T	T	T	F	T	**T**	T	**F**	T	**T**	F	T	F	T
V2	T	T	T	T	F	T	**T**	T	T	**T**	T	T	T	F	T	**F**	F	**F**	T	**F**	T	F	T	T
V3	T	T	T	F	T	T	**T**	T	T	**F**	T	F	F	F	T	**T**	T	**F**	T	**T**	F	T	F	F
V4	T	T	T	F	F	T	**T**	T	T	**F**	T	F	F	F	T	**F**	F	**F**	T	**T**	T	F	F	F
V5	T	T	F	T	T	T	**T**	T	T	**F**	F	F	T	F	T	**T**	T	**T**	F	**T**	F	T	F	T
V6	T	T	F	T	F	T	**T**	T	T	**F**	F	F	T	F	T	**F**	F	**T**	F	**F**	T	F	T	T
V7	T	T	F	F	T	T	**T**	T	T	**T**	F	T	F	F	T	**T**	T	**T**	F	**T**	F	T	F	F
V8	T	T	F	F	F	T	**T**	T	T	**T**	F	T	F	F	T	**F**	F	**T**	F	**T**	T	F	F	F
V9	T	F	T	T	T	T	**T**	F	T	**T**	T	T	T	T	F	**T**	T	**F**	T	**T**	F	T	F	T
V10	T	F	T	T	F	T	**T**	F	T	**T**	T	T	T	T	F	**T**	F	**F**	T	**F**	T	F	T	T
V11	T	F	T	F	T	T	**T**	F	T	**F**	T	F	F	T	F	**T**	T	**F**	T	**T**	F	T	F	F
V12	T	F	T	F	F	T	**T**	F	T	**F**	T	F	F	T	F	**T**	F	**F**	T	**T**	T	F	F	F
V13	T	F	F	T	T	T	**T**	F	T	**F**	F	F	T	T	F	**T**	T	**T**	F	**T**	F	T	F	T
V14	T	F	F	T	F	T	**T**	F	T	**F**	F	F	T	T	F	**T**	F	**T**	F	**F**	T	F	T	T
V15	T	F	F	F	T	T	**T**	F	T	**T**	F	T	F	T	F	**T**	T	**T**	F	**T**	F	T	F	F
V16	T	F	F	F	F	T	**T**	F	T	**T**	F	T	F	T	F	**T**	F	**T**	F	**T**	T	F	F	F
V17	F	T	T	T	T	F	**T**	T	F	**T**	T	T	T	F	T	**T**	T	**F**	T	**T**	F	T	F	T

V	*D*	*E*	*J*	*K*	*L*	*D*	**∨**	*E*	*D*	**→**	(*J*	↔	*K*)	¬	*E*	**∨**	*L*	**¬**	*J*	**¬**	(¬	*L*	∧	*K*)
V18	F	T	T	T	F	F	**T**	T	F	**T**	T	T	T	F	T	**F**	F	**F**	T	**F**	T	F	T	T
V19	F	T	T	F	T	F	**T**	T	F	**T**	T	F	F	F	T	**T**	T	**F**	T	**T**	F	T	F	F
V20	F	T	T	F	F	F	**T**	T	F	**T**	T	F	F	F	T	**F**	F	**F**	T	**T**	T	F	F	F
V21	F	T	F	T	T	F	**T**	T	F	**T**	F	F	T	F	T	**T**	T	**T**	F	**T**	F	T	F	T
V22	F	T	F	T	F	F	**T**	T	F	**T**	F	F	T	F	T	**F**	F	**T**	F	**F**	T	F	T	T
V23	F	T	F	F	T	F	**T**	T	F	**T**	F	T	F	F	T	**T**	T	**T**	F	**T**	F	T	F	F
V24	F	T	F	F	F	F	**T**	T	F	**T**	F	T	F	F	T	**F**	F	**T**	F	**T**	T	F	F	F
V25	F	F	T	T	T	F	**F**	F	F	**T**	T	T	T	T	F	**T**	T	**F**	T	**T**	F	T	F	T
V26	F	F	T	T	F	F	**F**	F	F	**T**	T	T	T	T	F	**T**	F	**F**	T	**F**	T	F	T	T
V27	F	F	T	F	T	F	**F**	F	F	**T**	T	F	F	T	F	**T**	T	**F**	T	**T**	F	T	F	F
V28	F	F	T	F	F	F	**F**	F	F	**T**	T	F	F	T	F	**T**	F	**F**	T	**T**	T	F	F	F
V29	F	F	F	T	T	F	**F**	F	F	**T**	F	F	T	T	F	**T**	T	**T**	F	**T**	F	T	F	T
V30	F	F	F	T	F	F	**F**	F	F	**T**	F	F	T	T	F	**T**	F	**T**	F	**F**	T	F	T	T
V31	F	F	F	F	T	F	**F**	F	F	**T**	F	T	F	T	F	**T**	T	**T**	F	**T**	F	T	F	F
V32	F	F	F	F	F	F	**F**	F	F	**T**	F	T	F	T	F	**T**	F	**T**	F	**T**	T	F	F	F

V	*D*	*E*	*J*	*K*	*L*	*K*	**→**	(*J*	→	*K*)	*L*	**∧**	¬	(*D*	→	*L*)
V1	T	T	T	T	T	T	**T**	T	T	T	T	**F**	F	T	T	T
V2	T	T	T	T	F	T	**T**	T	T	T	F	**F**	T	T	F	F
V3	T	T	T	F	T	F	**T**	T	F	F	T	**F**	F	T	T	T
V4	T	T	T	F	F	F	**T**	T	F	F	F	**F**	T	T	F	F
V5	T	T	F	T	T	T	**T**	F	T	T	T	**F**	F	T	T	T
V6	T	T	F	T	F	T	**T**	F	T	T	F	**F**	T	T	F	F
V7	T	T	F	F	T	F	**T**	F	T	F	T	**F**	F	T	T	T
V8	T	T	F	F	F	F	**T**	F	T	F	F	**F**	T	T	F	F
V9	T	F	T	T	T	T	**T**	T	T	T	T	**F**	F	T	T	T
V10	T	F	T	T	F	T	**T**	T	T	T	F	**F**	T	T	F	F
V11	T	F	T	F	T	F	**T**	T	F	F	T	**F**	F	T	T	T
V12	T	F	T	F	F	F	**T**	T	F	F	F	**F**	T	T	F	F
V13	T	F	F	T	T	T	**T**	F	T	T	T	**F**	F	T	T	T
V14	T	F	F	T	F	T	**T**	F	T	T	F	**F**	T	T	F	F
V15	T	F	F	F	T	F	**T**	F	T	F	T	**F**	F	T	T	T
V16	T	F	F	F	F	F	**T**	F	T	F	F	**F**	T	T	F	F
V17	F	T	T	T	T	T	**T**	T	T	T	T	**F**	F	F	T	T
V18	F	T	T	T	F	T	**T**	T	T	T	F	**F**	F	F	T	F
V19	F	T	T	F	T	F	**T**	T	F	F	T	**F**	F	F	T	T
V20	F	T	T	F	F	F	**T**	T	F	F	F	**F**	F	F	T	F
V21	F	T	F	T	T	T	**T**	F	T	T	T	**F**	F	F	T	T
V22	F	T	F	T	F	T	**T**	F	T	T	F	**F**	F	F	T	F
V23	F	T	F	F	T	F	**T**	F	T	F	T	**F**	F	F	T	T
V24	F	T	F	F	F	F	**T**	F	T	F	F	**F**	F	F	T	F
V25	F	F	T	T	T	T	**T**	T	T	T	T	**F**	F	F	T	T
V26	F	F	T	T	F	T	**T**	T	T	T	F	**F**	F	F	T	F

V	D	E	J	K	L	K	→	(J	→	K)	L	∧	¬	(D	→	L)
V27	F	F	T	F	T	F	**T**	T	F	F	T	**F**	F	F	T	T
V28	F	F	T	F	F	F	**T**	T	F	F	F	**F**	F	F	T	F
V29	F	F	F	T	T	T	**T**	F	T	T	T	**F**	F	F	T	T
V30	F	F	F	T	F	T	**T**	F	T	T	F	**F**	F	F	T	F
V31	F	F	F	F	T	F	**T**	F	T	F	T	**F**	F	F	T	T
V32	F	F	F	F	F	F	**T**	F	T	F	F	**F**	F	F	T	F

We are ready now to answer any question about the applicability of the logical concepts defined in 3.5.3–3.5.10 to Δ or members of Δ. We answer these questions by surveying the rows of the truth table above, searching for relevant truth valuations. Here is a sample of such questions and their answers.

(a) Is Δ consistent? No. None of V1–V32 makes the members of Δ all true (see 3.5.10b).

(b) Is *K*→(*J*→*K*) contingent? No, it is logically true. Each of V1–V32 assigns T to *K*→(*J*→*K*) (see 3.5.7 and 3.5.5a).

(c) Does Δ contain a logically false sentence? Yes. The last sentence, *L*∧¬(*D*→*L*), is logically false. It is false on all truth valuations, V1–V32 (see 3.5.6a).

(d) Is ¬(¬*L*∧*K*) logically false? No, it is contingent. The sentence is true on some truth valuations, such as V1, and false on others, such as V30 (see 3.5.6b and 3.5.7).

(e) Is ¬*J* a logical consequence of the set {*D*∨*E*, *D*→(*J*↔*K*), ¬*E*∨*L*}? No. V9, for instance, makes the members of the set all true and ¬*J* false (see 3.5.4).

(f) Is the following argument deductively valid?

D∨*E*
D→(*J*↔*K*)
¬*E*∨*L*
¬*J*
―――――
¬(¬*L*∧*K*)

Yes. None of V1–V32 makes the premises true and the conclusion false (see 3.5.3b).

(g) If we remove *D*∨*E* from the premises of the previous argument, is the resulting argument deductively valid? No. On V30 the premises are true and the conclusion is false (see 3.5.4).

(h) Are the sentences *D*→(*J*↔*K*) and ¬(¬*L*∧*K*) logically equivalent? No. V11, for instance, makes the first sentence false and the second true (see 3.5.8b).

(i) Is the set {*D*∨*E*, ¬*E*∨*L*, ¬*J*, ¬(¬*L*∧*K*)} consistent? Yes. There are truth valuations, such as V21, on which the members of the set are all true (3.5.9).

(j) Is ¬*K*∧¬*E* a logical consequence of Δ? Yes. Every SL sentence is a logical consequence of Δ because Δ is inconsistent. Let **X** be any SL sentence and Σ any inconsistent set of SL sentences. $\Sigma \vDash \mathbf{X}$ because there is no SL truth valuation on which the members of Σ are all true (Σ is inconsistent) and **X** is false. The first condition (Σ's members being all true) can never be satisfied. This is similar to our answer to 1.3.1k.

3.5.12 As in TL, **the SL worldview is incomplete**. This worldview identifies a certain type of states of affairs as basic ingredients of reality. But there are other types of states of affairs and of objects that are among the ingredients of reality. There are, for example, individuals, properties, relations, and *interdependent* states of affairs, which the SL worldview does not recognize as basic ingredients of reality. It is important to note that the SL worldview is controversial. Some philosophers do not believe that states of affairs are part of the ingredients of reality and others believe that they are not *basic* ingredients of reality. However, there are many philosophers who accept states of affairs as basic ingredients of reality and some philosophers go so far as to assert that they are the *only* basic ingredients of reality. In order to grant SL a possible place in the systems of classical logic, we will assume that states of affairs are at least part of the ingredients of reality (perhaps they are not basic ingredients). The incompleteness of the SL worldview results in there being SL truth valuations that fail to represent any relevant logical possibilities and relevant logical possibilities that are not represented by any SL truth valuations. As in TL, this conclusion implies that there are natural-language arguments that cannot be translated faithfully into SL. In order for the logical status of an SL argument to be indicative of the logical status of a natural-language argument of which the SL argument is a translation, the SL argument must be a faithful translation of the natural-language argument. An SL argument is a **faithful translation** of a natural-language argument just in case the interdependencies that exist between the sentences of the SL argument are precisely those interdependencies that exist between the sentences of the natural-language argument.

3.5.12:C COMMENTARY ON 3.5.12

3.5.12:C1 Note that we are talking about *relevant* logical possibilities. If we do not make this restriction, then every truth valuation represents some logical possibility. We gave in 1.2.1:C2 a general (and rough) description of the notion of a relevant logical possibility. We revisit this general description here. Later we will make our discussion more concrete by studying specific examples. Suppose that Σ is a set of natural-language declarative sentences. Let p be any logical possibility. p is a logical possibility *for* Σ, or that is *relevant to* Σ, if and only if every sentence in Σ makes a (true or false) assertion about some or all of the constituents of p. As stated in Chapters One and Two, the expressions that occur in the members of Σ must have the same meanings across all the logical possibilities for Σ. Let Σ^{SL} be the set consisting of the SL sentences that are translations of the natural-language sentences in Σ. Our conclusion asserts that there are cases of Σ and Σ^{SL}, such that some of the SL truth valuations for Σ^{SL} fail to represent any logical possibilities for Σ, and there are cases of Σ and Σ^{SL}, such that some of the logical possibilities for Σ are not represented by any SL truth valuations for Σ^{SL}.

It remains to define the relation of representation between SL truth valuations and informal logical possibilities. In 2.5.12:C1 we introduced certain notation, which we will extend to the case of SL. As above, Σ is a set of declarative sentences of some natural language L, and Σ^{SL} is a set consisting of SL translations of the sentences in Σ. Voc(Σ^{SL}) is the set of the extra-logical vocabulary that occurs in the members of Σ^{SL}, together with the logical vocabulary of SL. In other words, Voc(Σ^{SL}) consists of the atomic components of the sentences in Σ^{SL}, the five sentential connectives of SL, and the two parentheses. Voc(Σ) is the set of all the "basic" declarative L sentences that are translated into atomic sentences in Voc(Σ^{SL}), together with L expressions that correspond to all the logical symbols of SL. The atomic SL sentence X^{SL} in Voc(Σ^{SL}) is the translation of the "basic" declarative L sentence X in Voc(Σ), and the compound SL sentence Y^{SL} that is composed of Voc(Σ^{SL}) is the translation of the compound declarative L sentence Y that is composed of Voc(Σ). For the sake of simplicity, we will refer to any atomic sentence in Voc(Σ^{SL}) and to any compound sentence composed of Voc(Σ^{SL}) as a sentence composed of Voc(Σ^{SL}). Thus we consider an atomic sentence as composed of itself.

As we will see later, some SL sentences that syntactically mimic English sentences are unfaithful translations of the corresponding English sentences. There are three reasons for this semantical divergence between some English sentences and their SL translations: first, quite complex English sentences are translated as atomic SL sentences because these English sentences do not contain sentential connectives that correspond to SL connectives; second, SL truth-functional connectives may fail to capture important features of the sentential connectives in the English sentences; and, third, there are no interdependencies between the SL atomic sentences, while there might be important interdependencies between the English sentences that are translated into these atomic sentences. This semantical divergence could be so great that many of the SL truth valuations for the SL translations fail to represent any logical possibilities for the English sentences, and many logical possibilities for the English sentences fail to be represented by any truth valuations for the SL translations of the English sentences. Now we present the definition of the relation of representation.

Representation: An SL truth valuation V for Σ^{SL} **represents** a logical possibility p for Σ if and only if for every SL sentence **Z** that is composed of Voc(Σ^{SL}) and for every L sentence X that is composed of Voc(Σ), if **Z** is a translation of X (i.e., **Z** is X^{SL}), then **Z** is true (or false) on V if and only if X is true (or false) in p.

3.5.12:C2 We shall clarify the ideas and notation introduced in the preceding subsection by means of an example. We begin with a set Σ of declarative English sentences, and then produce a set Σ^{SL} of SL translations of these English sentences. As usual, we supply a translation key.

The set Σ

S1 Switch 1 is on and Switch 3 is off.
S2 Either Switch 2 or Switch 3 is on.
S3 If Switch 2 is off, then Switch 1 is off.
S4 At least two of Switches 1, 2, and 3 are on.

Translation Key

A: Switch 1 is on.
B: Switch 2 is on.
C: Switch 3 is on.

According to the notation above, 'Switch 1 is on'^{SL} is *A*, 'Switch 2 is on'^{SL} is *B*, and 'Switch 3 is on'^{SL} is *C*.

The set Σ^{SL}

$S1^{SL}$	$A \wedge \neg C$
$S2^{SL}$	$B \vee C$
$S3^{SL}$	$\neg B \rightarrow \neg A$
$S4^{SL}$	$((A \wedge B) \vee (A \wedge C)) \vee (B \wedge C)$

Voc(Σ^{SL}) = {A, B, C, ¬, ∧, ∨, →, ↔, (,)}
Voc(Σ) = {Switch 1 is on, Switch 2 is on, Switch 3 is on, it is not the case that, not, and, or, if-then, if and only if}

Below are more examples of English sentences that are constructed on the basis of Voc(Σ) and their SL translations that are composed of Voc(Σ^{SL}).

S5	Switches 1, 2, and 3 are on.	$S5^{SL}$	$(A \wedge B) \wedge C$
S6	Switch 3 is on if and only if Switch 2 is on.	$S6^{SL}$	$C \leftrightarrow B$
S7	Neither Switch 3 nor Switch 1 is on.	$S7^{SL}$	$\neg C \wedge \neg A$
S8	Switch 2 is on or Switch 1 is on, but not both.	$S8^{SL}$	$(B \vee A) \wedge \neg(B \wedge A)$
S9	If it is not the case that either Switch 2 or Switch 1 is on, then Switch 3 is off if and only if Switch 2 is on.	$S9^{SL}$	$\neg(B \vee A) \rightarrow (\neg C \leftrightarrow B)$

As is the case with TL, we allow ourselves a certain degree of freedom in constructing English sentences that are based on Voc(Σ). We expressed S5 as 'Switches 1, 2, and 3 are on', although 'are' is not in Voc(Σ); but we could have been more strict and expressed S5 as 'Switch 1 is on and Switch 2 is on and Switch 3 is on'. In S9, we wrote 'Switch 3 is off', where 'off' is not in Voc(Σ); we could have written instead 'Switch 3 is not on'. We stated S7 as 'Neither Switch 3 nor Switch 1 is on' in spite of the fact that the connective 'neither-nor' is not in Voc(Σ); but S7 could have been stated as 'Switch 3 is not on and Switch 1 is not on'. We expressed the English sentences in a more natural fashion by using words that are not in Voc(Σ). However, we can paraphrase all these sentences using only the expressions of Voc(Σ). This is a general rule: it is all right to reach beyond Voc(Σ) when constructing sentences that are composed on the basis of Voc(Σ) as long as it is possible to paraphrase these sentences solely, or almost solely, in terms of the sentences and connectives in Voc(Σ).

There are, up to equivalences, 8 different truth valuations for Σ^{SL}. Here is one of them: V(A) = T, V(B) = F, and V(C) = T. The reader can easily verify the following truth value assignments by consulting the truth conditions of the SL sentences.

V($S1^{SL}$) = V($A \wedge \neg C$) = F

V($S2^{SL}$) = V($B \vee C$) = T

V($S3^{SL}$) = V($\neg B \rightarrow \neg A$) = F

V($S4^{SL}$) = V($((A \wedge B) \vee (A \wedge C)) \vee (B \wedge C)$) = T

V($S5^{SL}$) = V($(A \wedge B) \wedge C$) = F

V($S6^{SL}$) = V($C \leftrightarrow B$) = F

$V(S7^{SL}) = (\neg C \wedge \neg A) = F$

$V(S8^{SL}) = V((B \vee A) \wedge \neg(B \wedge A)) = T$

$V(S9^{SL}) = V(\neg(B \vee A) \rightarrow (\neg C \leftrightarrow B)) = T$

Now we describe a logical possibility p that is represented by V. p consists of two facts and one negative fact that concern three switches connected in series as below.

We take O1, O2, and O3 to be the states of affairs that switch 1 is on, that switch 2 is on, and that switch 3 is on. It is clear from the diagram that O1 and O3 are facts of p and O2 is a negative fact of p. This logical possibility is relevant to the set Σ since every member of Σ makes an assertion about the constituents of p.

In order to show that V represents p, we need to argue that V and p satisfy the condition stated in the definition of representation (the "Representation Condition"). There are infinitely many sentences that can be constructed from the symbols in Voc(Σ^{SL}). How do we know that all these SL sentences have the same truth value on V as the English sentences that they translate have in p? In fact, we can know if we make two important assumptions. First, we need to assume that the operative conception of truth is the correspondence conception (see 3.1). Second, we have to assume that the English connectives in Voc(Σ) are truth-functional and have the same truth conditions as their counterparts in SL. These assumptions give us the resources to argue for the Representation Condition.

The correspondence conception of truth entails that the sentences 'Switch 1 is on' and 'Switch 3 is on' are true in p, since they correspond to the facts O1 and O2, respectively, and that the sentence 'Switch 2 is on' is false in p, since it corresponds to the negative fact O2. These are the same truth values that their SL translations *A*, *C*, and *B* have on V. Let X^{SL} be any SL sentence composed of Voc(Σ^{SL}) that is a translation of the English sentence X, which is composed of Voc(Σ). The truth value of X^{SL} on V is fully determined by the truth values of its atomic components and the truth conditions of the sentential connectives it contains. Since X^{SL} is a translation of X, we should expect that X consists of "basic" sentences that correspond to the atomic components of X^{SL} and of sentential connectives that correspond to the sentential connectives in X^{SL}. Since we assumed that the English sentential connectives in Voc(Σ) are truth-functional, the truth value of X is fully determined by the truth values of its "basic" sentential components and the truth conditions of the sentential connectives it contains. But we have established that the "basic" sentences and their SL translations have identical truth values, and we have assumed that the sentential connectives in X have the same truth conditions as their SL counterparts. Hence the truth value of X in p must be the same truth value of X^{SL} on V.

It is possible to weaken the Representation Condition. Rather than insisting that *all* X^{SL} and X have identical truth values, we could only require that the members of Σ^{SL} *and* their sentential components have the same truth values on V as their corresponding English translation have in p. It seems reasonable not to impose the demanding condition on V and p. After all we are only interested in Σ and Σ^{SL}. If p succeeds in preserving the truth values assigned by V to the members of Σ^{SL} and to their sentential components, then there seems to be good reason to claim that V represents p. All the examples we will study of SL truth valuations that do not represent any relevant logical possibilities and of logical possibilities that fail to be represented by any truth valuation violate the weaker condition. Of course, if a case violates the weaker condition, it automatically violates the stronger one.

3.5.12:C3 Using the terminology introduced in 2.5.12:C4, we call an SL truth valuation that represents a relevant logical possibility **genuine** and an SL truth valuation that represents no relevant logical possibility **superfluous**. To see an example of a superfluous truth valuation, consider the schematization of the first argument in Exercise 1.3.16.

P1 If Uranus is the last planet and if Newtonian physics is correct, then the theoretically calculated orbit of Uranus describes the motion of the planet correctly.
P2 The theoretically calculated orbit of Uranus differs significantly from the observational orbit.
P3 Newtonian physics is correct.
P4 The observational orbit of Uranus is the correct one.

C Uranus is not the last planet.

We translate this argument into SL using the familiar two-step procedure: first we give a translation key, and second we use the translation key to translate into SL the schematized argument.

Translation Key

U: Uranus is the last planet.
N: Newtonian physics is correct.
T: The theoretically calculated orbit of Uranus describes the motion of the planet correctly.
D: The theoretically calculated orbit of Uranus differs significantly from the observational orbit.
O: The observational orbit of Uranus is the correct one.

SL Argument

S1 $(U \wedge N) \rightarrow T$
S2 D
S3 N
S4 O

S5 $\neg U$

To test for deductive validity using truth analysis, we initially assign T to the premises and F to the conclusion and proceed with the analysis.

$(U^T \wedge^T N^T) \rightarrow^T T^T \quad D^T \quad N^T \quad O^T \quad / \quad \neg^F U^T$

We recover a truth valuation: V(*D*) = T, V(*N*) = T, V(*O*) = T, V(*T*) = T, V(*U*) = T. This proves that the SL argument is deductively invalid.

V is a superfluous truth valuation; it represents no logical possibility that is relevant to the English argument. If we read V in English, using the translation key above, we describe a situation in which Uranus is the last planet, Newtonian physics is correct, the theoretically calculated orbit of Uranus is correct, the theoretically calculated orbit differs significantly from the observational orbit, and the observational orbit is the correct one. This story describes an impossible situation. There is an implicit contradiction. The story says that the theoretically calculated orbit is correct. But the story also tells us that the theoretical and observational orbits differ significantly from each other and that the observational orbit is the correct one. This is impossible. If the theoretical and observational orbit differ significantly from each other, they cannot *both* be correct. At least one of them must be wrong. Thus we have the following contradiction: the theoretical and observational orbits are both correct *and* at least one of them is wrong (i.e., it is not the case that both of them are correct). The situation described by this story does not depict a logical possibility for the English argument. V, therefore, is a superfluous truth valuation. This demonstrates that there are cases in which an SL truth valuation fails to represent any relevant logical possibility.

In 2.5.12:C5 we described a procedure to deal with the case of a deductively invalid TL translation of an English argument, when there is a reason to suspect that the translation is not perfectly faithful. The same procedure is applicable to any deductively invalid SL translation of an English argument when there is a reason to believe that the SL translation is not totally faithful. We apply this procedure to the present case. The truth valuation V is the only counterexample to the SL argument. Since it is superfluous, the schematized argument is very likely deductively valid (in fact, it *is* deductively valid) in spite of the deductive invalidity of its SL translation. The reason is as follows: if the schematized argument had a counterexample, this logical possibility would very likely be represented by an SL truth valuation for the SL argument (see the analysis below). This truth valuation would be a genuine truth valuation that is a counterexample to the SL argument, but we said that the SL argument has only one counterexample, which is superfluous. Therefore it is very likely that there is no logical possibility that is a counterexample to the (schematized) English argument; hence it is very likely that the English argument is deductively valid.

In fact, in this case, we can make a stronger claim: because V is the only counterexample to the SL argument and it is superfluous, the English argument is indeed deductively valid. In this case, V's being superfluous supplies conclusive evidence, and not merely strong evidence, for the deductive validity of the English argument because the presence of V is due to the SL argument's failure to translate the relation of identity, since the SL worldview does not recognize relations as basic ingredients of reality. P2 of the English argument asserts that the theoretical orbit of Uranus is not identical with the observational orbit. P4 is equivalent to the assertion that the observational orbit of Uranus is identical with the actual orbit of the planet. Given the laws of identity, P2 and P4 imply that the theoretical orbit of Uranus is not identical with the actual orbit of the planet. Had the SL translation been able to capture this interdependency between P2 and P4, there would be no counterexample to the SL argument. This analysis shows that the failure of the SL argument to capture the relation of identity is the *only* reason that the SL argument is not a faithful translation of the English argument and it is the *only* reason for the presence of a counterexample to the SL argument. This in turn shows that all the relevant logical possibilities are represented by SL truth valuations for the SL argument.[1] Hence if there were a counterexample to the English argument, it would be represented by an SL truth valuation that is also a counterexample to the SL argument. But such a truth valuation would be genuine, yet we argued that the only counterexample to the SL argument is V, and V is superfluous. This example illustrates the point we made in 3.5.12—namely, that if an SL translation of a natural-language argument is not faithful, the deductive invalidity of the SL argument might not be indicative of the logical status of the natural-language argument; the latter might be valid.

We can improve the SL translation above by introducing into SL the information that if the theoretical and observational orbits are different, they cannot both be correct. This amounts to introducing a fifth premise into the SL argument: $D \rightarrow \neg(T \wedge O)$. As indicated in 2.5.12:C4, such a premise is usually called a **meaning-postulate**, because it helps the symbolic language capture some aspects of the meanings of the English expressions. '*D*' translates 'The theoretically calculated orbit of Uranus differs significantly from the observational orbit'. An aspect of the meaning of the last sentence is that the theoretical and observational orbits cannot both be correct, that is, if they differ, they cannot both be right, which is translated into $D \rightarrow \neg(T \wedge O)$. If we add this premise, the resulting SL argument is deductively valid. Here is the truth analysis.

$$(U^{T} \wedge^{T} N^{T}) \rightarrow^{T} T^{T} \qquad D^{T} \qquad N^{T} \qquad O^{T} \qquad D^{T} \rightarrow^{T} \neg^{T}(T^{T} \wedge^{<T,\,F>} O^{T}) \qquad / \qquad \neg^{F} U^{T}$$

The initial assignment of T to the premises and F to the conclusion is refuted. Therefore the SL argument is deductively valid. Because the system we shall consider in the next chapter, Predicate Logic (PL), contains the relation of identity, we will be able to capture in PL the Uranus argument without adding any meaning-postulate.

3.5.12:C4 Cases of deductively invalid English arguments whose SL translations are deductively valid are much less controversial than their counterparts in TL. These cases almost always arise from two features of SL: (1) the restriction the SL worldview places on permissible states of affairs, which is that there can be no interdependencies between these states of affairs, that is, each state of affairs can obtain or fail to obtain independently of the status of any other states of affairs; and (2) the truth conditions of the material conditional, that is, **X→Y** is true if and only if **X** is false or **Y** is true.

Philosophers gave many examples of inferences that are valid in SL but invalid under certain English interpretations. In this subsection, we consider one of them with some elaboration and we briefly discuss a second example.

SL1 $\quad (A \wedge B) \rightarrow C \models (A \rightarrow C) \vee (B \rightarrow C)$

The truth analysis below demonstrates that this inference is valid in SL.

SL1 $\quad (A^{T} \wedge^{T} B^{T}) \rightarrow^{<T,\,F>} C^{F} \quad / \quad (A^{T} \rightarrow^{F} C^{F}) \vee^{F} (B^{T} \rightarrow^{F} C^{F})$

1 Of course, a truth valuation might represent more than one logical possibility.

This analysis terminates with a contradiction: the initial truth-value assignment of T to the premise and F to the conclusion is refuted. SL1 is, therefore, valid. Now consider the following translation key.

A: We throw the first switch.
B: We throw the second switch.
C: The light turns on.

According to this translation key, we obtain the following English argument.

E1 If we throw the first switch and we throw the second switch, then the light turns on; therefore, either if we throw the first switch then the light turns on, or if we throw the second switch then the light turns on.

There is hardly any doubt that this English argument is deductively invalid. Imagine a situation in which the first and second switches are connected in series to the light: we must throw both switches in order for the light to turn on. In this situation the premise of E1 is clearly true regardless of the actual states of the switches. The premise describes a causal connection between the states of the switches and the state of the light. Whether one throws the switches or not, the causal connection remains true: if both switches are thrown, the light turns on. This causal sense of the English conditional is beyond the reach of the material conditional of SL. Thus if this electric circuit is functional, the premise of E1 is true. The conclusion of E1, however, *can be* false. There is a state of the circuit (call it 'S4') that makes both disjuncts of the conclusion false. Imagine that both switches are off. Now if one throws only the first switch, the light does not turn on, and if one throws only the second switch, the light does not turn on either. There is no explicit or implicit contradiction in the situation we just described. Our world contains many examples of similar electric circuits (and of other sorts of situations with the same general structure). Hence S4 is a logical possibility in which the premise of E1 is true and the conclusion of E1 is false. Therefore this logical possibility is a counterexample to the English argument. The diagram below depicts the logical possibility S4.

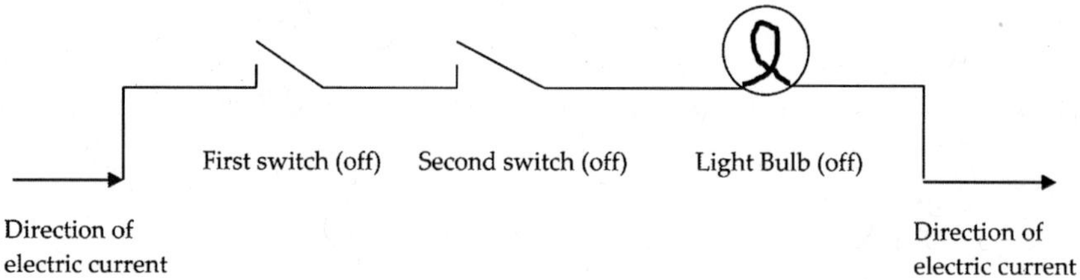

The state S4 of the electric circuit:
(1) If we throw both switches, the light turns on (true)
(2) If we throw the first switch, the light turns on (false)
(3) If we throw the second switch, the light turns on (false)
(4) Either if we throw the first switch then the light turns on, or if we throw the second switch then the light turns on (false)

The logical possibility S4 cannot be represented by any truth valuation for SL1, as SL1 is deductively valid, and hence it has no counterexamples. Let us examine the SL truth valuations for SL1 in order to see why none of them represents S4. To present matters clearly we draw two tables. The first table below is the standard truth table for the premise and conclusion of SL1. It displays, up to equivalence, all the SL truth valuations for SL1 and the truth values of its premise, conclusion, and their sentential components. The second table displays all the possible states of the electric circuit described above (assuming that the circuit is functional) and the truth values of the premise and the conclusion of E1 and of their sentential components. Let W1 stand for the sentence 'We throw the first switch', W2 for the sentence 'We throw the second switch', and L for the sentence 'The light turns on'. We will refer to the second table as the *circuit table*.

Truth Valuation	*A*	*B*	*C*	(*A*	∧	*B*)	→	*C*	(*A*	→	*C*)	∨	(*B*	→	*C*)
V1	T	T	T	T	T	T	**T**	T	T	T	T	**T**	T	T	T
V2	T	T	F	T	T	T	**F**	F	T	F	F	**F**	T	F	F
V3	T	F	T	T	F	F	**T**	T	T	T	T	**T**	F	T	T
V4	T	F	F	T	F	F	**T**	F	T	F	F	**T**	F	T	F
V5	F	T	T	F	F	T	**T**	T	F	T	T	**T**	T	T	T
V6	F	T	F	F	F	T	**T**	F	F	T	F	**T**	T	F	F
V7	F	F	T	F	F	F	**T**	T	F	T	T	**T**	F	T	T
V8	F	F	F	F	F	F	**T**	F	F	T	F	**T**	F	T	F

State of the Circuit	W1	W2	L	W1 and W2	If W1 and W2, L	If W1 then L	If W2 then L	If W1 then L, or if W2 then L
S1 (V1)	T	T	T	T	T	T	T	T
S2 (V4)	T	F	F	F	T	F	T	T
S3 (V6)	F	T	F	F	T	T	F	T
S4	F	F	F	F	T	F	F	F

The entries of the truth table require no explanations. We are by now quite familiar with standard truth tables and with the truth conditions of the SL connectives. But two things are worth noting. First, as expected, there is no truth valuation on which the premise is true and the conclusion is false. So none of V1–V8 is a counterexample to SL1. Second, if we invoke the translation key given above, V2, V3, V5, and V7 do not represent possible states of the electric circuit. Recall that SL presupposes that the atomic sentences are independent of each other. This is a reflection of the SL worldview, which rules out any interdependencies between the possible states of affairs. In our example, L is not independent of W1 and W2. If our circuit is functional, it is not possible for W1 and W2 to be true and L false and it is not possible for one of W1 and W2 to be false and L true. However, it does not follow that all of the other SL truth valuations represent possible states of the circuit. As we will see, V8 also does not represent a possible state of the circuit. An important point to note is that a state of the circuit consists of the states of the two switches and the causal features of the circuit. The state of the light is determined by the states of the switches and the causal features of the circuit.

The circuit table represents the four possible states of the circuit: (S1) both switches are thrown and the light is on; (S2) the first switch is thrown, the second is not, and the light is off;

(S3) the second switch is thrown, the first is not, and the light is off; and (S4) neither switch is thrown and the light is off. In all of these states, the causal connection between the switches and the light holds: if W1 and W2, then L. The premise of E1, therefore, is true in all possible states of the circuit. The truth value of the sentence 'If W1 then L' depends on the state of the second switch, that is, on the truth value of W2. It does not depend on the actual state of the first switch, that is, it does not depend on the truth value of W1, because, again, the sentence 'If W1 then L' describes a certain causal feature of the electric circuit. If the second switch is thrown (W2 is true), then throwing the first switch turns the light on ('If W1 then L' is true). One the other hand, if the second switch is off (W2 is false), then throwing the first switch does not turn the light on ('If W1 then L' is false). Similarly, the truth value of the sentence 'If W2 then L' depends on the state of the first switch, that is, on the truth value of W1. It does not depend on the actual state of the second switch, that is, it does not depend on the truth value of W2. If the first switch is thrown (W1 is true), throwing the second switch turns the light on ('If W2 then L' is true); and if the first switch is off (W1 is false), throwing the second switch does not turn the light on ('If W2 then L' is false). This analysis explains the entries of the circuit table.

The row headed by S4 is the counterexample to E1. In S4 neither switch is thrown; hence throwing either switch without throwing the other does not turn the light on. In S4 both sentences 'If W1 then L' and 'If W2 then L' are false. However, throwing both switches does turn the light on. So in S4 the sentence 'If W1 and W2, then L' is true. Thus in S4 the premise of E1 is true while the conclusion of E1 is false. No row of the truth table for SL1 represents S4. V8 assigns truth values to A, B, and C, that are identical with the truth values of W1, W2, and L in S4: neither switch is thrown and the light is off. However, on V8, $(A \wedge B) \rightarrow C$, $A \rightarrow C$, $B \rightarrow C$, and $(A \rightarrow C) \vee (B \rightarrow C)$ are all true; while in S4, the sentence 'If W1 and W2, then L' is true, and the sentences 'If W1 then L', 'If W2 then L', and 'If W1 then L, or if W2 then L' are all false. It is clear that V8 and S4 violate the weaker version of the Representation Condition. The truth values of the sentence $(A \rightarrow C) \vee (B \rightarrow C)$, which occurs in SL1, and of its sentential components $(A \rightarrow C)$ and $(B \rightarrow C)$ on V8 are different from the truth values of their English translations in S4. Since some of the truth values assigned to A, B, and C by each of V1–V7 are not identical with the truth values of W1, W2, and L in S4, the Representation Condition is violated. Hence none of the truth valuations for SL1 represents the logical possibility S4. This example, therefore, proves our claim in 3.5.12 that there are cases in which relevant logical possibilities are not represented by any relevant SL truth valuations.

Observe that V1 represents S1, V4 represents S2, and V6 represents S3. These SL truth valuations represent possible states of the circuit. Thus they are genuine truth valuations since the possible states of the circuit are logical possibilities that are relevant to E1. As stated previously, the truth valuations V2, V3, V5, V7, and V8 do not represent possible states of the circuit. However, it does not follow that they are superfluous truth valuations. The four states S1–S4 are only a few logical possibilities for the English argument E1. There might be other logical possibilities for E1 that some of these truth valuations represent. Consider the following set of four logical possibilities. Imagine that we have two parallel switches that are connected to a light bulb. These switches and light bulb are the first and second switches and the light that are mentioned in the premise and conclusion of E1. Thus the states of this electric circuit are logical possibilities that are relevant to E1. The diagram below depicts the state of this circuit when the first and second switches are not thrown and the light is off (call this state 'C4').

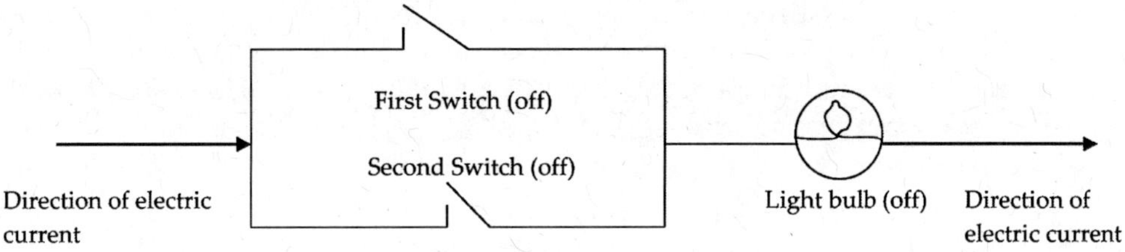

The state C4 of the electric circuit

Assuming that this electric circuit is functional, it has the following causal feature: if either or both switches are on, a current passes through the circuit, and hence the light turns on, but if both switches are off, no current passes through the circuit, and hence the light does not turn on. This circuit has the table below. As before we take W1, W2, and L to be the sentences 'We throw the first switch', 'We throw the second switch', and 'The light turns on', respectively. For the convenience of the reader we reproduce the truth table for SL1. Note that none of the logical possibilities C1–C4 is a counterexample to the English argument E1. This is part of the reason why, unlike the previous examples, all the states of this circuit are represented by relevant truth valuations.

Truth Valuation	*A*	*B*	*C*	(A	∧	*B*)	→	*C*	(A	→	C)	∨	(*B*	→	C)
V1	T	T	T	T	T	T	**T**	T	T	T	T	**T**	T	T	T
V2	T	T	F	T	T	T	**F**	F	T	F	F	**F**	T	F	F
V3	T	F	T	T	F	F	**T**	T	T	T	T	**T**	F	T	T
V4	T	F	F	T	F	F	**T**	F	T	F	F	**T**	F	T	F
V5	F	T	T	F	F	T	**T**	T	F	T	T	**T**	T	T	T
V6	F	T	F	F	F	T	**T**	F	F	T	F	**T**	T	F	F
V7	F	F	T	F	F	F	**T**	T	F	T	T	**T**	F	T	T
V8	F	F	F	F	F	F	**T**	F	F	T	F	**T**	F	T	F

State of the Circuit	W1	W2	L	W1 and W2	If W1 and W2, L	If W1 then L	If W2 then L	If W1 then L, or if W2 then L
C1 (V1)	T	T	T	T	T	T	T	T
C2 (V3)	T	F	T	F	T	T	T	T
C3 (V5)	F	T	T	F	T	T	T	T
C4 (V8)	F	F	F	F	T	T	T	T

As the reader can easily verify, in this case, the truth valuations V1, V3, V5, and V8 represent possible states of the circuit and the others do not. This means that these truth valuations are genuine because they represent logical possibilities that are relevant to E1. The previous example establishes that V1, V4, and V6 are genuine logical possibilities for SL1. This leaves V2 and V7. We leave it as an exercise for the reader to determine whether V2 and V7 are genuine or superfluous SL truth valuations for SL1.

SL1 is not a faithful translation of E1. The SL argument does not capture the dependency of L on W1 and W2, which is suggested by the premise of E1, since the material conditional of

SL cannot capture the causal sense of the English conditional.[1] This example demonstrates that when the SL translation of a natural-language argument is not faithful, the deductive validity of the SL argument might not be indicative of the logical status of the natural-language argument. The latter might be invalid.

We now consider briefly one more example of a deductively invalid English argument whose SL translation is valid.

SL2 $(D \rightarrow E) \vee (G \rightarrow H) \vDash (D \rightarrow H) \vee (G \rightarrow E)$

The truth analysis below demonstrates that this inference is valid in SL.

SL2 $(D^{T} \rightarrow^{F} E^{F}) \vee^{<T, F>} (G^{T} \rightarrow^{F} H^{F}) \quad / \quad (D^{T} \rightarrow^{F} H^{F}) \vee^{F} (G^{T} \rightarrow^{F} E^{F})$

Translation Key

D: Amanda is in London.
E: Amanda is in England.
G: Amanda is in Paris.
H: Amanda is in France.

According to this translation key, we obtain the following English argument.

E2 Either if Amanda is in London then she is in England, or if Amanda is in Paris then she is in France; therefore, either if Amanda is in London then she is in France, or if she is in Paris then she is in England.

We only mention, without further analysis, a plausible counterexample to the English argument. Our actual world is a counterexample to E2. London is in England and not in France, and Paris is in France and not in England. So it is true that if Amanda is in London, then she is in England, or if she is in Paris, then she is in France; but both disjuncts of the conclusion are false: if Amanda is in London, she is not in France, and if she is in Paris, she is not in England. Since the SL translation of E2 is deductively valid, none of its SL truth valuations is a counterexample. Hence the logical possibilities that are counterexamples to E2 are not represented by any of these truth valuations.

3.5.12:C5 As argued in 2.5.12:C7, it seems plausible to assume that the logical status of an argument is determined by the interdependencies that exist between its various parts. By definition, a faithful symbolic translation of a natural-language argument exhibits *all and only* the interdependencies that exist between the various parts of the natural-language argument. Therefore, since both arguments have exactly the same interdependencies between their *corresponding* parts, and since the logical status of an argument is determined by these dependencies, both arguments must have the same logical status. If this reasoning is correct (and we will assume that it is correct), then an SL argument that is a faithful translation of an English argument has the same logical status as the English argument: they are either both deductively valid or both deductively invalid.

---

1 An interesting aspect of the tables above is that all of them agree on the truth conditions for the conjunction and the disjunction.

3.6 Exercises

Note: All answers must be adequately justified.

3.6.1 Draw the construction tree of each of the SL sentences below.

3.6.1a* $((\neg(A \leftrightarrow B) \to \neg C) \vee (\neg A \wedge \neg(D \to B)))$
3.6.1b $(((A \vee \neg B) \wedge \neg(B \leftrightarrow D)) \wedge \neg((C \to E) \to (D \to A)))$
3.6.1c $((((H \to \neg\neg K) \leftrightarrow \neg(J \wedge L)) \vee (H \vee L)) \to \neg M)$
3.6.1d $(\neg N \leftrightarrow (((K \wedge \neg L) \vee \neg(M \to L)) \wedge (N \wedge K)))$

3.6.2 Draw the truth table for each of the following SL sentences and use it to determine the logical status of the sentence (i.e., contingent, logically true, or logically false) and of its immediate components.

3.6.2a* $\neg(A \wedge \neg(B \to C)) \leftrightarrow (A \to (\neg B \vee C))$
3.6.2b $\neg((D \leftrightarrow J) \vee (K \wedge \neg J))$
3.6.2c $((J \vee K) \wedge (J \leftrightarrow \neg H)) \to (H \to K)$
3.6.2d $((N \to E) \leftrightarrow (\neg N \vee E)) \wedge ((E \to M) \wedge (E \wedge \neg M))$

3.6.3 Let **X** be the following SL sentence: $(\neg(E \leftrightarrow D) \to ((D \wedge (A \vee \neg E)) \vee (\neg D \wedge (B \vee E))))$.

3.6.3a Draw the construction tree of **X**.
3.6.3b Use the truth-table method to determine the logical status of **X** and of its immediate components.

3.6.4 Establish the truth of the following claims.

3.6.4a* $\{\neg B \wedge (A \leftrightarrow B), D \to A\} \vDash \neg D$
3.6.4b $\{\neg L \vee M, M \to \neg(K \to K)\} \vDash \neg L$

3.6.5 Determine whether any of the following SL arguments is deductively valid.

3.6.5a*
$A \vee (B \to D)$
$C \to \neg E$
$B \wedge (E \leftrightarrow D)$

$D \leftrightarrow \neg C$

3.6.5b
$G \leftrightarrow (P \vee \neg(Q \wedge R))$
$(P \vee \neg Q) \to U$
$(P \vee \neg R) \to V$

$G \to (U \wedge V)$

3.6.6 Consider the following SL argument:

$(R \wedge \neg L) \to \neg N$
$N \leftrightarrow R$
$(M \wedge \neg L) \to N$

L

3.6.6a Give a counterexample to show that the argument above is deductively invalid.
3.6.6b Add the sentence $M \vee N$ to the premises of the argument and determine whether the resulting argument is deductively valid.

3.6.7 Show that each of the SL sentences below is logically true.

3.6.7a* $((A \vee B) \to C) \to (B \to C)$
3.6.7b $(((P \vee Q) \wedge R) \to (S \leftrightarrow W)) \to ((S \wedge \neg W) \to (Q \to \neg R))$
3.6.7c $((M \to \neg(\neg K \vee H)) \wedge (N \to (K \wedge \neg H))) \to ((M \vee N) \to \neg(K \to H))$

3.6.8 Determine whether each of the SL sentences below is logically true, logically false, or contingent.

3.6.8a* $(A \to (C \vee D)) \to ((A \to C) \vee (\neg A \vee (D \leftrightarrow B)))$
3.6.8b $((K \vee L) \to (M \leftrightarrow \neg N)) \wedge \neg((T \wedge \neg K) \to (M \vee N))$

3.6.9 Let **X**, **Y**, and **Z** be any SL sentences. Use the truth-table method to establish the logical equivalences below. (As in Chapter 1, the symbol '≅' stands for the relation of logical equivalence.)

3.6.9a	Double Negation:	$\neg\neg \mathbf{X}$	≅	$\mathbf{X}$
3.6.9b	Distribution:	$\mathbf{X} \wedge (\mathbf{Y} \vee \mathbf{Z})$	≅	$(\mathbf{X} \wedge \mathbf{Y}) \vee (\mathbf{X} \wedge \mathbf{Z})$
		$\mathbf{X} \vee (\mathbf{Y} \wedge \mathbf{Z})$	≅	$(\mathbf{X} \vee \mathbf{Y}) \wedge (\mathbf{X} \vee \mathbf{Z})$
3.6.9c*	De Morgan's Laws:	$\neg(\mathbf{X} \wedge \mathbf{Y})$	≅	$\neg \mathbf{X} \vee \neg \mathbf{Y}$
		$\neg(\mathbf{X} \vee \mathbf{Y})$	≅	$\neg \mathbf{X} \wedge \neg \mathbf{Y}$
3.6.9d	Material Conditional:	$\mathbf{X} \to \mathbf{Y}$	≅	$\neg \mathbf{X} \vee \mathbf{Y}$
3.6.9e	Negated Conditional:	$\neg(\mathbf{X} \to \mathbf{Y})$	≅	$\mathbf{X} \wedge \neg \mathbf{Y}$
3.6.9f	Contraposition:	$\mathbf{X} \to \mathbf{Y}$	≅	$\neg \mathbf{Y} \to \neg \mathbf{X}$
3.6.9g*	Exportation:	$\mathbf{X} \to (\mathbf{Y} \to \mathbf{Z})$	≅	$(\mathbf{X} \wedge \mathbf{Y}) \to \mathbf{Z}$
3.6.9h	Biconditional:	$\mathbf{X} \leftrightarrow \mathbf{Y}$	≅	$(\mathbf{X} \to \mathbf{Y}) \wedge (\mathbf{Y} \to \mathbf{X})$
3.6.9i*	Negated Biconditional:	$\neg(\mathbf{X} \leftrightarrow \mathbf{Y})$	≅	$\neg \mathbf{X} \leftrightarrow \mathbf{Y} \cong \mathbf{X} \leftrightarrow \neg \mathbf{Y}$

3.6.10 Are the following sets of SL sentences consistent?[1]

3.6.10a* $\{(R \vee K) \leftrightarrow \neg D, R \rightarrow \neg B, K \vee E, B \wedge \neg E\}$
3.6.10b $\{(S \vee Q) \leftrightarrow \neg C, S \rightarrow \neg B, Q \vee C, B \wedge \neg C\}$
3.6.10c $\{A \vee D, \neg(K \leftrightarrow M), (A \wedge K) \rightarrow R, (D \wedge \neg M) \rightarrow Q, \neg Q \wedge \neg R\}$
3.6.10d* $\{\neg(G \wedge H) \vee R, \neg R \rightarrow H, \neg(G \rightarrow R)\}$

3.6.11 Schematize the arguments in the following passages, translate the schematizations into SL, and show that the original arguments are deductively valid.

3.6.11a Sergio is bearded and bald if and only if he is not a philosopher. It is false that Sergio is bearded only if he is not a philosopher. Therefore, Sergio is not bald.

3.6.11b* If knowledge is impossible, then either we know that it is impossible or we don't. If we know that knowledge is impossible, then we know something. But if we know something, then knowledge is possible. Therefore, it is a necessary condition for knowledge to be impossible that we don't know that it is impossible.

3.6.12 Translate into SL the schematization of the argument in Exercise 1.3.17. Determine whether the SL argument is deductively valid or not. What can you conclude about the original English argument?

3.6.13 Schematize the arguments in the passages below and use the methods of SL to test for deductive validity.

3.6.13a* If beliefs are processes that occur in the brain, then they are physical processes. But if beliefs are physical processes, then the belief that God exists is a physical process. It is clear, however, that all physical processes can be duplicated by certain machines. Now we may assert that if the belief that God exists is a physical process and all physical processes can be duplicated by certain machines, then it is possible for a machine (say, a complex computer) to duplicate the belief that God exists. I take it to be obvious that it is possible for a machine to duplicate the belief that God exists only if it is possible for a machine to believe in God. Of course, we all know that this is impossible (it simply doesn't make sense to attribute religious beliefs to a machine).[2] Therefore, it is not the case that beliefs are processes that occur in the brain.

---

1 In the case of $\mathbf{X} \rightarrow^{\mathrm{T}} \mathbf{Y}$, if we assume that **X** is false, we don't need to analyze **Y** at all unless such an assumption leads to a contradiction in **X**. If it is not possible to assign F to **X**, the only available option is $\mathbf{X}^{\mathrm{T}} \rightarrow^{\mathrm{T}} \mathbf{Y}^{\mathrm{T}}$. But if it is possible for **X** to be false, **Y** need not be analyzed. If there are atomic sentences in **Y** that are not in **X**, we may assign to them any truth values. However in our answers to this question, we will, for the sake of completion and as an exercise, analyze **Y** in all cases.

2 The reader should not assume that we are committed to the truth of this premise. All the arguments that are given in the exercises of this book are constructed to illustrate the use of the symbolic systems to test for deductive validity and to give the reader some idea about interesting philosophical positions and possible justifications for them. Whether these arguments are sound or not is not a concern of this book.

3.6.13b Punishing a person for an act is never just unless the person is responsible for the act. But a person is responsible for an act just in case he or she has acted out of his or her free will. Whatever freedom of the will might involve, one thing, at least, is clear: a person would not have acted out of his or her free will if he or she couldn't have done otherwise. Modern behavioral science tells us, however, that human behavior is fully determined by one's environmental conditions and genetic makeup. If this is true, then any act is either an outcome of the person's environment or genes. If it is an outcome of the person's environment, then he or she could have done otherwise only if people can control their environmental conditions freely. On the other hand, if it is an outcome of the person's genes, then he or she could have done otherwise only if people can alter their genetic makeup by choice. It is clear, I think, that neither controlling one's environmental conditions *freely* nor altering one's genetic makeup by *choice* is possible if what modern behavioral science tells us is true. Therefore, punishing a person for an act is never just.

3.6.13c Philosophers use the adjective 'coreferential' to describe terms that refer to the same object. For example, the terms 'The first US President', 'The military leader of the American Revolution', and 'George Washington' are coreferential terms because they all refer to the same person. When terms have the same meaning, they are said to be *synonymous*. Many people would be inclined to affirm that coreferential terms are synonymous. Gottlob Frege attempted to show, however, that such an inclination is misguided. Here is a version of his argument. If coreferential terms are necessarily synonymous, then the terms '16+84' and '100' are synonymous; but it is clear that these terms are not synonymous unless the sentences '16+84 = 100' and '100 = 100' are saying exactly the same thing. However, the argument continues, if these two sentences are saying exactly the same thing, then the sentence '16+84 = 100' is informative if and only if the sentence '100 = 100' is informative. Since the sentence '16+84 = 100' is informative but '100 = 100' is not, Frege concluded that coreferential terms are not necessarily synonymous.

3.6.14 For each of the following passages, schematize the argument in the passage, translate the schematization into SL, and show that the schematization is deductively invalid. Explain why the counterexample to the schematization does not constitute a counterexample to the original argument (i.e., the argument in the passage). Supply the missing premise. Translate the revised schematization into SL and use it to establish the validity of the original argument.

3.6.14a The argument given in 3.6.13a attempts to show that beliefs, which are mental states, are not physical processes (precisely, they are not brain-processes). Here is an argument that aims at establishing that mental states *are* physical processes if the laws of physics hold. The argument runs as follows. If mental states (such as beliefs and desires) are not fully physical processes, then our mental states can cause our behavior (e.g., Elizabeth took aspirin *because* she *felt* pain, *had a desire* to get rid of it, and *believed* that aspirin would help her) only if nonphysical events can cause (at least in part) behavior. Behavior, of course, is a physical process. Now, if behavior is a physical process and can be caused by nonphysical events, then it is possible for a physical process to have a nonphysical cause. It is a physical fact, however, that causation must involve a transfer of energy from the cause to the effect.[1] We can thus assert that if it is possible for a physical process to have a nonphysical cause and if causation involves an energy transfer in the manner described above, then it is possible to transfer energy from a nonphysical event to the physical world. It is clear that a necessary condition for such a transfer to be possible is that it be possible to *add new* energy to the physical world. But unless the law of conservation of matter and energy is incorrect, it is not possible to add new energy to the physical world. Hence, we conclude that the correctness of the law of matter/energy conservation is sufficient for mental states to be fully physical processes.

3.6.14b* The French philosopher René Descartes (1596–1650) in his major philosophical work *Meditations on First Philosophy* attempted to show that a great deal of our ordinary beliefs about the world are genuine knowledge. He started by subjecting almost all of his previous beliefs to a severe test: a belief is knowledge just in case it is indubitable. He proceeded to show that statements such as 'I think', 'I exist', and 'I doubt' pass the test, and hence they are knowledge. He then gave a controversial argument attempting to prove the existence of God. From this, he concluded that everything that we perceive or understand *clearly and distinctly* is indubitable. Here is a modified version of his argument. We cannot help but believe in the truth of what we perceive (or understand) clearly and distinctly. It is clear, however, that we cannot help but have such beliefs only if it is in our nature to have such beliefs. But God created our nature. It is quite obvious that if God created our nature and if it is in our nature to have such beliefs, then God has created us such that we cannot help but believe in the truth of what we perceive clearly and distinctly. Now, either all of our beliefs in the truth of what we perceive clearly and distinctly are knowledge or some of such beliefs can be doubted. But if God has created us such that we must believe in the truth of what we perceive clearly and distinctly and yet some of such beliefs are not indubitable, then He has created us with a nature that deceives us.[2] Unless God is a deceiver, He would not have created us with such a nature. Therefore, we really know all those things that we perceive clearly and distinctly.

---

1 As stated in the preceding note, the reader should not assume that the premises of any argument in this book are true. For instance, it should not be assumed that it is *really* a physical fact that causation involves a transfer of energy from the cause to the effect. This premise might be true or it might be false. In this book, we are only concerned with the deductive validity and invalidity of arguments. The soundness of arguments lies outside the concerns of logical theory.

2 The intended justification of this premise is that if we find it inescapable to believe in the truth of what we perceive clearly and distinctly, then we find the belief that we *know* what we perceive clearly and distinctly to be inescapable as well, which means that it is in our nature to believe that we know those things. So, if it turns out that some of these beliefs can be doubted, then (given Descartes' criterion) they are not knowledge and, thus, our nature has deceived us in making us think that all of the beliefs formed by clear and distinct perception are knowledge. This line of reasoning is a modification of Descartes' original argument.

3.6.15 Schematize the argument in the passage below, translate it into SL, and test for validity. What is the status of the argument if the conclusion is replaced by the sentence "Essentialism is false only if biological species are natural kinds"?

> The *essence* of a natural kind is a collection of properties that are collectively sufficient and individually necessary for any object to be of that kind; furthermore, the essence of a natural kind explains why things of that kind have other relevant properties. For example, the natural kind *water* has the essence of being composed of H_2O molecules. Anything that is composed of such molecules is water (the essence is sufficient), and anything that is not composed of such molecules is not water (the essence is necessary). Also, the fact that something is composed of H_2O molecules explains why it has the typical properties of water, such as being colorless and odorless, freezing at 0° Centigrade, and boiling at 100° Centigrade. A philosophical doctrine that exerted substantial influence on science and philosophy alike is called *essentialism*. It is, on at least one reading, the view that every natural kind has essence. Consider, however, the following reasoning. If essentialism is true and if biological species are natural kinds, then they have essences. But if biological species have essences, then either no species contains deviant organisms or else the essence of a species contains vague properties (in order to allow for deviant members). However, every reasonably defined vague property must exclude many deviations. But if this is the case and if the essence of a biological species contains vague properties, then there are many deviations excluded from the species. A species would exclude many deviations only if its boundaries were defined arbitrarily. It is clear that a species cannot have essence if its boundaries are defined arbitrarily. Since at least some biological species could (and usually do) contain deviant organisms, essentialism is false unless biological species are not natural kinds.

3.6.16* Consider the following simple argument: all horses are animals; therefore, all heads of horses are heads of animals. Translate the argument into SL. Show that the SL argument is deductively invalid. Explain why the SL counterexamples cannot be used to construct counterexamples to the English argument.

3.6.17 The following SL argument is valid: $\neg(G \rightarrow H) \vDash G \wedge \neg H$. Produce an invalid English argument of which the SL argument is a translation. Explain the reason behind this divergence between the status of the English argument and the status of its SL translation.

Solutions to the Starred Exercises

SOLUTION TO 3.6.1

3.6.1a $((\neg(A\leftrightarrow B)\rightarrow\neg C)\vee(\neg A\wedge\neg(D\rightarrow B)))$

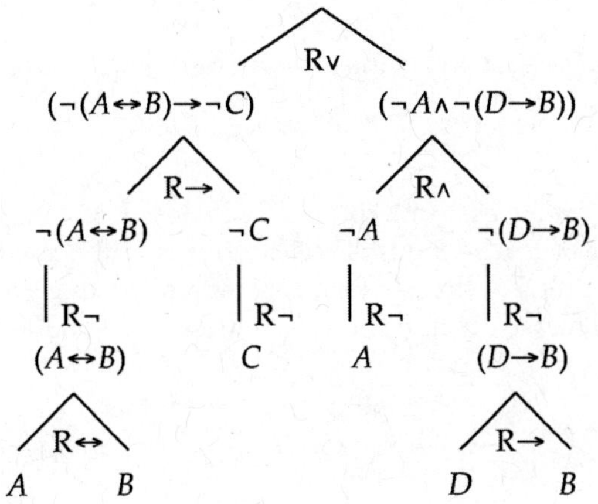

Because it is an SL sentence, its construction tree is complete. Recall that the syntactical decomposition of an ungrammatical expression of SL does not generate a complete construction tree.

SOLUTION TO 3.6.2

3.6.2a As usual we list the truth values of a sentence underneath its main connective. (The truth values of the main sentence and of its immediate components are in boldfaced font.)

V	A	B	C	¬	(A	∧	¬	(B	→	C))	↔	(A	→	(¬	B	∨	C))
V1	T	T	T	**T**	T	F	F	T	T	T	**T**	T	**T**	F	T	T	T
V2	T	T	F	**F**	T	T	T	T	F	F	**T**	T	**F**	F	T	F	F
V3	T	F	T	**T**	T	F	F	F	T	T	**T**	T	**T**	T	F	T	T
V4	T	F	F	**T**	T	F	F	F	T	F	**T**	T	**T**	T	F	T	F
V5	F	T	T	**T**	F	F	F	T	T	T	**T**	F	**T**	F	T	T	T
V6	F	T	F	**T**	F	F	T	T	F	F	**T**	F	**T**	F	T	F	F
V7	F	F	T	**T**	F	F	F	F	T	T	**T**	F	**T**	T	F	T	T
V8	F	F	F	**T**	F	F	F	F	T	F	**T**	F	**T**	T	F	T	F

The main sentence is logically true; it is true on every relevant truth valuation (see 3.5.5a). Each of its immediate components is contingent. They are true on V1 and V3–V8 and false on V2 (see 3.5.7). For the same reason they are logically equivalent (see 3.5.8a).[1]

1 It is a metatheorem of SL (i.e., a theorem *about* SL) that for all SL sentences **X** and **Y**, **X** ≅ **Y** if and only if **X**↔**Y** is logically true.

Solution to 3.6.4

3.6.4a We use truth analysis, which is our default method. Since we want to show that there is no truth valuation that makes the members of the set true and $\neg D$ false (see 3.5.3b), we initially assign T to the members of the set and F to $\neg D$, and then refute this assignment.

$$\neg^{T}B^{F}\wedge^{T}(A^{T}\leftrightarrow^{\langle T, F\rangle}B^{F}) \qquad D^{T}\rightarrow^{T}A^{T} \qquad / \qquad \neg^{F}D^{T}$$

We obtain a contradiction. This shows that the initial assignment of truth values is impossible.

Solution to 3.6.5

3.6.5a We assign T to the premises and F to the conclusion. If we recover a truth valuation, then the initial assignment of truth values is confirmed, and hence the argument is deductively invalid (see 3.5.4). But if all options lead to contradictions, then the initial assignment of truth values is refuted, and hence the argument is deductively valid (see 3.5.3b).

$$A\vee^{T}(B^{T}\rightarrow D) \qquad C\rightarrow^{T}\neg E \qquad B^{T}\wedge^{T}(E\leftrightarrow^{T}D) \qquad / \qquad D\leftrightarrow^{F}\neg C$$

Every sentence above presents us with options. The truth value of the conclusion can be satisfied by making D true and $\neg C$ false or D false and $\neg C$ true. Let us try the first of these options.

$$A\vee^{T}(B^{T}\rightarrow^{T}D^{T}) \qquad C^{T}\rightarrow^{T}\neg^{T}E^{F} \qquad B^{T}\wedge^{T}(E^{F}\leftrightarrow^{\langle T, F\rangle}D^{T}) \qquad / \qquad D^{T}\leftrightarrow^{F}\neg^{F}C^{T}$$

We reach a contradiction regardless of the truth value of A. This option is closed; so we try the second option.

$$A^{T}\vee^{T}(B^{T}\rightarrow^{F}D^{F}) \qquad C^{F}\rightarrow^{T}\neg^{T}E^{F} \qquad B^{T}\wedge^{T}(E^{F}\leftrightarrow^{T}D^{F}) \qquad / \qquad D^{F}\leftrightarrow^{F}\neg^{T}C^{F}$$

This truth analysis is consistent (i.e., it does not end with a contradiction). We recover the following truth valuation: V(A) = T, V(B) = T, V(C) = F, V(D) = F, and V(E) = F. The premises are true and the conclusion is false on V. Therefore the argument is invalid.

Solution to 3.6.7

3.6.7a An SL sentence is logically true (or false) if and only if there is no truth valuation for that sentence on which the sentence is false (or true). This definition presents a strategy to follow when applying the truth analysis method to show that a sentence is logically true (or false): assign F (or T) to the sentences and show that this assignment is impossible by refuting it.

$$((A\vee^{\langle T, F\rangle}B^{T})\rightarrow^{T}C^{F})\rightarrow^{F}(B^{T}\rightarrow^{F}C^{F})$$

The analysis produces a contradiction independently of the truth value of A. Therefore there is no relevant SL truth valuation that makes the sentence false, which means that the sentence is logically true (see 3.5.5b).

SOLUTION TO 3.6.8

3.6.8a In order to determine the logical status of a sentence using the truth analysis method, we assign to it a truth value and see whether this assignment is confirmed or refuted. If it is refuted, we know that the sentence is logically true or logically false, depending on the initial truth value. For instance, if we assign F to a sentence and this assignment is refuted, the sentence is logically true. On the other hand, if the initial truth-value assignment is confirmed, we must try the opposite truth value. Recovering one truth valuation shows that it is possible to assign the initial truth value to the sentence but we still need to determine whether it is possible to assign the opposite truth value to it.

Following this strategy, we now have a choice of assigning T or F to the sentence above. Since it is a conditional, assigning F to it produces fewer options. So we begin by assigning F to the sentence and performing truth analysis.

$$(A^{T}\rightarrow^{T}(C^{F}\vee^{T}D^{T}))\rightarrow^{F}((A^{T}\rightarrow^{F}C^{F})\vee^{F}(\neg^{F}A^{T}\vee^{F}(D^{T}\leftrightarrow^{F}B^{F})))$$

The analysis recovers the following SL truth valuation: V1(A) = T, V1(B) = F, V1(C) = F, and V1(D) = T. The assignment of F to the sentence has been confirmed, that is, it is possible to make it false. This implies that it is not logically true (see 3.5.5b). We still need to determine whether it is possible to make the sentence true. If it is, the sentence is contingent (see 3.5.7); if it is not, the sentence is logically false (see 3.5.6b). We assign T to it and analyze.

$$(A\rightarrow(C\vee D))\rightarrow^{T}((A\rightarrow C)\vee(\neg A\vee(D\leftrightarrow B)))$$

We have three options here: the antecedent and consequent are both true, they are both false, or the antecedent is false and the consequent is true. We should begin with the option that generates the fewest options. Thus we choose the second option.

$$(A^{T}\rightarrow^{F}(C^{F}\vee^{F}D^{F}))\rightarrow^{T}((A^{T}\rightarrow^{F}C^{F})\vee^{F}(\neg^{F}A^{T}\vee^{F}(D^{F}\leftrightarrow^{F}B^{T})))$$

This option also leads to a truth valuation: V2(A) = T, V2(B) = T, V2(C) = F, and V2(D) = F. The sentence, therefore, is contingent.

SOLUTIONS TO 3.6.9

3.6.9c De Morgan's Laws:

	X	Y	¬	(X	∧	Y)	¬	X	∨	¬	Y
1	T	T	**F**	T	T	T	F	T	**F**	F	T
2	T	F	**T**	T	F	F	F	T	**T**	T	F
3	F	T	**T**	F	F	T	T	F	**T**	F	T
4	F	F	**T**	F	F	F	T	F	**T**	T	F

The table above shows that there are at most four different ways of assigning truth values to **X** and **Y**. The table displays these assignments systematically. ¬(**X**∧**Y**) and ¬**X**∨¬**Y** have identical truth values on these assignments; therefore, they are logically equivalent (see 3.5.8a).

The table above is not a standard truth table and the formulas $\neg(\mathbf{X}\land\mathbf{Y})$ and $\neg\mathbf{X}\lor\neg\mathbf{Y}$ are not SL sentences. These formulas are part of the metalanguage of SL, which is English augmented with appropriate symbols. **X** and **Y** are metalinguistic variables that range over SL sentences. Thus $\neg(\mathbf{X}\land\mathbf{Y})$ and $\neg\mathbf{X}\lor\neg\mathbf{Y}$ are also metalinguistic variables, but they range over SL sentences that have specific forms. For instance, the SL sentences $\neg(A\land B)$ and $\neg(\neg K\land\neg\neg M)$ and $\neg((D\lor H)\land\neg(N\to G))$ are possible values for $\neg(\mathbf{X}\land\mathbf{Y})$. In the first sentence, A is substituted for **X** and B for **Y**. In the second sentence, $\neg K$ is substituted for **X** and $\neg\neg M$ for **Y**. In the third sentence, $(D\lor H)$ is substituted for **X** and $\neg(N\to G)$ for **Y**. No matter how complex the SL sentence is, if it has the right form, it is a possible value for $\neg(\mathbf{X}\land\mathbf{Y})$. Here is such a sentence.

$$\neg((\neg(G\to K)\land(A\leftrightarrow\neg C))\land((D\lor\neg E)\to(N\land\neg\neg M)))$$

In this sentence, $(\neg(G\to K)\land(A\leftrightarrow\neg C))$ is substituted for **X** and $((D\lor\neg E)\to(N\land\neg\neg M))$ is substituted for **Y**. Similar remarks are applicable to $\neg\mathbf{X}\lor\neg\mathbf{Y}$. It is a metalinguistic variable that ranges over SL sentences that have the form displayed by the variable. The following SL sentences are possible values for $\neg\mathbf{X}\lor\neg\mathbf{Y}$: $\neg A\lor\neg B$, $\neg\neg K\lor\neg\neg G$, $\neg(M\lor N)\lor\neg(H\to L)$, and $\neg((P\to\neg Q)\leftrightarrow\neg(S\lor R))\lor\neg(\neg(U\land V)\to(R\leftrightarrow Q))$.

The metalinguistic variables **X** and **Y** have no structure; hence any SL sentence may be substituted for **X** and **Y**. $\neg(\mathbf{X}\land\mathbf{Y})$ and $\neg\mathbf{X}\lor\neg\mathbf{Y}$, on the other hand, have specific structures (i.e., they exhibit specific forms). Because of this, metalinguistic variables such as $\neg(\mathbf{X}\land\mathbf{Y})$ and $\neg\mathbf{X}\lor\neg\mathbf{Y}$ are called **structural variables**. The form displayed by $\neg(\mathbf{X}\land\mathbf{Y})$ is that of a negated conjunction. Thus any SL sentence formed by negating the conjunction of two (not necessarily distinct) SL sentences is a possible value for $\neg(\mathbf{X}\land\mathbf{Y})$.

The table above is not a standard truth table for two specific SL sentences. This table establishes the logical equivalence of infinitely many pairs of SL sentences. Any two SL sentences obtained by substituting a specific SL sentence for the **X** in $\neg(\mathbf{X}\land\mathbf{Y})$ and $\neg\mathbf{X}\lor\neg\mathbf{Y}$ and a specific SL sentence for the **Y** in $\neg(\mathbf{X}\land\mathbf{Y})$ and $\neg\mathbf{X}\lor\neg\mathbf{Y}$ are logically equivalent. For example, $\neg((A\lor B)\land(C\to\neg D))$ is logically equivalent to $\neg(A\lor B)\lor\neg(C\to\neg D)$. We do not need to construct the truth table for these sentences (which consists of 16 rows) in order to establish that they are logically equivalent. We know that they are logically equivalent because we know that for any SL sentences **X** and **Y**, $\neg(\mathbf{X}\land\mathbf{Y})$ and $\neg\mathbf{X}\lor\neg\mathbf{Y}$ are logically equivalent. The four-row table we constructed above is sufficient for establishing the logical equivalence of $\neg((A\lor B)\land(C\to\neg D))$ and $\neg(A\lor B)\lor\neg(C\to\neg D)$. This is greatly convenient, for $\neg(\mathbf{X}\land\mathbf{Y})$ and $\neg\mathbf{X}\lor\neg\mathbf{Y}$ may range over very complex SL sentences with many atomic components. Proving the logical equivalence of such sentences may require constructing a huge truth table or performing a complicated truth analysis. But if two SL sentences can be shown to instantiate the metalinguistic formulas $\neg(\mathbf{X}\land\mathbf{Y})$ and $\neg\mathbf{X}\lor\neg\mathbf{Y}$ by substituting the same SL sentence for **X** in both formulas and the same SL sentence for **Y** in both formulas, then these sentences are logically equivalent. Here is an example of such pair of logically equivalent sentences (observe the complexity of these sentences).

$$\neg(\mathbf{X}\land\mathbf{Y})\text{: } \neg(((E\to\neg G)\land\neg(K\land(M\lor N)))\land\neg((L\leftrightarrow\neg R)\to\neg(P\to Q)))$$
$$\neg\mathbf{X}\lor\neg\mathbf{Y}\text{: } \neg((E\to\neg G)\land\neg(K\land(M\lor N)))\lor\neg\neg((L\leftrightarrow\neg R)\to\neg(P\to Q))$$

The truth-value assignments displayed in our four-row table do not necessarily represent SL truth valuations. **X** and **Y** may be compound sentences. This implies that a distribution of truth values to **X** and **Y** might not produce a unique distribution of truth values to their atomic components.[1] Consider, for instance, the sentence $\neg((D \vee G) \wedge K)$, which is of the form $\neg(\mathbf{X} \wedge \mathbf{Y})$. Assigning F to $(D \vee G)$ produces a unique truth valuation, V(D) = F and V(G) = F. But assigning T to $(D \vee G)$ gives us three possible truth valuations, V1(D) = T and V1(G) = T, V2(D) = T and V2(G) = F, and V3(D) = F and V3(G) = T. If the truth-value assignments displayed in the table above are not necessarily SL truth valuations, then what are they? They represent all the possible distributions of truth values that can be assigned to **X** and **Y** by the various truth valuations for them. This does not mean that every pair of SL sentences may receive all of these distributions of truth values (e.g., the assignment of T to $K \wedge H$ and F to $P \rightarrow K$ is not possible). But it means that every truth valuation for every pair, **X** and **Y**, of SL sentences must assign to **X** and **Y** one of the distributions of truth values listed (as 1–4) in the table above.

	X	Y	¬	(X	∨	Y)	¬	X	∧	¬	Y
1	T	T	**F**	T	T	T	F	T	**F**	F	T
2	T	F	**F**	T	T	F	F	T	**F**	T	F
3	F	T	**F**	F	T	T	T	F	**F**	F	T
4	F	F	**T**	F	F	F	T	F	**T**	T	F

The table above is the truth table for the second De Morgan's Law. We follow the same reasoning as above. There are at most four different ways of assigning truth values to **X** and **Y**, on each of which $\neg(\mathbf{X} \vee \mathbf{Y})$ and $\neg\mathbf{X} \wedge \neg\mathbf{Y}$ have identical truth values. Hence, for all SL sentences **X** and **Y**, $\neg(\mathbf{X} \vee \mathbf{Y})$ and $\neg\mathbf{X} \wedge \neg\mathbf{Y}$ are logically equivalent (see 3.5.8a).

3.6.9g Exportation:

	X	Y	Z	X	→	(Y	→	Z)	(X	∧	Y)	→	Z
1	T	T	T	T	**T**	T	T	T	T	T	T	**T**	T
2	T	T	F	T	**F**	T	F	F	T	T	T	**F**	F
3	T	F	T	T	**T**	F	T	T	T	F	F	**T**	T
4	T	F	F	T	**T**	F	T	F	T	F	F	**T**	F
5	F	T	T	F	**T**	T	T	T	F	F	T	**T**	T
6	F	T	F	F	**T**	T	F	F	F	F	T	**T**	F
7	F	F	T	F	**T**	F	T	T	F	F	F	**T**	T
8	F	F	F	F	**T**	F	T	F	F	F	F	**T**	F

There are at most eight assignments of truth values to **X**, **Y**, and **Z**. As explained in the Solution to 3.6.9c above, the formulas $\mathbf{X} \rightarrow (\mathbf{Y} \rightarrow \mathbf{Z})$ and $(\mathbf{X} \wedge \mathbf{Y}) \rightarrow \mathbf{Z}$ are structural metalinguistic variables that range over all the SL sentences that exhibit the same forms as these formulas. This table proves that any two SL sentences that are obtained by substituting the same sentence for **X**, the same sentence for **Y**, and the same sentence for **Z** in the first and second formulas are logically equivalent.

1 In fact, it might not produce any truth valuation. For example, assigning F to $\neg(A \wedge \neg A)$ leads to a contradiction. F is not a possible truth-value assignment for $\neg(A \wedge \neg A)$.

3.6.9i Negated Biconditional:

	X	Y	¬	(X	↔	Y)	¬	X	↔	Y	X	↔	¬	Y
1	T	T	**F**	T	T	T	F	T	**F**	T	T	**F**	F	T
2	T	F	**T**	T	F	F	F	T	**T**	F	T	**T**	T	F
3	F	T	**T**	F	F	T	T	F	**T**	T	F	**T**	F	T
4	F	F	**F**	F	T	F	T	F	**F**	F	F	**F**	T	F

SOLUTIONS TO 3.6.10

3.6.10a The strategy to test for consistency is to assign T to every member of the set and then to see whether the truth analysis confirms or refutes the initial assignment. If the initial assignment is confirmed, then the set is consistent (see 3.5.9), but if it is refuted, then the set is inconsistent (see 3.5.10b).

$(R^{F} \vee^{T} K^{T}) \leftrightarrow^{T} \neg^{T} D^{F} \qquad R^{F} \rightarrow^{T} \neg^{F} B^{T} \qquad K^{T} \vee^{T} E^{F} \qquad B^{T} \wedge^{T} \neg^{T} E^{F}$

We recover the following SL truth valuation: V(*B*) = T, V(*D*) = F, V(*E*) = F, V(*K*) = T, and V(*R*) = F. The set, therefore, is consistent.

3.6.10d As explained in 3.6.10a, we assign T to every sentence in the set and see whether we confirm or refute the initial assignment.

$\neg^{F}(G^{T} \wedge^{T} H^{T}) \vee^{\langle T, F\rangle} R^{F} \qquad \neg^{T} R^{F} \rightarrow^{T} H^{T} \qquad \neg^{T}(G^{T} \rightarrow^{F} R^{F})$

The analysis leads to a contradiction. The set, therefore, is inconsistent (see 3.5.10b).

SOLUTION TO 3.6.11

3.6.11b

Schematization of the argument

P1 If knowledge is impossible, then either we know that it is impossible or we don't.
P2 If we know that knowledge is impossible, then we know something.
P3 If we know something, then knowledge is possible.

C It is a necessary condition for knowledge to be impossible that we don't know that it is impossible.

Translation Key

P: Knowledge is possible.
K: We know that knowledge is impossible.
S: We know something.

SL Argument

S1	$\neg P \rightarrow (K \vee \neg K)$
S2	$K \rightarrow S$
S3	$S \rightarrow P$
S4	$\neg P \rightarrow \neg K$

As usual we test for deductive validity by assigning T to the premises and F to the conclusion and then performing truth analysis.

$$\neg^{T} P^{F} \rightarrow^{T} (K^{T} \vee^{T} \neg^{F} K^{T}) \qquad K^{T} \rightarrow^{<T, F>} S^{F} \qquad S^{F} \rightarrow^{T} P^{F} \qquad / \qquad \neg^{T} P^{F} \rightarrow^{F} \neg^{F} K^{T}$$

We reach a contradiction, which shows that the SL argument is deductively valid. The schematization and translation are straightforward. We conclude that the original argument is deductively valid too.

There is an interesting observation to make about this truth analysis: the contradiction is reached independently of the analysis of the truth value of the first premise, that is, the truth of the second and third premises, together with the falsity of the conclusion, are sufficient for the derivation of a contradiction. This implies that the first premise is logically superfluous. We can remove the first premise and the resulting argument would still be valid. The first premise, perhaps, is serving a rhetorical purpose. It makes the reasoning clearer but it is not needed for the conclusion to follow.

In fact, the first premise is logically true. We can show this by constructing the truth table for this sentence or by performing truth analysis. But there is an easier approach. It is obvious that the consequent of the first premise—namely the disjunction $K \vee \neg K$—is logically true. One can easily verify, from the truth conditions of the conditional, that a conditional with a logically true consequent (or a logically false antecedent) is logically true. A logically true premise has no influence on the logical status of an argument, that is, removing or adding a logically true premise does not alter the deductive validity or invalidity of an argument. We will show this for an SL argument, but similar reasoning may be given for a TL, a PL, or an English argument.

Let **Γ/X** be an SL argument and **Y** a logically true SL sentence. We prove first that if **Γ/X** is deductively valid and **Y** is a premise in **Γ**, then removing **Y** from **Γ** does not make the argument **Σ/X** deductively invalid,[1] where **Σ** is the set obtained from **Γ** by removing **Y** from it. Take V to be an SL truth valuation for the argument **Σ/X**, such that V is a model of **Σ**. If V is not relevant to **Y**, that is, if there are atomic components of **Y** that do not occur in **X** or in any member of **Σ**, we extend V to V* by assigning arbitrary truth values to these atomic sentences of **Y**. V* does not change any of the truth-value assignments that V makes. Since **Y** is logically true, it is true on every truth valuation for it; thus it is true on V*. **Γ** is the set **Σ** plus **Y**; therefore V* is a model of **Γ** as well. But **Γ/X** is deductively valid; so **X** is true on V* (see 3.5.3a). This entails

1 If **Y** is not in **Γ**, adding **Y** to **Γ** does not make the new argument invalid whether **Y** is logically true or not. We discussed this case for informal arguments in the Solution to 1.3.1k. The reasoning is similar for symbolic arguments of classical logic.

that **X** is true on V because, as we said above, V* does not change any of the truth-value assignments that V makes. Hence every truth valuation for Σ/**X** that satisfies Σ makes **X** true. We conclude that the argument Σ/**X** is deductively valid.

Now we prove that if Γ/**X** is deductively invalid and **Y** is not a member of Γ, then adding **Y** to Γ does not make the argument Σ/**X** deductively valid,[1] where Σ is the set obtained from Γ by adding **Y** to it. Since Γ/**X** is deductively invalid, there is a truth valuation V that satisfies Γ and makes **X** false (see 3.5.4). If **Y** contains atomic sentences that do not occur in **X** or in any member of Γ, we extend V to V* by assigning any truth values to those atomic sentences without changing any of the truth-value assignments that V makes. V* is an SL truth valuation for **Y**. But **Y** is logically true. It follows that **Y** is true on V* (see 3.4.5a). Since the set Σ is Γ plus **Y**, V* satisfies Σ. Given that **X** is false on V* (because it is false on V and because V* does not change the truth-value assignments made by V), it follows that V* is a counterexample to the argument Σ/**X**, which means that Σ/**X** is deductively invalid.

SOLUTION TO 3.6.13

3.6.13a

Schematization of the argument

P1 If beliefs are processes that occur in the brain, then they are physical processes.
P2 If beliefs are physical processes, then the belief that God exists is a physical process.
P3 All physical processes can be duplicated by certain machines.
P4 If the belief that God exists is a physical process and all physical processes can be duplicated by certain machines, then it is possible for a machine to duplicate the belief that God exists.
P5 It is possible for a machine to duplicate the belief that God exists only if it is possible for a machine to believe in God.
P6 It is impossible for a machine to believe in God.

C It is not the case that beliefs are processes that occur in the brain.

Translation Key

B: Beliefs are processes that occur in the brain.
P: Beliefs are physical processes.
G: The belief that God exists is a physical process.
M: All physical processes can be duplicated by certain machines.
R: It is possible for a machine to duplicate the belief that God exists.
L: It is possible for a machine to believe in God.

---

1 If **Y** is in Γ, removing **Y** from Γ does not make the new argument valid whether **Y** is logically true or not. 1.3.11 deals with this case for informal arguments. Similar reasoning applies to symbolic arguments.

SL Argument

S1 $B \rightarrow P$
S2 $P \rightarrow G$
S3 M
S4 $(G \wedge M) \rightarrow R$
S5 $R \rightarrow L$
S6 $\neg L$

S7 $\neg B$

As usual, we test for deductive validity by assigning T to the premises and F to the conclusion and performing truth analysis.

$$B^{T} \rightarrow^{T} P^{T} \quad P^{T} \rightarrow^{T} G^{T} \quad M^{T} \quad (G^{T} \wedge^{T} M^{T}) \rightarrow^{T} R^{T} \quad R^{T} \rightarrow^{<T,\,F>} L^{F} \quad \neg^{T} L^{F} \quad / \quad \neg^{F} B^{T}$$

We reach a contradiction. This shows that the SL argument is deductively valid. The schematization is clearly correct and the SL translation is *reasonably* faithful. Hence the original English argument is valid as well (see 3.5.12).

One might ask how the SL translation can be considered reasonably faithful when there are interdependencies between the sentences of the English argument, yet the atomic sentences of the SL argument are independent of each other. For instance, the consequent of P4 is a logical consequence of the conjuncts of its antecedent. The point can be made clear by translating the conjuncts of the antecedent of P4 and its consequent into TL and representing this conditional as a standard TL argument. Thus let '*g*' stand for 'the belief that God exists', '*P*' for 'physical processes', and '*D*' for 'things that can be duplicated by certain machines'. The conditional P4 could be rendered as the following TL argument.

S8 *g* is *P*
S9 all *P* are *D*

S10 *g* is *D*

S10 says in English that the belief that God exists can be duplicated by a certain machine; this is a paraphrase of the consequent of P4. TL captures the interdependency between the consequent of P4 and the conjuncts of its antecedent, while SL translates these three sentences as three atomic sentences (*G*, *M*, and *R*), which are logically independent of each other. The SL argument allows for the possibility that *R* is false and *G* and *M* are true, but the English argument does not allow for the possibility that the consequent of P4 is false and the conjuncts of its antecedent are true. The TL argument agrees with the English argument: S10 cannot be false while S8 and S9 are true.

This question raises an important point. However, we can still maintain that the SL argument is a reasonably faithful translation of the English argument. In a certain sense it can be said that SL "captures" the dependency of the consequent of P4 on its antecedent by simply translating P4 as the material conditional S4. It is true that the consequent of P4 logically follows from its antecedent, and it is true that the consequent of S4 does not logically follow from its antecedent;

but observe that any truth valuation that makes the premises of the SL argument true can never assign T to the antecedent of S4 and F to its consequent. The logical status of the SL argument is determined solely by the SL truth valuations that are relevant to the argument and that make its premises true. An SL truth valuation on which some of the premises of the SL argument are false is irrelevant to the logical status of the SL argument. If there is an SL truth valuation on which S1–S6 are true and S7 is false, the SL argument would be deductively invalid (see 3.5.4); if there is no such truth valuation, the SL argument is valid (see 3.5.3b). Thus SL guarantees that there are no circumstances that are relevant to the logical status of the SL translation and under which *G* and *M* are true and *R* is false. SL truth valuations that make use of the lack of interdependency between the consequent and antecedent of S4 are truth valuations that make *R* false and *G* and *M* true. These truth valuations make S4 false; hence they are irrelevant to the logical status of the SL argument, because they fail to make all the premises true. Thus while indeed SL renders the consequent of S4 logically independent from its antecedent, this independence is not allowed to play a role in determining the logical status of the SL argument.

The important point that the question raises is really an objection to the English argument itself. The argument contains logically true premises, which, as explained in the solution to 3.6.11b, are irrelevant to the logical status of the argument. P4 is one such premise and P2 is another. We demonstrated that the consequent of P4 is a logical consequence of its antecedent. This shows that P4 is logically true; for it is a theorem of (informal) classical logic (and of many other logical systems) that {X, Y} ⊨ Z if and only if the conditional 'If (X and Y) then Z' is logically true. To see that P2 is also logically true, we can render it as an argument: all beliefs are physical processes; therefore the belief that God exists is a physical process (since, by definition, the belief that God exists is a belief). TL can capture this argument with the help of a meaning-postulate. Let '*g*' stand for 'the belief that God exists', '*B*' for 'beliefs', and '*P*' for 'physical processes'. The TL translation of the argument based on P2 is: *g* is *B*; all *B* are *P*; therefore, *g* is *P*. So the important point raised by the question is that the English argument is not economical: it contains superfluous premises, which can be deleted from the argument without altering the logical status of the argument.

If we keep only the essential premises, the English argument would look like this.

P1 If beliefs are processes that occur in the brain, then they are physical processes.
P3 All physical processes can be duplicated by certain machines.
P5 It is possible for a machine to duplicate the belief that God exists only if it is possible for a machine to believe in God.
P6 It is impossible for a machine to believe in God.

C It is not the case that beliefs are processes that occur in the brain.

We can informally demonstrate the deductive validity of this argument by using an indirect proof. We assume that the argument is deductively invalid and show that this assumption leads to a contradiction. So we assume that there is a relevant logical possibility q in which the premises are all true and the conclusion is false. This is the Reductio Assumption. Since C is false in q, beliefs are processes that occur in the brain. Given that P1 is true in q, it follows that beliefs are physical processes. But the belief that God exists is, by definition, a belief; hence in q this belief is a physical process. According to P3, all physical processes can be duplicated by certain machines; therefore in q the belief that God exists can be duplicated by a certain machine. P5 entails that it is possible for a machine to believe in God, since in q a certain machine can duplicate the belief that God exists. This contradicts P6, which asserts that it is impossible for a machine to believe in God. Therefore we reject the Reductio Assumption and conclude that there is no logical possibility in which P1, P3, P5, and P6 are all true and C is false. This proves that the English argument is deductively valid.

The real divergence between the original English argument and its SL translation is that the SL argument contains no superfluous premises: all of S1–S6 are needed for the conclusion (S7) to follow. Unlike the original argument, none of the SL premises is logically true, and this is due to SL's inability to see certain interdependencies that exist between some of the sentences in the original English argument. But given the original formulation of the English argument, SL produced an argument that captured the deductive validity of that argument. However, no amount of analysis in SL can reveal that peculiar feature of the English argument—namely that it is not economical. Since the SL argument fails to capture fully all the interdependencies that exist in the English argument, we did not describe the SL translation as "totally faithful"; but since this failure does not influence the logical status of the SL argument, we described the SL translation as "reasonably faithful." Thus there is no divergence between the logical status (valid or invalid) of the SL argument and the logical status of the original English argument. On the other hand, an SL translation of the second economical argument is invalid, because it severely distorts the logical structure of the argument. As we will see later, Predicate Logic (PL) will be a suitable logic for capturing almost all the logical features of the English argument. We say "almost all" because PL, like TL, also needs the aid of the meaning-postulate stated above.

Some of the English arguments in this chapter contain logically true premises. These premises are irrelevant to the deductive validity of the English arguments but they are essential for the validity of their SL translations. In fact, they were included in the English arguments in order to render their SL translations suitable for determining the deductive validity of the original arguments. When we encounter such arguments, we will not reiterate this analysis but we will describe the SL translations as "reasonably faithful." This should alert the reader that the English argument contains superfluous premises.

Solution to 3.6.14

3.6.14b

Schematization of the Argument

P1 We cannot help but believe in the truth of what we perceive (or understand) clearly and distinctly.

P2 We cannot help but believe in the truth of what we perceive clearly and distinctly only if it is in our nature to believe in the truth of what we perceive clearly and distinctly.

P3 God created our nature.

P4 If God created our nature, and if it is in our nature to believe in the truth of what we perceive clearly and distinctly, then God has created us such that we cannot help but believe in the truth of what we perceive clearly and distinctly.

P5 Either all of our beliefs in the truth of what we perceive clearly and distinctly are knowledge, or some of such beliefs can be doubted.

P6 If God has created us such that we must believe in the truth of what we perceive clearly and distinctly, and yet some of our beliefs in the truth of what we perceive clearly and distinctly can be doubted, then He has created us with a nature that deceives us.

P7 Unless God is a deceiver, He would not have created us with a nature that deceives us.

C All of our beliefs in the truth of what we perceive clearly and distinctly are knowledge.

We paraphrased some of the premises to make clear the logical relations between them. None of our rewordings is problematic; for example, we spelled out what 'such beliefs' stands for in P2 and we substituted 'can be doubted' for 'are not indubitable' in P6.[1] Our paraphrase of the conclusion, perhaps, calls for some explanation. If we know something, then our belief in the truth of that thing is knowledge, and vice versa. We are using the word 'thing' here to mean "statement"; thus, for instance, to know that blood is red is to know that the statement that blood is red is true. Thus our paraphrase of the conclusion presupposes that saying that one knows something is equivalent to saying that one's belief in the truth of that thing is knowledge. This is a reasonable presupposition, since knowledge is nearly universally assumed to require belief. Our paraphrase brings out the logical relation between the conclusion and the fifth premise.

Translation Key

B: We cannot help but believe in the truth of what we perceive clearly and distinctly.
N: It is in our nature to believe in the truth of what we perceive clearly and distinctly.
G: God created our nature.
L: God has created us such that we cannot help but believe in the truth of what we perceive clearly and distinctly.
K: All of our beliefs in the truth of what we perceive clearly and distinctly are knowledge.
S: Some of our beliefs in the truth of what we perceive clearly and distinctly can be doubted.
D: God has created us with a nature that deceives us.
R: God is a deceiver.

SL Argument

S1 B
S2 $B \rightarrow N$
S3 G
S4 $(G \wedge N) \rightarrow L$
S5 $K \vee S$
S6 $(L \wedge S) \rightarrow D$
S7 $\neg R \rightarrow \neg D$

S8 K

---

1 'Are not indubitable' literally means "are not incapable of being doubted," which has the same meaning as 'are capable of being doubted', which, in turn, is a paraphrase of 'can be doubted'.

We follow the standard strategy of assigning T to the premises and F to the conclusion and performing truth analysis.

$$B^{T} \quad B^{T}\rightarrow^{T}N^{T} \quad G^{T} \quad (G^{T}\wedge^{T}N^{T})\rightarrow^{T}L^{T} \quad K^{F}\vee^{T}S^{T} \quad (L^{T}\wedge^{T}S^{T})\rightarrow^{T}D^{T} \quad \neg^{F}R^{T}\rightarrow^{T}\neg^{F}D^{T} \quad / \quad K^{F}$$

We recover, up to equivalence, only one truth valuation that makes the premises true and the conclusion false. Here is this truth valuation: V(B) = T, V(D) = T, V(G) = T, V(K) = F, V(N) = T, V(R) = T, V(S) = T, and V(L) = T. The SL argument is deductively invalid (see 3.5.4).

Since V is, up to equivalence, the only counterexample to the SL argument, the original English argument is most likely valid if V fails to generate a counterexample to it (see 3.5.12:C3). To see what sort of story V tells in English, we use our translation key to read V in English. Here is the story based on V: some of our beliefs in the truth of what we perceive clearly and distinctly can be doubted, and hence, according to the Cartesian test described in the passage, not all of our beliefs in the truth of what we perceive clearly and distinctly are knowledge; however, it is true that we cannot help but believe in the truth of what we perceive clearly and distinctly and, this implies, it is in our nature to have such beliefs; furthermore, God created our nature and so God has created us such that we cannot help but believe in the truth of what we perceive clearly and distinctly; but this means that God has created us with a nature that deceives us, and hence God is a deceiver.

The problem with this story is that it flies in the face of the traditional monotheistic conception of God. According to this conception, God is the holder of all perfections: He is benevolent, omnipotent, omniscient, and so on. It contradicts this conception to say that God is a deceiver. But there is no premise in the schematization that suggests we are presupposing the traditional monotheistic conception of God. The story told by V in English is a counterexample to the schematized argument. The V story, however, is *not* a counterexample to the original argument (i.e., the argument in the passage). The passage supplies an historical context for the argument. It is a version of an argument proposed by the French philosopher René Descartes. Descartes is not an obscure figure. Much about him, his ideas, and his life are well known. We know, for instance, that he was a theist and a Catholic. There is nothing in his written work that suggests that his conception of God radically differs from the standard monotheistic conception. The historical context supplied in the passage is evidence that the argument presupposes a traditional conception of God. Given this evidence, the Principle of Charity directs us to add to the schematized argument the following missing premise:

P8: God is not a deceiver.

The revised SL translation has a new premise, S8, and its conclusion is now S9.

S8: $\neg R$
S9: K

We said above that the original argument is most likely deductively valid if V fails to generate a counterexample to the original argument, because V is the only counterexample to the SL argument. We show that the original argument is indeed deductively valid by showing that its revised SL translation is valid. We assign T to the premises and F to the conclusion and perform truth analysis.

$B^{T} \quad B^{T} \rightarrow^{T} N^{T} \quad G^{T} \quad (G^{T} \wedge^{T} N^{T}) \rightarrow^{T} V^{T} \quad K^{F} \vee^{T} S^{T} \quad (V^{T} \wedge^{T} S^{T}) \rightarrow^{T} D^{T}$

$\neg^{T} R^{F} \rightarrow^{\langle T, F \rangle} \neg^{F} D^{T} \quad \neg^{T} R^{F} \quad / \quad K^{F}$

We obtain a contradiction. This shows that there is no relevant SL truth valuation on which the premises of the (revised) SL argument are true and the conclusion is false. The SL argument is deductively valid (see 3.5.3b). We argued for the correctness of the revised schematization, and there is no reason to doubt the faithfulness of the SL translation. The original English argument, therefore, is deductively valid.

SOLUTION TO 3.6.16

Schematization of the Argument

P All horses are animals.

C All heads of horses are heads of animals.

Translation Key

H: Something is a horse.
A: Something is an animal.
R: Something is a head of a horse.
M: Something is a head of an animal.

SL Argument

S1: $H \rightarrow A$

S2: $R \rightarrow M$

Translating 'All horses are animals' as $H \rightarrow A$ and 'All heads of horses are heads of animals' as $R \rightarrow M$ severely distorts the original argument, but it is the best SL can do. There is no logical relation between H and A and between R and M. But according to the English argument the "thing" that is a horse is the same "thing" that is an animal and the "thing" that is a head of a horse is the same "thing" that is a head of an animal. SL cannot capture these interdependencies because SL has no linguistic means to express general terms such as 'horse' and 'animal' and quantifiers such as 'all' and 'some'. Recall that the SL worldview does not recognize individuals and properties as basic ingredients of reality. Furthermore, there is no relation between R and H and between M and A. But the English expressions 'horses' and 'heads of horses' and the expressions 'animals' and 'heads of animals' are related. SL cannot express relations such as the relation of something being the head of something. Again, the SL worldview does not include relations among its basic ingredients of reality.

It is obvious that the English argument is deductively valid and that the SL argument is deductively invalid (consider the truth valuation that assigns T to H, A, and R and F to M). However, no counterexample to the SL argument represents a counterexample to the English argument. All the SL counterexamples, by definition, make the premise, S1, true and the conclusion, S2, false. Since S2 is the conditional, $R \rightarrow M$, it can be made false only by making R true and M false. Using the translation key to read these counterexamples in English, we obtain stories in which there is a head α of a horse β that is not the head of any animal. But since the SL counterexamples make S1 (the premise of the SL argument) true, we expect the English stories based on these SL counterexamples to make P (the premise of the English argument) true as well.[1] But P says that all horses are animals. Since β is a horse, β must be an animal too. α, therefore, is a head of an animal after all. Any such story, thus, contains a contradiction: α is not the head of any animal and α is the head of an animal (namely, β). It follows that none of the stories based on the SL counterexamples is a logical possibility. In other words, none of the truth valuations that are counterexamples to the SL argument represents a logical possibility that is relevant to the English argument; these SL truth valuations are all superfluous.

One might suggest adding a meaning-postulate to the SL argument in order to improve the translation. This meaning-postulate would have to be $(H \rightarrow A) \rightarrow (R \rightarrow M)$. But this is quite unreasonable. This meaning-postulate is simply the argument itself, stated as a conditional. It seems that the only way to improve the SL translation is to express the deductive validity of the argument itself as a premise of the SL argument.

As we saw in the Solution to 2.6.10, TL is also inadequate to establish the deductive validity of this simple English argument. This argument requires a logical system that has the linguistic wherewithal to express properties, relations, and quantifications. The symbolic system we study in Chapters Four and Five has such resources. It is the system of Predicate Logic (PL).

1 If any such story makes P false, then it does not constitute a counterexample to the English argument.

Chapter Four

Predicate Logic (PL)

4.1 The PL Worldview

Predicate Logic (PL)[1] is the most important and the most studied modern symbolic system. In some sense, it includes Term Logic and extends Sentence Logic. Although it has, when compared to natural languages, severe expressive limitations, its expressive power far exceeds those of TL and SL, and its resources for constructing complex, yet extremely precise, declarative sentences go far beyond anything typically available for making similar sentences in natural languages. The **PL worldview** specifies **individuals**, **properties**, and **relations** as the **basic ingredients of reality**. As in the TL and the SL worldviews, every combination of basic ingredients of reality constitutes a **logical possibility** according to this worldview. However, there is a restriction: every combination of basic ingredients of reality must contain at least one individual and the relation of identity. According to the PL worldview, the relation of identity is **token identity**—every individual bears this relation to itself and no distinct individuals bear this relation to each other. In 2.1, we defined the **extension of a property** in some logical possibility as the set that consists of all the individuals that instantiate the property in that logical possibility, and we said that an individual **instantiates** a property if and only if it has this property. Relations relate certain numbers of individuals to each other. For example, the relation "being a sister of" relates pairs of individuals to each other in a specific order. A relation that relates n individuals to each other is referred to as an **n-place relation**. A sequence of n individuals in some specific order is called **n-tuple**. For instance, the 3-tuple ⟨a, b, c⟩ is an ordered triple consisting of the first three letters of the English alphabet in their standard order. The **extension of an n-place relation** in some logical possibility is the set that consists of all the n-tuples of individuals that bear this relation to each other in that logical possibility. Unlike the TL worldview, the PL worldview permits properties and relations to have empty extensions. However, the relation of identity cannot have an empty extension in any logical possibility, since every logical possibility must contain at least one individual and every individual bears the relation of identity to itself. The PL worldview imposes two restrictions on the language that is used to make assertions about the constituents of a logical possibility: first, every singular term must refer to a unique individual,[2] and

---

1 Predicate Logic is also called "First-Order Logic" and "Quantificational Logic."

2 As defined in 2.3.2, a singular term of some natural language is a word or a phrase that stands for exactly one individual and could not stand for more than one individual without being ambiguous. Proper names, such 'America', and definite descriptions, such as 'The country that dropped two nuclear bombs on Japan', are, according to this definition, singular terms *if they have referents*. Proper names, such as 'Pegasus', and definite descriptions, such as 'The present king of France', that have no referents are not classified, on the basis of this definition, as singular terms. If such non-referring expressions are allowed to count as singular terms, they could not be translated into PL, since PL worldview presupposes that every singular term refers to a unique individual.

second, every individual must have a name. The first restriction is a standard feature of PL and the second is a feature of the particular approach to PL that we will follow in this book.[1]

4.1:C COMMENTARY ON 4.1

4.1:C1 Because of the richness of the PL worldview, possible situations could be analyzed at varied levels of detail. We will illustrate this type of analysis with an example and then will address how to interpret these different levels of analysis.

Consider the example of the two cats that was discussed in 3.1:C. We have two sentences:

S1 Tom is on the mat.
S2 Sylvester is *not* on the mat.

where Tom and Sylvester are two tomcats. According to the SL worldview, the relevant possible situations are analyzed in terms of states of affairs. Let us revisit the SL analysis briefly in order to contrast it with the types of analyses PL provides. We may take the states of affairs to be Tom's being on the mat and Sylvester's being on the mat. Four logical possibilities could be constructed from these two states of affairs.[2]

q1 Tom is on the mat and Sylvester is on the mat.
q2 Tom is on the mat and Sylvester is not on the mat.
q3 Tom is not on the mat and Sylvester is on the mat.
q4 Tom is not on the mat and Sylvester is not on the mat.

It is obvious that S1 is true in q1 and q2 and false in q3 and q4, and S2 is true in q2 and q4 and false in q1 and q3.[3] We have seen in Chapter Three that the extra-logical vocabulary of SL con-

---

1 This particular approach will be explained when the semantics of PL is discussed. To look ahead, however, the approach is called *substitutional quantification* and will be explained in 4.4.1:C5.

2 In all the descriptions of logical possibilities that we will consider in this chapter, if we do not explicitly affirm or implicitly presuppose a certain *relevant* fact about the constituents of a logical possibility, then we are assuming that it is not the case. For instance, since we do not explicitly affirm or implicitly presuppose that the mat is on Tom or that Sylvester is on Tom, then we are assuming that neither the mat nor Sylvester is on Tom. (However, for the sake of clarity, we will explicitly describe the relations that obtain or fail to obtain between the cats and the mat.) The preceding general claim cannot be stated precisely or accurately. It is not clear at all what it means to say that something is implicitly presupposed or what a relevant fact is. Tom's being a cat is presupposed and its being on the mat is explicitly stated. But what about Tom's having long hair? It is neither explicitly affirmed nor implicitly presupposed; should we conclude that Tom is a shorthaired cat? But the latter is also neither explicitly affirmed nor implicitly presupposed. Perhaps we should not make any assumptions about the length of Tom's hair, since his being longhaired or shorthaired is an irrelevant fact to the situation at hand. But, again, the notion of a relevant fact is extremely vague. We shall not attempt to resolve any of these difficulties here. The situation will relatively improve in PL, when we invoke a "universe of discourse" and extensions in the description of a logical possibility. The universe of discourse determines what the relevant individuals are and the extensions of properties and relations clearly specify which individuals have the properties or relation and which do not.

3 In all the groups of logical possibilities listed in this section, S1 is true in the first and second logical possibilities and false in the third and fourth, and S2 is true in the second and fourth logical possibilities and false in the first and third.

sists of sentence letters since its worldview posits states of affairs as the basic ingredients of reality. So let the sentence letter *A* translate the English sentence 'Tom is on the mat' and *B* translate 'Sylvester is on the mat'. Logical possibilities in SL are characterized as truth valuations. Since there are two sentence letters, there are, up to equivalence, four truth valuations, each of which represents one of the logical possibilities q1–q4.

V1 V1(*A*) = T and V1(*B*) = T
V2 V2(*A*) = T and V2(*B*) = F
V3 V3(*A*) = F and V3(*B*) = T
V4 V4(*A*) = F and V4(*B*) = F

In PL, we can give "deeper" analyses. For instance, we can analyze the possibilities that are relevant to {S1, S2} in terms of individuals and a property. The relevant individuals are the two cats and the mat, and the relevant property is "being on the mat." We use the letter '*M*' to refer to this property. This analysis also gives us (at least) four logical possibilities.[1]

p1 Tom and Sylvester have the property *M*.
p2 Tom has the property *M* but Sylvester does not.
p3 Sylvester has the property *M* but Tom does not.
p4 Neither Tom nor Sylvester has the property *M*.

As we will see later, logical possibilities in PL are characterized as PL *interpretations*. We will study the notion of a PL interpretation with some detail in 4.4.1. However, we give a brief description of it here. Every PL interpretation comes with a *universe of discourse*, which is a set of individuals, and *semantical assignments* that attach extensions to PL predicates. PL predicates are symbols that stand for properties and relations. The four logical possibilities p1-p4 can be represented by infinitely many PL interpretations.[2] We consider only four of them, which closely resemble the four logical possibilities described above. These four interpretations have the same universe of discourse; it consists of the two cats, Tom and Sylvester, and the mat, because these are the individuals mentioned in S1 and S2. Each one of these interpretations assigns a certain extension to the predicate '*M*' (it is a predicate because it designates a property). The symbolic expression 'J(*M*)' stands for the extension that the PL interpretation J assigns to the predicate '*M*'. In general terms, the extension of a predicate is the extension of the property or the relation that is designated by this predicate. Here are the four interpretations. (We use ':' as our metalinguistic symbol for the relation of identity in order to reserve the traditional identity predicate '=' for the language of PL.)

J_1 $J_1(M)$: {Tom, Sylvester}
J_2 $J_2(M)$: {Tom}
J_3 $J_3(M)$: {Sylvester}
J_4 $J_4(M)$: ∅ (the empty set)

1 As stated in note 2, page 180, by not explicitly affirming or implicitly presupposing that the mat has the property *M*, we are assuming that this mat does not have the property *M*.
2 In 4.5.13:C1, we will give a relatively precise definition of the relation of representation that holds between PL interpretations and logical possibilities.

We can give an even "deeper" analysis. We can analyze the situation in terms of individuals and a relation. The relevant individuals, again, are the two cats and the mat, and the relevant relation is the relation "being on." We use '*O*' to designate this relation. So to say that Tom is on the mat is to say that Tom bears the relation *O* to the mat and to say that Sylvester is not on the mat is to say that Sylvester does not bear the relation *O* to the mat. This analysis, too, gives us (at least) four logical possibilities.

p5 Each of Tom and Sylvester bears the relation *O* to the mat.
p6 Tom bears the relation *O* to the mat but Sylvester does not.
p7 Sylvester bears the relation *O* to the mat but Tom does not.
p8 Neither Tom nor Sylvester bears the relation *O* to the mat.

O is a *binary relation*, that is, it can relate two things to each other. The extension of a binary relation, if it is nonempty, consists of *ordered pairs* (i.e., 2-tuples). As its name indicates, an ordered pair is made of two things in a certain order. An ordered pair whose components are x and y, in this order, is designated as $\langle x, y\rangle$. For example, the ordered pair ⟨Bethany, Anna⟩ is in the extension of the binary relation "taller than" if and only if Bethany is taller than Anna. In general, the ordered pair $\langle x, y\rangle$ is in the extension of the binary relation R if and only if x bears the relation R to y. Using this terminology, we can represent these logical possibilities as four PL interpretations each of which consists of the universe of discourse {Tom, Sylvester, the mat} and an extension for the predicate '*O*'.

J_5 $J_5(O)$: {⟨Tom, the mat⟩, ⟨Sylvester, the mat⟩}
J_6 $J_6(O)$: {⟨Tom, the mat⟩}
J_7 $J_7(O)$: {⟨Sylvester, the mat⟩}
J_8 $J_8(O)$: ∅

Could we analyze further? We can identify other *relevant* ingredients. We omitted one obvious relation—the relation of identity, which is traditionally designated by the symbol '='. But there are others; for example, there are the properties "being a cat" and "being a mat." Let '*C*' and '*D*' stand for these two properties, respectively. The logical possibilities p9–p12 are described as p5–p8, respectively, with the addition of the following clauses to each description: every individual is identical only with itself; Tom and Sylvester have the property *C*; and the mat has the property *D*. We represent the logical possibilities p9–p12 as the PL interpretations J_9–J_{12}, which are obtained from J_5–J_8, respectively, by adding to each of them the appropriate extensions for '=', '*C*', and '*D*'. If n is 9–12, we have:

$J_n(=)$: {⟨Tom, Tom⟩, ⟨Sylvester, Sylvester⟩, ⟨the mat, the mat⟩}
$J_n(C)$: {Tom, Sylvester}
$J_n(D)$: {the mat}

4.1:C2 It seems warranted to make explicit these additional ingredients—namely, =, *C*, and *D*. The sentences S1 and S2 are about two cats and a mat and about whether these cats are on the mat or not. But couldn't we also include other properties, such as "being an animal" and "having eyes," that the cats have, and properties, such as "being on the floor" and "being made of fiber," that the mat might have, and relations, such as "being heavier than," that might hold between these individuals? And since the PL worldview allows properties and relations to have empty extensions, couldn't we include, in the relevant logical possibilities, properties and relations, such as the property "being a vegetable" and the relation "being a sister of," that do not hold for any individuals in these logical possibilities, and then assign to them the empty extension ∅?[1]

We explained in 1.2.1:C2 that logical possibilities are almost always entertained in relation to some set of declarative sentences. These logical possibilities contain objects, such as individuals, properties, and relations, about which the sentences make assertions. We said there that if the set of sentences is Σ, these logical possibilities are described as *relevant to* Σ, or simply as *for* Σ. This entails that we should consider only logical possibilities that contain individuals, properties, and relations that are mentioned in the sentences S1 and S2. (Recall that S1 is the sentence 'Tom is on the mat' and S2 is the sentence 'Sylvester is not on the mat'.) Thus the constituents of the logical possibilities that are relevant to {S1, S2} must contain the following objects: the two cats Tom and Sylvester, the mat, an extension for the relation of identity, and either an extension for the property "being on the mat" or an extension for the relation "being on," or both.[2] However, according to the PL worldview, not every logical possibility for {S1, S2} must consist of only these objects. This point is clear, since the definition of a logical possibility for some set of declarative sentences requires that every sentence in the set make an assertion about *some* or all of the constituents of the logical possibility. So the answer to the questions above is that logical possibilities that are relevant to the set {S1, S2} could contain a large variety of constituents in addition to the objects mentioned in S1 and S2. However, most of these constituents are irrelevant to the truth values of S1 and S2 in these logical possibilities. What matters is whether Tom or Sylvester is on the mat or not. Having short or long hair, being fat or skinny, being old or young, and so on have no bearing on the truth values of S1 and S2. Thus in considering logical possibilities that are relevant to a certain set of sentences, we typically focus on those possibilities whose constituents are precisely the objects mentioned in the members of the set.

The constituents identified above are not independent of each other. For instance, the property "being on the mat" is analyzable in terms of the relation "being on" and the mat. From another perspective, however, whether a cat bears the relation "being on" to the mat is reducible to whether this cat has the property "being on the mat." Let S3 be the biconditional 'Tom has the property "being on the mat" if and only if Tom bears the relation "being on" to the mat'. Any logical possibility that is relevant to the set {S1, S2, S3} must include at least the two cats, the mat, the property "being on this mat", and the relation "being on," since these are the constituents about which the sentences S1–S3 make assertions. In all of these logical

1 As indicated in Chapter Two, there is only one empty set; hence there is only one empty extension.

2 Strictly speaking, this claim is not true. It is possible to "carve" the sentences S1 and S2 in multiple ways. For instance, we can think of S1 and S2 as making assertions about the two cats, the mat, and the relation that can be expressed as "x has y on it." Describing this relation is a little awkward, but it is a genuine relation, which may hold between the mat and the cats. A possible extension of this relation is {⟨the mat, Tom⟩, ⟨the mat, Sylvester⟩}. According to this extension, S1 is true and S2 is false. Another possible way of "carving" these sentences is to think of them as making assertions about the two cats, the mat, and two properties "having Tom on it" and "having Sylvester on it." If we want S1 and S2 to be true, we should assign the extension {the mat} to the first property and the empty extension to the second property. This shows that in most cases there is no *unique* set of objects about which a sentence is making an assertion.

possibilities, S3 is true. Hence S3 is logically true. It is not hard to see why it is logically true. Each of the left-hand and the right-hand sides of the biconditional S3 is a paraphrase of the sentence that Tom is on the mat. So S3 is equivalent to the biconditional 'Tom is on the mat if and only if Tom is on the mat', and the latter is clearly logically true.

As explained in 2.1:C3, combinations of basic ingredients of reality should be constrained by interdependencies between properties. The example we discussed in 2.1:C3 concerns the extensions of three properties: "being a Frenchman," "being a man," and "being French." (We are assuming here that the term 'Frenchman' is not used in its old sense meaning a French person.) We said there that there is no possible combination that includes these properties, and in which the extension of the property "being a Frenchman" is not the intersection of the extensions of the properties "being a man" and "being French." However, we also noted there that the TL worldview and TL semantics, which reflects that worldview, take a very liberal attitude towards combinations of basic ingredients of reality: *all* such combinations, without any restrictions, constitute logical possibilities. Similarly, in SL there is no restriction on combinations of basic ingredients of reality, since according to the SL worldview states of affairs, which are the basic ingredients of reality, are assumed to be independent of each other. PL is no exception: *all* combinations of individuals, properties, and relations constitute logical possibilities. This, of course, is counterintuitive. For instance, there should be no possible combination that includes the individuals Tom the cat and the mat, the property "being on the mat," and the relation "being on," and in which Tom is in the extension of the property "being on the mat," but the ordered pair ⟨Tom, the mat⟩ is not in the extension of the relation "being on." In other words, there can be no possible combination of basic ingredients of reality in which Tom the cat instantiates the property "being on the mat" but fails to bear the relation "being on" to the mat. It seems that combinations of basic ingredients of reality must be constrained by certain interdependencies between properties and relations. However, the PL worldview and PL semantics do not impose any such restrictions. As is the case with TL and SL, this liberal attitude toward permissible combinations of basic ingredients of reality will be a source of trouble for PL.

Each of p1–p8 represents a different combination of basic ingredients of reality; thus they are different logical possibilities for {S1, S2}. Different analyses describe different logical possibilities for a given set of declarative sentences. It is clear that there are infinitely many logical possibilities that are relevant to {S1, S2}. However, there are common features to these logical possibilities: (1) they must contain constituents about which S1 and S2 make assertions, (2) they must determine which cat, if any, is on the mat and which cat, if any, is not on the mat. In terms of the truth values of S1 and S2, these logical possibilities divide into four types: those that make S1 and S2 true, those that make S1 true and S2 false, those that make S1 false and S2 true, and those that make S1 and S2 false.

4.1:C3 If *all* the logical possibilities that are relevant to the set {S1, S2} must contain three distinct individuals—Tom the cat, Sylvester the cat, and the mat—then we have some interesting consequences. The following sentences are logically true: (S3) 'Tom is a cat', (S4) 'Sylvester is a cat', and (S5) 'Tom, Sylvester, and the mat are all distinct individuals'. There is a philosophical position that renders all of these sentences logically true. Not all philosophers accept this position, but it should be discussed.

We said in 1.2.1:C2 that in considering the logical possibilities that are relevant to a certain set Σ of declarative sentences, the language of Σ must be held fixed, that is, the linguistic expressions that occur in the members of Σ should have the same meanings across all the logical possibilities that are relevant to Σ. Some philosophers consider the referent of a *proper name*, such as 'Tom' and 'Sylvester', to be part of the meaning of the name (remember that the *referent* of a singular term, whether it is a proper name or not, is the individual to which the term refers). Since the meanings of the expressions that occur in S1 and S2 are invariant across all the relevant logical possibilities, the referents of these proper names must also be held fixed in these possibilities. This entails that Tom the cat and Sylvester the cat inhabit every logical possibility that is relevant to {S1, S2}.[1]

Rendering S3–S5 logically true is not the only implication of this position. If Tom the cat inhabits every logical possibility for {S1, S2}, then the sentence that Tom exists is also logically

---

1 Some philosophers believe that there is nothing to the meaning of a referring proper name other than its referent. Others believe that the meaning of any referring singular term, whether it is a name, such as 'George Washington', or a definite description, such as 'The first President of the United States', is different from its referent. We say 'referring' because there are names, such as 'Santa Claus', and definite descriptions, such as 'The present king of France', that have no referents. The logic of non-referring singular terms is complex and controversial. Luckily, neither Term Logic nor Predicate Logic permits non-referring singular terms. The branch of logical theory that deals with non-referring singular terms is called "free logic." To give a flavor of the difficulties that surround the issue of non-referring singular terms, consider the sentence 'Pegasus exists' (call this sentence G). The name 'Pegasus' does not have a referent in our world but it seems reasonable to assume that it could have a referent in another logical possibility. Intuitively, we want to say that G is *contingently* false in our world. In other words, using the definition of contingency (see 1.2.8), we want to assert that since G is contingent, there are two logical possibilities, p and q, such that G is false in p (Pegasus does not exist) and true in q (Pegasus exits). If we require G to have the same meaning in both p and q, the name 'Pegasus' cannot have different meanings in p and q. But if 'Pegasus' is a proper name and if the referent of a proper name is part of its meaning, then the name 'Pegasus' has a meaning in q that is different from its meaning in p because 'Pegasus' has a referent in q but not in p. This entails that G does not have the same meaning in p and q, which contradicts our previous requirement. There are several responses to this conundrum. One common response is to say that non-referring names are not really proper names but definite descriptions disguised as names, and that the meaning of a definite description has nothing to do with the presence or the type of a referent. For example, 'Pegasus' could be thought of as the definite description 'The winged horse described in Greek mythology'. This view implies that this definite description has a meaning that is independent of whether there is such a winged horse or not. Hence G would have the same meaning, but different truth values, in p and q. A variation on this view holds that definite descriptions have meanings *only* within sentences and, as before, non-referring names are not proper names but definite descriptions. Thus, for example, G should be reformulated as the following sentence: 'There is one and only one individual that is a winged horse as described in Greek mythology'. This sentence has the same meaning in p and q; but it is false in p and true in q. A problem with these two views is that if the term 'Pegasus' has a referent in q, then it is not clear why it cannot be considered a proper name there. But if it is considered a proper name in q, then the semantical category of the term 'Pegasus' is different in p and q: in p it is a definite description and in q it is a proper name. We are no longer sure, therefore, that G has the same meaning in p and q. Difficulties of this sort led certain philosophers to propose that all names, referring and non-referring, must be considered as abbreviations of collections of definite descriptions, and that whether such collections designate certain individuals or not is irrelevant to the meanings of these names. According to this view, the referent of a name is not part of its meaning. So the fact that 'Pegasus' has a referent in q but not in p has nothing to do with the meaning of this name. This ensures that G has the same meaning in p and q. There are other views as well, though the brief discussion here is sufficient to give a feel for the sorts of complexities that surround the issue of non-referring singular terms.

true. In fact, we can assert something stronger. If '*e*' is a referring proper name, the sentence '*e* exists' is logically true, since we assumed that the referent of a proper name is part of its meaning and that the meanings of linguistic expressions are invariant across all the relevant logical possibilities. As we will see later, in PL every sentence that affirms the existence of *s*, where '*s*' is a proper name or a definite description is logically true. This is a consequence of the PL worldview, which requires that every singular term refer to an individual. However, PL does not treat existence as a property. So in PL we cannot attribute existence to *s*, and hence we cannot say '*s* exists' verbatim. We will be able to say something that is close to '*s* exists'. We will be able to say that there exists something that is identical with *s*. The formal semantics of PL, as we will see in due course, attributes a content to the sentence 'There exists something that is identical with *s*' that is different from its content according to the natural semantics described here. Ordinarily, when we say that there exists something that is identical with Tom, we mean, among other things, that there exists a cat that is identical with Tom. But the content that PL formal semantics attributes to the sentence 'There exists something that is identical with *s*' (more precisely, to the PL translation of this sentence) is something akin to saying that there is an individual that is named '*s*'; the individual need not be a cat.

Although this philosophical position requires that Tom and Sylvester have the property "being a cat" in every logical possibility that is relevant to S1 and S2, it allows certain properties and relations of Tom and Sylvester to vary from one logical possibility to another. For example, in p5 Sylvester is on the mat while in p6 Sylvester is *not* on the mat. Why should we, then, assume that Tom and Sylvester have the property "being a cat" in all the relevant logical possibilities? The reason, which many of the philosophers who subscribe to this position give, is that if we allow the cat Tom to be replaced, say, with a dog in a certain logical possibility, then it seems that we have changed altogether the referent of the proper name 'Tom'. By changing the referent of the proper name 'Tom', then, according to this position, we change its meaning. Meanings, though, must be held fixed across all the relevant logical possibilities; thus Tom and Sylvester must have the property "being a cat" in every logical possibility.

This analysis suggests that we distinguish between two types of properties and relation an individual may have. The first type comprises properties and relations that are not part of the "definition" of the individual: their alteration does not change the identity of the individual. The second type comprises properties and relations that are part of the "definition" of the individual: their alteration does change the identity of the individual. Philosophers call the first type of properties and relations "accidental" and the second type "essential." Although not all philosophers believe that there is such a distinction,[1] the distinction between accidental and essential properties and relations has an intuitive appeal. For example, almost all of us would agree that whether I grow a beard or not is irrelevant to my identity. I would be the same individual either way. But almost all of us would also agree that the property of being human is different from the property of being bearded. It seems that it is not possible for me to lose this property and remain the same individual; a logical possibility in which I am replaced with a camel called 'Aladdin Mahmud Yaqub' is a logical possibility in which I

1 In fact, many philosophers think the distinction between essential and accidental properties applies to types of things rather than to individuals. The standard example is that a human being is a rational animal. Thus being rational is an essential property of the type "human." We will discuss types later in this section. But in this passage we are talking about essential and accidental properties of individuals.

ceased to exist and in which the proper name 'Aladdin Mahmud Yaqub' has a different referent. Hence "being bearded" is one of my accidental properties while "being human" is one of my essential properties.

It is a coherent, though restrictive, philosophical position to presuppose that individuals have essential as well as accidental properties and relations, that altering some of the essential properties and relations of an individual changes its identity, and that the referent of a (referring) proper name is part of the meaning of the name. This entails that if α is an individual that is the referent of a *proper name* occurring in a member of Σ, then none of the essential properties or relations of α may be altered in any of the logical possibilities that are relevant to Σ.

Let us now return to the set {S1, S2}, where, as before, S1 is the sentence 'Tom is on the mat' and S2 is the sentence 'Sylvester is not on the mat'. Given the philosophical position stated above, the essential qualities of Tom, Sylvester, and this mat are maintained across the logical possibilities for the set {S1, S2}. If there are essential properties, then surely "being a cat" is an essential property of Tom and Sylvester and "being a mat" is an essential property of the mat; also the fact that the relation of identity does not hold between any two of these individuals is also essential to them. However, the relation of "being on" is an accidental relation between the cats and the mat. Whether a cat is on a mat or not does not usually alter the identity of the cat. Given the definitions of logically true and contingent sentences (see 1.2.6 and 1.2.8), the sentences (S3) 'Tom is a cat', (S4) 'Sylvester is a cat', (S5) 'The mat is a mat', and (S6) 'Tom, Sylvester, and the mat are all distinct individuals' are logically true, while the sentences S1 and S2 are contingent.

Some philosophers would dispute these claims. They might either deny the distinction between essential and accidental properties and relations, or argue that altering essential properties is *metaphysically impossible* but not necessarily *logically impossible*. For example, if one considers "being a mammal" an essential property of whales and "not being a mammal" an essential property of fish, then it is *impossible* that some whales are fish. The standard position is that this impossibility is not logical but *metaphysical*. We need not be concerned with the precise meaning of 'metaphysical' here. We only need to have a general understanding of the distinction between logical and metaphysical possibility. Both notions presuppose that the language of the sentences in question is held fixed across all relevant possibilities. We said in Chapter One that something is logically possible if and only if the supposition of its truth or of its existence does not lead to a contradiction. On the other hand, something is metaphysically possible if and only if the supposition of its truth or of its existence does not lead to a contradiction and does not alter any of the essential properties and relations that the relevant individuals have. Thus, by definition, something is metaphysically possible only if it is logically possible. In other words, every metaphysical possibility is a logical possibility. Philosophers who maintain the distinction between logical and metaphysical possibilities deny the converse; they say that not every logical possibility is a metaphysical possibility, that is, there are things that are logically possible but metaphysically impossible. A defender of this view might say that the sentence 'Some whales are fish', while metaphysically impossible, is not logically impossible; for there is no contradiction which results from assuming that it is true. However, there are philosophers who collapse the notions of logical possibility and metaphysical possibility into one notion. We will discuss this position in the following subsection.

This is another place where we have to make a philosophical commitment. We need to decide whether the referent of a proper name is part of its meaning, and whether logical possibility is the same as metaphysical possibility or if it is a less restrictive notion. We will follow the traditional view of distinguishing between logical and metaphysical possibility: all metaphysical possibilities are logical possibilities but not all logical possibilities are metaphysical possibilities. We also adopt the less restrictive position of not including the referent of a proper name in the meaning of the name.

4.1:C4 Finally, we will take a closer look at the requirement that the meanings of the expressions occurring in the members of some set Σ of natural-language declarative sentences must be held fixed across all the logical possibilities that are relevant to Σ. If the notion of meaning invoked in this requirement is not adequately clarified, the distinction between logical and metaphysical possibility might be lost and the condition of consistency for logical possibility might not be satisfied. In order to explain these claims, we need to examine first the scope of the meaning of an expression and second the *conceptual content* of an expression.

Philosophers who believe that individuals have essential and accidental properties and relations usually hold the further view that *types* also have essential and accidental properties and relations. It is not easy to define what a type is; but we can give a rough account of this notion. A type *T* may be thought of as an ordered pair ⟨Q, E⟩, where Q is a property or a relation and E is the extension of Q. We say that Q is the property or relation that *defines* the type *T* and E is the extension of *T*. Since in this book we deal only with bivalent languages, we extend bivalence to all the properties and relations we allow. A property is bivalent if and only if every individual either has this property or lacks it but it cannot both have it and lack it at the same time. An n-place relation is bivalent if and only if every sequence of n individuals either has this relation or lacks it but it cannot both have it and lack it at the same time. In 4.1, we gave a brief explanation of what we mean by 'n-place relation'. Let us revisit this issue here. Relations can hold between two individuals as when we say that Mary is taller than John, or between three individuals as when we say that four is greater than three but less than five, or between four individuals as when we say that New York is closer to Philadelphia than San Francisco is to Dallas, or, in general, between any number of individuals. If a relation R can hold between n individuals, we say that R is an n-place relation. In this book, we will require the property or relation that defines a type to be bivalent. Here are a few examples of types: *W* is the type whose defining property is "being a whale" and whose extension is the set of all whales; *S* is the type whose defining relation is "being a sister of" and whose extension is the set of all ordered pairs ⟨*x*, *y*⟩ where *x* is a sister of *y*; and *M* is the type whose defining relation is "sharing borders with country __ and country ..." and whose extension is the set of all ordered triples ⟨*x*, *y*, *z*⟩ where *x*, *y*, and *z* are countries such that *x* shares borders with *y* and *z*.

When we say that a type *T* has a certain property, we mean that every member of the extension of *T* has this property, and when we say that *T* has a certain relation, we mean that the members of the extension of *T* have this relation to each other or to members of another type. Many philosophers affirm that individuals *and* types have essential and accidental properties and relations. We argued previously that attributing essential and accidental properties to individuals is intuitively appealing. The same is true of types. Consider an example that is commonly used by philosophers to argue that water is *essentially* composed of H_2O molecules. Water has many properties. One of them is that it is the main ingredient in carbonated beverages. What would happen to water if the manufacturers of carbonated beverages were to change the ingredients of these drinks, and water would no longer be an ingredient in carbonated beverages? Most likely nothing important would happen to water. People would go on drinking water in one form or another, our bodies would still be made mostly of water, water would continue to fall down as rain, it would remain to be the substance that fills the oceans and the rivers, and, most important, its composition of H_2O molecules would not change. It seems intuitively clear that the property "being the main ingredient in carbonated beverages" is accidental to water. Water is water, whether it is used in the making of carbonated beverages or not. However, the property "being composed of H_2O molecules" seems to be a different kind of property. If a small lake in Iceland was discovered to be filled with a substance that appears to have almost the same properties as water but, upon further testing, it was determined that this substance is not composed of H_2O molecules but, say, of XYZ molecules, most likely we would say that this substance is *not* water—that it is extremely similar to water but, nevertheless, not water. This intuitive judgment suggests that we consider the property "being composed of H_2O molecules" an essential property of water: any substance that does not have this property is not water.

We said in 4.1:C3 that some philosophers take the referent of a proper name to be part of its meaning. If this assumption is extended to "type names," such as the terms 'water', 'gold', 'whales', and 'humans', the distinction between metaphysical and logical possibility would be mostly lost. If being composed of H_2O molecules is an essential property of water, then there is no metaphysical possibility in which water does not have this property. But if the type water, or at least its essential properties, are part of the meaning of the term 'water', then since the meaning of 'water' must be held fixed across all the relevant logical possibilities, it would be logically impossible for water not to be composed of H_2O molecules. Intuitively, it seems too strong to suppose that it is logically impossible for water not to be composed of H_2O molecules. If we do not want to nullify the distinction between metaphysical and logical possibility regarding types, we should not assume that types, or at least their essential properties, must be part of the meanings of their names. If we want to maintain this distinction *in general*, we should not assume that objects or their essential properties (if they have any) must be part of the meanings of the expressions that stand for these objects. In other words, if a linguistic expression, including a proper name or a "type name," stands for an object, then this object and its essential properties need not be within the scope of the meaning of the expression. After all, it seems reasonable to assume that there is a logical possibility in which water is not composed of H_2O molecules.

However, some philosophers, especially of the medieval era, do not accept that there is a clear distinction between logical and metaphysical possibility. Those philosophers typically affirm that every existent, whether it is an individual or a type, has essential properties. They further argue that if the essential properties are not, at least, part of the meaning of any expression that designates that existent, then it is no longer clear what the meaning of this expression is. For example, we surely mean something by the term 'water'; if no properties of water were allowed to be part of the meaning of this term, then there would be no justification for saying that the term 'water' designates the substance water. Therefore, in order for the term 'water' to stand for water, it is necessary that we include in the meaning of this term some of the properties of water. But if we include in the meaning of the term 'water' accidental properties of water, say, that it is the main ingredient in carbonated beverages, and if the meaning of this term is to be invariant across all the relevant logical possibilities, then it would be logically impossible for water not to be the main ingredient in carbonated beverages, which is absurd. This leaves us with one of two options: (1) include at least some essential properties of an existent in the meaning of any expression that designates this existent, or (2) allow the meaning of the term to change across some relevant logical possibilities. If we take the second option, the notions of logical truth and of logical falsehood would lose much of their force. For instance, we would not be able to affirm that it is logically true for bachelors to be unmarried, since 'bachelor' might mean "married male" in some logical possibility; and that it is logically false for a triangle to have four angles, since 'triangle' might mean "square" in some logical possibility. These examples show that the second option leads to absurdities. Hence our only alternative is to accept the first option, which is to allow essential properties of an object to be part of the meaning of any expression that stands for that object. This entails that the distinction between logical and metaphysical possibility cannot be maintained in most cases.

We will not engage this argument in depth. We will hold to the positions that essential properties of objects (if there are any) need not be part of the meanings of the expressions that designate these objects, and that the meanings of the expressions must remain unchanged across all relevant logical possibilities. But we accept the initial motivation of the argument above—namely, that some properties and relations of objects must be included in the meanings of the expressions that stand for these objects, for if no such properties are included in the meaning of an expression, it would be difficult to claim that this expression designates this object. Some philosophers say that these properties are what can be termed 'conceptual properties', that is, properties that are part of the concept (or perhaps, concepts) of the object. So, for example, the property of having three angles must be included in the meaning of the expression 'triangle' because this is a conceptual property of triangles, and the property of being a physical substance must be included in the meaning of the expression 'water' because being a physical substance is part of the concept of water. This may be right, and for the most part we will follow this suggestion. Occasionally, however, we will let the context help determine some of the properties that must be included in the meaning of an expression that stands for the object that has these properties.

When we say that the meanings of linguistic expressions must be held fixed across the relevant logical possibilities, we usually assume that the notion of meaning invoked here is that of the "conceptual content" of an expression. A meaningful linguistic expression, except, perhaps, for a proper name, typically involves a concept; it could be the concept of a physical object such as "water" and "man," of an abstract object such as "triangle" and "justice," of a mental object such as "pain" and "belief," of a property such as "being red" and "being rational," of a relation such as "being taller than" and "being a brother of," or what have you. The conceptual content of a meaningful linguistic expression is the concepts that are conveyed by this expression.

We will not attempt to give a definition of 'concept' here. The issue of defining 'concept' is philosophically involved and beyond the scope of this book. However, we can illustrate the idea with a couple of examples. The expression 'Euclidean triangle' conveys the concept of a three-sided figure that satisfies the standard axioms of Euclidean geometry. This is a "logically deep" concept: many obvious and non-obvious things logically follow from this concept. For instance, a Euclidean triangle has three interior angles whose sum is 180 degrees; the three lines that connect the vertices of a Euclidean triangle and the midpoints of its corresponding sides intersect in a single point; the side of a Euclidean triangle that is opposite to a right angle (i.e., an interior angle measuring 90 degrees) is the longest side of the triangle, and the square of the length of that side is the sum of the squares of the lengths of the other two sides. An object that satisfies the concept of a Euclidean triangle must have all of these properties. If an object lacks one of these properties in some logical possibility, it would fail to satisfy the concept of a three-sided figure that fulfills the axioms of Euclidean geometry, and hence it would not be a Euclidean triangle in that logical possibility. A simpler example is the expression 'bachelor'. It conveys the concept of an unmarried adult male. Unlike the concept of a Euclidean triangle, this concept is "logically shallow": only a few, mostly obvious, things logically follow from the concept of a bachelor. For instance, we can affirm that no bachelor is a female (assuming that there are no gender ambiguities), that no bachelor has a spouse, that there are no married bachelors, and other things of this sort.

A linguistic expression whose conceptual content is logically shallow does not usually present us with a problem. The logical reach of such content is quite limited and hence it is unlikely to run into conflict with other assumptions. For example, if the relevant logical possibilities have inhabitants who are bachelors, we most likely would not make the mistake of assuming that one of these possibilities contain a married bachelor. It is clear that the concept of a bachelor excludes the possibility of a married bachelor. On the other hand, a linguistic expression whose conceptual content is logically deep may present us with a problem. For instance, if a relevant logical possibility contains a Euclidean triangle, it is not immediately obvious that this logical possibility cannot allow the sum of the triangle's interior angles to be 181 degrees. We need to know some geometry in order to know that a story in which a Euclidean triangle has interior angles whose sum is 181 degrees is inconsistent, and hence it does not depict a logical possibility.

These examples suggest that in holding the meanings of the linguistic expressions fixed across the relevant logical possibilities, we need to restrict the meanings of these expressions to the "most immediate" level of their conceptual contents. Consider the expression 'Euclidean triangle' again. The meaning of this expression should be restricted to the concept of a three-sided figure that satisfies the standard axioms of Euclidean geometry. We should be very careful not to include in the meaning of this expression "deeper" assumptions, for such assumptions might run into conflict with some of the things that logically follow from the concept of a Euclidean triangle. Here is another mathematical example. The expression 'even integer' conveys the concept of an integer that can be divided by two without a remainder. This is the most immediate level of the conceptual content of this expression. (Of course, the notion of an integer needs to be defined as well.) Nothing more should be included in the meaning of the expression 'even integer'; and it is this "minimal" meaning that must be held fixed across the relevant logical possibilities. Even integers have many properties, some of them are known and others remain unknown.[1] If one were to present a more elaborate meaning for the expression 'even integer', this more elaborate meaning might attribute to the even integers certain properties that they do not have. In such a case the expression 'even integer' would be *incoherent*, that is, it would express a contradictory concept. Since a logical possibility must be consistent, no logical possibility can satisfy an incoherent concept, i.e., no logical possibility can contain an object to which the concept applies. In 1.2.1:C1 we encountered an incoherent concept—namely, the concept of the barber of Russell Town. No logical possibility can contain such a barber. Thus in the case of an even integer, we must hold fixed its conceptual content,

---

1 For instance, it is not known yet whether every even integer greater than two is the sum of two prime numbers: no one has found an even integer greater than two that is not the sum of two prime numbers but, also, no one has proved that this is true of *all* even integers greater than two.

which is of an integer that is divisible by two, and let this content determine the other properties that even integers have.

The moral of all this discussion is that the requirement that the meanings of the linguistic expressions must be held fixed across all the relevant logical possibilities presupposes an "austere" notion of meaning: if an expression E stands for an object, then we do not insist that the meaning of E must include the object for which E stands or its essential properties; if E involves a concept, we restrict the meaning of E to the most immediate level of the conceptual content of E. These two conditions ensure that the sphere of relevant logical possibilities is not made too narrow. If the meanings of the expressions are inflated by building more content into them, the sphere of relevant logical possibilities would be made too limited, because this additional content would place more demands on what can be allowed as a relevant logical possibility. For instance, if we build into the meaning of the expression 'water' the additional content that water is composed of H_2O molecules, then no logical possibility in which water is not composed of H_2O molecules would be allowed into the sphere of relevant logical possibilities. The more content we build into the meanings of the expressions, the fewer logical possibilities would be allowed into the sphere of relevant logical possibilities. This might collapse the notion of logical possibility into more restricted notions, such as the notion of metaphysical possibility. An extreme case obtains when the conceptual content of an expression is expanded to the extent that the content becomes incoherent. In this case the sphere of logical possibilities would be empty, because logical possibilities must be consistent, and hence they cannot satisfy an incoherent concept.

4.2 The Syntax of PL

4.2.1 The **basic vocabulary** of PL consists of six categories.[1]

- 4.2.1a **Names**, which are the following lowercase italic letters: $a, b, c, \ldots, r, s, t$ (excluding 'f', 'g', and 'h'); with numeric subscripts if needed.
- 4.2.1b **Function symbols**, which are the following lowercase italic letters with numeric superscripts: $f^1, g^1, h^1, f^2, g^2, h^2, f^3, g^3, h^3, \ldots$; with numeric subscripts if needed.
- 4.2.1c **Predicates**, which are uppercase italic letters with numeric superscripts: A^1, B^1, $C^1, \ldots, X^1, Y^1, Z^1; A^2, B^2, C^2, \ldots, X^2, Y^2, Z^2; A^3, B^3, C^3, \ldots, X^3, Y^3, Z^3; \ldots$; with numeric subscripts if needed.
- 4.2.1d **Variables**, which are the following lowercase italic letters: u, v, w, x, y, z; with numeric subscripts if needed.
- 4.2.1e Eight **logical symbols**: $\neg, \wedge, \vee, \rightarrow, \leftrightarrow, \forall, \exists, =$
- 4.2.1f **Parentheses**: '(' and ')'

The variables, logical symbols, and parentheses are referred to as the **logical vocabulary** of PL, and the names, function symbols, and predicates are referred to as the **extra-logical vocabulary** of PL.

---

1 In Chapters Four and Five, we will follow the convention we employed in the previous two chapters: unless there is a cause for misunderstanding, we will not enclose the symbols of the object language between single quotes when they are mentioned in the metalanguage.

4.2.1:C COMMENTARY ON 4.2.1

As is the case with TL and SL, the PL worldview guides the syntax and semantics of PL. We have names to stand for individuals, predicates to stand for properties and relations, and function symbols to stand for relations of a special type, which we call **functions**.[1] The term 'individual' is used here in its philosophical sense. Individuals can be any kind of entities or objects. The logical symbols allow us to form complex expressions that describe relations between these ingredients of reality. The variables have a syntactical role that pertains to the use of quantifiers such as 'every' and 'some.' To illustrate the point we translate the English sentence 'all whales are mammals' into a hybrid PL-English sentence. First, we reformulate the sentence as 'if something is a whale, then it is a mammal'. The latter could be paraphrased, using variables, as 'for every *x*, if *x* is a whale, then *x* is a mammal'. The variable '*x*' here assumes the role of the pronoun 'it' in the English sentence. In general, PL variables in some ways behave like pronouns in English. As we will explain later, in PL sentences variables are placeholders: they indicate which quantifier applies to which place.

We designate the boldfaced letters '**P**', '**Q**', and '**R**', sometimes with numeric superscripts, to be metalinguistic variables ranging over PL predicates, the boldfaced letters '**f**' and '**g**', possibly with numeric superscripts, to be metalinguistic variables ranging over function symbols, and the boldfaced letters '**x**', '**y**', and '**z**' to be metalinguistic variables ranging over PL variables.

PL predicates come with superscripts. A superscript indicates the number of "places" the predicate has. For example, the predicate A^1 is a 1-place predicate, meaning that it applies to single individuals. 1-place predicates stand for properties because properties apply to single individuals. The predicate A^1 might stand for the property "being a man." As with any property, this one applies to single individuals. A given individual either has this property or does not (recall that in this book we deal only with bivalent languages, properties, and relations). Socrates has this property but the Statue of Liberty does not. We normally express this fact by saying that Socrates is a man but the Statue of Liberty is not. The English expression 'is a man' is a 1-place predicate: if we put a singular term, such as 'Socrates' or 'The Statue of Liberty', to the left of 'is', we obtain a declarative sentence. You can think of a 1-place predicate as an expression with a "blank"; if you fill the blank with a singular term, the resulting (complete) expression is a declarative sentence. The singular term that completes the 1-place predicate is usually referred to as the subject of the resulting declarative sentence. Using this terminology, we say that the sentence 'Socrates is a man' consists of the subject 'Socrates' and the predicate 'is a man'. The same analysis applies to PL. However, the "blanks" of any PL predicate, except the identity predicate '=', are always to the right of the predicate. So if the PL name '*s*' stands for Socrates, the English sentence 'Socrates is a man' is translated into the PL sentence A^1s. 1-place predicates are also called **unary predicates** or **monadic predicates**.

A predicate that has more than one place stands for a relation with the same number of places. Thus, for instance, 2-place predicates stand for 2-place relations, which are commonly called "binary relations," and 3-place predicates stand for 3-place relations. In general, an n-place predicate (where n is greater than one) stands for an n-place relation. These predicates are called **relational predicates**. In order to improve the readability of n-place English predi-

1 Functions will be explained when we deal with PL semantics.

cates, we use variables instead of blanks. We will refer to the language that results from augmenting English with PL variables as **PL-English**. For example, instead of writing '— shares borders with ... and ***' we will write '*x* shares borders with *y* and *z*'. This 3-place relational predicate may be used to express the 3-place relation of a country's sharing borders with two other countries, or of a region's sharing borders with two regions on a certain map, or other similar 3-place relations. Rather than saying that the English sentence 'The USA shares borders with Canada and Mexico' is obtained from the predicate '— shares borders with ... and ***' by filling its blanks with the names 'The USA', 'Canada', and 'Mexico', respectively, we will say that this English sentence is obtained from the (PL-English) predicate '*x* shares borders with *y* and *z*' by substituting the names 'The USA', 'Canada', and 'Mexico' for the variables '*x*', '*y*', and '*z*', respectively. We will find that PL-English is quite useful in translating English sentences into PL and conversely. We should note that English is part of PL-English. Hence every natural English sentence is a PL-English sentence.

We extend the use of variables to PL predicates. In order to simplify the notation and improve the readability of n-place PL predicates, we use variables instead of superscripts to indicate the number of places. For instance, instead or writing B^3 we write *Bxyz*. We can let the PL predicate *Bxyz* translate the PL-English predicate '*x* shares borders with *y* and *z*', and the PL names *s*, *c*, and *m* translate the English names 'The USA', 'Canada', and 'Mexico', in this order. As we will explain later, there are several ways of generating sentences in PL. One simple way is substituting PL names for the variables of a PL predicate. Thus *Bscm* is a PL sentence, and it translates the English sentence 'The USA shares borders with Canada and Mexico'.

Function symbols also have places. The superscript of a function symbol indicates the number of places this function symbol has. As with PL predicates, we will employ the convention of using variables instead of superscripts to mark the number of places a function symbol has. For instance, we might write *gxyz* instead of g^3 and *hvw* instead of h^2. We abandon this convention where there is cause for ambiguities. Function symbols are used to generate complex terms. Later we will define what we mean by "a PL term." For now it suffices to consider the case of names. If we apply a function symbol to a name, we obtain a complex name that refers to some individual. A complex PL name is similar to a **definite description**. In natural languages, definite descriptions are complex expressions that refer to unique individuals, if they refer at all. In English, definite descriptions are typically generated with the aid of the definite article 'the'. For example, 'The first president of the United States' is a definite description that refers to one and only one individual—George Washington. Definite descriptions can be analyzed syntactically as being composed of *functional descriptions* and *singular terms*. Functional descriptions are similar to predicates in that they have "blanks." The expression 'The first president of ___' is a 1-place functional description. If we allow the use of variables in English, we can paraphrase this functional description as 'The first president of *x*'. Substituting a name for *x* generates a definite description. Depending on the name we substitute for *x*, the resulting definite description might or might not have a referent. For instance if we substitute 'Iraq' for *x*, the definite description 'The first president of Iraq' has a unique referent: General Qasim (in this case, we say that the definite description is *referring*). But if we replace *x* with the name 'England', the resulting definite description has no referent (we say in this case that this definite description is *non-referring*). It is possible to replace *x* with another definite description. For instance, instead of using the name 'The USA', we may replace *x* with the definite description 'The country that dropped a nuclear bomb on Japan'. In this case we generate the definite description 'The first president of the country that dropped a nuclear bomb on Japan'. Again, this definite description refers to George Washington.

We call names and definite descriptions collectively **singular terms**. We said above that a definite description is composed of a functional description and a singular term. This assertion is not exactly correct. A functional description might have more than one "blank." A 2-place functional description has two "blanks," a 3-place functional description has three "blanks," and, in general, an n-place functional description has n blanks. The functional description 'The sum of *x*, *y*, and *z*' is 3-place. If we substitute three numerals for the variables *x*, *y*, and *z*, we obtain a definite description that refers to exactly one number. For example, the definite description 'The sum of 7, 11, and 16' refers to the number 34. Here too it is possible to replace a variable with another definite description. Consider the definite description 'The square of 4'. We can substitute this definite description for the variable *z* in the original functional description. In this case, we obtain the definite description 'The sum of 7, 11, and the square of 4', which refers to 34. Thus we can assert more accurately that a definite description is composed of a functional description and one or more singular terms. As we have seen, definite descriptions might be referring or non-referring.

Natural-language functional descriptions are translated into PL as function symbols with the same number of places. For example, *fx* and *hxyz* may be the PL translations of the English functional descriptions 'The first president of *x*' and 'The sum of *x*, *y*, and *z*', respectively. If we let the name *e* translate 'The USA', the PL term *fe* would translate the definite description 'The first president of the USA', and if we let *a*, *b*, *c* translate the numerals '7', '11', and '16', respectively, then the PL term *habc* would translate the definite description 'The sum of 7, 11, and 16'. Just as the case with English definite description, a PL term may be substituted for a variable in another PL term. Consider the following example. Say we let the function symbol *gv* translate the functional description 'The square of *v*' and the name *s* translate the numeral '4'. We obtain the PL term *habgs*, which translates the definite description 'The sum of 7, 11, and the square of 4'. PL terms that contain no variables can never fail to refer: every such PL term has exactly one referent on each PL interpretation (we will define this notion later).

Variables are powerful syntactical devices. They can serve two functions: they can indicate the number of places in PL predicates and formulas, the order of these places, and their types, and they can serve as placeholders for the quantifiers. We illustrate by means of a PL-English example their first function. The sentence 'John is a brother of Mary and a son of William, who is a student in a class taught by Sarah, who is a friend of John' expresses a complex 4-place relation that holds between John, Mary, William and Sarah. If we want to express this relation in general, without specifying these particular individuals, we can use the 4-place PL-English predicate '*x* is a brother of *y* and a son of *z*, who is a student in a class taught by *w*, who is a friend of *x*'. Notice that although this predicate has five "blanks," it is really a 4-place predicate because two of these "blanks" are represented by the "same" variable, that is, they are represented by two occurrences of the same variable—namely '*x*'. Also the order of these variables is essential for expressing the relation correctly. We can say that the English sentence above is generated by substituting the names 'John', 'Mary', 'William', and 'Sarah', for the variables '*x*', '*y*', '*z*', and '*w*', in this order. Using the same variables, we may translate the complex PL-English predicate into the PL formula $(((Bxy \wedge Sxz) \wedge Czw) \wedge Fwx)$, where *Bxy* translates '*x* is a brother of *y*', *Sxz* translates '*x* is a son of *z*', *Czw* translates '*z* is a student in a class taught by *w*', *Fwx* translates '*w* is a friend of *x*', and, as in SL, $\wedge$ is the conjunction sign. If we let the PL names *j*, *m*, *l*, and *s* stand for John, Mary, William, and Sarah, respectively, the original English sentence can be translated into the PL sentence $(((Bjm \wedge Sjl) \wedge Cls) \wedge Fsj)$.

To illustrate the role of variables in the use of quantifiers, we consider two examples. We paraphrase the English sentence 'Someone who was the teacher of Plato was condemned to death by an Athenian jury' into PL-English as 'There is *x* such that *x* was a teacher of Plato and *x* was condemned to death by an Athenian jury'. This sentence may be analyzed as being composed of a 2-place predicate, a 1-place predicate, and a proper name. They are, respectively: '*x* was a teacher of *y*', '*x* was condemned to death by an Athenian jury', and 'Plato'. If we let Txy translate '*x* was a teacher of *y*', Ax translate '*x* was condemned to death by an Athenian jury', and p translate 'Plato', we obtain the PL sentence $(\exists x)(Txp \wedge Ax)$, where the quantifier '$(\exists x)$' is a translation of the PL-English expression 'There is *x*'. For our second example, we modify the English sentence in the preceding paragraph to read 'John is a brother of Mary and a son of William, who is fond of all John's friends'. First, we paraphrase this sentence as the following PL-English sentence 'John is a brother of Mary and a son of William, and for every *v*, if *v* is a friend of John, then William is fond of *v*'. The last clause is another way of saying that William is fond of everyone who is a friend of John. Now we can extract a 3-place complex predicate from this sentence: '*x* is a brother of *y* and a son of *z*, and for every *v*, if *v* is a friend of *x*, then *z* is fond of *v*'. This predicate is 3-place even though it contains occurrences of four variables. The three occurrences of the variable '*v*' indicate the applicability of the universal quantifier 'every'. These places are not available for substitution; they are not really "blanks." To see the point clearly, try to substitute 'John' for '*x*', 'Mary' for '*y*', William for '*z*', and 'Sarah' for '*v*'. We get 'John is a brother of Mary and a son of William, and for every Sarah, if Sarah is a friend of John, then William is fond of Sarah'. This sentence is infelicitous, because the expression 'every Sarah' makes no sense—we are not referring here to all individuals who are named 'Sarah'; rather we are referring to a single specific individual named 'Sarah'. With the aid of the PL-English paraphrase of the original English sentence, we now translate the English sentence into PL. The expression 'for every *v*' is translated into PL as '$(\forall v)$'. We let Dzv translate '*z* is fond of *v*', and Bxy, Sxz, Fwx, j, m, and l be as above. The English sentence may be translated into PL as $((Bjm \wedge Sjl) \wedge (\forall v)(Fvx \rightarrow Dzv))$, where $\rightarrow$ is the material conditional sign of SL. We will study PL translation with elaboration later.

The logical symbols of PL consist of the standard sentential connectives of SL, $\neg$, $\wedge$, $\vee$, $\rightarrow$, and $\leftrightarrow$, two quantifier symbols, $\forall$ and $\exists$, which are used to express the quantifiers 'all' and 'some', respectively, and a symbol, =, for the relation of token identity. Since the quantifier symbols are not really sentential connectives, we will refer to the logical symbols of PL (except the symbol =) as 'logical operators'. Thus one of these logical operators is a unary connective, $\neg$, four of them are binary connectives, $\wedge$, $\vee$, $\rightarrow$, and $\leftrightarrow$, and two of them are quantifier symbols, $\forall$ and $\exists$. When we speak of the PL connectives we refer to the standard five sentential connectives. When we speak of the logical operators of PL we mean the connectives and the quantifier symbols. In 4.2.2 below, we will be more specific regarding the quantifiers. According to the syntactical approach we adopt in this book, the quantifiers of PL are expressions of the forms $(\forall \mathbf{z})$ and $(\exists \mathbf{z})$, where **z** is any PL variable. So precisely speaking, the logical operators of PL are the sentential connectives and all the expressions of the forms $(\forall \mathbf{z})$ and $(\exists \mathbf{z})$. In 4.2.3 we will study how to use these logical operators to form complex grammatical expressions of PL.

4.2.2 The PL connectives $\neg$, $\wedge$, $\vee$, $\rightarrow$, and $\leftrightarrow$ are the same as the SL connectives. There are three more logical symbols, which we have not yet discussed, that appear in PL. The symbol '=' is called the **identity predicate**. It is a 2-place predicate that stands for the relation of token identity. As explained in 4.1, this relation holds between each individual and itself and does not hold between distinct individuals. A PL expression of the form (∀**z**), where **z** is a PL variable, is called a **universal quantifier** and of the form (∃**z**) is called an **existential quantifier**. We also refer to these quantifiers as **z-quantifiers**. The universal and existential quantifiers of PL are similar in meaning to the universal and existential quantifiers of TL, 'all' and 'some'. Thus a PL universal *x*-quantifier (∀*x*) is read in PL-English as 'for all *x*', 'for every *x*', or 'for each *x*'; and a PL existential *x*-quantifier (∃*x*) is read in PL-English as 'for some *x*' or 'there is (exists) *x*' (in the sense that there is at least one *x*). 1-place PL predicates play the role of TL general terms. However, there is an important difference. TL general terms may not name the empty set, while PL predicates (except for the identity predicate) may name the empty set. In other words, no TL general term may have an empty extension, while PL predicates may have empty extensions. A PL **term** is either a PL name, a PL variable, or a PL expression that is generated from the names and variables by applying the following formation rule any finite number of times: if $\mathbf{f}^n$ is an n-place function symbol and $\mathbf{t}_1$, $\mathbf{t}_2$, …, $\mathbf{t}_n$ are any PL terms, then the PL expression $\mathbf{f}^n\mathbf{t}_1\, \mathbf{t}_2 \ldots \mathbf{t}_n$ is a PL term. (We use the boldfaced letters '**r**', '**s**', and '**t**', with numeric subscripts if needed, as metalinguistic variables ranging over PL terms.) If a PL term contains function symbols and variables, we call it a **functional term**; and if it is a name or contains function symbols but no variables, we call it a **singular term**. We note that, by definition, PL variables are neither functional nor singular terms; they are only PL terms. PL functional terms correspond to functional descriptions and PL singular terms to singular terms in natural languages. It is intuitively clear, from our previous discussion, that a PL functional term does not designate any individual (it is an "incomplete" expression) while a PL singular term designates a single individual. (We will state these facts with precision when we define the notion of a PL interpretation.) PL singular terms also correspond to the singular terms of TL; and they share the same restriction imposed on TL singular terms—namely that every PL singular term must refer to exactly one individual.

4.2.2:C COMMENTARY ON 4.2.2

4.2.2:C1 Predicate Logic **includes Term Logic** and **extends Sentence Logic**. The connectives of SL are part of the vocabulary of PL and, as we will see below, the formation rules of SL are also formation rules of PL. Furthermore the truth conditions of these connectives are the same in SL and PL. PL is said to be a **proper extension** of SL because every inference that is valid in SL is also valid in PL but not every inference that is valid in PL is valid in SL (we will explain this assertion in 4.5.12).

The relation between PL and TL is not as straightforward. The vocabulary of TL is not part of the vocabulary of PL, but every TL sentence can be translated into one or more PL sentences. TL singular terms correspond to PL singular terms, and TL general terms correspond to 1-place PL predicates. The TL universal quantifier 'all' corresponds to a PL universal quantifier (∀**z**) (the symbol '∀' is an upside-down 'A' for 'All') and the TL existential quantifier 'some' corresponds to a PL existential quantifier (∃**z**) (the symbol '∃' is a backward 'E' for 'Exists'). The 'is' of predication of TL, such as the 'is' in '*c* is *A*', corresponds to the concatenation of a 1-place predicate and a PL singular term (e.g., *Ac*). The 'is' of identity of TL, such as the 'is' in '*c* is *e*', corresponds to the identity predicate '=' (e.g., *c* = *e*). The list below shows how TL sentences may be translated into PL.

(a) TL: all *A* are *B*	PL: $(\forall x)(Ax \rightarrow Bx)$	PL-English: for every *x*, if *x* is *A*, then *x* is *B*.
(b) TL: no *A* is *B*	PL: $(\forall x)(Ax \rightarrow \neg Bx)$	PL-English: for every *x*, if *x* is *A*, then *x* is not *B*.
	Or: $\neg(\exists x)(Ax \wedge Bx)$	PL-English: there is no *x*, such that *x* is both *A* and *B*.
(c) TL: some *A* are *B*	PL: $(\exists x)(Ax \wedge Bx)$	PL-English: there is *x*, such that *x* is both *A* and *B*.
(d) TL: some *A* are not *B*	PL: $(\exists x)(Ax \wedge \neg Bx)$	PL-English: there is *x*, such that *x* is *A* but not *B*.
(e) TL: *c* is *A*	PL: Ac	
(f) TL: *c* is not *A*	PL: $\neg Ac$	
(g) TL: *c* is *e*	PL: $c = e$	
(h) TL: *c* is not *e*	PL: $c \neq e$ (we conventionally write '$c \neq e$' instead of '$\neg c = e$')	

Using these translations, we can construct valid TL arguments that are invalid in PL. The source of the problem is that PL permits predicates to have empty extensions while TL does not permit general terms to have empty extensions. We will explain later that in PL a sentence of the form $(\forall x)(Ax \rightarrow Bx)$ is true if *A* has an empty extension. The following TL argument is deductively valid: all *A* are *B*; therefore, some *A* are *B*. However, the corresponding PL argument is invalid: $(\forall x)(Ax \rightarrow Bx)$ does not logically imply $(\exists x)(Ax \wedge Bx)$. The situation can be remedied. If a TL argument makes use of the requirement that the general terms do not name the empty set, we would translate 'all *A* are *B*', in this case, as the PL sentence $(\forall x)(Ax \rightarrow Bx) \wedge (\exists x)Ax$ (every *A* is *B*, and there is something that is *A*). The following PL argument is deductively valid: $(\forall x)(Ax \rightarrow Bx) \wedge (\exists x)Ax$; therefore $(\exists x)(Ax \wedge Bx)$. This argument captures the previous TL argument. However, even if we add an existential clause, such as $(\exists x)Ax$, to every inference that invokes the assumption that *A* has a nonempty extension, there would still be inferences that are valid in TL but cannot be translated in a natural way into valid PL inferences. For instance, the TL sentence 'some *A* are *A*' is logically true in TL. Its PL counterpart $(\exists x)(Ax \wedge Ax)$, which reduces to $(\exists x)Ax$, is not logically true in PL, since in PL it is possible for *A* to have an empty extension. There is no natural way of translating 'some *A* are *A*' into a logically true PL sentence. Because of this, PL is not an extension of TL. On the other hand, if we restrict PL sentences to those that have the forms of the sentences (a)–(h) listed above, and if we restrict PL interpretation to those that do not assign the empty extension to any 1-place predicate, the resulting (restricted) logic is equivalent to TL. We call this logic the "PL counterpart of TL" and we denote it as "PL(TL)." To say that TL and PL(TL) are equivalent systems is to say that an inference is valid in TL if and only if it is valid in PL(TL). Furthermore, we will see later that every TL diagram can be converted into a PL interpretation. These facts allow us to say that PL *includes* TL but it is not an extension of it. We will return to these issues with much more elaboration in 4.5.12.

4.2.2:C2 When we need to indicate that the quantifiers $(\forall \mathbf{z})$ and $(\exists \mathbf{z})$ apply to the variable **z**, we refer to these quantifiers as **z**-quantifiers. (The symbol **z** is a metalinguistic variable that stands for any PL variable.) For instance, the PL formula $((\forall v)(D^2vj \rightarrow (\exists w)H^1w) \wedge (\forall v)R^4vbxy)$ contains two *v*-quantifiers, $(\forall v)$ twice, and one *w*-quantifier, $(\exists w)$. This terminology is useful in stating the quantifier formation rules.

As we said previously, PL names will be interpreted as referring to individuals. PL variables, on the other hand, are purely syntactical entities; they receive no semantical interpretation in PL. This does not mean that the use of variables has no semantical significance. In fact, variables and their orders have important roles in the interpretation of PL sentences. For example, if the 2-place PL predicate L^2xy stands for the relation "x loves y", $(\forall x)(\exists y)L^2xy$ has a very different interpretation from $(\forall y)(\exists x)L^2xy$. The first sentence says that everyone loves someone (in PL-English, for every x, there is y, such that x loves y), and the second says that everyone is loved by someone (in PL-English, for every y, there is x, such that x loves y). When we say that variables in PL have no semantical content, we do not mean that they are semantically insignificant; we only mean that they do not receive any interpretation. They are not treated as names of individuals or of properties and relations. This is a feature of the type of PL semantics we adopt in this book, though there are other semantical approaches to PL that, in some sense, treat variables as names of individuals.

According to the definitions in 4.2.2, a PL term is either a variable, a functional term, or a singular term. PL singular terms are either names or terms that consist *only* of function symbols and names. PL singular terms correspond to English singular terms, which can be either proper names, such as 'Dr. Seuss', or definite descriptions, such as 'The author of the children's classic *The Cat in the Hat*'. A PL functional term is a PL term that has "blanks." It must contain at least one function symbol and at least one variable. PL functional terms correspond to English functional descriptions. Both of these categories are "incomplete" expressions, which can be transformed into singular terms by filling the "blanks" with singular terms. For instance, h^4cdeg^1x is a functional term. It is composed of the 4-place function symbol h^4, the three names c, d, and e, and the functional term g^1x, which, in turn, is composed of the 1-place function symbol g^1 and the variable x. This functional term can be transformed into a singular term by replacing the variable x with a singular term. If we replace x with the name b, we obtain the singular term h^4cdeg^1b; and if we replace x with the singular term f^2sr, we get the singular term $h^4cdeg^1f^2sr$. Just as in English we can iterate functional descriptions in the formation of definite description, PL functional terms can be iterated in the formation of singular terms. For instance, in English we can write 'The father of the father of Nagham', and in PL we can write g^1g^1s. It is clear that the formation rule of PL terms can generate infinitely many terms from a finite set of vocabulary and can allow the formation of very complex terms.

PL terms can be used to fill the "blanks" of PL predicates. The number of terms immediately to the right of a predicate indicates the number of places this predicate has. This allows us to simplify our notation by dropping the superscripts from the predicates. So when we write, for instance, $Gxycs$, we know that G is a 4-place predicate and the expression's precise form is G^4xycs. Note that x and y are variables and c and s are names. The use of function symbols could engender ambiguities if we insist on following the convention of dropping the superscripts. To avoid misreading PL terms or PL formulas, we need to do two things. First, we should keep in mind that only the letters 'f', 'g', and 'h', with or without numeric subscripts or superscripts, may be used as function symbols, that only the letters 'a', 'b', 'c', 'd', 'e', 'i', 'j', …, and 's', with or without subscripts, may be used as PL names, and that only the letters 'u', 'v', 'w', 'x', 'y', and 'z', with or without subscripts, may be used as PL variables. Second, if there is any possibility of ambiguity, we do not follow the convention of dropping the superscripts of function symbols. Let us illustrate by means of an example. The expression $Pavcgt$ is not ambiguous. It can only be an abbreviation of the expression P^4avcg^1t. However, the expression $Pavgtc$ is ambiguous; it can be an abbreviation of one of the following two PL expressions: P^4avg^1tc or P^3avg^2tc. So we need to be careful with applying the convention of dropping superscripts.

4.2.3 The **formulas** of PL are either atomic or compound. The **atomic formulas** of PL are all the expressions of the form **r** = **s** where **r** and **s** are any PL terms, and all the expressions of the form $\mathbf{Q}^n\mathbf{t}_1\mathbf{t}_2\ldots\mathbf{t}_n$ where $\mathbf{Q}^n$ is any n-place PL predicate (except the identity predicate) and $\mathbf{t}_1$, $\mathbf{t}_2$, ..., and $\mathbf{t}_n$ are any PL terms. The **compound formulas** of PL are those expressions constructed from the atomic ones by applying, some finite number of times, one or more of the seven **formation rules** listed below. Let **X** and **Y** be any PL formulas.

R¬: ¬**X** is a PL formula.
R∧: (**X**∧**Y**) is a PL formula.
R∨: (**X**∨**Y**) is a PL formula.
R→: (**X**→**Y**) is a PL formula.
R↔: (**X**↔**Y**) is a PL formula.
R∀: If **X** contains occurrences of the variable **z** but no **z**-quantifiers, then (∀**z**)**X** is a PL formula.
R∃: If **X** contains occurrences of the variable **z** but no **z**-quantifiers, then (∃**z**)**X** is a PL formula.

4.2.3:C COMMENTARY ON 4.2.3

The first five rules are the standard formation rules of SL. The only difference is that they apply here to sentences and formulas. In the next subsection we will explain the difference between PL formulas and PL sentences. The only thing that concerns us now is to understand how to construct PL formulas. We begin with atomic formulas. If we put a PL term to the left of the identity predicate and a PL term to its right, we obtain an atomic formula. Here are a few examples: $x = y$, $t = g^1c$, and $h^3xer = w$. The metalinguistic variable $\mathbf{Q}^n$ stands for any n-place PL predicate, other than the identity predicate. For example, $\mathbf{Q}^1$ stands for any 1-place predicate, such as A^1 and P^1, $\mathbf{Q}^3$ stands for any 3-place predicate, such as B^3 and L^3, and $\mathbf{Q}^5$ stands for any 5-place predicate, such as L^5 and E^5. Note that the predicates L^3 and L^5 are independent of each other. The symbols $\mathbf{t}_1$, $\mathbf{t}_2$, ..., and $\mathbf{t}_n$ are metalinguistic variables that stand for at most n PL terms. These terms could be variables, functional terms, singular terms, or any combination of them. Since some of these metalinguistic variables may stand for the same term, the number of the distinct terms may be less than n. Consider, as examples, two PL predicates L^3 and E^5. The following expressions are atomic formulas of PL: L^3xya, L^3uuf^1e, L^3cg^1vc, E^5uvwwl, E^5axbg^2xda, and E^5ach^1zbd.

Once we have atomic formulas, we can construct compound ones from these formulas by using one or more of the seven formation rules any finite number of times. Here are a few examples of compound formulas produced by using the first five rules. The formulas are listed on the left and the rules used in constructing the formulas are listed on the right in the order of their application.

$(\neg A^1z \wedge B^3xah^1y)$	R¬ R∧
$((\neg A^1z \wedge B^3xah^1y) \rightarrow K^2zg^2we)$	R¬ R∧ R→
$\neg(((\neg A^1z \wedge B^3xah^1y) \leftrightarrow K^2zg^2we) \wedge D^1y)$	R¬ R∧ R↔ R∧ R¬

The last two formation rules are more involved. There are two conditions that must be met in order to apply the quantifier rules: (1) the variable of the quantification must occur in the formula, and (2) no quantifier of *that variable* occurs in the formula. For example, the variable x occurs in the formula $(A^1z \wedge B^3xah^1y)$ and no x-quantifier occurs in that formula. Thus we can apply either R∀ or R∃. Say, we apply R∀; we obtain the compound formula $(\forall x)(A^1z \wedge B^3xah^1y)$. We say that we have *quantified over x*. Once we quantify over a variable, we cannot quantify over it again. The second condition prevents us from quantifying over the same variable twice. So the expression $(\exists x)(\forall x)(A^1z \wedge B^3xah^1y)$ is ungrammatical, according to our rules.[1] However, the variables z and y in the formula $(\forall x)(A^1z \wedge B^3xah^1y)$ are available for quantification. Applying R∃ to z and y, we get $(\exists z)(\exists y)(\forall x)(\neg A^1z \wedge B^3xah^1y)$. No more quantification may be applied to the last formula. We cannot quantify over x, y, or z since the formula already contains an x-quantifier, a y-quantifier, and a z-quantifier, and we cannot quantify over any variable that does not occur in the formula (the first condition). For instance, the expression $(\exists v)(\forall x)(A^1z \wedge B^3xah^1y)$ is ungrammatical because the formula $(\forall x)(A^1z \wedge B^3xah^1y)$ contains no occurrences of the variable v.

These formation rules allow for the construction of very complex formulas with *nested* quantifiers. Consider the following sequence of constructions.

(a) We begin with $(\forall x)(A^1z \wedge B^3xah^1y)$ and $(z = y \rightarrow (\exists v)D^3evy)$.

(b) The variable z occurs in the second formula, which contains no occurrences of any z-quantifiers. We can use R∃ to quantify over z. We obtain $(\exists z)(z = y \rightarrow (\exists v)D^3evy)$.

(c) We apply R¬ to (b). We get $\neg(\exists z)(z = y \rightarrow (\exists v)D^3evy)$.

(d) We apply R∨ to (c) and the first formula in (a). We obtain the formula

$$((\forall x)(A^1z \wedge B^3xah^1y) \vee \neg(\exists z)(z = y \rightarrow (\exists v)D^3evy)).$$

(e) Since y occurs in (d) and no y-quantifier occurs in it, we can use R∀ to quantify over y. We get $(\forall y)((\forall x)(A^1z \wedge B^3xah^1y) \vee \neg(\exists z)(z = y \rightarrow (\exists v)D^3evy))$.

No more quantification can be applied to the formula in (e). There is quantification over every variable that occurs in this formula. As defined in 4.2.4 below, the scope of a quantifier is the formula immediately to its right. Thus the scope of $(\exists v)$ is the formula D^3evy; the scope of $(\exists z)$ is the formula $(z = y \rightarrow (\exists v)D^3evy)$, which already contains the quantifier $(\exists v)$; the scope of $(\forall x)$ is the formula $(A^1z \wedge B^3xah^1y)$; and the scope of $(\forall y)$ is the formula $((\forall x)(A^1z \wedge B^3xah^1y) \vee \neg(\exists z)(z = y \rightarrow (\exists v)D^3evy))$, which contains all the previous quantifiers. When a quantifier occurs inside the scope of another quantifier, we say that these quantifiers are **nested**. The formula $(\forall y)((\forall x)(A^1z \wedge B^3xah^1y) \vee \neg(\exists z)(z = y \rightarrow (\exists v)D^3evy))$ contains several nested quantifiers, such as $(\exists v)$ of $(\exists v)D^3evy$ and $(\exists z)$ of $(\exists z)(z = y \rightarrow (\exists v)D^3evy)$ (the former is nested inside the scope of the latter).

The *immediate component* of a PL formula of the form ¬**X** is **X** and its *main operator* is the unary connective ¬. The immediate components of PL formulas of the forms (**X**∧**Y**), (**X**∨**Y**), (**X**→**Y**), and (**X**↔**Y**) are **X** and **Y** and their main operators are the binary connectives ∧, ∨, →, and ↔, respectively. The immediate components of (∀**z**)**X** and (∃**z**)**Y** are **X** and **Y** and their

---

1 There are approaches to PL syntax that allow for multiple quantifications over the same variables. Mathematicians usually favor a PL syntax that allows multiple quantifications. According to this syntax, the formula $(\exists x)(\forall x)(\neg A^1z \wedge B^3xay)$ would be grammatical.

main operators are the quantifiers (∀**z**) and (∃**z**), respectively. If a formula **X** occurs in a formula **Y**, we say that **X** is a *subformula* of **Y**. Technically speaking, every formula is a subformula of itself, because every formula "occurs" in itself. We say that **X** is a *proper subformula* of **Y** just in case **X** is a subformula of **Y** and it is not identical with **Y**. An *atomic component* of a PL formula **X** is a subformula of **X** that is an atomic formula.

4.2.4 The **scope** of a quantifier is the PL formula that immediately follows the quantifier. An occurrence of a variable **z** in a PL formula is **bound** if and only if it is an occurrence inside a quantifier in the formula or inside the scope of a **z**-quantifier in the formula. An occurrence of a variable in a PL formula is **free** if and only if it is not bound. The **sentences** of PL are precisely those PL formulas that contain no free occurrences of any variable.

4.2.4:C COMMENTARY ON 4.2.4

4.2.4:C1 We already talked about the scope of a quantifier when we discussed nested quantifiers above. It is important to note that there is only one formula that immediately follows any quantifier. Hence every quantifier in a PL formula has only one scope. Consider for example the formula:

$$(\forall y)((\forall x)(\neg A^1c \wedge B^3xag^1y) \vee \neg(\exists z)(z = y \rightarrow (\exists v)D^3evy))$$

The formula that is immediately to the right of $(\forall x)$ is $(\neg Ac \wedge B^3xag^1y)$. If we begin at the left-hand parenthesis that follows the quantifier $(\forall x)$ and proceed by examining the PL expressions that occur to the right of $(\forall x)$, we find that the only formula that occurs immediately to the right of $(\forall x)$ is $(\neg Ac \wedge B^3xag^1y)$. No other expression that immediately follows $(\forall x)$ is a grammatical PL formula. One might think that the PL expression $(\neg Ac \wedge B^3xag^1y) \vee \neg(\exists z)(z = y \rightarrow (\exists v)D^3evy)$, which also occurs immediately to the right of $(\forall x)$, is a PL formula. It is not, if the rules are applied correctly. The most likely formation rule that might generate this expression is R∨; but this rule requires the compound formula to be enclosed between two parentheses. Thus this rule would generate $((\neg Ac \wedge B^3xag^1y) \vee \neg(\exists z)(z = y \rightarrow (\exists v)D^3evy))$, which is different from the previous expression. The complete PL expression that occurs to the right of $(\forall x)$ is $(\neg Ac \wedge B^3xag^1y) \vee \neg(\exists z)(z = y \rightarrow (\exists v)D^3evy))$. This expression is also not a grammatical PL formula: its parentheses are not *balanced*, that is, not every right-hand parenthesis is matched with an opposing left-hand parenthesis. The parentheses of a PL formula must be balanced.

Both occurrences of the variable y in the formula $\neg(\exists z)(z = y \rightarrow (\exists v)D^3evy)$ are free, since neither is an occurrence inside a quantifier or inside the scope of a y-quantifier (there are no y-quantifiers in this formula). On the other hand, the two occurrences of the variable z in the formula $\neg(\exists z)(z = y \rightarrow (\exists v)D^3evy)$ are bound because the first is inside the quantifier $(\exists z)$ and the second is inside the scope of a z-quantifier. The occurrences of the variable y in the previous formula are inside the scope of a quantifier—namely, $(\exists z)$. However, since the quantifier is a z-quantifier and not a y-quantifier, these occurrences are free and not bound. An occurrence of a variable **x** inside the scope of a quantifier whose variable is not **x** need not be bound. A bound occurrence of **x** must be inside a quantifier or inside the scope of an **x**-quantifier. The first occurrence of the variable v in the formula $\neg(\exists z)(z = y \rightarrow (\exists v)D^3evy)$ is bound because it is inside the quantifier $(\exists v)$, and the second occurrence of v is bound not because it is inside the scope of $(\exists z)$ but because it is inside the scope of the v-quantifier $(\exists v)$.

The notion of a free occurrence of a variable is important. A PL formula that has no free occurrences of any variable is a PL sentence, and a PL formula that has at least one free occurrence of a variable is not a PL sentence. PL formulas that are not sentences are called **proper formulas**. Every PL formula is either a proper formula or a sentence. The formula $((\forall x)(\neg A^1c \wedge B^3xag^1y) \vee \neg(\exists z)(z = y \rightarrow (\exists v)D^3evy))$ is a proper formula, since all the occurrences of the variable y are free. However, if we quantify over these occurrences, we obtain a PL sentence; the formula $(\forall y)((\forall x)(\neg A^1c \wedge B^3xag^1y) \vee \neg(\exists z)(z = y \rightarrow (\exists v)D^3evy))$ is a PL sentence, since it contains no free occurrences of any variable. The importance of a free occurrence of a variable is that it supplies a place over which a quantifier might be applied.

The notion of a bound occurrence of a variable is also significant. Syntactically, a bound occurrence of a variable is not available for quantification. For instance, the occurrences of the variable v in the formula $\neg(\exists z)(z = y \rightarrow (\exists v)D^3evy)$ are not available for quantification. The first occurrence is in $(\exists v)$, which is not suitable for quantification, and the second is inside the scope of $(\exists v)$, which means that it is already quantified over, and hence no further quantification is possible over this occurrence. Bound variables have also semantical significance, even though they have no semantical content. They serve as *placeholders* for the quantifiers: they indicate the places to which the quantifiers apply. For instance, the existential quantifier in the sentence $(\exists z)(\forall x)S^2zx$ applies to the first place of the predicate S^2zx and the universal quantifier applies to the second place of the predicate S^2zx. If we interchange the variables of the quantifiers, we obtain the sentence $(\exists x)(\forall z)S^2zx$. In the latter the existential quantifier applies to the second place of the predicate S^2zx and the universal quantifier to the first place. These two sentences are syntactically and semantically different from each other. To see this, let S^2zx translate the PL-English predicate "z hates x." The first sentence says in English that there is someone who hates everyone. The second says that there is someone such that everyone hates him or her, that is, there is someone who is hated by everyone.[1]

The scope of a quantifier has semantical significance because it determines the range of the applicability of the quantifier. Occurrences of the same variable in different scopes of quantifiers are independent of each other, in the sense that the interpretations of the quantified clauses need not be dependent on each other. Let us consider an example. The sentences $(\exists x)A^1x \wedge (\exists x)B^1x$ and $(\exists x)(A^1x \wedge B^1x)$ are syntactically and semantically very different. In the first sentence the second occurrence of x is inside the scope of the first $(\exists x)$ and the fourth occurrence of x is inside the scope of the second $(\exists x)$. These occurrences are independent of each other. We could have used different variables for these clauses. The formula $(\exists z)A^1z \wedge (\exists y)B^1y$ has the same semantical significance as the first formula above: any PL interpretation that makes one of them true (or false) makes the other true (or false) as well. However, in the sentence $(\exists x)(A^1x \wedge B^1x)$ the second and third occurrences of the variable x are within the scope of the same quantifier, $(\exists x)$. These occurrences are not independent of each other. If we use different variables, we might change the syntax and semantics of the sentence. For example, the formula $(\exists z)(A^1z \wedge B^1y)$ is not a sentence; it is a proper formula because the occurrence of y is free. Furthermore, if we interpret the sentence $(\exists x)(A^1x \wedge B^1x)$ in a natural language, we obtain a declarative sentence, but if we interpret $(\exists z)(A^1z \wedge B^1y)$ in a natural language, we do not obtain a declarative sentence; rather we get a complex predicate.

1 Since the quantifier 'everyone' covers all people, the first sentence implies that the one who hates everyone also hates himself or herself, and the second sentence implies that the one who is hated by everyone is also hated by himself or herself.

In order to make our previous discussion concrete, let us interpret the sentences and formula in the preceding paragraph. Suppose A^1 stand for the property of being a whale and B^1 stand for the property of being a fish. The sentence $(\exists x)A^1x \wedge (\exists x)B^1x$ says, on this interpretation, that there is a whale and there is a fish, which is true in our world. The sentence $(\exists z)A^1z \wedge (\exists y)B^1y$ says the same thing on this interpretation. On the other hand, the sentence $(\exists x)(A^1x \wedge B^1x)$ says that there is something that is both a whale and a fish, which is false in our world. The formula $(\exists z)(A^1z \wedge B^1y)$ expresses the English predicate 'there is something, such that this thing is a whale and __ is a fish'. The presence of the blank indicates that this is not a complete sentence but only a complex predicate. We can fill this blank with a name, obtaining a complete declarative sentence. For example, we can say that there is something, such that this thing is a whale and Goldie is a fish. If 'Goldie' is indeed a name of a fish, this sentence would be true in our world (though it is an odd sentence).

4.2.4:C2 In Subsection 3.2.2:C3 we discussed two **bookkeeping rules**: the Balance Rule and the Main-Connective Rule. These rules can be modified to apply to PL expressions. The Balance Rule for PL is the same as the Balance Rule for SL with one modification: the parentheses of the quantifiers are not included in the count. It is easy to verify whether a quantifier has balanced parentheses or not. A PL expression has balanced parentheses when and only when all its quantifiers have balanced parentheses, the rightmost parenthesis receives the number zero, and no parenthesis in the expression receives a negative number. Recall that we begin the count at the leftmost parenthesis with the number one and we proceed to the right adding one for every left parenthesis and subtracting one for every right parenthesis, ignoring the parentheses of the quantifiers. Here are three examples, the first and second expressions have balanced parentheses and the third does not.

E1 $(\forall y)({}_1(\forall x)({}_2\neg A^1c \wedge (\exists y)B^3xag^1y)_1 \vee \neg(\exists z)({}_2({}_3z = y \leftrightarrow G^1h)_2 \rightarrow (\exists v)D^3evy)_1)_0$

E2 $({}_1\neg A^1c \wedge ({}_2(\exists y)B^3xag^1y)_1)_0 \vee ({}_1({}_2z = y \leftrightarrow G^1h)_1 \rightarrow (\exists v)D^3evy)_0$

E3 $({}_1(\forall x)({}_2\neg A^1c \wedge (\exists y)B^3xag^1y)_1)_0 \vee \neg(\exists z)({}_1({}_2({}_3z = y \leftrightarrow G^1h)_2 \rightarrow (\exists v)D^3evy)_1)_0)_{-1}$

None of these expressions is a grammatical PL formula but, nevertheless, the first two have balanced parentheses. The Balance Rule may be applied to any PL expression not only to PL formulas. All PL formulas have balanced parentheses. The converse, however, is not true. There are many expressions that have balanced parentheses, but they are ungrammatical PL expressions. E1 and E2 have balanced parentheses because the counting process meets the criterion above. E3 does not have balanced parentheses because one of the parentheses (the rightmost) is assigned a negative number. E3 is not grammatical, for grammatical expressions have balanced parentheses. E1 is not a formula; the third occurrence of y is quantified over twice: one with $(\exists y)$ and one with $(\forall y)$. E2 is not a formula because the expression $((\exists y)B^3xag^1y)$, which is the second conjunct in E2, is ungrammatical. The quantifier rules do not introduce parentheses. A correct application of R$\exists$ to B^3xag^1y would generate $(\exists y)B^3xag^1y$, without outer parentheses. It is worth noting that the expression $(\exists y)(B^3xag^1y)$ is also not a formula because atomic formulas are not enclosed between parentheses.

In PL the Main-Connective Rule is called "the Main-Operator Rule," and, as in SL, it applies only to grammatical expressions, that is, to formulas. We speak of operators in PL instead of connectives because the quantifiers are not really connectives. It is important to note that this rule does not produce the correct results if the syntactical rules used in constructing the PL formula are relaxed according to any convention. The same applies to construction trees, which we will cover in the following subsection. There are two modifications to this rule in PL: (1) if the formula begins with a quantifier, then this quantifier is the main operator of this sentence; and (2) the parentheses of the quantifiers are not included in the count. We apply the same counting process as in the Balance Rule. As in SL, if we are dealing with a PL formula, we will reach the number one once, twice, or thrice. We will not restate this rule here. The reader is referred to 3.2.2:C3 for a complete description of this rule. But we will consider a few examples of PL sentences and proper formulas.

E4 $(\exists y)((T^2ca \vee \neg B^1y) \wedge (\forall x)(T^2xy \leftrightarrow \neg(\exists w)B^1w))$

E5 $({}_1({}_2T^2ca \vee ({}_3\neg B^1y \rightarrow H^1a)_2)_1 \wedge (\forall x)({}_2T^2xy \leftrightarrow \neg(\exists w)B^1w)_1)_0$

E6 $\neg((T^2ca \vee (\neg B^1y \rightarrow H^1a)) \wedge (\forall x)(T^2xy \leftrightarrow \neg(\exists w)B^1w))$

E7 $({}_1(\exists v)R^1v \vee \neg(\forall x)({}_2(\exists y)({}_3L^2xk \vee L^3yka)_2 \rightarrow L^2h^1xa)_1)_0$

E8 $({}_1\neg({}_2(\exists w)P^1w \wedge ({}_3B \leftrightarrow L^2ab)_2)_1 \leftrightarrow C)_0$

E9 $({}_1\neg\ P^1e \leftrightarrow C^1g^2sr)_0$

The main operator of E4 is the first quantifier, $(\exists y)$, because the formula begins with it. E4 is a sentence, since all the occurrences of variables are bound. We do not need to apply the counting method for E4. The same is true for E6. The formula begins with $\neg$, so $\neg$ is the main operator. The counting method is not needed. E5 is a proper formula (it is not a sentence) because all the occurrences of y in it are free. E5 begins with a parenthesis. Hence we need to apply the counting method. Since we reach 1 thrice, the main operator is the binary connective that immediately follows the second occurrence of 1. This is the connective $\wedge$. In E7 we reach 1 only twice. The symbol that is immediately to the right of the second occurrence of 1 is the rightmost parenthesis of the formula. According to the Main-Operator Rule, the main operator is the first binary connective that occurs in this formula. This is the connective $\vee$. E7 is a PL sentence. All the occurrences of variables in it are bound. In E8, 1 is also reached twice. This time, however, the second occurrence of 1 is immediately followed by a binary connective; thus this binary connective, $\leftrightarrow$, is the main operator of the formula. E8 is also a PL sentence: the two occurrences of w are bound. In the last formula, E9, 1 is reached only once. This indicates that the formula contains only one binary connective, which is the main operator of the formula. E9 is a sentence because it contains no free variable. In fact, it contains no variable at all.

4.2.5 The **construction tree** of a PL formula **X** displays the main logical operators and the immediate components of **X** and of its subformulas. The tree terminates with the atomic components of **X**.

4.2.5:C COMMENTARY ON 4.2.5

PL construction trees are very similar to SL construction trees. The main differences are that in PL we have formation rules for the quantifiers and that the grammatical expressions of PL are formulas and not only sentences. We have already introduced the notions of a main logical operator, immediate component, subformula, and atomic component (see 4.2.3:C). A construction tree all of whose branches terminate with atomic formulas is called a **complete construction tree**. Construction trees are used to determine whether a PL expression is a PL formula, and, if it is a formula, whether it is a PL sentence. A PL expression is a formula if and only if it has a complete construction tree. A PL formula is a sentence if and only if every variable that occurs in one of its atomic components receives quantification on the branch that connects the top node with that atomic formula. No occurrence of a variable can receive more than one quantification. When generating construction trees, conventions must not be used to relax the syntactical rules. The reader may wish to invoke the Main-Operator Rule in order to determine the main operators of the formulas below.

Here is the construction tree of the PL expression $((E^1m \wedge P^1n) \rightarrow (\exists y)(E^1y \wedge P^1y))$.

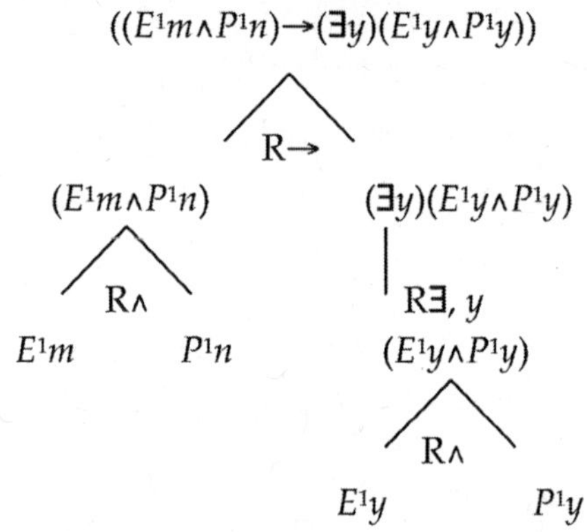

Every branch of this construction tree terminates with an atomic formula. Hence this is a complete construction tree, which proves that the original expression is a PL formula. Every node that is a compound formula is decomposed into its immediate components, and the formulation rule that is used to form this compound formula is indicated immediately below the node. All the formulas displayed on this construction tree are the subformulas of the original formula. The atomic formulas at the bottom of the tree are the atomic components of the formula at the top. When a quantifier rule is applied, we indicate the variable of the quantification. This is why we wrote 'R∃, y'. Care must be applied when a quantifier rule is used. We must make sure that the rule is applied correctly. So we must examine the immediate component of the quantified formula and make sure that the variable of the quantification occurs in this component, and that no quantifier of that variable occurs in this component. We observe that the formula $(E^1y \wedge P^1y)$ satisfies these conditions for the variable y.

This formula is also a sentence. There are two occurrences of the variable y in the atomic components. Both occurrences are quantified over: 'R∃, y' appears on a branch above these atomic formulas. In practice, if we see that an atomic formula contains a variable, we follow the branch upward and see if there is quantification over that variable. If there is, then that occurrence of that variable is bound in the original formula. If all the occurrences of the variables in the atomic components are quantified over, then the original formula is a PL sentence. If no atomic component contains a variable, then we know that the original formula is a sentence.

Let us consider another example.

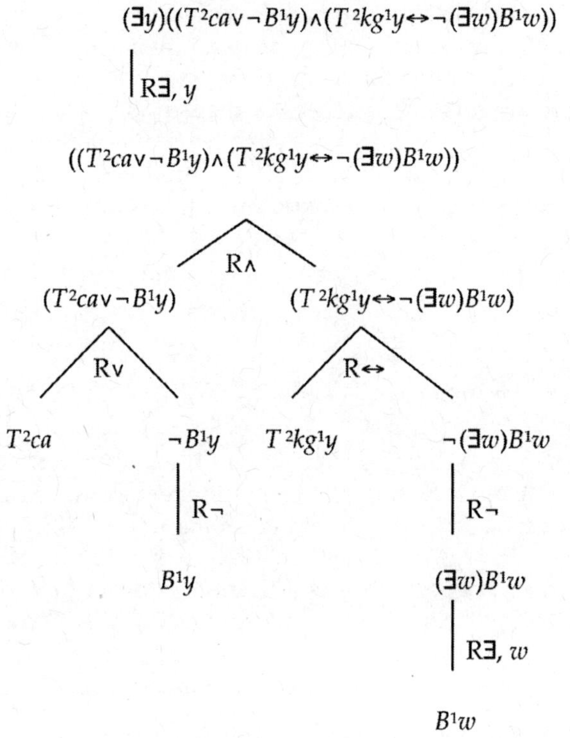

All the branches terminate with atomic formulas. This is a complete construction tree, and hence the original expression is a PL formula. It is also a sentence. The y in B^1y and in T^2kg^1y are quantified over by $(\exists y)$, which is the main operator of the original formula. The w in B^1w is quantified over by $(\exists w)$.

The expression below is a proper PL formula.

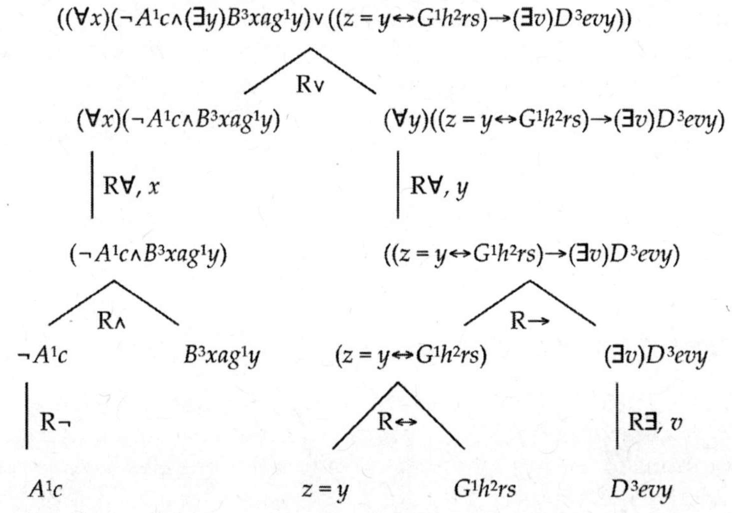

Since the construction tree is complete, the original PL expression is a formula. The variables x in B^3xag^1y, y in $z = y$ and in D^3evy, and v in D^3evy are quantified over. However, the variables y in B^3xag^1y and z in $z = y$ are free. Thus the original formula is not a sentence; it is a proper formula. Observe that there are three occurrences of the variable y. The first occurrence (in B^3xag^1y) is free while the second (in $z = y$) and the third (in D^3evy) are bound.

If we try to produce a construction tree for an ungrammatical expression, the tree would not be complete. The decomposition would halt before reaching all atomic formulas. Here is an example

$$((\forall x)(A^1c \wedge B^3xag^1y) \vee (\forall y)((z = y \leftrightarrow G^1h^2rs) \rightarrow (\exists y)D^3evy))$$

R∨

$$(\forall x)(A^1c \wedge B^3xag^1y) \qquad (\forall y)((z = y \leftrightarrow G^1h^2rs) \rightarrow (\exists y)D^3evy)$$

The subformula $(\forall x)(A^1c \wedge B^3xag^1y)$ can be decomposed fully into its atomic components A^1c and B^3xag^1y. However, there is no reason to proceed with the decomposition because the expression on the right is not a PL formula—it is ungrammatical according to our formation rules. The universal y-quantifier, $(\forall y)$, is applied to the expression $((z = y \leftrightarrow G^1h^2rs) \rightarrow (\exists y)D^3evy)$. But the quantifier rule R∀ does not permit this application. In order for the rule to apply, the expression $((z = y \leftrightarrow G^1h^2rs) \rightarrow (\exists y)D^3evy)$ must satisfy two conditions: (1) it contains occurrences of y, and (2) no y-quantifier occurs in it. The first condition is satisfied but the second is not: a y-quantifier—namely, $(\exists y)$—occurs in the expression. So this construction tree is not complete.

4.2.6 In most cases we shall employ the **conventions** of dropping the outermost parentheses of formulas, of writing predicates without superscripts, and of writing '$\mathbf{s} \neq \mathbf{t}$' instead of '$\neg \mathbf{s} = \mathbf{t}$'. Occasionally we will write function symbols without superscripts.

4.2.6:C COMMENTARY ON 4.2.6

We can always tell the number of places a predicate has by counting the PL terms that immediately follow the predicate. Some readers might think that the expression $\neg \mathbf{s} = \mathbf{t}$ is ambiguous; for it is not clear whether the negation sign applies to **s** or to the formula $\mathbf{s} = \mathbf{t}$. They might think that if we want the negation sign to apply to the formula, we should write $\neg(\mathbf{s} = \mathbf{t})$. This is incorrect. The negation sign cannot apply to **s**, since **s** is a PL term, and it makes no sense to negate a term. Only formulas can be negated. So the negation must apply to the formula $\mathbf{s} = \mathbf{t}$. Furthermore, the expression $\neg(\mathbf{s} = \mathbf{t})$ is ungrammatical, according to our syntactical rules. No atomic formula, such as $\mathbf{s} = \mathbf{t}$, may be enclosed between parentheses. Traditionally, the negation of an identity statement is written as '$\mathbf{s} \neq \mathbf{t}$', and we will follow this traditional notation. Dropping the outermost parentheses simplifies the forms of many formulas. It is a convention we adopted for SL, and we will adopt for PL as well. However, when working with construction trees, we abandon all conventions and strictly follow the syntactical rules of PL.

4.2.7 The type of syntax described above is usually called **generative recursive grammar**. It is so called because every compound formula of PL can be generated from the basic vocabulary by repeated applications of one or more of the seven formation rules listed in 4.2.3. The first five rules can be iterated indefinitely—they are *iterative* or *recursive* (these words are synonyms). This feature implies that the number of PL formulas that can be generated from a finite list of basic vocabulary is infinite. The quantifier rules, however, cannot be iterated indefinitely.

4.2.7:C COMMENTARY ON 4.2.7

We said in 3.2.6:C that the formation rules of SL are indefinitely iterative, that is, they can be repeated an arbitrary number of times. The first five formation rules of PL are virtually the same as the formation rules of SL. The difference is that the formation rules of PL may be applied to proper formulas as well as to sentences. Thus the first five PL formation rules are indefinitely iterative. If any of these rules is applied to PL formulas, the outcome is also a PL formula. Furthermore, these five rules can be applied to any PL formulas without restrictions. These two features entail that these rules can generate an infinite number of PL formulas from a finite set of basic vocabulary.

The last two formation rules of PL—the quantifier rules—can neither be iterated indefinitely nor can they be applied to any PL formulas indiscriminately. In order for a quantifier $(\forall \mathbf{z})$ or $(\exists \mathbf{z})$ to be placed at the left of a PL formula **X**, the variable **z** must occur in **X** and **X** must not contain any **z**-quantifiers. These two conditions restrict greatly the applicability of the quantifier rules. For instance, these rules cannot be applied to any of the following PL formulas.

F1 $((Red \leftrightarrow Msj) \lor (Gr \land \neg Mabgc))$
F2 $((\exists u)(Auk \land \neg Bnn) \leftrightarrow (Gr \land \neg Casu))$
F3 $((\forall z)z = a \rightarrow (\exists x)(Gh^2xte \leftrightarrow \neg Jax))$

F1 contains no variables at all. The only variable that occurs in F2 is u. It occurs three times; the first two are bound and the third is free. However, F2 contains a u-quantifier, which inhibits the applicability of R$\forall$ and R$\exists$. The quantifier $(\forall u)$ or $(\exists u)$ may be placed in front of the formula $(Gr \land \neg Casu)$. But no formulation rule would generate this construction, since this formula is a proper subformula, yet every formation rule of PL can be applied only to whole formulas and not to parts of formulas. F3 contains occurrences of two variables z and x but it also contains a z-quantifier and an x-quantifier. In fact, all the occurrences of variables in F3 are bound.

On the other hand, the quantifier rules can be applied to the PL formula below generating only a finite number of PL formulas.

F4 $((Rxy \leftrightarrow Mxj) \lor (Gz \land \neg Mxf^2by))$

Here is a sample of the PL formulas that can be generated.

F4.1 $(\exists x)((Rxy \leftrightarrow Mxj) \lor (Gz \land \neg Mxf^2by))$
F4.2 $(\forall z)((Rxy \leftrightarrow Mxj) \lor (Gz \land \neg Mxf^2by))$
F4.3 $(\exists x)(\forall y)((Rxy \leftrightarrow Mxj) \lor (Gz \land \neg Mxf^2by))$
F4.4 $(\exists z)(\forall y)((Rxy \leftrightarrow Mxj) \lor (Gz \land \neg Mxf^2by))$
F4.5 $(\forall y)(\forall x)(\forall z)((Rxy \leftrightarrow Mxj) \lor (Gz \land \neg Mxf^2by))$
F4.6 $(\exists z)(\forall y)(\exists x)((Rxy \leftrightarrow Mxj) \lor (Gz \land \neg Mxf^2by))$

No further iterations of the quantifier rules can be applied to F4.5 and F4.6.

4.3 Translating PL into English and English into PL

The translation guides for TL and SL are applicable to PL. The translations of the **English idioms** listed in 3.3.2 are the same for PL, because all of the SL connectives and formation rules are part of PL syntax. The translations of the English idioms listed in 2.3.4 are *expressible* in PL, because PL basic vocabulary contains counterparts to the basic-vocabulary categories of TL. We explained in 4.2.2:C1 how to express TL sentences in PL. In many cases it is helpful to translate first into PL-English and then into the target language. Given the relative richness of PL vocabulary and the relative complexity of its syntax, translating into and from PL is far more involved than translating into and from SL and TL. It is a skill that is developed through practice.

4.3:C COMMENTARY ON 4.3

4.3:C1 We will begin with translating PL into English. We will be given translation keys, universes of discourse, and PL sentences. The translation key indicates the English translations of the extra-logical vocabulary used in the PL sentences. The universe of discourse (UD) determines the set of the individuals over which the quantifiers range. For example, if UD is the set of all people, then the quantifiers $(\exists x)$ and $(\forall x)$ should be read as "there is someone" and "for everyone," but if the UD is the set of all things, then $(\exists x)$ and $(\forall x)$ should be read as "there is something" and "for everything." We will use the translation key and the universe of discourse to convert the PL sentences into idiomatic English sentences. The English translations of the PL sentences below need not be true, they must only be faithful translations.

(a)

Translation Key

e: Elizabeth
j: James
s: Samantha
f^1z: The father of z
Lxy: x loves y
UD: The set of all people

PL Sentences

S1 Lej
S2 Lje
S3 Lf^1ss
S4 Lf^1sf^1s
S5 $(\exists x)(\exists y)Lxy$
S6 $(\forall x)(\forall y)Lxy$
S7 $(\forall x)(\exists y)Lxy$
S8 $(\forall y)(\exists x)Lxy$
S9 $(\exists x)(\forall y)Lxy$
S10 $(\exists y)(\forall x)Lxy$

English Translations

S1 *Elizabeth loves James.*
S2 *James loves Elizabeth.*
S3 *Samantha's father loves Samantha.*
S4 *Samantha's father loves himself.*
S5 *Someone loves someone.*
S6 *Everyone loves everyone.*
S7 *Everyone loves someone.*
S8 *Everyone is loved by someone.*
S9 *There is someone who loves everyone.*
S10 *There is someone who is loved by everyone.*

(b)

Translation Key

a: Adam
Suv: u is a son of v
UD: The set of all people

PL Sentences

S1 $(\forall x)\neg Sax$
S2 $\neg(\exists x)Sax$
S3 $(\forall x)(x \neq a \rightarrow (\exists u)Sxu)$
S4 $(\forall x)(\exists u)(Sux \wedge (\forall z)(Szx \rightarrow z = u))$

English Translations

S1 *Adam is not a son of anyone.*
S2 *Adam is not a son of anyone.*
S3 *Anyone who is not Adam is a son of someone.*
S4 *Everyone has exactly one son.*

Note: The clause $(\forall z)(Szx \rightarrow z = u)$ is usually referred to as *the uniqueness clause*. The first clause, Sux, says that u has the relation S to x. The uniqueness clause says that any individual that has the relation S to x must be identical with u. Thus none but u has the relation S to x.

(c)

Translation Key

s: Simsim
d: Fido
Hx: x is a cat
Fx: x is a dog
UD: The set of all animals

PL Sentences

S1 $Hs \wedge Fd$
S2 $(\forall v)(Hv \rightarrow \neg Fv)$
S3 $(\exists v)Hv \wedge (\exists v)Fv$

English Translations

S1 *Simsim is a cat and Fido is a dog.*
S2 *No cat is a dog.*
S3 *There is a cat and there is a dog.*

(d)

Translation Key

r: Ranah
Nz: *z* wins a gold medal
Ry: *y* receives $1000
UD: The set that consists of the members of the US Olympic Team

PL Sentences

S1 $Nr \rightarrow (\forall x)Rx$
S2 $(\forall u)(Nu \rightarrow (\forall x)Rx)$
S3 $(\forall u)Nu \rightarrow (\forall x)Rx$

English Translations

S1 *If Ranah wins a gold medal, every member of the US Olympic Team receives* $1000.
S2 *If any member of the US Olympic Team wins a gold medal, every member of the team receives* $1000.
S3 *If all members of the US Olympic Team win gold medals, then every member of the team receives* $1000.

(e)

Translation Key

a: Johanna
b: Margaret
Hw: *w* is a granddaughter of Diana
Ky: *y* is a great grandmother of Diana
Dzy: *z* is a descendent of *y*
Sy: *y* is a human
UD: The set of all people

PL Sentences

S1 $(\neg Ha \vee \neg Kb) \rightarrow \neg Dab$
S2 $(Ha \wedge Kb) \rightarrow (\exists x)(Dax \wedge Dxb)$
S3 $(\forall x)(\forall y)(Dxy \rightarrow Sy)$

English Translations

S1 *If Johanna is not a granddaughter of Diana or Margaret is not a great grandmother of Diana, then Johanna is not a descendent of Margaret.*
S2 *If Johanna is a granddaughter of Diana and Margaret is a great grandmother of Diana, then there is someone of whom Johanna is a descendent and who is a descendent of Margaret.*
S3 PL-English: For all *x* and *y*, if *x* is a descendent of *y*, then *y* is a human.
English: *If a person is a descendent of any individual, then that individual is a human.*

Note on the use of variables in a translation key: The variables that are used in expressing PL predicates and their English translations are "dummy" variables and are independent of each other. For instance, if we write '*x* loves *y*' as a translation of *Lxy*, these occurrences of *x* and *y* are independent of each other: each one can be replaced by any name, and it can serve as a placeholder for its own quantifier. However, the variables *x* and *y* themselves are dummy variables, that is, they can be replaced by any other variables as long as they are different letters. For example, we could have written '*Lyx*' or '*Lwz*', but we cannot write '*Lvv*' because, in this case, this predicate expresses the property of self-love and not the relation of love between two different people. However, it is permissible to replace independent variables with the same name. For instance, if *j* stands for Johanna, we can replace *w* and *z* in '*Lwz*' with *j*; we obtain '*Ljj*'. On the other hand, if the predicate is '*Lzz*', then these occurrences of the variable *z* are interdependent: either they are to be replaced with the same name, or they are to be quantified over by the same quantifier. We can write '*Ljj*' and we can write '(∀z)*Lzz*', but we cannot use '*Lzz*' to produce '*Lae*', or to produce '(∃z)(∃z)*Lzz*'. In fact, '(∃z)(∃z)*Lzz*' is ungrammatical; it can never be generated by our syntactical rules. However, *z* itself is a dummy variable. We could have written '*Luu*', '*Lyy*', or, in general, '*L***xx**' for any PL variable **x**.

(f)

Translation Key

t: The number three
a: The number four
b: The number two
g^2xy: The sum of *x* and *y*
Bxy: *x* is greater than or equal to *y*
UD: The set of the natural numbers: {0, 1, 2, 3, 4, ...}

PL Sentences

S1 Bg^2abt
S2 $(\forall v)(\forall u)((Bvu \wedge Buv) \rightarrow v = u)$
S3 $(\forall u)(\forall v)(\forall z)((Buv \wedge Bvz) \rightarrow Buz)$

English Translations

S1 *The sum of four and two is greater than or equal to three.*
S2 *If two natural numbers are greater than or equal to each other, then they are identical.*
S3 PL-English: For all natural numbers *u*, *v*, and *z*, if *u* is greater than or equal to *v* and *v* is greater than or equal to *z*, then *u* is greater than or equal to *z*.
English: *If a natural number is greater than or equal to a second natural number and if the second number is greater than or equal to a third natural number, then the first number is greater than or equal to the third number.*

Note on the use of PL-English in mathematics: The technical language of mathematics makes extensive use of PL-English. The use of variables is a standard practice in mathematics. It is nearly impossible to carry out mathematical computations or mathematical proofs without the use of variables. Almost always if one tries to avoid variables, the resulting expressions either have erroneous connotations, are awkward, ambiguous, or even unintelligible. The last sentence above illustrates some of these aspects. The PL-English translation of the PL sentence is clearer than its English counterpart. The use of the words 'first', 'second', and 'third' in the English translation suggests that these numbers cannot be identical. This is an erroneous connotation. These numbers can all be equal to each other. Mathematical expressions are also very compact due to the use of elaborate symbolism. For instance the relation "greater than or equal to" is denoted in mathematics by the single symbol '≥'. Thus the PL sentence could be translated as follows: for all natural numbers *u*, *v*, and *z*, if $u \geq v$ and $v \geq z$, then $u \geq z$.

4.3:C2 We have gained some experience in translating PL into English. Now we will practice translating English into PL. We will be given English sentences, and our task is to supply a translation key and PL translations of the English sentences. Of course, we may use the logical symbols of PL, PL variables, and parentheses. Unlike translating PL into English, no specific universe of discourse may be assumed. We take the quantifiers as ranging over all things. Sometimes we will find it useful to restate the English or PL sentences in PL-English. Neither the English sentences nor their PL translation need be true.

(a) We translate the English sentences below into PL using only the identity predicate and the name *j* for John.

English Sentences

S1 *Every object is identical with itself.*
S2 *Some object is not identical with itself.*
S3 *No object is identical with itself.*

S4 *Every object is identical with some object.*
S5 *John exists.*
S6 *Only John exists.*
S7 *There is something other than John.*
S8 *John does not exist.*
S9 *There is something.*
S10 *There is only one object.*
S11 *There are at least two objects.*
S12 *There are at most two objects.*
S13 *There are exactly two objects.*

PL Translations

S1 $(\forall v)v = v$
S2 $(\exists z)z \neq z$
S3 $(\forall v)v \neq v$
S4 $(\forall v)(\exists u)v = u$
S5 $(\exists x)x = j$ (There is something that is John.)
S6 $(\forall x)x = j$ (Everything is John.)
S7 $(\exists x)x \neq j$
S8 $\neg(\exists x)x = j$ (There is no individual that is John.)
S9 $(\exists y)y = y$ (PL-English: There is *y*, such that *y* is identical with itself.)
S10 $(\exists y)(\forall z)z = y$ (PL-English: There is *y*, such that everything is *y*.)
S11 $(\exists w)(\exists z)w \neq z$ (There are two things that are not identical with each other.)
S12 $(\exists w)(\exists z)(\forall x)(x = w \vee x = z)$ (PL-English: There are *w* and *z*, such that everything is either *w* or *z*. Note that *w* and *z* could be identical.)
S13 $(\exists w)(\exists z)(w \neq z \wedge (\forall x)(x = w \vee x = z))$ (PL-English: There are *w* and *z*, such that *w* is not *z* and everything is either *w* or *z*.)

(b)

English Sentences

S1 *Everyone who believes in God also believes in the afterlife.*
S2 *Some people believe in God but not in the afterlife.*
S3 *There are those who believe neither in God nor in the afterlife.*

Translation Key

o: God
l: The afterlife
Px: *x* is a person
Bzy: *z* believes in *y*

PL Translations

S1 $(\forall x)((Px \wedge Bxo) \rightarrow Bxl)$

S2 $(\exists z)(Pz \wedge (Bzo \wedge \neg Bzl))$

S3 $(\exists x)(Px \wedge (\neg Bxo \wedge \neg Bxl))$

(c)

English Sentences

S1 *Every animal that lives in the sea is either a whale or a fish.*
S2 *If an animal lives in the sea and is not a whale, then it is a fish.*

Translation Key

Az: *z* is an animal
Sx: *x* lives in the sea
Wy: *y* is a whale
Fx: *x* is a fish

PL Translations

S1 $(\forall y)((Ay \wedge Sy) \rightarrow (Wy \vee Fy))$

S2 $(\forall z)(((Az \wedge Sz) \wedge \neg Wz) \rightarrow Fz)$, or equivalently, $(\forall z)((Az \wedge Sz) \rightarrow (\neg Wz \rightarrow Fz))$

(d)

English Sentence

S *No one who does all the homework for the Symbolic Logic class will fail this class.*

Translation Key

s: Symbolic Logic class
Pv: *v* is a person
Dxy: *x* does *y*
Hxy: *x* is a homework for *y*
Fwz: *w* will fail *z*

PL Translation

S $(\forall w)((Pw \wedge (\forall x)(Hxs \rightarrow Dwx)) \rightarrow \neg Fws)$

PL-English: For every *w*, if *w* is a person such that *w* does all the homework for the Symbolic Logic class, then *w* will not fail the Symbolic Logic class.

Note: The clause '$(\forall x)(Hxs \rightarrow Dwx)$' may be read more literally in PL-English as 'for every *x*, if *x* is a homework for the Symbolic Logic class, then *w* does *x*'.)

(e)

English Sentence

S *No one who has a daughter tolerates acts of sexism.*

Translation Key

Px: *x* is a person
Dw: *w* has a daughter
Sw: *w* is an act of sexism
Txy: *x* tolerates *y*

PL Translation

S $(\forall w)((Pw \wedge Dw) \rightarrow \neg(\exists x)(Sx \wedge Twx))$

Note: In PL the clause '$\neg(\exists x)(Sx \wedge Twx)$' is equivalent to '$(\forall x)(Sx \rightarrow \neg Twx)$', which reads in PL-English as "for every *x*, if *x* is an act of sexism, then *w* does not tolerate *x*." Observe that we interpreted the sentence as saying "No one who has a daughter tolerates *any* act of sexism." This reading is warranted. It is clear that the sentence is not meant to suggest that someone who has a daughter does not tolerate some acts of sexism but he or she might tolerate other acts of sexism.

(f)

English Sentence

S *You can fool some people all of the time or you can fool all people some of the time but you cannot fool all people all of the time.*

Translation Key

Pz: *z* is a person.
Axy: *x* can fool *y* all of the time
Sxy: *x* can fool *y* some of the time

PL Translation

S $(\forall z)(Pz \rightarrow (((\exists x)(Px \wedge Azx) \vee (\forall x)(Px \rightarrow Szx)) \wedge \neg(\forall x)(Px \rightarrow Azx)))$

PL-English: For every *z*, if *z* is a person, then either *z* can fool some person for all of the time or *z* can fool every person for some of the time, but it is not the case that *z* can fool every person for all of the time.

Note: The clause '$(\forall x)(Px \rightarrow Szx)$' could be given a more literal reading as "for every x, if x is a person, then z can fool x for some of the time." The clause '$\neg(\forall x)(Px \rightarrow Azx)$' could be read more literally as "it is not that case that for every x, if x is a person, then z can fool x for all of the time." The last clause is equivalent to the PL-English expression 'there is at least one person x, such that z cannot fool x for all of the time'. Observe that we treated the second-person pronoun, 'You', as a universal quantifier in this context. 'You' could be used in English as a universal quantifier. For example when I say, "If you smoke, you will get sick," I most likely mean that if anyone smokes, he or she will get sick. Similarly, if a warning label on an over-the-counter medicine says, "You should not use this medicine for more than a week," the meaning is not that one or more specific people should not use this medicine for more than a week, but rather that *no one* should. In PL-English this may be expressed as "For every x, if x is a person, then x should not use this medicine for more than a week." The reader might have observed that we elected to build time into our predicates; for instance, Axy translates 'x can fool y all of the time'. But the expressions 'all of the time' and 'some of the time' are quantifiers ranging over time. Thus one could offer a "deeper" translation by making these quantifications explicit in PL. In this case, instead of Axy and Sxy, we would have one 3-place predicate $Fxyz$, which translates 'x can fool y for the time period z'. I leave it as an exercise for the reader to produce a PL translation using $Fxyz$.

(g)

English Sentence

S *The square of an even number is even and the square of an odd number is odd.*

Translation Key

gx: The square of x
Ez: z is an even number
Oy: y is an odd number

PL translation

S $(\forall z)(Ez \rightarrow Egz) \wedge (\forall y)(Oy \rightarrow Ogy)$

4.4 The Semantics of PL

4.4.1 The semantics of PL is much more complex and natural than either the semantics of TL or SL. In TL we have diagrams and in SL truth valuations; in PL we have interpretations. A PL interpretation for some set of PL sentences gives the meanings of these sentences. It tells us what the sentences assert about the individuals in a given collection. In other words, a PL interpretation for some PL sentences is a possible situation that is described (correctly or incorrectly) by these sentences. Here is the precise definition of this notion. A PL **interpretation** I for a set Γ of PL sentences consists of a universe of discourse, a list of names, and semantical assignments. The **universe of discourse** (UD) is simply a *nonempty* (finite or infinite) collection of individuals: people, birds, numbers, planets, geometric figures, linguistic expressions, ideas, or what have you. The **list of names** (LN) consists of names for all the individuals in the universe of discourse, and it must contain all the names that occur in the members of Γ. The **semantical assignments** made by I are as follows.

4.4.1a To every name in LN, I assigns exactly one individual in UD; and every individual in UD is assigned by I to at least one name in LN. The individual I assigns to the name **s** is called the **referent of s on I** and is denoted as 'I(**s**)'.

4.4.1b To every n-place function symbol $\mathbf{f}^n$ that occurs in a sentence in Γ, I assigns exactly one **n-place function** on UD. The function I assigns to the n-place function symbol $\mathbf{f}^n$ is denoted as 'I($\mathbf{f}^n$)'.

4.4.1c To every singular term $\mathbf{f}^n\mathbf{t}_1\mathbf{t}_2\mathbf{t}_3 \ldots \mathbf{t}_n$, where $\mathbf{f}^n$ is an n-place function symbol and $\mathbf{t}_1$, $\mathbf{t}_2$, $\mathbf{t}_3$, …, and $\mathbf{t}_n$ are PL singular terms, I assigns the individual $F(\alpha_1, \alpha_2, \alpha_3, \ldots, \alpha_n)$, where F is the function I assigns to the n-place function symbol $\mathbf{f}^n$, α_1 is the referent I assigns to $\mathbf{t}_1$, α_2 is the referent I assigns to $\mathbf{t}_2$, α_3 is the referent I assigns to $\mathbf{t}_3$, …, and α_n is the referent I assigns to $\mathbf{t}_n$. Symbolically, $I(\mathbf{f}^n\mathbf{t}_1\mathbf{t}_2\mathbf{t}_3, \ldots, \mathbf{t}_n) = I(\mathbf{f}^n)(I(\mathbf{t}_1), I(\mathbf{t}_2), I(\mathbf{t}_3), \ldots, I(\mathbf{t}_n)) = F(\alpha_1, \alpha_2, \alpha_3, \ldots, \alpha_n)$.[1]

4.4.1d I assigns to the identity predicate, '=', the binary relation of **token identity**, which holds between every individual in UD and itself and does not hold between different individuals in UD.

4.4.1e To every 1-place predicate that occurs in a sentence in Γ, I assigns exactly one **bivalent property** on UD. The property I assigns to the 1-place predicate $\mathbf{P}^1$ is denoted as 'I($\mathbf{P}^1$)'.

4.4.1f To every n-place predicate (n is greater than 1) that occurs in a sentence in Γ, I assigns exactly one **bivalent n-place relation** on UD. The relation I assigns to the n-place predicate $\mathbf{R}^n$ is denoted as 'I($\mathbf{R}^n$)'.

I is said to be an interpretation for a PL sentence if it is an interpretation for a set containing that sentence. If I is a PL interpretation *for* a set Γ of PL sentences, it is also described as *relevant to* Γ. We call the individuals, functions, properties, and relations of a PL interpretation "the constituents of the interpretation."

4.4.1:C Commentary on 4.4.1

4.4.1:C1 Before discussing PL interpretations, we need to explain some of the notions mentioned in the definitions above. For the sake of clarity, we will restate some of the definitions and explanations we mentioned previously. In Chapter One, we defined the notions of a bivalent sentence and a bivalent language. We said there that a bivalent sentence is a declarative sentence that is either true or false but not both in every logical possibility, and a bivalent language is a language whose declarative sentences are all bivalent. The notions of bivalence can be extended to properties and relations. A bivalent property is such that every individual in UD either has or lacks this property, but it cannot both have this property and lack it. To see

1 The concept of a function and the notation used in 4.4.1b and 4.4.1c will be fully explained in the commentary.

the relation between bivalent sentences and bivalent properties, consider the following example. A sentence that was proposed in 1.2.2:C1 as an example of a non-bivalent sentence was 'Pat gave up smoking'. We said there that the truth and falsity of this sentence seem to presuppose that Pat has smoked during her life. This entails that the property "*x* gave up smoking" (call this property *Gx*) is not bivalent. If Pat smoked during her life and then stopped smoking, then she has this property, and hence the sentence *Gp*, where *p* stands for Pat, is true. If Pat continues to smoke, then she lacks this property, and hence the sentence *Gp* is false. However, if Pat never smoked in her life, then it seems that it is not the case that she either has this property or lacks it; it seems that she neither has nor lacks this property. In this case the sentence *Gp* is neither true nor false. The point is that if we allow non-bivalent properties and relations, then we have to allow non-bivalent sentences. Since we do not allow non-bivalent sentences, we should not allow non-bivalent properties and relations.

As we explained previously, the set of all the individuals that have a property P is called **the extension of** P. The extension of a 1-place PL predicate on any relevant PL interpretation I is the extension of the property that this predicate designates on I. A sequence of n individuals in some specific order is called **n-tuple.** For instance, the 2-tuple ⟨0, 1⟩ is an ordered pair consisting of the first two natural numbers ordered by the relation "less than," and the 3-tuple ⟨a, b, c⟩ is an ordered triple consisting of the first three letters of the English alphabet in their standard order. In general, an n-tuple of individuals $\mathbf{a}_1, \mathbf{a}_2, \mathbf{a}_3, \ldots, \mathbf{a}_n$ (not necessarily all distinct) is denoted as $\langle \mathbf{a}_1, \mathbf{a}_2, \mathbf{a}_3, \ldots, \mathbf{a}_n \rangle$. The individuals that constitute an n-tuple are called **coordinates.** A bivalent n-place relation is such that every n-tuple of individuals in UD either bears this relation or it does not; it cannot bear and not bear this relation. The **extension of an n-place relation** is the set that consists of all the n-tuples of individuals that bear this relation to each other. For example, the ordered pairs ⟨0, 1⟩, ⟨5, 7⟩, and ⟨13, 22⟩ are all in the extension of the binary relation "less than," and the ordered pairs ⟨Turkey, Iraq⟩ and ⟨China, India⟩ are in the extension of the binary relation "to the north of"; of course, the extensions of these relations contain many more ordered pairs.

An n-place function on UD may be roughly defined as a rule that assigns to *every* n-tuple of individuals in UD a *unique* individual also in UD. For example, if UD is the set of all the natural numbers, {0, 1, 2, 3, ...}, and A is defined as the rule "the sum of *x* and *y*," then A is a 2-place function on UD. A satisfies the definition of a function: to *every* ordered pair ⟨n, m⟩ of numbers in UD, A assigns a *unique* (i.e., one and only one) number in UD that is the sum of n and m. Let k be the sum of n and m. We express this fact as "A(n, m) = k." We refer to the numbers n and m as the **arguments** of A and to the number k as the **value** of A *at* these arguments. Using the standard mathematical symbol for addition, we may express this function by writing "A(n, m) = n+m." Since PL interpretations assign functions to PL function symbols and since PL function symbols are meant as counterparts of functional descriptions, we will in most cases use PL-English functional descriptions to define the functions that are assigned to function symbols. For instance, rather than defining the function A above as the rule that assigns to every ordered pair of natural numbers the sum of these numbers, we will define A by the functional description 'The sum of *x* and *y*'.

Precisely speaking, functions are relations of a special type; and hence they too have extensions. The extension of the function A consists of ordered triples ⟨n, m, k⟩, where the first two coordinates of the triples are the arguments of A and the last coordinate is the value of A at n and m. For instance, the ordered triples ⟨5, 7, 12⟩, ⟨13, 13, 26⟩, and ⟨0, 21, 21⟩ are all in the extension of A. As another example, consider the 3-place function on the positive integers, {1, 2, 3, ...}, that is defined by the functional description "The greatest common devisor of *x*, *y*, *z*."

This is a function because any three positive integers have a unique greatest common divisor. For example, the greatest common divisor of 21, 35, and 49 is 7, the greatest common divisor of 7, 13, and 15 is 1, and the greatest common divisor of 6, 12, and 18 is 6. Using the notation introduced in the preceding paragraph, we may define this function as "F(n, m, k) = the greatest common divisor of n, m, and k, where n, m, and k are any positive integers." The numbers n, m, and k are the arguments of F and F(n, m, k) is the value of F at these arguments. The extension of F consists of 4-tuples ⟨n, m, k, j⟩ such that the first three coordinates are the arguments of the function F and the last coordinate is its value at those arguments. In general, the extension of an n-place function consists of (n+1)-tuples such that the first n coordinates of each tuple are the arguments of the function and the n+1st (i.e., the last) coordinate of the tuple is the value of the function at those arguments.

As we said above, functions are relations of a special type. We explain this type of relation by means of an example. Let I be a PL interpretation whose UD consists of all human males. Assume for the sake of example that every human male has exactly one biological father. Further assume that the constituents of I includes the 1-place function "The biological father of *x*" and the 2-place relation "*y* is the biological father of *x*." Let us refer to the function as F*x* and to the relation as R*xy*. This function and this relation have exactly the same extension on I. The extension of F*x* on I consists of ordered pairs $\langle\alpha, \beta\rangle$ such that α is the argument of this function and β is the value of the function at this argument. In simpler terms, α is the biological father of β. The extension of R*xy* on I consists of precisely the same ordered pairs in the extension of F*x*. It is clear from the definition of R*xy* that an ordered pair $\langle\alpha, \beta\rangle$ is in the extension of R*xy* if and only if α is the biological father of β. Since PL is an extensional system, functions, properties, and relations are identified with their extensions. Hence, on I, both the function F*x* and the relation R*xy* are identified with the unique extension that consists of all the ordered pairs whose second coordinate is the biological father of the first coordinate. It follows that F*x* and *Rxy* are identical on I. In general, a 1-place function on some set D is a binary relation Q*xy* on D that satisfies the following condition: for every member α of D, there is exactly one member β of D such that the ordered pair $\langle\alpha, \beta\rangle$ is in the extension of Q*xy*. This definition can be generalized in a straightforward way to any n-place function.

Function: An n-place function F on some set D is an (n+1)-place relation such that for every n-tuple $\langle\alpha_1, \alpha_2, \alpha_3, \ldots, \alpha_n\rangle$ of coordinates in D, there is a unique individual β in D where the (n+1)-tuple $\langle\alpha_1, \alpha_2, \alpha_3, \ldots, \alpha_n, \beta\rangle$ belongs to the extension of F. The individuals $\alpha_1, \alpha_2, \alpha_3, \ldots,$ and α_n are called "the arguments of F" and β "the value of F at the arguments $\alpha_1, \alpha_2, \alpha_3, \ldots,$ and α_n." β is typically denoted as $F(\alpha_1, \alpha_2, \alpha_3, \ldots, \alpha_n)$.

As defined previously, a PL singular term is a PL name or a PL term that contains a function symbol but no variables. Given that function symbols are interpreted as functions on UD, singular terms behave like names in that they also refer to unique individuals in UD. To see this, consider the following example. Suppose that we have the singular term f^2ag^1b, and that a PL interpretation I, whose UD is the set of all positive integers, {1, 2, 3, …}, assigns the function "the sum of *x* and *y*" to f^2xy, the function "the square of *z*" to g^1z, and the numbers 11 and 7 to the names *a* and *b*, respectively. The interpretation of the singular term f^2ag^1b on I is the definite description "the sum of 11 and the square of 7" (i.e., $11 + 7^2$), which is the number 60. Thus on I, the singular term f^2ag^1b refers uniquely to the number 60. This is true of all PL singular terms. We state this fact as a theorem.

Theorem 4.1: Every PL singular term has one and only one referent (in UD) on every PL interpretation that is relevant to it (i.e., that interprets it).

If **t** is a PL singular term and I is a PL interpretation that is relevant to **t**, we designate the referent of **t** on I as I(**t**). This theorem follows from 4.4.1a–4.4.1c, and the definition of a function. Although the proof is straightforward, it invokes a certain mathematical principle called the "Principle of Mathematical Induction," which we do not intend to discuss in this book. However, it is not hard to see why this theorem is true. If the term **t** is a PL name, the theorem is obvious: 4.4.1a of the definition of a PL interpretation stipulates that every name must be assigned to a unique referent in UD. Now consider the case of a singular term of the form $\mathbf{g^n s_1 s_2 s_3 \ldots s_n}$, where $\mathbf{g^n}$ is an n-place function symbol and $\mathbf{s_1}$, $\mathbf{s_2}$, $\mathbf{s_3}$, ... and $\mathbf{s_n}$ are PL *names*. By 4.4.1b of the definition of a PL interpretation, $\mathbf{g^n}$ is assigned an n-place function on UD. Let that function be G. Also let $\langle \alpha_1, \alpha_2, \alpha_3, \ldots, \alpha_n \rangle$ be an n-tuple whose coordinates are members of UD such that α_1 is the referent of $\mathbf{s_1}$, α_2 of $\mathbf{s_2}$, α_3 of $\mathbf{s_3}$, and so on. We know that all of these α's exist and are unique since $\mathbf{s_1}$, $\mathbf{s_2}$, $\mathbf{s_3}$, ... and $\mathbf{s_n}$ are PL names. According to the definition of a function (above) there is a *unique* individual in UD that is the value of G at the arguments α_1, α_2, α_3, ..., and α_n. This unique individual is denoted as $G(\alpha_1, \alpha_2, \alpha_3, \ldots, \alpha_n)$. According to 4.4.1c of the definition of a PL interpretation, $G(\alpha_1, \alpha_2, \alpha_3, \ldots, \alpha_n)$ is the unique referent of $\mathbf{g^n s_1 s_2 s_3 \ldots s_n}$. However, in general, PL singular terms admit degrees of complexity that far exceed the structure of $\mathbf{g^n s_1 s_2 s_3 \ldots s_n}$. We briefly discussed the structure of PL terms in 4.2.2:C2. We will revisit this issue here with more elaboration. The general case is that of a PL term that consists of an n-place function symbol followed by n singular terms, many of which could contain several function symbols followed by other singular terms. As an example consider the PL singular term $f^4cg^2h^1bf^3ang^1esf^1r$. This singular term consists of the 4-place function symbol f^4 followed by four singular terms: (1) the name *c*, (2) the singular term $g^2h^1bf^3ang^1e$, (3) the name *s*, and (4) the singular term f^1r. The singular term $g^2h^1bf^3ang^1e$ consists of the 2-place function symbol g^2 followed by two singular terms: (1) the singular term h^1b, and (2) the singular term f^3ang^1e. The last singular term consists of the 3-place function symbol f^3 followed by three singular terms: (1) the name *a*, (2) the name *n*, and (3) the singular term g^1e. Each of the singular terms h^1b, g^1e, and f^1r consists of a 1-place function symbol followed by a name. We can describe the structure of this complex term diagrammatically using a sort of construction tree.

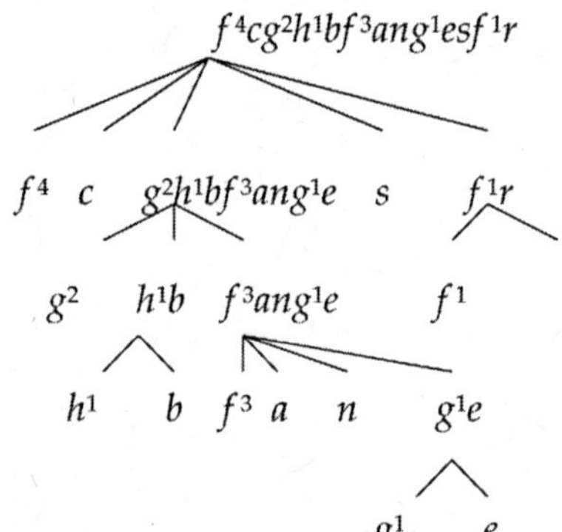

As this example suggests, all PL singular terms, no matter how complex they may be, can be decomposed into names or function symbols followed by names or both; and we have established above that each one of these cases generates a unique referent. This is not exactly a proof of Theorem 4.1 but it is an informal explanation of why it is true. For our purposes, such an explanation is sufficient. Theorem 4.1 shows that we are entitled to our definition of the semantics of singular terms as stated in 4.4.1c.

4.4.1:C2 A PL interpretation I can be represented as an ordered triple ⟨UD, LN, SA⟩, where UD is the universe of discourse, LN is the list of names, and SA are the semantical assignments. There is something peculiar about the way we defined a PL interpretation. We required an interpretation to bring its own names for the individuals in its universe of discourse. In other words, the individuals of an interpretation come labeled. In practice the list of names seems extraneous to the function of an interpretation. In reality these names are essential for the *formal* truth conditions of the quantifiers that are given in this book. Once we give an interpretation for a PL sentence, the sentence acquires a specific meaning on that interpretation. We describe this meaning by translating the PL sentence into the language of the interpretation. This gives the sentence a "natural reading" on that interpretation. In most cases, we can rely on this natural reading to determine the truth value of the interpreted sentence without invoking the formal truth conditions, which we will discuss with elaboration later.

The semantical assignments of names to individuals ensure three things: (1) every name has a referent, that is, there are no non-referring names; (2) no name has more than one referent, that is, there are no ambiguous names; and (3) every individual has *at least one* name, that is, there are no unnamed individuals. Every function symbol and every predicate is assigned, on a PL interpretation, exactly one function and exactly one property or a relation, respectively. This ensures that no function symbol and no predicate are ambiguous.

4.4.1:C3 The best way to explain the notion of PL interpretation is to give a few examples of it. In the next subsection, we will be given sets of PL sentences and we will be asked to give interpretations for them. Some problems place certain conditions on these interpretations. Unless the problem requires an interpretation to make certain sentences true, the truth and falsity of the interpreted sentences are not a condition on the proposed interpretations. A PL interpretation gives semantical contents (i.e., meanings) to the relevant PL sentences; it need not make them true.

When given a set of PL sentences and asked to give an interpretation for this set, we might wonder how to begin constructing an interpretation. The best approach is to begin by specifying a certain universe of discourse. We usually specify a UD that "makes sense" to us, i.e., we choose individuals some of whose properties and relations are familiar to us. If you know some biology, you might be comfortable with a UD that consists of biological species. If you are interested in certain mathematical structures, you might want to specify a UD that consists of certain types of mathematical objects, say, numbers or geometric figures. Perhaps, you like pets and would like the PL sentences to make assertions about some pets you know. All of us are comfortable with talking about people and their properties and relations.

Once we specify a UD, we make sure that we have enough names to label all the individuals in our UD. Our list of names must include the names that occur in the PL sentences that we want to interpret. Then we assign to each 1-place predicate and each n-place predicate that occurs in a given PL sentence a bivalent property or a bivalent n-place relation that the individuals in UD might have. We need to make clear which individuals, if any, have these properties and relations and which individuals do not. We can accomplish this in one of three ways:

(1) state explicitly "facts" that indicate the individuals that have this property or relation and the individuals that do not; (2) give the extension of this property or relation; and (3) rely on common knowledge to determine the individuals that have this property or relation (mathematical interpretations are frequently of this sort).

In interpreting the predicates that occur in the given PL sentences, we can assign properties and relations, or we can assign the extensions of properties and relations without specifying the properties and relations that have these extensions. In fact, PL semantics is extensional: a property or a relation is reduced to its extension. Thus, according to this semantics, specifying the extension of a property or a relation is the same thing as specifying a property or a relation. However, we will find in practice that specifying a "natural-sounding" property or a relation, instead of just giving an extension, improves the natural readings of the PL sentences.

Function symbols are interpreted by assigning functions to them. Since the extension of a 1-place function consists of ordered pairs and, in general, the extension of an n-place function consists of (n+1)-tuples, almost always defining functions by means of their extensions is a cumbersome exercise. As stated previously, we will define functions by stating functional descriptions. We usually use PL-English for this purpose.

When all the tasks described above are accomplished, the construction of a PL interpretation is completed. We do not need to determine the truth values of the given PL sentences on this interpretation, unless we are asked to do so. However, in order to gain some experience, we will always *try* to determine the truth values of the PL sentences on the interpretations we construct. We say that we will *try*, because, unlike the case of SL and TL, determining the truth values of PL sentences on certain interpretations might be very difficult or even impossible given the state of our current knowledge. So it is advisable that we avoid complex interpretations that might require "deep" knowledge in order to determine the truth values of the given PL sentences on these interpretations. Although we have not yet stated the truth conditions of the PL sentences, it is usually sufficient to rely on the natural readings of the interpreted PL sentences and the truth conditions of the sentential connectives, with which we are familiar from Chapter Three, in order to determine the truth values of these PL sentences.

If the problem requires that a PL interpretation make certain sentences true and others false, we, after determining the truth values of the interpreted PL sentences, try to manipulate various aspects of the interpretation in order to meet these requirements. A PL sentence might come false on our interpretation, but the problem requires that the sentence be true; so we "tinker" with the semantical assignments of properties and relations, or with the semantical assignments of referents, or with the UD itself until we manage to make the sentence true on the modified interpretation. We should note that *every* change in an aspect of an interpretation, no matter how minor it might be, produces a new interpretation.

4.4.1:C4 We now demonstrate how to implement the procedure described in the preceding subsection. As usual, any semantical assignment that I makes is denoted as I(#), where # is either a name, a function symbol, or a predicate.

(a) We will construct a PL interpretation for the following set of PL sentences on which the first five sentences are true and the sixth sentence is false.

S1 $Hc \vee De$
S2 $(\exists x)Hx \wedge (\exists y)Dy$
S3 $(\exists x)Hx \rightarrow Pc$
S4 $(\exists x)Dx \rightarrow Pe$
S5 $(\exists x)(Hx \wedge Dx) \rightarrow c \neq e$
S6 $c \neq e$

I am going to describe the family of one of my brothers-in-law as it was in the year 2000. I will not pay attention first to the requirement stated in the problem of making the first five sentences true and the sixth false. After I construct the interpretation, I will manipulate certain aspects of the interpretation in order to meet that requirement. We use the colon, instead of the identity sign '=', to indicate identity in the metalanguage in order to reserve the sign '=' for the object language.

The PL Interpretation I

UD: {Jim, Peggy, Andrea, Tom}
LN: *c*, *e*, *j*, *p*

Note: We have to include the names *c* and *e*, which occur in the set above, and add two more names in order to have enough names for all the individuals in the UD.

Semantical assignments

I(*c*): Andrea; I(*e*): Tom; I(*j*): Jim; and I(*p*): Peggy
I(*Dx*): the property "*x* is a parent": {Jim, Peggy}
I(*Hx*): the property "*x* is a child": {Andrea, Tom}
I(*Px*): the property "*x* was born in Iowa": {Jim, Peggy}
I($x = y$): the relation "*x* is identical with *y*": {⟨Jim, Jim⟩, ⟨Peggy, Peggy⟩, ⟨Andrea, Andrea⟩, ⟨Tom, Tom⟩}

Note: For the sake of simplicity, we will omit in future interpretations the phrases 'the property' and 'the relation'. We determine which individuals have these properties and which do not by giving the extensions of the properties. We could, instead, state explicitly facts that indicate the individuals that have these properties and the individuals that do not. It is important to note that **the quantifiers range over UD**. Thus if we say "every individual" and "there is an individual" we mean "every individual in UD" and "there is an individual in UD." In the future, we will indicate the interpretation of '=' by simply stating that it denotes the relation of token identity on UD. We describe the properties and relations that are assigned to PL predicates by giving PL-English predicates that interpret the PL predicates.

The truth values of the PL sentences

Natural reading of (S1) $Hc \vee De$ on I: *Andrea is a child or Tom is a Parent.*
S1 is true on I because the first disjunct is true (the second disjunct is false).

Natural reading of (S2) $(\exists x)Hx \wedge (\exists y)Dy$ on I: *There is a child and there is a parent.*
S2 is true on I because the extensions assigned to *H* and *D* are not empty.

Natural reading (S3) $(\exists x)Hx \rightarrow Pc$ on I: *If there is a child, then Andrea was born in Iowa.*
S3 is false on I since the antecedent is true and the consequent is false.

Natural reading of (S4) $(\exists x)Dx \rightarrow Pe$ on I: *If there is a parent, then Tom was born in Iowa.*
S4 is false on I because the antecedent is true and the consequent is false.

Natural reading of (S5) $(\exists x)(Hx \wedge Dx) \rightarrow c \neq e$ on I: *If there is a child who is also a parent, then Andrea is not Tom.*
S is true on I because the antecedent is false and the consequent is true on I.

Natural reading of (S6) $c \neq e$ on I: *Andrea is not Tom.*
S6 is true on I.

We need to modify I in order to meet the requirement stated in the problem. To make S6 false, we should assign the same referent to *c* and *e*. We use I* to designate the modified interpretation.

First modification I*(*c*): Andrea; I*(*e*): Andrea

But now we need a name for Tom. So we add a fifth name to LN, say, *t*.

Second modification I*(*t*): Tom

Since $c \neq e$ is false on I*, the antecedent and consequent of $(\exists x)(Hx \wedge Dx) \rightarrow c \neq e$ are false on I*; hence S5 is still true on I*. It remains to make S3 and S4 true. It suffices to make *Pe* true on I*. In order to make *Pe* true, the referent of *e* must be in the extension of *P*. So we add Andrea to the extension of *P*.

Third modification I*(*P*): "*x* was born in Iowa" = {Jim, Peggy, Andrea}

In fact, I* no longer describes the actual situation: Andrea was born in Minnesota and not in Iowa. This is immaterial. PL interpretations describe actual or possible situations. We may design our interpretations as we please, so long as we adhere strictly to the definition of PL interpretation in order to ensure consistency. The description of I and the modifications listed above supply a complete description of the new interpretation, I*.

S1–S6 have the same natural readings on I* as they do on I. We explained above that S1–S5 are true and S6 is false on I*. Therefore the requirement of the problem has been met. We can look ahead and make an interesting observation. If we consider the PL argument whose premises are the sentences S1–S5 and whose conclusion is S6, that is, the argument {S1, S2, S3, S4, S5} / S6, the interpretation I* demonstrates that this PL argument is deductively invalid because on I* the premises are true and the conclusion is false. Although we have not yet given the definition of deductive validity in PL, we can anticipate what this definition is: a PL argument is deductively valid if and only if there is no PL interpretation on which the premises of the argument are all true and the conclusion is false.

(b) We will give two interpretations for the PL sentence S, one of which makes S true and the other makes it false. We will begin with *any* interpretation for S; S is bound to be either true or false on this interpretation. We then modify the interpretation in order to switch the truth value of S.

S $(\forall x)(Qx \rightarrow (Kx \wedge Dx)) \leftrightarrow (\exists v)(Dv \wedge Kv)$

The PL Interpretation I

UD: {Samantha, Christopher, Lisa, Darius}
LN: *s*, *c*, *l*, *d*

Semantical Assignments

I(*s*): Samantha; I(*c*): Christopher; I(*l*): Lisa; I(*d*): Darius
I(*Qx*): "*x* is an English major": {Samantha, Darius}
I(*Dx*): "*x* plays tennis": {Samantha, Darius, Lisa}
I(*Kx*): "*x* plays the piano": {Samantha, Darius, Christopher}

Natural reading of $(\forall x)(Qx \rightarrow (Kx \wedge Dx)) \leftrightarrow (\exists v)(Dv \wedge Kv)$ on I:
All English majors play the piano and play tennis if and only if there is someone who plays tennis and plays the piano.
The left-hand side of the biconditional is true on I, since every individual in the extension of *Q*—namely, Samantha and Darius—is also in the extensions of *D* and *K*. The right-hand side of the biconditional is also true on I, since there is an individual—e.g., Samantha—who is in the extensions of *D* and *K*. Thus S, which is the biconditional sentence, is true on I.

In order to make S false we need to make one of the sides of the biconditional true and the other false. We will keep the right-hand side true and try to switch the truth value of the left-hand side. If there is an English major who does not play tennis or does not play the piano, the left-hand side, $(\forall x)(Qx \rightarrow (Kx \wedge Dx))$, would be false because this side asserts that every English major (*Q*) plays tennis (*D*) and plays the piano (*K*). Thus it is sufficient to remove, say, Samantha from the extension of *D*. In this case, Samantha, who is an English major, does not play tennis. We let I* be the modified interpretation.

The natural reading of S on I* is the same as its reading on I. We explained above that the left-hand side of S is false and the right-hand side is true on I*. Therefore S is false on I*. Although we have not yet defined the logical concept of contingency in PL, we can see that this sentence is contingent since it is true on one PL interpretation and false on another.

(c) We want to construct an interpretation for the following set of PL sentences, such that it has an infinite universe of discourse and every sentence is true on it.

S1 $(\forall x)\, o \neq gx$
S2 $(\forall x)(\forall y)(gx = gy \rightarrow x = y)$
S3 $(\forall x)(x \neq o \rightarrow (\exists z)\, x = gz)$

The PL Interpretation I

UD: The set of all natural numbers: {0, 1, 2, 3, …}
LN: $o, a_1, a_2, a_3, a_4, \ldots, a_n, \ldots$

Semantical Assignments

I(o): 0; I (a_1): 1; I(a_2): 2; I(a_3): 3; …; in general, I(a_n): n
I(gx): "The successor of x" (i.e., I(gx) is x+1)

Natural reading of (S1) $(\forall x)\ o \neq gx$ on I:
0 is not the successor of any natural number.
True on I.

Natural reading of (S2) $(\forall x)(\forall y)(gx = gy \rightarrow x = y)$ on I:
Any two natural numbers that have the same successor are identical.
True on I.

Natural reading of (S3) $(\forall x)(x \neq o \rightarrow (\exists z)\ x = gz)$ on I:
Every natural number that is not 0 is the successor of some natural number.
True on I.

Since S1–S3 are true on I and UD is an infinite set, the requirement of the problem is met. We will see later that, according to the definition of the logical concept of consistency in PL, the set {S1, S2, S3} is consistent because there is a PL interpretation on which the members of the set are all true.

4.4.1:C5 Let us look more closely at what we termed "natural readings" of some of the PL sentences we interpreted in the previous subsection. Consider the sentence $(\forall x)(Qx \rightarrow (Kx \wedge Dx))$, which is the left-hand side of the biconditional $(\forall x)(Qx \rightarrow (Kx \wedge Dx)) \leftrightarrow (\exists v)(Dv \wedge Kv)$ interpreted in 4.4.1:C4(b). We said that the natural reading of $(\forall x)(Qx \rightarrow (Kx \wedge Dx))$ on I is: All English majors play the piano and play tennis. According to this reading, the universal quantifier is interpreted as ranging over the members of UD. To see the point clearly, consider the PL-English reading of this sentence: For every x, if x is an English major, then x plays the piano and plays tennis. Thus 'For every x, …' is interpreted as making an assertion about all the individuals in UD. This way of interpreting the quantifiers is referred to as **objectual quantification** ('objectual' from 'object').

However, the formal truth conditions we will give for the quantifiers in the next section specify an interpretation of the quantifiers on which the variable of the quantification ranges over the names in LN. This way of interpreting the quantifiers is called **substitutional quantification**. In order to state this sort of quantification precisely, we need to introduce the notion of *substitutional instance* of a quantified sentence. Consider the following definition.

Substitutional Instance: The PL sentence **X[t]** is a substitutional instance of the quantified sentence (Δ**z**)**X**, where Δ is either the universal quantifier symbol ∀ or the existential quantifier symbol ∃, if and only if **X[t]** is obtained from **X** by replacing all the occurrences of the variable **z** in the formula **X** with the singular term **t**. If **t** is a name, we refer to **X[t]** as a **basic substitutional instance** of (Δ**z**)**X**.

This definition makes sense since once we remove the quantifier (Δ**z**) the occurrences of the variable **z** in **X** become free, and hence they become available for substitution. For example, $(Ql \rightarrow (Kl \wedge Dl))$ is a substitutional instance of the universally quantified sentence $(\forall x)(Qx \rightarrow (Kx \wedge Dx))$. This substitutional instance is obtained by removing the universal quantifier from $(\forall x)(Qx \rightarrow (Kx \wedge Dx))$, resulting in the formula $(Qx \rightarrow (Kx \wedge Dx))$, and then substituting the name l for x in all its occurrences in $(Qx \rightarrow (Kx \wedge Dx))$.

On the interpretation described in Problem (b) in the previous subsection, $(\forall x)(Qx \rightarrow (Kx \wedge Dx))$ has four substitutional instances because there are only four singular terms that are interpreted by I. These are the four names, *s*, *c*, *l*, and *d*, that are listed in LN. There are no other singular terms because there are no function symbols interpreted by I. In other words, the substitutional instance of $(\forall x)(Qx \rightarrow (Kx \wedge Dx))$ on I are precisely its *basic* substitutional instances. Thus these substitutional instances are:

$(Qs \rightarrow (Ks \wedge Ds))$; $(Qc \rightarrow (Kc \wedge Dc))$; $(Ql \rightarrow (Kl \wedge Dl))$; and $(Qd \rightarrow (Kd \wedge Dd))$

If the universal quantifier $(\forall x)$ is interpreted substitutionally, the sentence $(\forall x)(Qx \rightarrow (Kx \wedge Dx))$ asserts that all its basic substitutional instances are true. This entails that in order for $(\forall x)(Qx \rightarrow (Kx \wedge Dx))$ to be true on I, the substitutional instances stated above must all be true on I. Since they are true on I, the sentence $(\forall x)(Qx \rightarrow (Kx \wedge Dx))$ is true on I.

The objectual reading of the sentence $(\exists v)(Dv \wedge Kv)$ on I is the natural reading we stated in 4.4.1:C4(b)—namely, that there is an individual in UD who plays tennis and plays the piano. The quantifier 'There is *v*, such that ...' is interpreted objectually as asserting that there is an individual in UD, such that ... . The sentence is true on I because there is such an individual in UD (e.g., Samantha). Substitutional quantification interprets $(\exists v)(Dv \wedge Kv)$ as asserting that there is a basic substitutional instance that is true on I. Just as $(\forall x)(Qx \rightarrow (Kx \wedge Dx))$, this sentence has four substitutional instances on I. Since the substitutional instance $(Ds \wedge Ks)$ is true on I, the sentence $(\exists v)(Dv \wedge Kv)$ is also true on I.

In general, the **objectual interpretation** of a PL sentence of the form (∀**z**)**X** is that "for every individual in UD, this individual *satisfies* the formula **X**";[1] and the objectual interpretation of (∃**z**)**X** is that "there is an individual in UD, such that this individual satisfies the formula **X**." The **substitutional interpretation** of (∀**z**)**X** is that "every *basic* substitutional instance of (∀**z**)**X** is true"; and the substitutional interpretation of (∃**z**)**X** is that "there is at least one *basic* substitutional instance of (∃**z**)**X** that is true." When we discuss the truth conditions of the quantifiers, we will explain that, as long as every individual in UD has a name in LN, both

1 The relation of *satisfaction*, which can hold between individuals in UD and PL formulas, is a technical notion. We will not discuss this notion in this book. We only speak of an interpretation *satisfying* a set of PL sentences, in the sense that every member of the set is true on that interpretation. This sense is different from the technical notion of satisfaction.

interpretations of the quantifiers deliver the same truth values for the same quantified sentences. Having said this, there are logical and philosophical reasons for preferring one interpretation to the other. We will discuss some of these reasons in 4.4.3:C2.[1]

4.4.2 The **size of a PL interpretation** is the **cardinality** of its universe of discourse. Intuitively, the cardinality of a set is the number of its members. Thus the cardinality of a set consisting of n members, where n is a non-negative integer, is n. Such a set is finite. For example, the sets ∅ (the empty set), {*a*, *b*, *c*}, {Abraham, Sarah, Hagar, Ishmael, Isaac}, and {*x*: *x* is a planet in our solar system} (i.e., the set of all *x*, such that *x* is a planet in our solar system) are finite and their cardinalities are, respectively, 0, 3, 5, and 8. A **finite interpretation** is an interpretation whose universe of discourse is a finite set. A set whose cardinality is at least as great as the cardinality of the set of the natural numbers, {0, 1, 2, 3, …} (call it ℕ), is an infinite set. An **infinite interpretation** is an interpretation whose universe of discourse is an infinite set.

4.4.2:C COMMENTARY ON 4.4.2

The interpretations described in 4.4.1:C4(a) and (b) are finite. Since there are four individuals in each of their UDs, their size is four. On the other hand, the interpretation of 4.4.1:C4(c) is infinite: its UD is the set of all natural numbers, which is an infinite set. An interpretation might be finite, even though we have no idea what its size is. For example, an interpretation whose UD is the set of all the bright stars in the universe is most likely a finite interpretation even though we have no idea what its size is, since we have no idea how many bright stars there are in the universe. Similarly, it is common to have interpretations whose UD is the set of all people. If one means "all people who have ever lived or will ever live on Earth," then this interpretation is definitely finite even though it is extremely unlikely that its size will ever be known.

Infinite interpretations come in many sizes. This might sound counterintuitive, since many people are inclined to say that the infinite is infinite whether we add to it or subtract from it. Indeed, we do not change the cardinality of an infinite set by adding to it or subtracting from it finitely many things. Similarly, sometimes an infinite cardinality does not change even if we add to it or subtract from it infinitely many things. For instance, if we add to the set of positive even integers, {2, 4, 6, 8, 10, …} (call it **E**), the *infinitely many* positive odd integers, {1, 3, 5, 7, 9, …}, we obtain the set of all positive integers, {1, 2, 3, 4, 5, …} (call it **PI**). Even though **E** is

---

1 A philosophically informed reader might have noticed that our discussion of objectual and substitutional quantification presupposes "the truth-conditional theory of meaning." This theory says that the meaning of a declarative sentence is given by the truth conditions of this sentence. According to this theory, since substitutional quantification attributes truth conditions to the quantified sentences that are based on the truth values of their substitutional instances, it is assumed that these sentences *make assertions about* their substitutional instances. Similarly, the standard objectual semantics attributes truth conditions to the quantified sentences that are based on the relation of satisfaction between individuals and formulas; hence we treated these sentences as making assertions about the relation of satisfaction, individuals, and formulas. It is not essential to presuppose the truth-conditional theory of meaning in order to explain the difference between the objectual and substitutional interpretations of the quantifiers. This distinction could be cashed out solely in terms of the truth conditions that these interpretations attribute to the quantified sentences. In this case, these different interpretations of the quantified sentences would say nothing about the assertions made by these sentences. However, we will proceed under the supposition that the truth-conditional theory of meaning is true. We will revisit this issue later.

a proper subset of **PI**, they are demonstratively of the same cardinality. A compelling criterion for the identity of cardinalities is the existence of a **one-to-one correspondence** between two sets, that is, two sets have the same cardinality if and only if there is a one-to-one correspondence between them. For instance, the sets {*s*, *c*, *l*, *d*} and {Samantha, Christopher, Lisa, Darius} are of the same cardinality because the correspondence between every individual in the latter set and the first letter of his or her name is a one-to-one correspondence between the two sets. Similarly, the correspondence between a natural number and its double is a one-to-one correspondence between **PI** and **E** (1 corresponds to 2, 2 to 4, 3 to 6, 4 to 8, 5 to 10, and so on). However, if we subtract from **PI** all the natural numbers greater than 5, the resulting set is a finite set of cardinality 5. The general rule is that adding to an infinite set or removing from it finitely many things does not change its cardinality but adding to it or removing from it infinitely many things *might* change its cardinality.

There are infinite cardinalities that are larger than other infinite cardinalities. A lot of things are known about infinite cardinalities. For instance, we know that the cardinality of $\mathbb{N}$ is the *smallest infinite cardinality*. Set theorists denote this cardinality as $\aleph_0$ (aleph-null). Hence every infinite subset of $\mathbb{N}$ has the cardinality $\aleph_0$. $\aleph_1$ denotes the *least* infinite cardinality that is greater than $\aleph_0$, $\aleph_2$ the least infinite cardinality that is greater that $\aleph_1$, $\aleph_3$ the least infinite cardinality that is greater that $\aleph_2$, and in general, $\aleph_{n+1}$ denotes the least infinite cardinality that is greater than $\aleph_n$, where n is a nonnegative integer. This gives us an *infinite* sequence of infinite cardinalities: $\aleph_0 < \aleph_1 < \aleph_2 < \aleph_3 < \aleph_4 < \ldots$, where '<' denotes the relation of less than.

Is there a familiar set that has a cardinality greater than $\aleph_0$? Indeed, the set of all real numbers $\mathbb{R}$ is such a set. The real numbers are the numbers that can be assigned to all the points on an infinite straight line. Intuitively, every real number measures the distance that a point on a line has from the zero-point. Distances on the right of the zero-point are considered positive and distances on the left of the zero-point are considered negative.[1] It is clear that $\mathbb{N}$ is a proper subset of $\mathbb{R}$. It is a mathematical theorem that the cardinality of $\mathbb{R}$, which is usually denoted as C, is greater than the cardinality of $\mathbb{N}$, $\aleph_0$.

We will not present a proof of this theorem here, but there is a point of special significance to our approach to PL semantics, which concerns cardinalities larger than $\aleph_0$. A set whose cardinality is less than or equal to $\aleph_0$ is called **countable**. All sets whose cardinalities are larger than $\aleph_0$ are called **uncountable**. The distinction is intuitive enough, so we can explain it here. A countable set is a set that permits a finite or an infinite process of *counting* its members. Every finite set is clearly countable, since the process of counting its members is finite, meaning that it terminates after finitely many steps. For example, the set {Abraham, Sarah, Hagar Ishmael, Isaac} can be counted via a 5-step counting process. This counting process reveals the cardinality of the set. The fact that the process terminates at the number 5 shows that the cardinality of the set is 5. The set of all natural numbers, $\mathbb{N}$, can also be counted via an infinite process. We count the members of the set {0, 1, 2, 3, 4, 5, ...} as follows: we assign the number one to 0, the number two to 1, the number three to 2, the number four to 3, and so on. The process never terminates, but it has a nice feature: every natural number, no matter how big

1 To be technically correct, we should have used the term 'displacement' instead of 'distance'. Distance is the length of the line that a point traversed; it has no direction, and hence it cannot be negative. Displacement, roughly speaking, is distance with direction, and hence it can be positive or negative.

it is, will eventually be counted via this process. The facts that the process never terminates and that it will eventually reach any arbitrary natural number allow us to say that this set is **countably infinite.** All finite sets, that is, sets whose cardinalities are less than $\aleph_0$, and all sets whose cardinality is $\aleph_0$, admit such a process, and hence they are countable. Every set whose cardinality is greater than $\aleph_0$ does not admit such counting process, and hence it is uncountable. Recall that all sets whose cardinalities are at least as great as $\aleph_0$ are infinite. Since C, the cardinality of the set of real numbers $\mathbb{R}$, is greater than $\aleph_0$, $\mathbb{R}$ is uncountable. This means that no matter how we might arrange and rearrange the order of the points on a line, there is no infinite counting process for this set. Thus the size of an interpretation whose UD is $\mathbb{R}$ is not only infinite but also uncountable. We will revisit this issue when we consider the truth conditions for the quantifiers in the following subsection.

4.4.3 The **truth conditions** of a PL sentence **X** are the conditions that determine its truth value on any given PL interpretation for it. In other words, **X** is true on a PL interpretation I if and only if I satisfies its truth conditions, and **X** is false on I if and only if I is relevant to **X** and does not satisfy its truth conditions. Here is a complete list of those truth conditions. Let **X** and **Y** be any PL sentences and I be any PL interpretation for them (i.e., I is an interpretation for a set containing both **X** and **Y**).

4.4.3a If **X** is of the form '**r** = **s**' where **r** and **s** are PL singular terms, then **X** is true on I if and only if the referents of **r** and **s** on I are identical, that is, they are the same individual (symbolically, I(**s**) = I(**r**)).

4.4.3b If **X** is of the form '$\mathbf{P}^1\mathbf{s}$' where $\mathbf{P}^1$ is a 1-place predicate of PL and **s** a PL singular term, then **X** is true on I if and only if the referent of **s** on I has the property that I assigns to $\mathbf{P}^1$ (symbolically, $I(\mathbf{s}) \in I(\mathbf{P}^1)$).

4.4.3c If **X** is of the form '$\mathbf{R}^n\mathbf{t}_1\mathbf{t}_2\ldots\mathbf{t}_n$' where $\mathbf{R}^n$ is an n-place predicate (n is greater than 1) and $\mathbf{t}_1$, $\mathbf{t}_2$, …, and $\mathbf{t}_n$ are PL singular terms, then **X** is true on I if and only if the referents of those singular terms, in the order indicated, are related to each other according to the relation that I assigns to $\mathbf{R}^n$ (symbolically, $\langle I(\mathbf{t}_1), I(\mathbf{t}_2), \ldots, I(\mathbf{t}_n)\rangle \in I(\mathbf{R}^n)$).

4.4.3d

¬**X** is true on I	if and only if	**X** is false on I.
(**X**∧**Y**) is true on I	if and only if	**X** is true on I and **Y** is true on I.
(**X**∨**Y**) is true on I	if and only if	**X** is true on I or **Y** is true on I or both.
(**X**→**Y**) is true on I	if and only if	**X** is false on I or **Y** is true on I or both.
(**X**↔**Y**) is true on I	if and only if	**X** and **Y** are both true on I or both false on I.

4.4.3e If **X** is of the form (∀**y**)**Z**, then **X** is true on I if and only if *all* its basic substitutional instances are true on I, that is, for *every* name **s** in LN, the sentence **Z**[**s**] is true on I, where **Z**[**s**] is formed by replacing all the occurrences of **y** in **Z** by **s**.

4.4.3f If **X** is of the form (∃**y**)**Z**, then **X** is true on I if and only if *at least one* of its basic substitutional instances is true on I, that is, for *some* name **s** in LN, the sentence **Z**[**s**] is true on I, where **Z**[**s**] is formed by replacing all the occurrences of **y** in **Z** by **s**.

If every PL sentence in a set Γ is true on I, we say that I **satisfies** Γ or that I is a PL **model** of Γ.

4.4.3:C Commentary on 4.4.3

4.4.3:C1 If $\mathbf{P}^1$ is a 1-place predicate and **s** is a PL singular term, I assigns a property to $\mathbf{P}^1$, which is denoted as I($\mathbf{P}^1$), and a referent to **s**, which is denoted as I(**s**). The atomic sentence $\mathbf{P}^1\mathbf{s}$ is true on I if and only if the individual I(**s**) has the property I($\mathbf{P}^1$). Since PL is an extensional system the property I($\mathbf{P}^1$) is identified with its extension on I. Thus to say that I(**s**) has the property I($\mathbf{P}^1$) is to say that I(**s**) belongs to the extension I($\mathbf{P}^1$). As indicated in Chapter Two, the relation "belong to" is designated by the symbol '$\in$'. So 'I(**s**) $\in$ I($\mathbf{P}^1$)' means that the individual I(**s**) belongs to (is a member of) the extension I($\mathbf{P}^1$). Using this symbolism and the extensionality of PL, we can state the truth conditions of the atomic sentence $\mathbf{P}^1\mathbf{s}$ as follows: $\mathbf{P}^1\mathbf{s}$ is true on I if and only if I(**s**) $\in$ I($\mathbf{P}^1$).

The truth conditions of the atomic sentences of the form '$\mathbf{R}^n\mathbf{t}_1\mathbf{t}_2...\mathbf{t}_n$' are slightly more involved. $\mathbf{R}^n$ is an n-place predicate (n is greater than 1) and $\mathbf{t}_1$, $\mathbf{t}_2$, $\mathbf{t}_3$, ..., and $\mathbf{t}_n$ are (not necessarily all distinct) PL singular terms. I assigns an n-place relation to $\mathbf{R}^n$; we denote this relation as I($\mathbf{R}^n$). I($\mathbf{t}_1$), I($\mathbf{t}_2$), ..., and I($\mathbf{t}_n$) are the individuals (i.e., the referents) in UD that I assigns to the singular terms $\mathbf{t}_1$, $\mathbf{t}_2$, $\mathbf{t}_3$, ..., and $\mathbf{t}_n$. The sentence $\mathbf{R}^n\mathbf{t}_1\mathbf{t}_2...\mathbf{t}_n$ is true on I if and only if the relation I($\mathbf{R}^n$) holds between the individuals I($\mathbf{t}_1$), I($\mathbf{t}_2$), ..., and I($\mathbf{t}_n$), in the order indicated. For instance, if I(T^2) is the relation "taller than," I(a) is Anna, and I(b) is Bethany, then T^2ba is true on I if and only if Bethany is taller than Anna. Observe that the order is very important: T^2ba says that Bethany is taller than Anna; it does not say that Anna is taller than Bethany. Again, since PL is an extensional system, I(T^2) is identified with its extension on I. I(T^2) is a binary relation; therefore its extension consists of all ordered pairs $\langle\alpha, \beta\rangle$, such that α and β are individuals in UD and α has the relation I(T^2) to β, that is, α is taller than β. Using this notation and the extensionality of PL, we may say that T^2ba is true on I if and only if the ordered pair $\langle$I(b), I(a)$\rangle$ $\in$ I(T^2), that is, the ordered pair $\langle$Bethany, Anna$\rangle$ belongs to the extension of T^2 on I. In general, the atomic sentence $\mathbf{R}^n\mathbf{t}_1\mathbf{t}_2...\mathbf{t}_n$ is true on I if and only if the ordered n-tuple $\langle$I($\mathbf{t}_1$), I($\mathbf{t}_2$), ..., I($\mathbf{t}_n$)$\rangle$ $\in$ I($\mathbf{R}^n$).

The truth conditions of the compound sentences $\neg\mathbf{X}$, $(\mathbf{X}\wedge\mathbf{Y})$, $(\mathbf{X}\vee\mathbf{Y})$, $(\mathbf{X}\rightarrow\mathbf{Y})$, and $(\mathbf{X}\leftrightarrow\mathbf{Y})$ are the same as their truth conditions in SL. As we explained in 4.4.1:C5, the formal truth conditions adopted in this book for the quantified sentences are based on a substitutional interpretation of the quantifiers. Truth conditions of the quantified sentences that are based on objectual interpretation of the quantifiers would be: $(\forall\mathbf{y})\mathbf{Z}$ is true on I if and only if every individual in UD satisfies the formula **Z**, and $(\exists\mathbf{y})\mathbf{Z}$ is true on I if and only if there is at least one individual in UD that satisfies the formula **Z**.[1] We stated in note 1, p. 229 of this chapter that the notion of *satisfaction* invoked in objectual quantification is rather a technical notion, which we would like to avoid. In fact, one of the main reasons for choosing substitutional over objectual quantification in this book is to avoid the technicalities of the relation of satisfaction, which can hold between individuals in UD and PL formulas.

1 These are simplified statements of the objectual truth conditions of quantified sentences. The precise definition of the relation of satisfaction describes a certain relation between *sequences* of individuals in UD and PL formulas. However, our simplified definitions capture the essence of this relation.

4.4.3:C2 Substitutional quantification is less technically involved, and most readers find it easier to understand. But there is a logical and philosophical price for it. In order for substitutional quantification to deliver the right truth values for the quantified sentences, all the individuals in UD must be named. To see the point clearly, consider the sentence $(\forall x)(Qx \rightarrow (Kx \wedge Dx))$ and the PL interpretation I of 4.4.1:C4(b). Whether we interpret the quantifier $(\forall x)$ objectually or substitutionally, the sentence comes true on I. The objectual reading of $(\forall x)(Qx \rightarrow (Kx \wedge Dx))$ on I asserts that for every individual x in UD, x satisfies the formula $(Qx \rightarrow (Kx \wedge Dx))$. Given the semantical assignments of I and the nature of the relation of satisfaction, this assertion may be interpreted on I as saying that for every individual in UD, if this individual is an English major, then he or she plays the piano and plays tennis. Since this assertion is true on I, the sentence $(\forall x)(Qx \rightarrow (Kx \wedge Dx))$ is also true on I. On the other hand, the substitutional reading of $(Qx \rightarrow (Kx \wedge Dx))$ on I asserts that for every name **t** in LN, the substitutional instance $(Q\mathbf{t} \rightarrow (K\mathbf{t} \wedge D\mathbf{t}))$ is true on I. As explained previously, there are four substitutional instances of $(\forall x)(Qx \rightarrow (Kx \wedge Dx))$ on I: $(Qs \rightarrow (Ks \wedge Ds))$, $(Qc \rightarrow (Kc \wedge Dc))$, $(Ql \rightarrow (Kl \wedge Dl))$, and $(Qd \rightarrow (Kd \wedge Dd))$. The interpretations of these substitutional instances on I are all true; therefore the sentence $(\forall x)(Qx \rightarrow (Kx \wedge Dx))$ is also true on I. It is important to note that when we interpret PL sentences (whether we invoke objectual or substitutional quantification), we require that the English connectives that appear in these interpretations have the truth conditions of the corresponding PL connectives. This requirement is *not* observed when we deal with translations of PL into English and conversely.[1]

Both interpretations of the quantifiers, objectual and substitutional, deliver the same truth value for $(\forall x)(Qx \rightarrow (Kx \wedge Dx))$ on I. In fact we can construct an even closer parallel between these two interpretations. The interpretations on I of the four substitutional instances are stated in English. The resources of English allow certain conjunctions to be *converted into* universally quantified sentences and certain disjunctions to be converted into existentially quantified sentences. For instance, if we say that Lindsey, Vanessa, Halston, *and* Martha are smart, and we know that these are the only female students in Advanced Logic, then we can reformulate our conjunction as a universally quantified statement by saying that *every* female student in Advanced Logic is smart. And if we say that Lindsey, Vanessa, Halston, *or* Martha is smart, we can reformulate this disjunction as an existentially quantified statement by saying that at least one female student in Advanced Logic is smart. Thus the conjunction of the four interpretations of the substitutional instances above can be converted in English into the single sentence 'For every individual in UD, if this individual is an English major, then he or she plays the piano and plays tennis'. This sentence makes the same assertion obtained from interpreting the quantifiers objectually.

The point of the forgoing discussion is that if every individual in UD is named, substitutional and objectual quantifications deliver the same truth values for the same quantified sentences, and in certain cases their interpretations can be *converted into* very similar assertions. However, if we allow for unnamed individuals, objectual quantification gets things right while substitutional quantification might not. Let us modify the interpretation I of 4.4.1:C4(b) in

1 When we give PL translations of English sentences, we aim at PL translations that are *faithful* to the original. 'Faithfulness' here means "capturing correctly the interdependencies that might exist between the various components of the English sentences." Therefore we cannot impose on the English sentences *formal* truth conditions, which might be alien to the natural semantics of the English sentences.

order to allow for unnamed individuals. We will call this "interpretation" J. The reader must be warned, however, that J does not satisfy the definition of a PL interpretation stated in 4.4.1; it violates the condition 4.4.1a. Nevertheless, we will consider J in order to see what might happen if unnamed individuals are permitted.

The "interpretation" J

UD: {Samantha, Christopher, Lisa, Darius}
LN: s, c, l

Semantical Assignments

J(s): Samantha; J(c): Christopher; J(l): Lisa
J(Qx): "x is an English major": {Samantha, Darius}
J(Dx): "x plays tennis": {Samantha, Darius, Lisa}
J(Kx): "x plays the piano": {Samantha, Christopher}

On J, no name in LN refers to Darius, and Darius is removed from the extension of K. So in the world of J, Darius is an unnamed individual, and he does not play the piano.

The objectual reading of $(\forall x)(Qx \rightarrow (Kx \wedge Dx))$ asserts, as before, that for every individual x in UD, x satisfies the formula $(Qx \rightarrow (Kx \wedge Dx))$. This reading may be interpreted on J as saying that for every individual in UD, if this individual is an English major, then he or she plays the piano and plays tennis. This assertion, of course, is false on J, since Darius is an English major who does not play the piano. Therefore, according to the objectual interpretation of $(\forall x)(Qx \rightarrow (Kx \wedge Dx))$, this sentence is false on J.

The substitutional reading of $(\forall x)(Qx \rightarrow (Kx \wedge Dx))$ makes the same assertion as stated previously: for every name **t** in LN, the substitutional instance $(Q\mathbf{t} \rightarrow (K\mathbf{t} \wedge D\mathbf{t}))$ is true on J. However, we now only have the three names s, c, and l; so we have the following substitutional instances: $(Qs \rightarrow (Ks \wedge Ds))$, $(Qc \rightarrow (Kc \wedge Dc))$, and $(Ql \rightarrow (Kl \wedge Dl))$. All of these substitutional instances are true on J. Hence $(\forall x)(Qx \rightarrow (Kx \wedge Dx))$ is true on J. Therefore substitutional quantification delivers the wrong verdict regarding the truth value of the quantified sentence $(\forall x)(Qx \rightarrow (Kx \wedge Dx))$ on J. The problem is that the individual who is the sole counterexample to the universal claim made by $(\forall x)(Qx \rightarrow (Kx \wedge Dx))$ is Darius—he is the only individual in UD who is in the extension of Q but not in the intersection of the extensions of K and D. However, since Darius has no name in LN, there is no substitutional instance of $(\forall x)(Qx \rightarrow (Kx \wedge Dx))$ that applies to Darius. In other words, the individual in UD that demonstrates the falsity of $(\forall x)(Qx \rightarrow (Kx \wedge Dx))$ is not "disclosed" at the level of language. Any individual that has no name in LN is "hidden" from substitutional quantifications, and hence is not considered in determining the truth values of quantified sentences.

4.4.3:C3 There are several objections to substitutional quantification. One of them is that the requirement that every individual in UD have a name in LN creates (1) a linguistic aspect of every PL interpretation and (2) a dependency of the given PL language on its PL interpretations. The first point seems artificial and unrealistic. It seems artificial to require actual and possible realities to contain names for their individuals. Why should the planets of a distant solar system be labeled in order for such a system to count as a possible situation? Furthermore, there seem to be many situations in which the requirement is practically impossible to fulfill. How do we name all the stars in the universe, all the grains of sand on the beaches of Florida, and all the bees in Africa? However, an advocate of substitutional quantification might respond to these objections in this way. The requirement of including names for all the individuals of a PL interpretation is a formal requirement that is neither meant to be authentic nor realistic. Also, we do not assume that there is a "practical" method of assigning names to all the individuals of an interpretation. One is free to consider an interpretation whose UD is the set of all stars in the universe. Naming these stars is simply an *in-principle* requirement. There is nothing contradictory (and hence, logically impossible) about assigning names to all these stars in the universe, and there is no shortage of names: PL allows the use of subscripts to generate as many names as needed. PL only requires that *in principle* there is a way of naming all these individuals. In practice, when such an interpretation is described, it is enough to state that there is a name for every star in the universe. Even if UD is infinite, the use of subscripts ensures that there are plenty of names to label all the individuals in UD.

The second point, which is the dependency of the given PL language on its PL interpretations, is a more serious problem for substitutional quantification. Assume that Γ is a set of PL sentences. These sentences consist of basic PL vocabulary, which defines the PL language in which the members of Γ are constructed. A PL interpretation I for Γ brings its own set of names. Almost always many of these names do not occur in members of Γ. These names are, therefore, new items that are to be added to the basic vocabulary of the language of Γ. The language of Γ would be expanded by adding these names to it. This means that every PL interpretation for Γ might change the language of Γ in a particular way. The language of Γ would not be a fixed language but a "dynamic" language that changes with almost every PL interpretation for it. The situation becomes more troubling if we consider an interpretation whose UD is an uncountable set, say, the set $\mathbb{R}$ of all the real numbers. In this case, LN would also be uncountable. Therefore the PL language of Γ would have to be expanded beyond any countable limit.

Again, an advocate of substitutional quantification might be able to respond to these objections. First, we do not need to consider the language of Γ dynamic, changing with every PL interpretation for Γ. The language of Γ consists of the extra-logical vocabulary that occurs in the members of Γ and of the logical vocabulary of PL (i.e., the eight logical symbols, the variables, and the parentheses). This language does not change if an interpretation I for Γ brings with it additional names. The new names do not belong to the language of Γ, they belong to the language of I. Every PL interpretation has its own language, whose vocabulary is the PL logical vocabulary together with the extra-logical vocabulary that I interprets. In fact, the definition of PL interpretation given in this book permits an interpretation for Γ to interpret not only new names but even new function symbols and predicates, which are not part of the language of Γ. So there are two languages: the language of the set Γ, which is fixed and does not vary with its PL interpretations, and the language of a PL interpretation for Γ. Γ has its own fixed vocabulary and every interpretation for Γ has its own vocabulary, which is the vocabulary that is interpreted by that interpretation. There is a relation between those two vocabularies: the vocabulary of Γ is always a subset of the vocabulary of any of its PL interpretations. The vocabulary of an interpretation of uncountable size is also uncountable, since its LN consists of uncountably many names in order to name all the individuals in its UD. There is nothing in principle wrong about having languages whose vocabulary is uncountable. The same formation rules apply for terms and formulas. The only difference is that the set of formulas of such a language is uncountable as well. As for finding enough subscripts to index the uncountable vocabulary of a PL interpretation whose size is uncountable, we can use UD itself as the source of these subscripts. For instance, if the set $\mathbb{R}$ is the UD, we let LN consist of all the names that can be generated by indexing the letter '*c*' with a real number r, that is, for every real number r, c_r is a name in LN.

There are other objections to substitutional quantification, but we will mention only one last objection. Substitutional quantification interprets the quantified sentence as making an assertion about the language itself. (∀**z**)**X** is interpreted as saying that for every name **s** in LN, the substitutional instance **X**[**s**] is true on I, and (∃**z**)**X** as saying that there is at least one name **s** in LN, such that the substitutional instance **X**[**s**] is true on I. But this is counterintuitive. For example, when we assert (∀*x*)*Px*, we ordinarily mean that every individual in UD has the property I(*P*). We make an assertion about the individuals of an interpretation and the property that the predicate *P* designates on I. We do not intend to make an assertion about the language in question—about its names and the substitutional instances of its formulas.

Two responses might be offered to this objection. First, there is no reason to take the truth conditions of a sentence as determining the assertion made by the sentence. The truth conditions of (∀**z**)**X** and (∃**z**)**X** involve the truth status of their basic substitutional instance. However, these truth conditions need not constitute the assertions made by these sentences. We may assume that (∀**z**)**X** and (∃**z**)**X** make assertions about the individuals in UD, but their truth conditions are given in terms of the truth status of their basic substitutional instances. Many philosophers would not find this response very persuasive, since it is typically assumed that there is a close relation between the truth conditions of a sentence and the assertion made by the sentence. An influential philosophical position, which was championed by the American Philosopher Donald Davidson (1917–2003), affirms that the meaning of a sentence is given by its truth conditions. The position is known in the philosophical literature as 'the truth-conditional theory of meaning'. If we equate the meaning of a sentence with the assertion it makes, we can say that the truth conditions of a sentence constitute the assertion it makes. According to this position, the assertion that (∀**z**)**X** makes is that every basic substitutional instance of it is true, and the assertion (∃**z**)**X** makes is that at least one of its basic substitutional instances is true. If this influential position is correct, then the first response is misguided.

The second response accepts this position—namely that the truth conditions of a sentence constitute its meaning; and hence it accepts that the assertions made by (∀**z**)**X** and (∃**z**)**X** are about the truth status of their basic substitutional instances. However, this response argues that objectual quantification fares no better. In order to appreciate the full force of this response, some knowledge of the technical relation of satisfaction is required. At any rate, we can describe the main idea. We only need to be aware that this relation is defined to hold between sequences of individuals in UD and PL formulas. So a defender of substitutional quantification might respond that the objectual readings of (∀**z**)**X** and (∃**z**)**X** do not escape the complaint of this objection; for the standard account of objectual quantification interprets (∀**z**)**X** as asserting that every sequence of individuals in UD that meets a certain condition *satisfies the formula* **X**, and (∃**z**)**X** as saying that there is a sequence of individuals in UD that meets a certain condition and which *satisfies the formula* **X**. These readings invoke the syntactical notion of a PL formula, and hence the assertions they make are, in part, about the language of PL.

A possible rebuttal to this response is that once the details of the relation of satisfaction are spelled out and once we invoke the interpretations, supplied by I, of the basic vocabulary, an assertion that a quantified sentence makes is ultimately reduced to assertions solely about the individuals in UD and their properties and relations. However, a defender of substitutional quantification might say the same thing about a substitutional reading of a quantified sentence. She might say that, similarly, once the semantical contents of the basic substitutional instances are fully cashed out in terms of the interpretations supplied by I, the assertion that the quantified sentence makes is reduced to assertions about the individuals of I and the properties and relations they might have.

We will not attempt to settle the debate between advocates of substitutional quantification and advocates of objectual quantification. This debate is rather philosophically and logically involved. Our goal, for the sake of intellectual honesty, is to make clear that the semantical approach we adopted is controversial and to give a feel for some of the issues that are involved in the controversy. However, we should stress that even though we gave substitutional truth conditions for the quantifiers, in practice almost always we determine the truth values of quantified sentences by examining the assertions they make about the individuals in UD and their properties and relations.

We end this section by stating precisely the relationship between the *basic* substitutional instances of a quantified sentence (Δ**z**)**X** (where 'Δ' stands for '∀' or '∃') and its substitutional instances in general. Recall that a *basic* substitutional instance of (Δ**z**)**X** is obtained by substituting a PL *name* for the variable **z** in all its occurrences in **X**. On the other hand, a substitutional instance of (Δ**z**)**X** is obtained by substituting a PL *singular term* for **z** in all its occurrences in **X**. Since every name is a singular term, it immediately follows that every basic substitutional instance is a substitutional instance. The converse, however, is not true. If there are function symbols, then there are many singular terms that are not names; and hence there are many substitutional instances that are not basic. For instance, the PL sentences $Pe \rightarrow (Rea \wedge Te)$ and $Pg^1e \rightarrow (Rg^1ea \wedge Tg^1e)$ are both substitutional instances of the sentence $(\forall x)(Px \rightarrow (Rxa \wedge Tx))$; but the first is basic and the second is not. The first is obtained by substituting the name *e* for *x* and the second by substituting the singular term g^1e for *x*. So we have two trivial implications: (1) if all the substitutional instances of (Δ**z**)**X** are true (or false) on some PL interpretation I, then all its basic substitutional instances are true (or false) on I; and (2) if one of the basic substitutional instances of (Δ**z**)**X** is true (or false) on I, then one of its substitutional instances is true (or false) on I. But are the converse of these claims true? The answer is "Yes." We state these facts as a theorem.

Theorem 4.2: If all the basic substitutional instances of (Δ**z**)**X** are true (or false) on a PL interpretation I, then all its substitutional instances are true (or false) on I; and hence if one of the substitutional instances of (Δ**z**)**X** is true (or false) on I, then one of its basic substitutional instances is true (false) on I.

It is not hard to see why this theorem is true. Informally speaking, if all the basic substitutional instances of (Δ**z**)**X** are true on I, then the formula **X** is true of all the individuals in UD (since all the individuals in UD have names in LN). But according to Theorem 4.1, all singular terms refer to unique individuals in UD. Thus if we substitute a singular term for **z** in **X**, we obtain a true sentence on I. On the other hand if there is a true substitutional instance of (Δ**z**)**X** on I, then there is an individual in UD of which the formula **X** is true (because every singular term has a unique referent on I). But this individual must have a name in LN. This implies that if we substitute this name for **z** in **X**, we obtain a true sentence on I. The same informal reasoning works for the case of falsehood. This is not a precise proof, since we never defined what we mean by a formula being true of an individual. However, the idea is intuitive enough. The informal reasoning we just gave conveys the central idea of the proof.

4.4.4 **Bivalence** is one of the central features of **classical logic**. A notion of truth that supports bivalence is usually described as **classical truth**. The notions of truth that are defined in TL, SL, and PL are classical notions. By a notion of truth in a logical system L, we intend the definition of the property "being true on M," where M is L's surrogate for the informal notion of logical possibility. In TL, M is a TL diagram; in SL, an SL truth valuation; and in PL, a PL interpretation. Thus articulating a notion of truth in a classical logical system L requires a definition of M, which is L's representation of a logical possibility, and a specification of truth conditions for all the sentences of L in a manner that ensures that every sentence of L is either true or false, but not both, on every M that is relevant to that sentence.

4.4.4:C COMMENTARY ON 4.4.4

The reader might have noticed that we are operating in this chapter with an implicit assumption: every PL sentence that is composed of the PL vocabulary of a certain interpretation is either true or false on that interpretation, where the PL vocabulary of an interpretation consists of the PL basic vocabulary that is interpreted by the semantical assignments of that interpretation, the eight logical symbols of PL, the PL variables, and the two parentheses. We made similar assumptions for TL and SL. These assumptions entail that the languages of TL, SL, and PL are bivalent. As defined in 1.2.2, a bivalent language is a language of which every declarative sentence is bivalent; and a bivalent declarative sentence is a sentence that is either true or false, but not both, in every logical possibility.

It is not hard to see that this claim is true for TL and SL. The definition of a TL diagram and the truth conditions of the TL sentences are designed so that every TL sentence has a classical truth value (i.e., T or F, but not both) on every TL diagram for that sentence. Similarly, the definition of SL truth valuation and the truth conditions of the SL sentences ensure that every SL sentence has a classical truth value on every truth valuation for that sentence. This is not as obvious in the case of PL. Nevertheless, it is true. Given the definition of PL interpretation and the truth conditions of the PL sentences, the notion of truth in PL is classical, that is, every PL sentence is either true or false, but not both, on every PL interpretation for that sentence.

Let us state this claim in full generality. Suppose that Σ is any set of PL sentences and H is any PL interpretation *for* Σ. Let Voc(Σ) be the PL basic vocabulary that occurs in the members of Σ, that is, the extra-logical vocabulary of which the sentences in Σ are composed, together with the logical vocabulary of PL. Similarly, we denote the vocabulary of the interpretation H as Voc(H). Voc(H) consists of all the extra-logical vocabulary that is interpreted by H, together with the logical vocabulary of PL. It is clear that Voc(Σ) is a subset of Voc(H), since H interprets all the function symbols and predicates in Voc(Σ) and all the names in LN, which might contain additional names that are not in Voc(Σ). Also there is no requirement that every predicate interpreted by H be one of the predicates in Voc(Σ). Hence every PL sentence that is composed of Voc(Σ) is a sentence that is composed of Voc(H), while the converse might not be true. H can interpret all the sentences composed of Voc(H). The semantics of PL is designed such that every sentence composed of Voc(H) (and hence, every sentence composed of Voc(Σ)) is either true or false, but not both, on H. This is true no matter what Σ or H is, so long as H is a PL interpretation for Σ. The language of PL, therefore, is bivalent. It is possible to prove this claim; but any proof of this claim is beyond the scope of this book.

4.5 Logical Concepts in PL

4.5.1 As in 1.1.1, a PL **argument** is defined as a nonempty collection of PL sentences; one of these sentences is the conclusion of the argument, and the others are its premises.

4.5.2 In PL, **logical possibilities** are characterized as **PL interpretations**. The truth conditions stated in 4.4.3 define the notion **true-on-an-interpretation** which is intended to capture the informal notion **true-in-a-logical-possibility**. Thus the **logical concepts** introduced in 1.2.4–1.2.11 can now be defined in PL by simply replacing the expression 'logical possibility' with 'PL interpretation' in those definitions and making the necessary adjustments. The resulting statements are listed in 4.5.3–4.5.10.

4.5.2:C COMMENTARY ON 4.5.2

The concepts of deductive validity, logical truth, logical falsehood, logical equivalence, and inconsistency are **universal concepts**, that is, their definitions make assertions about *all* PL interpretations. Hence in order to establish that one of these concepts applies to a given set of PL sentences, we need to prove something about all PL interpretations. For example, to show that a PL argument is deductively valid, we need to show that *every* PL interpretation that makes the premises true also makes the conclusion true; in order to prove that two PL sentences are logically equivalent, we must prove that there is *no* PL interpretation on which the sentences have different truth values; and in order to establish that a PL sentence is logically true, we must establish that it is true on *every* PL interpretation for it.

In TL and SL, there are finitely many TL diagrams or SL truth valuations for every finite set of sentences. Thus, as we explained in 2.5.11:C and 3.5.11:C, one can *in principle* survey all the relevant TL diagrams or SL truth valuations to determine whether a certain logical concept applies or not. This approach is not available to us in PL, since the number of PL interpretations for any set of PL sentences is infinite. We have to give a proof establishing that every PL interpretation meets a certain condition or that there is no PL interpretation that has a certain feature. In the subsections below, we will give examples of such proofs. However, in the next chapter, we will develop precise and formal methods for constructing proofs that establish the applicability of universal concepts to a given set of PL sentences.

The denials of the universal concepts listed above are **existential concepts**, that is, their definitions assert the existence of PL interpretations that meet certain conditions. So in order to show that one of these concepts applies to a certain set of PL sentences, we must demonstrate the existence of a PL interpretation that fulfills certain conditions. For example, in order to prove that a PL argument is deductively invalid, we must prove the existence of a PL interpretation that makes the premises true and the conclusion false; in order to show that a PL sentence is contingent, we need to show that there are two PL interpretations, one of which makes the sentence true and the other makes it false; and in order to establish that a set of PL sentences is consistent, we must establish the existence of a PL interpretation that satisfies the set, that is, on which all the members of the set are true. The most common approach to demonstrating the existence of a PL interpretation that meets certain conditions is to construct an appropriate PL interpretation and show that it actually meets the desired conditions. In this chapter, we will only give exercises (in 4.6) that require the construction of PL interpretations of certain types. Exercises that require establishing something about *all* PL interpretations will be given in Chapter Five, when the concept of formal proof is introduced.

4.5.3a A PL argument **Γ/X** is **deductively valid** (or **Γ**$\vDash$**X**) if and only if on every PL interpretation for **Γ/X** on which all the members of **Γ** are true **X** is true as well.

4.5.3b A PL argument **Γ/X** is **deductively valid** (or **Γ**$\vDash$**X**) if and only if there is no PL interpretation on which all the members of **Γ** are true and **X** is false.

4.5.3:C COMMENTARY ON 4.5.3

We will show that the argument below is deductively valid.

S1 $(Raa \wedge Pc) \rightarrow (\exists x)Ax$
S2 $(\forall y)\neg Ay$
S3 Pc

S4 $(\exists x)\neg Rxx$

We will prove that every PL interpretation for S1–S4 that makes S1–S3 true makes S4 true as well. We will use a style of proof that we may term 'arbitrary-object proof'.[1] The strategy of this proof style is to take an object as an *arbitrary* representative of a certain type *T*, and to show that this arbitrary object meets a certain condition C. It is essential that in this proof no assumption about this arbitrary object be invoked other than that it is of type *T*. An arbitrary-object proof allows us to conclude that *all* objects of type *T* meet the condition C. In the following proof the arbitrary object of type *T* is a PL interpretation for S1–S4 on which the premises, S1–S3, are true. We invoke only the assumption that I is a PL interpretation for S1–S4 that makes S1–S3 true, the definition of PL interpretation given in 4.4.1, and the formal truth conditions listed in 4.4.3.

Proof

1. Assume that I is an *arbitrary* PL interpretation for the sentence S1–S4, such that S1–S3 are true on I, that is, the following sentences are true on I:
    1.1. $(Raa \wedge Pc) \rightarrow (\exists x)Ax$
    1.2. $(\forall y)\neg Ay$
    1.3. Pc
2. From 1.2 by the truth conditions of the universal quantifier: every basic substitutional instance of $(\forall y)\neg Ay$ is true on I. This implies that every sentence of the form $\neg A$**s**, where **s** is any name in LN, is true on I.

---

1 The technical name of this proof style is 'universal generalization'. We will formalize this type of proof in Chapter Five.

3. From 2 by the truth conditions of the negation: for every name **s** in LN, the substitutional instance *A***s** is false on I.
4. From 3 by the definition of basic substitutional instance: there is no basic substitutional instance of $(\exists x)Ax$ that is true on I.
5. From 4 by the truth conditions of the existential quantifier: $(\exists x)Ax$ is false on I.
6. From 1.1 and 5 by the truth conditions of the conditional: $Raa \wedge Pc$ is false on I.
7. From 6 by the truth conditions of the conjunction: either *Raa* is false on I or *Pc* is false on I.
8. From 1.3 and 7: *Raa* must be false on I.
9. From 8 by the truth conditions of the negation: $\neg Raa$ is true on I.
10. From 9 by the truth conditions of the existential quantifier: $(\exists x)\neg Rxx$ is true on I.
11. From 1 through 10: if I is a PL interpretation for S1–S4 that makes S1–S3 true, then I makes S4 true as well.
12. From 11: given that I is *arbitrary*, it follows that *every* PL interpretation for S1–S4 that makes the premises, S1-S3, true also makes the conclusion, S4, true.
13. From 12 by Definition 4.5.3a: the PL argument above is deductively valid.

4.5.4 A PL argument **Γ/X** is **deductively invalid** (or $\Gamma \not\models \mathbf{X}$) if and only if there is a PL interpretation on which the members of **Γ** are all true and **X** is false.

4.5.4:C COMMENTARY ON 4.5.4

We said previously that in order to establish the applicability of an existential concept, it is sufficient to present a PL interpretation that meets the conditions of the concept. Since the concept of deductive invalidity is an existential concept, we will prove that the following PL argument is deductively invalid by constructing a PL interpretation I on which the premises S1–S5 are true and the conclusion S6 is false.

S1 $(\neg Ha \vee \neg Kb) \rightarrow a = gb$

S2 $a \neq gb$

S3 $(Ha \wedge Kb) \rightarrow (\exists x)(a = gx \vee x = gb)$

S4 $(\forall x)((x = b \vee x = c) \rightarrow Sagx)$

S5 $(\forall z)(\forall w)(z = gw \rightarrow Swb)$

S6 $\neg Saa \vee \neg Sbb$

The PL Interpretation I

UD: The set of all positive even numbers: {2, 4, 6, ...}
LN: $a, b, c, d, e_2, e_4, e_6, \ldots, e_n, \ldots$; where n is any positive even number

Semantical Assignments

> I(a): 16; I(b): 2; I(c): 4; I(d): 8; I(e_n) = n
> I(gz): "The square of z"
> I(Hx): "x is even"
> I(Kv): "v is prime"
> I(Sxy): "x is divisible by y"

Natural reading of (S1) $(\neg Ha \vee \neg Kb) \rightarrow a = gb$ on I:
If 16 *is not even or* 2 *is not prime, then* 16 *is the square of* 2.
The sentence is true on I, because the antecedent and consequent are false on I. (2 is a prime number.)

Natural reading of (S2) $a \neq gb$ on I:
16 *is not the square of* 2.
The sentence is true on I.

PL-English reading of (S3) $(Ha \wedge Kb) \rightarrow (\exists x)(a = gx \vee x = gb)$ on I:
If 16 *is even and* 2 *is prime, then there is an even number x, such that* 16 *is the square of x and x is the square of* 2.
The sentence is true on I, since the antecedent and the consequent are true on I. The consequent is true on I because 16 is the square of 4 and 4 is the square of 2.

Natural reading of (S4) $(\forall x)((x = b \vee x = c) \rightarrow Sagx$ on I:
16 *is divisible by the square of* 2 *and by the square of* 4.

PL-English reading of (S5) $(\forall z)(\forall w)(z = gw \rightarrow Swb)$ on I:
For any even numbers z and w, if z is the square of w, then w is divisible by 2.
The sentence is true on I, since all the numbers in UD are divisible by 2.

Natural reading of (S6) $\neg Saa \vee \neg Sbb$ on I:
16 is not divisible by 16 or 2 is not divisible by 2.
The sentence is false on I because both disjuncts are false on I.

According to Definition 4.5.4, since there is a PL interpretation that makes the premises of the PL argument true and its conclusion false, the argument is deductively invalid.

4.5.5a A PL sentence is **logically true** if and only if it is true on every PL interpretation for that sentence.

4.5.5b A PL sentence is **logically true** if and only if there is no PL interpretation on which it is false.

4.5.5:C COMMENTARY ON 4.5.5

We will show that the PL sentence $(\forall z)Gzd \rightarrow Gdd$ is logically true by showing that it satisfies Definition 4.5.5b. In previous chapters we introduced a style of proof that we called an "indirect proof." We invoke this proof style here. We will assume that there is a PL interpretation I on which the sentence is false, and then derive a contradiction from this assumption (the Reductio Assumption). This allows us to conclude that the Reductio Assumption is false. In this proof we only invoke the Reductio Assumption, the definition of PL interpretation, and the truth conditions of PL sentences.

Proof

1. Reductio Assumption: there is a PL interpretation I on which the sentence $(\forall z)Gzd \rightarrow Gdd$ is false.
2. From 1 by the truth conditions of conditional: $(\forall z)Gzd$ is true on I, and *Gdd* is false on I.
3. From 2 by the truth conditions of the universal quantifier: every basic substitutional instance of $(\forall z)Gzd$ is true on I, that is, for every name **s** in LN, G**s***d* is true on I.
4. From 3: *Gdd* is true on I.
5. From 2 and 4: *Gdd* is both true and false on I, which is a contradiction.
6. From 1 through 5: the Reductio Assumption is false; hence, there is no PL interpretation on which $(\forall z)Gzd \rightarrow Gdd$ is false.
7. From 6 by Definition 4.5.5b: $(\forall z)Gzd \rightarrow Gdd$ is logically true.

4.5.6a A PL sentence is **logically false** if and only if it is false on every PL interpretation for that sentence.

4.5.6b A PL sentence is **logically false** if and only if there is no PL interpretation on which it is true.

4.5.6:C COMMENTARY ON 4.5.6

We will demonstrate that the PL sentence $\neg(\exists v)Rvv \wedge (Red \wedge e = d)$ is logically false. We use an indirect proof to establish that there is no PL interpretation on which $\neg(\exists v)Rvv \rightarrow (Red \wedge e = d)$ is true. We only invoke the Reductio Assumption, the definition of PL interpretation, and the formal truth conditions.

Proof

1. Reductio Assumption: there is a PL interpretation I on which the sentence $\neg(\exists v)Rvv \wedge (Red \wedge e = d)$ is true.
2. From 1 by the truth conditions of the conjunction: the following sentences are true on I:
    2.1. $\neg(\exists v)Rvv$
    2.2. *Red*
    2.3. $e = d$
3. From 2.1 by the truth conditions of the negation: $(\exists v)Rvv$ is false on I.
4. From 3 by the truth conditions of the existential quantifier: every basic substitutional instance of $(\exists v)Rvv$ is false on I, that is, for every name **s** is LN, R**ss** is false on I.

5. From 4: *Ree* is false on I.
6. From 2.3 by the truth conditions of the identity predicate: I(e) is the same individual as I(d).
7. From 2.2 by the truth conditions of '*Red*': $\langle I(e), I(d)\rangle \in I(R)$.
8. From 6 and 7: $\langle I(e), I(e)\rangle \in I(R)$.
9. From 8 by the truth conditions of '*Ree*': *Ree* is true on I.
10. From 5 and 9: *Ree* is both true and false on I, which is a contradiction.
11. From 1 through 10: the Reductio Assumption is false; hence we conclude that there is no PL interpretation on which the sentence $\neg(\exists v)Rvv \wedge (Red \wedge e = d)$ is true.
12. From 11 by Definition 4.5.6b: $\neg(\exists v)Rvv \wedge (Red \wedge e = d)$ is logically false.

4.5.7 A PL sentence is **contingent** if and only if it is true on some PL interpretations and false on other interpretations.

4.5.7:C COMMENTARY ON 4.5.7

Almost every PL sentence we write without much thought is contingent. Contingency is an existential concept. Thus we establish the contingency of a sentence by presenting two PL interpretations one of which makes the sentence true and the other makes it false. We will show that $(\forall x)(Bx \vee Kx) \leftrightarrow (\exists v)(Bv \wedge Kv)$ is contingent.

The PL Interpretation I

UD: {John, Mary, Bill, Sarah}
LN: j, m, b, s

Semantical Assignments

I(j): John; I(m): Mary; I(b): Bill; I(s): Sarah
I(Bx): "x is male": {John, Bill}
I(Kx): "x is female": {Mary, Sarah}

Natural reading of $(\forall x)(Bx \vee Kx)) \leftrightarrow (\exists v)(Bv \wedge Kv)$ on I:
Everyone is either male or female if and only if there is someone who is both male and female.
This is false on I because the left-hand side of the biconditional is true and the right-hand side is false on I.

The PL Interpretation I*

UD: {2, 3, 4, 5}
LN: a, b, c, d

Semantical Assignments

> I*(*a*): 2; I*(*b*): 3; I*(*c*): 4; I*(*d*): 5
> I*(*Bx*): "*x* is prime": {2, 3, 5}
> I*(*Kx*): "*x* is even": {2, 4}

Natural reading of $(\forall x)(Bx \vee Kx) \leftrightarrow (\exists v)(Bv \wedge Kv)$ on I*:
Every number is either prime or even if and only if there is a number that is both prime and even.
This is true on I* since both sides of the biconditional are true on I.

According to Definition 4.5.7, since the sentence $(\forall x)(Bx \vee Kx) \leftrightarrow (\exists v)(Bv \wedge Kv)$ is false on I and true on I*, it is contingent.

4.5.8a Two PL sentences are **logically equivalent** if and only if on every PL interpretation for these sentences they have identical truth values.

4.5.8b Two PL sentences are **logically equivalent** if and only if there is no PL interpretation on which they have different truth values.

4.5.8:C COMMENTARY ON 4.5.8

We will show that the PL sentences

> S1 $\neg(\forall v)(Dv \rightarrow Ce)$, and
> S2 $(\exists z)(Dz \wedge \neg Ce)$

are logically equivalent. We first prove that every PL interpretation for S1 and S2 that makes S1 true makes S2 true as well. We use the strategy of arbitrary-object proof, introduced in 4.5.3:C, to establish this claim. We only invoke the assumption that I is a PL interpretation for S1 and S2 on which $\neg(\forall v)(Dv \rightarrow Ce)$ is true, the definition of PL interpretation, and the formal truth conditions.

Proof 1

1. Assume that I is an arbitrary PL interpretation for S1 and S2 on which $\neg(\forall v)(Dv \rightarrow Ce)$ is true.
2. From 1 by the truth conditions of the negation: $(\forall v)(Dv \rightarrow Ce)$ is false on I.
3. From 2 by the truth conditions of the universal quantifier: there is a basic substitutional instance of $(\forall v)(Dv \rightarrow Ce)$ that is false on I, that is, there is a name **s** in LN such that the substitutional instance $D\mathbf{s} \rightarrow Ce$ is false on I.
4. From 3 by the truth conditions of the conditional: $D\mathbf{s}$ is true on I and Ce is false on I.
5. From 4 by the truth conditions of the negation: $D\mathbf{s}$ and $\neg Ce$ are true on I.
6. From 5 by the truth conditions of the conjunction: $D\mathbf{s} \wedge \neg Ce$ is true on I.
7. From 6 by the definition of a basic substitutional instance: there is a basic substitutional instance of $(\exists z)(Dz \wedge \neg Ce)$ that is true on I.
8. From 7 by the truth conditions of the existential quantifier: $(\exists z)(Dz \wedge \neg Ce)$ is true on I.

9. From 1 through 8: if I is a PL interpretation for S1 and S2 on which S1 is true, S2 is true on I as well.
10. From 9: since I is arbitrary, every PL interpretation for S1 and S2 that makes S1 true makes S2 true as well.

We now prove that every PL interpretation for S1 and S2 that makes S1 false also makes S2 false.

Proof 2

1. Assume that I is an arbitrary PL interpretation for S1 and S2 on which $\neg(\forall v)(Dv \to Ce)$ is false.
2. From 1 by the truth conditions of the negation: $(\forall v)(Dv \to Ce)$ is true on I.
3. From 2 by the truth conditions of the universal quantifier: every basic substitutional instance of $(\forall v)(Dv \to Ce)$ is true on I, i.e., for every name **s** in LN, the substitutional instance $D\mathbf{s} \to Ce$ is true on I.
4. From 3 by the truth conditions of the conditional: for every name **s** in LN, either $D\mathbf{s}$ is false on I or Ce is true on I.
5. From 4 by the truth conditions of the negation: for every name **s** in LN, either $D\mathbf{s}$ is false on I or $\neg Ce$ is false on I.
6. From 5 by the truth conditions of the conjunction: for every name **s** in LN, $D\mathbf{s} \wedge \neg Ce$ is false on I.
7. From 6 by the definition of basic substitutional instance: all basic substitutional instances of $(\exists z)(Dz \wedge \neg Ce)$ are false on I.
8. From 7 by the truth conditions of the existential quantifier: $(\exists z)(Dz \wedge \neg Ce)$ is false on I.
9. From 1 through 8: if I is a PL interpretation for S1 and S2 on which S1 is false, S2 is false on I as well.
10. From 9: since I is arbitrary, every PL interpretation for S1 and S2 that makes S1 false also makes S2 false.
11. From 10 and the conclusion of Proof 1 by Definition 4.5.8a: S1 and S2 are logically equivalent.

4.5.9 A set of PL sentences is **consistent** if and only if there is a PL interpretation on which every member of the set is true, that is, there is a PL interpretation that satisfies the set (or simply, if and only if the set has a PL model).

4.5.9:C COMMENTARY ON 4.5.9

We will demonstrate that the set of PL sentences below is consistent.

S1 $(\forall x)(\forall y)(Bxy \vee Byx)$
S2 $(\exists x)(\exists y)(Bxy \wedge x \neq y)$
S3 $(\forall x)(\forall y)((Bxy \wedge Byx) \to x = y)$
S4 $(\forall x)(\forall y)(\forall z)((Bxy \wedge Byz) \to Bxz)$

As indicated in 4.5.2:C, consistency is an existential concept. So we will establish the consistency of the set above by presenting a model for the set.

The PL Interpretation I

UD: {Anna, Margaret, Patricia, Jennifer}
LN: a, m, p, j

Semantical Assignments

I(a): Anna; I(m): Margaret; I(p): Patricia; I(j): Jennifer
I(Bxy): "x likes y": {⟨Anna, Anna⟩, ⟨Margaret, Margaret⟩, ⟨Patricia, Patricia⟩, ⟨Jennifer, Jennifer⟩, ⟨Anna, Margaret⟩, ⟨Margaret, Patricia⟩, ⟨Patricia, Jennifer⟩, ⟨Anna, Patricia⟩, ⟨Anna, Jennifer⟩, ⟨Margaret, Jennifer⟩}
I($x = y$): the relation of token identity

Natural reading of (S1) $(\forall x)(\forall y)(Bxy \vee Byx)$ on I:
For any two people, either the first likes the second or the second likes the first. (The "two" people could be one person.)
PL-English reading: For all people x and y, either x likes y or y likes x.
True on I.

Note: The PL-English reading is more precise, since it does not convey the connotation that the people x and y are distinct. On the other hand, the English expression 'two people' has the connotation that the two people are distinct. However, it is a common usage in natural languages to use number-words, such as 'two' and 'three', without excluding the possibility that these individuals are identical. For instance, we may say that the two planets that we call "the Morning Star" and "the Evening Star" are actually the same planet. Philosophers have a term for this type of usage, *semantic pretense*. The idea is that we "pretend" that there are two planets, and then we abandon the pretense when we assert that these are the same planet. PL-English does not require semantic pretense; its use of variables provides it with an adequate resource to eliminate any need for semantic pretense. However, mathematicians, who rely extensively on PL-English, make frequent use of semantic pretense. It is not unusual to see in mathematics a proof that begins with the assumption that there are two mathematical objects, such as numbers, satisfying a certain condition and then ends with the conclusion that these mathematical objects are really one object. We will sometimes make use of semantic pretense when we give natural readings of PL sentences. If our usage might engender misunderstanding, we will also give a PL-English reading of the sentence. When the natural readings are quite awkward, we will only give PL-English readings.

Natural reading of (S2) $(\exists x)(\exists y)(Bxy \wedge x \neq y)$ on I:
There are two distinct people such that the first likes the second.
PL-English reading: There is a person x and a person y such that x likes y and x is not identical with y.
True on I.

Natural reading of (S3) $(\forall x)(\forall y)((Bxy \wedge Byx) \rightarrow x = y)$ on I:
For any two people, if the first likes the second and the second likes the first, then they are the same person.
True on I.

Natural reading of (S4) $(\forall x)(\forall y)(\forall z)((Bxy \wedge Byz) \rightarrow Bxz)$ on I:
For any three people, if the first likes the second and the second likes the third, then the first likes the third.
True on I.

4.5.10a A set of PL sentences is **inconsistent** if and only if on every PL interpretation for that set at least one member of the set is false.

4.5.10b A set of PL sentences is **inconsistent** if and only if there is no PL interpretation on which all the members of the set are true, that is, there is no PL interpretation that satisfies the set (or simply, if and only if the set has no PL model).

4.5.10:C COMMENTARY ON 4.5.10

We will prove, using indirect proof, that the following set of PL sentences is inconsistent. The Reductio Assumption is that there is a PL interpretation I that satisfies the set. We will show that this assumption leads to a contradiction; hence the assumption is false. This establishes that there is no PL interpretation on which S1–S3 are true, which means that the set is inconsistent.

S1 Ec
S2 $(\forall w)(Ew \rightarrow Le)$
S3 $\neg(\exists v)Lv$

Proof

1. Reductio Assumption: there is a PL interpretation I on which S1–S3 are true, that is, the following sentences are true on I:
    1.1. Ec
    1.2. $(\forall w)(Ew \rightarrow Le)$
    1.3. $\neg(\exists v)Lv$
2. From 1.2 by the truth conditions of the universal quantifier: every basic substitutional instance of $(\forall w)(Ew \rightarrow Le)$ is true on I, i.e., for every name **s** in LN, the substitutional instance $E\mathbf{s} \rightarrow Le$ is true on I.
3. From 2: $Ec \rightarrow Le$ is true on I.
4. From 1.1 and 3 by the truth conditions of the conditional: Le is true on I.
5. From 4 by the definition of a basic substitutional instance: there is a basic substitutional instance of $(\exists v)Lv$ that is true on I.
6. From 5 by the truth conditions of the existential quantifier: $(\exists v)Lv$ is true on I.
7. From 1.3 by the truth conditions of the negation: $(\exists v)Lv$ is false on I.

8. From 6 and 7: $(\exists v)Lv$ is both true and false on I, which is a contradiction.
9. From 1 through 8: the Reductio Assumption is false; hence there is no PL interpretation on which S1–S3 are all true.
10. From 9 by Definition 4.5.10b: the set above is inconsistent.

4.5.11 The number of PL interpretations for a set of PL sentences is infinite; hence we cannot list all of them. In a *typical* case, it is not possible to survey all of the PL interpretations for a given set of PL sentences in order to determine whether one of the eight logical concepts, defined in 4.5.3–4.5.10, applies or not. Furthermore, there is no mechanical procedure for generating the PL interpretations for every set of PL sentences, and there is no mechanical procedure for determining the truth value of every PL sentence on every PL interpretation for that sentence. This suggests that these logical concepts in PL are **undecidable**. In fact, they are demonstratively undecidable. The proofs of their undecidability are logically and mathematically advanced. They are beyond the scope of this book. However, we can give an informal explanation of their undecidability.

4.5.11:C COMMENTARY ON 4.5.11

4.5.11:C1 As defined in 2.5.11:C and 3.5.11:C, a concept is **decidable** if and only if there is an **effective decision procedure** for determining whether something is subsumed under the concept or not; and a decision procedure is effective if and only if it can, in principle, be followed mechanically and after a finite number of deterministic steps leads to the correct answer, whether it is "Yes" or "No." We also explained there that not all **effective procedures** are decision procedures. We call a procedure "effective" just in case it is mechanical (i.e., it involves no creative steps) and generates the desired result after finitely many deterministic steps. There are effective procedures that produce the answer "Yes" when and only when the correct answer is "Yes," but that might not produce any answer when the correct answer is "No," and others that produce the answer "No" when and only when the correct answer is "No," but that might not produce any answer when the correct answer is "Yes." The first type of procedure is called an **effective Yes-procedure**, and the second type of procedure is called an **effective No-procedure**. Our definition of an effective *decision* procedure makes it clear that such a procedure must produce *both* answers. A concept that has only an effective Yes-procedure is called **semidecidable**.

We said in Chapters Two and Three that the methods of TL diagrams and truth tables provide us with effective decision procedures for determining the applicability of any of the eight logical concepts, defined in 2.5.3–2.5.10 and 3.5.3–3.5.10, to any *finite* set of TL or SL sentences. We described this fact by saying that these logical concepts are *decidable* in TL and SL. We illustrated the methods of TL diagrams and truth tables in 2.5.11:C and 3.5.11:C, respectively. The situation is different in PL. There are infinitely many PL interpretations for each set of PL sentences, whether it is finite or infinite. The interpretations of some sets of PL sentences are "manageable." They can be classified as belonging to a finite collection of classes of interpretations. In other words, there are cases in which infinitely many PL interpretations can be reduced to a finite number of "representative" interpretations. Consider as a very simple example the following set: $\Gamma = \{Pa \wedge \neg(Qb \wedge Rc), (Rc \vee Pa) \leftrightarrow \neg Se, \neg Rc \rightarrow Se, \neg Qb\}$. Γ has infinitely many PL interpretations. There are infinitely many universes of discourse, infinitely many lists of names, infinitely many referents that can be assigned to the names *a*, *b*, *c*, and *e*, and infinitely many extensions that can be assigned to the 1-place predicates *P*, *Q*, *R*, and *S*. However, so far as the applicability of the logical concepts to Γ and its members is concerned, these infinitely many PL interpretations can be reduced to 16 "representative" interpretations. Any PL interpretation I for Γ must include the names *a*, *b*, *c*, and *e* in its LN, assign referents to these names (which are I(*a*), I(*b*), I(*c*), and I(*e*)), and assign extensions to the predicates *P*, *Q*, *R*, and *S* (which are I(*P*), I(*Q*), I(*R*), and I(*S*)). Of course, I might include many more names in its LN and assign extensions to many more predicates. Regardless of the individuals and names that I has and the semantical assignments that it makes, there are only 16 different combinations of how the membership relation obtains between I(*a*) and I(*P*), I(*b*) and I(*Q*), I(*c*) and I(*R*), and I(*e*) and I(*S*). Here are two of these combinations: (1) I(*a*) ∈ I(*P*), I(*b*) ∈ I(*Q*), I(*c*) ∉ I(*R*), and I(*e*) ∉ I(*S*); (2) I(*a*) ∉ I(*P*), I(*b*) ∈ I(*Q*), I(*c*) ∉ I(*R*), and I(*e*) ∈ I(*S*). ('*x* ∈ *Z*' is read "*x* is a member of Z," and '*x* ∉ *Z*' is read "*x* is not a member of Z.") Each one of these combinations defines a "type" of PL interpretation for Γ. Each one of these 16 types consists of infinitely many PL interpretations for Γ. However, the members of each of these types all agree on the truth values of the members of Γ and of their sentential components.

The members of the set Γ are extremely simple PL sentences. They contain only sentential connectives, which make them similar to SL sentences. Such cases are the exception and not the norm in PL. The infinitely many PL interpretations for a *typical* set of PL sentences cannot be reduced to a finite number of representative interpretations. Moreover, there is no effective procedure for generating all the PL interpretations for every set of PL sentences, and there is no effective procedure for determining the truth value of every PL sentence on every PL interpretation for that sentence. This suggests that there is no effective *decision* procedure for answering *all* the questions of the form: "Is this PL sentence, or is this set of PL sentences, subsumed under this concept?" It is known, however, that even finite sets of PL sentences or sets that have effective decision procedures for determining membership in them typically fail to have an effective decision procedure for determining the applicability of all the eight logical concepts to them or to their members.[1] This entails that these logical concepts are undecidable in PL.[2]

Let us illustrate the point with an example. Consider the following set Γ of PL sentences.

S1 $(\forall z)(\exists x)(\exists y)Txyz$

S2 $(\forall z)((Ez \wedge Gcz) \rightarrow (\exists x)(\exists y)((Px \wedge Py) \wedge Szxy))$

There are infinitely many PL interpretations for this set. These interpretations vary in their universes of discourse, sizes, and semantical assignments. We would be surprised if there is a mechanical procedure for generating all of these interpretations. Given the great variety of the interpretations for Γ, the process of generating these interpretations seems "essentially" creative. We will not give a proof of this claim; its truth seems sufficiently intuitive. However,

1 A set for which there is an effective decision procedure for determining whether an object is a member of the set or not is said to be *decidable*.

2 Some of these concepts are not decidable but they are semidecidable. The concept of logical truth in PL is such a concept. There is an effective procedure that always produces the answer "Yes" when and only when the given PL sentence is logically true, but it does not *always* produce an answer if the sentence is contingent (if the sentence is logically false, its negation is logically true). This procedure, however, is only *in-principle* effective. It is not practical at all.

we will give a few PL interpretations for Γ to illustrate the degrees of similarity and difference that such interpretations can have to each other. We will also illustrate that there is no effective procedure for determining the truth value of *every* PL sentence on *every* PL interpretation for that sentence.

The PL Interpretation I

UD: {Simsim, Rumie, Gabby, Pickle, Catherine the Great, Murjana, Dena, Amy}
LN: *s*, *r*, *b*, *k*, *c*, *m*, *d*, *a*

Semantical Assignments

I(*s*): Simsim; I(*r*)Rumie; I(*b*): Gabby; I(*k*): Pickle; I(*c*): Catherine the Great; I(*m*): Murjana; I(*d*): Dena; I(*a*): Amy
I(*Ez*): "*z* is a kitten"
I(*Px*): "*x* is a person"
I(*Gzy*): "*z* is the mother of *y*"
I(*Txyz*): "*x*, *y*, and *z* are members of the same species, and *x* and *y* are siblings"
I(*Szxy*): "*z* is bred by *x* and *y*"

Facts of the interpretation I

F1. The only kittens are Simsim, Rumie, Gabby, and Pickle.
F2. Murjana, Dena, and Amy are people.
F3. Murjana and Amy, but not Dena, are siblings.
F4. Catherine the Great is a full-grown cat, which is the mother of the kittens Simsim, Rumie, and Gabby only.
F5. Murjana and Dena bred the kittens Simsim, Rumie, and Gabby, but not Pickle. Amy has nothing to do with breeding cats.

PL-English reading of (S1) $(\forall z)(\exists x)(\exists y)Txyz$ on I:
For every individual z, there are two individuals x and y, such that x, y, and z are members of the same species, and x and y are siblings.
True on I.

Natural reading of (S2) $(\forall z)((Ez \wedge Gcz) \rightarrow (\exists x)(\exists y)((Px \wedge Py) \wedge Szxy))$ on I:
Any kitten whose mother is Catherine the Great is bred by two people.
True on I.

The PL Interpretation I*

UD: {Simsim, Rumie, Gabby, Pickle, Catherine the Great, Murjana, Dena, Amy}
LN: *s*, *r*, *b*, *k*, *c*, *m*, *d*, *a*

Semantical Assignments

I(s): Simsim; I(r)Rumie; I(b): Gabby; I(k): Pickle; I(c): Catherine the Great; I(m): Murjana;
I(d): Dena; I(a): Amy
I*(Ez): "z is a kitten"
I*(Px): "x is a person"
I*(Gzy): "z is the mother of y"
I*($Txyz$): "x, y, and z are members of the same species, and x and y are siblings"
I*($Szxy$): "z is bred by x and y"

Facts of the interpretation I*

F1. The only kittens are Simsim, Rumie, Gabby, and Pickle.
F2. Murjana, Dena, and Amy are people who are not related to each other.
F3. Catherine the Great is a full-grown cat, which is the mother of all the kittens.
F4. Murjana and Dena bred the kittens Simsim, Rumie, and Gabby. Pickle is from a different litter, which was not bred by Murjana or Dena. Amy has nothing to do with breeding cats.

PL-English reading of (S1) $(\forall z)(\exists x)(\exists y)Txyz$ on I*:
For every individual z, there are two individuals x and y, such that x, y, and z are members of the same species, and *x and y are siblings.*
False on I*.
Any of the people in UD demonstrates that this sentence is false, since none of these people is a sibling of another.

Natural reading of (S2) $(\forall z)((Ez \wedge Gcz) \rightarrow (\exists x)(\exists y)((Px \wedge Py) \wedge Szxy))$ on I*:
Any kitten whose mother is Catherine the Great is bred by two people.
False on I*.
Pickle is a kitten whose mother is Catherine the Great, but Pickle was not bred by anyone in UD.

Note: A slight change in I produced a new interpretation, and the truth values of S1 and S2 also changed. I and I* have the same UD (and hence the same size, 8), the same LN, the same semantical assignments except for T and G (they have different extension on I and I*).

The reader must have noticed that we determined the truth values of S1 and S2 on I and I* mechanically and with remarkable ease. We simply examined small finite extensions of five predicates and the referent of one name, and searched these extensions (most likely mentally) for appropriate individuals or ordered tuples of individuals. It is essential to note that we did *not* claim that there is no PL sentence whose truth value can be determined by an effective procedure on certain PL interpretations for that sentence. Our claim is rather that there is no effective procedure, or even a finite collection of effective procedures, that can determine the truth value of *every* PL sentence on *every* PL interpretation for that sentence. We will see an illustration of this claim below.

The PL Interpretation N

UD: : {0, 1, 2, 3, …}
LN: c, n_0, n_1 n_2, n_3, …

Semantical Assignments

N(c): 2; N(n_0): 0; N(n_1): 1; N(n_2): 2; N(n_3): 3;…; in general, N(n_k): k, where k is any natural number.
N(Ex): "x is an even number": {2, 4, 6, 8, 10, …}
N(Pz): "z is a prime number": {2, 3, 5, 7, 11, 13, 17, …}
N(Gzy): "y is greater than z" (observe the order of the variables)
N($Txyz$) "x and y are twin primes that are greater than z" (x and y are *twin primes* if and only if x and y are prime numbers and $x = y+2$; for example, 3 and 5, 5 and 7, 11 and 13, 17 and 19, and 29 and 31)
N($Szxy$): "z is the sum of x and y" (in symbols, $z = x+y$)

Natural reading of (S1) $(\forall z)(\exists x)(\exists y)Txyz$ on N:
For every natural number, there are twin primes greater than that number.
In other words, *there are infinitely many pairs of twin primes.*[1]
The truth value of this sentence on N is unknown. No one has yet been able to prove it or disprove it.

Natural reading of (S2) $(\forall z)((Ez \wedge Gcz) \rightarrow (\exists x)(\exists y)((Px \wedge Py) \wedge Szxy))$ on N:
For every even number greater than 2, there are two prime numbers, such that the even number is the sum of the two prime numbers.
In other words, *every even number greater than 2 is the sum of two prime numbers.*[2]
The truth value of this sentence on N is unknown. No one has yet been able to prove it or disprove it.

To prove that S1 is true on N we need to give a proof, based on the *standard axioms of arithmetic*, whose conclusion is that there are infinitely many pairs of twin primes. On the other hand, in order to prove that S1 is false on N we need to show, also on the basis of the standard axioms of arithmetic, that there is a natural number such that there are no twin primes greater than that number. As for proving that S2 is true on N, we must prove that the standard axioms of arithmetic logically imply that every even number greater than two is the sum of two prime

1 The assertion that there are infinitely many pairs of twin primes is known as the "Twin-Prime Conjecture."
2 The assertion that every even number greater than two is the sum of two prime numbers is known as the "Goldbach Conjecture."

numbers, while to prove that S2 is false we ought to show that these axioms logically imply that there is an even number greater than two that is not the sum of any two prime numbers.

Based on our current state of mathematical knowledge we do not know what these proofs are nor do we know whether it is even possible to prove or disprove these statements. However, for the sake of argument let us suppose that there are such proofs and that we simply do not know what they are. The point here is that there does not *seem* to be any mechanical procedure for finding these proofs. If this is true, and if finding these proofs is our *only* means for determining the truth values of S1 and S2 on N, then there would be no effective decision procedure for determining the truth values of these sentences on N. This would establish our claim that there is no effective decision procedure for determining the truth value of every PL sentence on every interpretation for that sentence.

4.5.11:C2 Perhaps, we are wrong. Perhaps, this is only an appearance, and in actuality there is a deterministic mechanical procedure for finding these proofs; and hence there exists an effective decision procedure for determining the truth values of S1 and S2 on N. We will try to describe such a procedure, and we will see whether this procedure is really an effective decision procedure for determining the truth values of S1 and S2 on N.

First, we need to designate an appropriate set of PL sentences as the set of axioms that all proofs of arithmetical propositions, such as S1 and S2, may invoke. Mathematicians have produced a set of PL sentences that are widely considered as the *standard axioms of arithmetic*. These axioms are sometimes called the "Dedekind-Peano Axioms." Let DP stand for the set of these axioms. It is not important for our purposes here what these axioms are. The only thing that matters is that these axioms are almost universally accepted as the standard axioms of arithmetic. Let us also assume that there is a PL model of DP. As described above, the interpretation N is not rich enough to interpret all of the DP axioms. For instance, one of these axioms may be expressed as $(\forall x)n_0 \neq gx$, where g stands for the function of succession. Thus gx is interpreted as "The successor of x." If we take the universal quantifier as ranging over the natural numbers, the interpretation of $(\forall x)n_0 \neq gx$ is that 0 is not the successor of any natural number (recall that n_0 stands for the number 0). But N, as it stands, does not contain a semantical assignment for the 1-place function symbol g. So we have to expand the semantical assignments of N in order to accommodate *all* the extra-logical PL vocabulary that occurs in the DP axioms. Let us suppose that we accomplished this task and that the resulting PL interpretation, which we will also call N, is a model of DP.

We will refer to any PL sentence that is composed of Voc(DP) (i.e., the vocabulary of DP) as an arithmetical sentence. Let Th(N) ("the theory of N") be the set that consists of all the arithmetical sentences that are true on the interpretation N. In other words, for every PL sentence **X** that is composed of Voc(DP), **X** $\in$ Th(N) if and only if **X** is true on N. Th(N) is standardly called "Arithmetic." We will follow this terminology here. We also define Th(DP) ("the theory of DP") as the set of all the arithmetical sentences that are logical consequences of DP. Said differently, for every PL sentence **X** that is composed of Voc(DP), **X** $\in$ Th(DP) if and only if DP $\models$ **X**. Th(DP) is traditionally called "Peano Arithmetic." It is clear that DP is a subset of Th(DP) (every sentence is a logical consequence of itself). If N is a model of DP, then it is a model of Peano Arithmetic. Since every sentence of Peano Arithmetic is an arithmetical sentence that is a logical consequence of DP and since, by definition, all the logical consequences of DP are true on every model of DP, all the members of Th(DP) are true on N. It follows that Peano Arithmetic is a subset of Arithmetic. As we will explain below, Peano Arithmetic is a *proper subset* of Arithmetic.

The set DP is an infinite set but it is *decidable*. A **set is decidable** just in case there is an effective decision procedure that determines whether any object (of a certain type) is a member of the set or not. For example, the set of the PL formulas is decidable, since the method of construction trees is an effective decision procedure that determines whether any PL expression is a PL formula or not. Similarly, there is an effective decision procedure that determines whether any PL sentence is a DP axiom or not. This is very important if we want to construct proofs that invoke these axioms. We would want to know whether a PL sentence we intend to invoke is an axiom of Peano Arithmetic or not. Since we have an effective decision procedure, we can apply this procedure to this PL sentence and determine whether it is indeed one of the DP axioms or not.

Second, we must describe precisely a set of *rules of inference*. A **PL rule of inference** is a rule that permits the inference of a PL sentence of a certain form from a finite set of PL sentences also of certain forms. We will study PL rules of inference with some detail in Chapter Five. For now, it suffices to give a few examples of inference rules. The rule *Modus Ponens* allows us to infer **Y** from two sentences whose forms are **X→Y** and **X**; the rule *Hypothetical Syllogism* licenses the inference of a sentence of the form **X→Z** from two sentences whose forms are **X→Y** and **Y→Z**; and the rule *Negated Quantifiers* permits the inference of a formula of the form ¬(∀**z**)**X** from a formula of the form (∃**z**)¬**X**, and conversely. As we will explain in Chapter Five, PL has a set of inference rules, which we call the "Natural Deduction System" (NDS), that is both *sound* and *complete*. The rules of NDS are **sound** in the sense that they *only* permit the inference of logical consequences: if **Γ** is a set of PL sentences and **X** is a PL sentence that can be derived from **Γ** by applying one or more of the rules of NDS, then **X** is a logical consequence of **Γ**, that is, the PL argument **Γ/X** is deductively valid. On the other hand, the rules of NDS are **complete** in the sense that they allow the inference of *all* logical consequences: if **Γ** is a set of PL sentences and **X** is a logical consequence of **Γ**, then **X** can be derived from **Γ** by applying one or more of the rules of NDS. The soundness of the rules of NDS entails that these rules are *truth-preserving*, which means that if a PL sentence **X** is inferred, by applying NDS rules, from a set of PL sentences that are true on some PL interpretation I, then **X** is also true on I.

Now we are ready to define the notion of arithmetical proof. π is an arithmetical proof of some PL sentence **X** if and only if π is a finite sequence of PL sentences, such that every sentence in π is either a DP axiom or is obtained by one of the NDS rules of inference from earlier sentences in π, and the terminal (i.e., last) sentence in π is **X**. We will construct many PL proofs in Chapter Five. However, it is sufficient for our purposes here to illustrate this notion with one example. The following sequence of seven PL sentences is an arithmetical proof of the sentence $(\exists x)(\exists y)\neg(n_0 \neq gy \rightarrow n_0 = gx)$. This proof invokes the arithmetical axiom stated above—namely, $(\forall x)n_0 \neq gx$—and four NDS rules. We will give informal descriptions of these rules when they are applied first.

1	$(\forall x)\ n_0 \neq gx$	Arithmetical axiom
2	$n_0 \neq ga$	From 1 by *Universal Instantiation*: from a universally quantified sentence, we may infer any of its substitutional instances.
3	$n_0 \neq gb$	From 1 by *Universal Instantiation.*
4	$n_0 \neq ga \land n_0 \neq gb$	From 2 and 3 by *Conjunction*: from two sentences, we may infer their conjunction.

5	$\neg(n_0 \neq ga \rightarrow n_0 = gb)$	From 4 by *Negated Conditional*: $\mathbf{X} \wedge \neg \mathbf{Y}$ may be replaced with $\neg(\mathbf{X} \rightarrow \mathbf{Y})$, and conversely.
6	$(\exists y)\neg(n_0 \neq gy \rightarrow n_0 = gb)$	From 5 by *Existential Generalization*: from a substitutional instance, we may infer an existentially quantified sentence.
7	$(\exists x)(\exists y)\neg(n_0 \neq gy \rightarrow n_0 = gx)$	From 6 by *Existential Generalization.*

The arithmetical proof itself is only the sequence of the seven PL sentences above, not including the numbers on the left or the explanations on the right. These numbers and explanations are for our benefit in order to understand how these sentences are obtained.

It is known that there is a mechanical procedure for listing all of these arithmetical proofs in some order. You may reasonably doubt that there is really such a mechanical procedure, but in this book you have to take my word that there is *in principle* such a mechanical procedure. So using this mechanical procedure we produce a list whose members are *all* the proofs that are based on the axioms of Peano Arithmetic (i.e., the DP axioms) and correct applications of the NDS rules of inference. Let this list be $\pi_1, \pi_2, \pi_3, \pi_4, \pi_5, \ldots$. We remark three things about this list. First, it is infinite, since we can always repeat one of the premises of a proof or one of its intermediate conclusions as many times as we would like (it is surely a valid inference rule that infers **Z** from **Z**). Second, since every π_k is a finite sequence of PL sentences, we can in principle mechanically examine the terminal sentence of π_k in order to see what the conclusion of π_k is (we say "in principle" because the proof could be extremely long). Third, the terminal sentence of any π_k is true on N. The reason for the last claim is this. By the definition of an arithmetical proof, π_k does not make use of anything other than DP axioms, which are true on N, and NDS rules of inference, which are truth-preserving; thus when these rules are applied to DP axioms, they yield conclusions that are also true on N; it follows that the terminal sentence of π_k (i.e., the conclusion of π_k) is true on N.

It seems that we now have all the components of an effective procedure for determining whether S1 and S2 are true on N or not. The procedure works as follows.

1. We use the mechanical procedure to obtain the first arithmetical proof π_1. We examine the last sentence of π_1; if this sentence is S1 or S2, we have *effectively* determined that S1 or S2 is true on N; if this sentence is ¬S1 or ¬S2, we have effectively determined that S1 or S2 is false on N.
2. If the terminal sentence of π_1 is neither S1, S2, ¬S1, nor ¬S2, we use the same mechanical procedure to locate the next proof, π_2, in the sequence of all arithmetical proofs. We repeat the same procedure as in 1 above, but this time for the proof π_2. Again if the terminal sentence of π_2 is neither S1, S2, ¬S1, nor ¬S2, we repeat this step; this time we examine π_3.
3. We repeat step 2 for subsequent proofs in the list of all arithmetical proofs. Since we assumed that there is either a proof of S1 or of ¬S1, and either of S2 or of ¬S2, this procedure will eventually locate these proofs. All the steps of this procedure are mechanical and deterministic; therefore we have an effective procedure for determining the truth values of S1 and S2 on N, *if* indeed there are proofs of S1 or of ¬S1, and of S2 or of ¬S2.

Even if the procedure we described above is indeed successful in recovering the desired proofs, no one uses this procedure. It is practically impossible to use. Mathematicians employ a great deal of creativity and ingenuity in discovering proofs of mathematical propositions. A more serious problem with this procedure is that it is not really a decision procedure. It is possible that there is no arithmetical proof of S1, ¬S1, S2, or ¬S2. It is established demonstratively that the axioms of Peano Arithmetic are *incomplete*. The relevant notion of completeness here is different from the notion of completeness we defined above for the NDS rules of inference. The DP axioms are incomplete in the sense that there are arithmetical sentences that can neither be proved nor disproved on the basis of these axioms. Let G be such a PL sentence, and assume that we do not know whether G or ¬G is provable or not. Due to the principle of bivalence, we know that G is either true or false on N. Since there is no arithmetical proof of G or of ¬G, the effective procedures described above for finding arithmetical proofs will not be able to yield any proof of G or ¬G, no matter how long we apply the procedure. Thus this procedure will not be able to determine the truth value of G on N. Since we do not know in advance whether G or ¬G is provable or not, we will not be able to tell whether the procedure has not been applied long enough to locate a proof of G or of ¬G, or whether there is no such a proof, and hence the procedure would continue to run forever without being able to locate the desired proof. This entails that the effective procedure we described above is not a decision procedure. It cannot generate a decisive answer to *every* question of the form: "Is this PL sentence true on N?" If the sentence is provable, the procedure will return the answer "Yes"; if the negation of the sentence is provable, the procedure will return the answer "No." However, if neither the sentence nor its negation is provable, the procedure will not return any answer; it will not halt.

The point of this discussion is that since there are arithmetical PL sentences that are neither provable nor disprovable on the basis of the DP axioms, the effective procedure described above is not a decision procedure that works for all arithmetical PL sentences. If this procedure does not work for all arithmetical PL sentences, is there another effective procedure that is a decision procedure for determining the truth values of all the arithmetical PL sentences on N? The answer is negative. It is a metatheorem of PL that the set of all the arithmetical PL sentences that are true on N is not a decidable set. In the language of concepts, the concept of arithmetical truth is not a decidable concept, where by 'arithmetical truth' we mean "an arithmetical sentence that is true on N."

The incompleteness of the DP axioms entails that Peano Arithmetic is a *proper* subset of Arithmetic. Recall that Peano Arithmetic is Th(DP), which is the PL theory whose members are all the arithmetical sentences that are logical consequences of DP. We argued previously that if the PL interpretation N is a model of DP, it is also a model of Th(DP). This means that Peano Arithmetic is a subset of Arithmetic (symbolically, Th(DP) $\subseteq$ Th(N)). Since the set DP is incomplete, there is an arithmetical sentence G such that neither G nor ¬G is provable from DP. Given that the NDS rules of inference are complete, it follows that neither G nor ¬G is a logical consequence of DP. Hence, by the definition of Th(DP), neither G nor ¬G is a member of Th(DP). However, by the principle of bivalence, either G or ¬G is true on N. Therefore, Th(N) contains an arithmetical sentence that does not belong to Th(DP). This shows that Peano Arithmetic is a proper subset of Arithmetic.

We have not yet explained why the eight logical concepts defined in 4.5.3–4.5.10 are undecidable in PL. We have explained that the concept of arithmetical truth is undecidable, that Peano Arithmetic is incomplete, and that it is a proper subset of Arithmetic. The undecidability of arithmetical truth demonstrates what we said earlier that there is no effective decision procedure that determines whether any PL sentence is true or false on any given PL interpretation for that sentence. This is already very different from SL and TL. The truth conditions of the TL and SL sentences allow us to determine *effectively* the truth value of any TL or SL sentence on any TL diagram or SL truth valuation that is relevant to this sentence. One might argue, however, that just because the concept "true-on-an-interpretation" is not decidable in general in PL, there might still be effective methods for determining the applicability of all or some of the eight logical concepts. After all, we initially determined that the sentences S1 and S2, mentioned in 4.5.11:C1, are contingent because we were able to determine their truth values on two interpretations I and I*. We did not need the interpretation N to reach a conclusion about the logical status of S1 and S2. The existence of I and I* is sufficient for determining that they are contingent.

This is true. Nevertheless, these logical concepts are undecidable in PL. It is enough to explain that the relation of logical consequence is undecidable in PL. All the other logical concepts can be defined on the basis of the relation of logical consequence. Deductive validity is just another way of talking about logical consequence, and deductive invalidity is simply the negation of deductive validity. A sentence is logically false if and only if it logically implies a contradiction (i.e., a sentence of the form $\mathbf{X} \wedge \neg \mathbf{X}$ is a logical consequence of it); it is logically true if and only if its negation is logically false; and it is contingent just in case it is neither logically true nor logically false. Two sentences are logically equivalent if and only if each one is a logical consequence of the other. A set is inconsistent just in case it logically implies a contradiction, and it is consistent if and only if it is not inconsistent. This shows that if the relation of logical consequence is proved to be undecidable in PL, then none of these logical concepts is decidable in PL.

At any rate, what does it mean to say that the relation of logical consequence is undecidable? Let us explain first what it means to say that it is decidable. The relation of logical consequence is decidable when and only when there is an effective decision procedure such that for any *decidable* set Γ and any sentence **X**, this decision procedure can determine the answer to the question "Is **X** a logical consequence of Γ?" It is important that Γ is decidable; otherwise membership in Γ would not be determined effectively, which means we might not be able, even in principle, to determine whether any sentence is a member of Γ or not. Thus to show that the relation of logical consequence is undecidable in PL, it suffices to describe a decidable PL set Δ for which there is no effective decision procedure for determining whether or not any given sentence is a logical consequence of it. In other words, if we take Th(Δ) ("the theory of Δ") to be the collection of all the logical consequences of Δ, then to say that there is no effective decision procedure for determining whether or not any sentence is a logical consequence of Δ is simply to say that there is no effective decision procedure for determining membership in the set Th(Δ). Given the definition of a decidable set, the previous assertion amounts to saying that Th(Δ) is an undecidable set. In other words, to show that the relation of logical consequence is undecidable in PL is simply to show that there is a decidable set Δ such that Th(Δ) is undecidable. The incompleteness of Peano Arithmetic together with its relative richness as a mathematical theory entails that it is undecidable. Since Peano Arithmetic is Th(DP) and DP is a decidable set of PL sentences, the undecidability of Th(DP) implies that the relation of logical consequence is undecidable in PL. The fact that Peano Arithmetic is undecidable is a theorem of the metatheory of PL. Its proof lies far beyond the scope of this book. So we accept it without proof. However, as explained above, this fact is sufficient to demonstrate that the relation of logical consequence is undecidable in PL; and hence the eight logical concepts defined in 4.5.3–4.5.10 are also undecidable in PL.

4.5.12 At the outset of this chapter we said that PL includes TL and extends SL. In this subsection we explain this statement. First, every TL sentence X and every set Γ of TL sentences can be transformed via an effective procedure into a PL sentence **Z** and a set Σ of PL sentences, respectively. We call the sentence **Z** a **PL counterpart of** X and denote it as PL(X), and we call the set Σ a PL **counterpart of** Γ and denote it as "PL(Γ)." There might be many PL counterparts of the same TL sentence and of the same TL set. If Γ/X is a TL argument, its PL counterparts are denoted as "PL(Γ/X)." The notation 'PL(Γ/X)' makes sense since an argument Γ/X is, by definition, merely a set whose members are the premises (i.e., the members of Γ) and the conclusion (i.e., X). Every TL diagram for a TL set Γ can be converted into a PL interpretation for PL(Γ), such that the truth values of the TL sentences on the TL diagram are identical with the truth values of their PL counterparts on the PL interpretation. If we restrict the PL sentences to those that are counterparts of TL sentences, and restrict the PL interpretations to those that assign *nonempty* extensions to the 1-place predicates, then we call the resulting logic the PL **counterpart of** TL and denote it as "PL(TL)." A TL argument Γ/X is valid in TL if and only if any of its PL counterparts PL(Γ/X) are valid in PL(TL). We say that PL **includes** TL because PL includes a restricted portion—namely PL(TL)—that is equivalent to TL. Second, every SL sentence **W** and every set Φ of SL sentences can be transformed via an effective procedure into a PL sentence **Y** and a set Ψ of PL sentences, such that there is a one-to-one correspondence between the two sets, and every pair of corresponding sentences has the same **sentential form**. We call the sentence **Y** a PL **counterpart of W** and denote it as "PL(**W**)" and we call the set Ψ a PL **counterpart of** Φ and denote it as "PL(Φ)." There are many PL counterparts of the same SL sentence and of the same set of SL sentences. If Φ/**Z** is an SL argument, any of its PL counterparts is denoted as "PL(Φ/**Z**)." PL is a **proper extension** of SL because for every valid SL argument Φ/**Z**, any of its PL counterparts PL(Φ/**Z**) is also valid, but the converse is not true.

4.5.12:C COMMENTARY ON 4.5.12

4.5.12:C1 In 4.2.2:C1 we described how to translate TL sentences into PL. In this subsection we will describe a portion of PL that is equivalent to TL. We impose two restrictions on PL: one syntactical and one semantical. The syntactical restriction is that only PL sentences that are counterparts (i.e., translations) of TL sentences are permitted into this logic. The semantical restriction is that PL interpretations for these restricted sentences are not permitted to assign the empty extension to any 1-place predicate. We call the resulting portion of PL **the PL counterpart of TL** and denote it as "PL(TL)." We now describe a procedure that transforms every TL sentence W into a PL(TL) sentence **V**, such that W and **V** can be considered as having the "same" truth conditions.[1] We call **V** a PL counterpart of W and we denote it as PL(W). We will

1 Precisely speaking, W and **V** do not have the *same* truth conditions. As a TL sentence, the truth conditions of W have to do with circles and dots and their relations to each other. As a PL sentence, the truth conditions of **V** have to do with extensions and individuals and their relations to each other. Nevertheless, the truth conditions of W and **V** are structurally similar to each other: their structural relations mirror each other. The tables below make this correspondence between the truth conditions of W and the truth conditions of **V** clear.

consider one example of each TL sentence-form. The table below describes this procedure when applied to all the TL sentence-forms, it also suggests a way of transforming every TL diagram into a PL(TL) interpretation and vice versa. (See 2.1:C1 for the meaning of the set-theoretic symbols used in the table below.)

No.	TL sentence W	PL(TL) sentence PL(W)	Truth conditions of W	Truth Conditions of PL(W)
1	all *A* are *B*	$(\forall x)(Ax \rightarrow Bx)$	*B*-circle includes *A*-circle	$I(Ax) \subseteq I(Bx)$
2	no *A* is *B*	$\neg(\exists x)(Ax \wedge Bx)$ Or, $(\forall x)(Ax \rightarrow \neg Bx)$	*A*-circle and *B*-circle are disjoint	$I(Ax) \cap I(Bx) = \varnothing$
3	some *A* are *B*	$(\exists x)(Ax \wedge Bx)$	*A*-circle and *B*-circle are not disjoint	$I(Ax) \cap I(Bx) \neq \varnothing$
4	some *A* are not *B*	$(\exists x)(Ax \wedge \neg Bx)$	*B*-circle does not include *A*-circle	$I(Ax) \not\subseteq I(Bx)$[1]
5	*s* is *A*	As	*s*-dot is inside *A*-circle	$I(s) \in I(Ax)$
6	*s* is not *A*	$\neg As$	*s*-dot is outside *A*-circle	$I(s) \notin I(Ax)$[2]
7	*s* is *t*	$s = t$	*s*-dot and *t*-dot are identical	$I(s) = I(t)$
8	*s* is not *t*	$s \neq t$	*s*-dot and *t*-dot are not identical	$I(s) \neq I(t)$

This procedure can be extended to any set of TL sentences. We simply replace every TL sentence in the set with *exactly one* of its PL counterparts provided that we use the same PL names and 1-place predicates to replace the same singular and general TL terms, and that we use different PL names and 1-place predicates to replace different singular and general TL terms. Thus if the TL set is {all *A* are *B*, some *B* are not *C*}, then we may transform this set into $\{(\forall x)(Ax \rightarrow Bx), (\exists z)(Bz \wedge \neg Cz)\}$. It is not permissible, though, to use two different PL predicates, say 'Bx' and 'Dz', for the '*B*' in 'all *A* are *B*' and for the '*B*' in 'some *B* are not *C*', respectively, or to use one PL predicate, say Gx, for both the '*A*' in 'all *A* are *B*' and for the '*C*' in 'some *B* are not *C*'. The rule is that the same letter is replaced with the same name or predicate and that two different letters are replaced with two different names or predicates. There is an easy way to ensure that this rule is never violated. We simply use the singular and general terms of the TL sentences as our PL names and 1-place predicates. If Γ is a set of TL sentences, then a PL counterpart of Γ, which is denoted as "PL(Γ)," is a set that consists of PL counterparts of all the members of Γ such that for every X in Γ, there is exactly one PL counterpart of X in PL(Γ).

If Γ is a set of TL sentences, then every TL diagram D for Γ can be converted, in a natural way, into a PL(TL) interpretation I for PL(Γ), and every PL(TL) interpretation I for PL(Γ) can be converted, also in a natural way, into a TL diagram D for Γ. Given that TL sentences and their PL(TL) counterparts have the "same" truth conditions, the truth value on D of any sentence in Γ is identical with the truth value on I of its counterpart in PL(Γ). This fact explains why TL and PL(TL) are equivalent systems. The equivalence of TL and PL(TL) entails that a set of TL sentences is subsumed under a logical concept if and only if any of its PL counter-

1 The set-theoretic sentence '$I(Ax) \not\subseteq I(Bx)$' means that the extension $I(Ax)$ is *not a subset of* the extension $I(Bx)$, that is, there is at least one individual in $I(Ax)$ that is not in $I(Bx)$.

2 The set-theoretic sentence '$I(s) \notin I(Ax)$' means that the individual $I(s)$ is not in the extension $I(Ax)$.

parts are subsumed under the same logical concept in PL(TL). The following paragraph and theorem articulate in precise terms the relationship between TL and its PL counterpart, PL(TL).

TL and PL(TL) are equivalent systems. For every TL argument Σ/W, for all TL sentences X and Y, and for every set Γ of TL sentences, (1) Σ/W is valid (or invalid) if and only if PL(Σ/W) is valid (or invalid) in PL(TL); (2) X is logically true (or logically false, or contingent) if and only if PL(X) is logically true (or logically false, or contingent) in PL(TL); (3) X and Y are logically equivalent if and only if PL(X) and PL(Y) are logically equivalent in PL(TL), where PL(X) and PL(Y) are the members of any PL counterpart of the set {X, Y}; and (4) Γ is consistent (or inconsistent) if and only if PL(Γ) is consistent (or inconsistent) in PL(TL). All of these assertions are corollaries of a single theorem. In order to state this theorem precisely, we need to introduce a simple notation. If Σ is a set of TL sentences and W a TL sentence, then PL(Σ, W) is a PL counterpart of the TL set that consists of W and of all the members of Σ. Really, PL(Σ, W) is just another notation for PL(Σ/W). We are now ready to state the relevant theorem.

Theorem 4.3: For every TL sentence W and every TL set Σ (possibly empty), Σ ⊨ W in TL if and only if PL(Σ) ⊨ PL(W) in PL(TL), where PL(Σ) is a subset of PL(Σ, W) and PL(W) is a member of PL(Σ, W).

We illustrate this theorem and some of its corollaries with a few examples.

(a) The TL sentence 'some *E* are not *H*' is a logical consequence of the TL set {all *E* are *A*, all *A* are *G*, no *G* is *H*}. The non-similar TL diagrams that make the premises all true are D below and its standard variants. In all of them the conclusion is true.

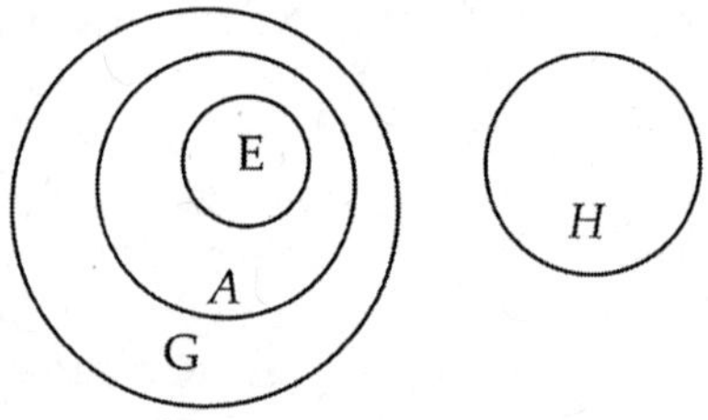

D: all *E* are *A* (true); all *A* are *G* (true); no *G* is *H* (true) / some *E* are not *H* (true)

The relation of logical consequence holds, in PL(TL), between the PL counterparts of these TL sentences, that is,

$$\{(\forall x)(Ex \to Ax), (\forall x)(Ax \to Gx), (\forall z)(Gz \to \neg Hz)\} \vDash (\exists z)(Ez \wedge \neg Hz)$$

Proof that the logical consequence relation above holds in PL(TL).

1. Let I be a PL(TL) interpretation such that
   1.1. $(\forall x)(Ex \to Ax)$ is true on I,
   1.2. $(\forall x)(Ax \to Gx)$ is true on I, and
   1.3. $(\forall z)(Gz \to \neg Hz)$ is true on I.
2. From 1 by the definition of a PL(TL) interpretation: I(*Ex*) is not empty, that is, there is an individual that belongs to I(*Ex*).

3. From 2 by the definition of a PL(TL) interpretation: there is a name **s** in LN such that I(**s**) belongs to I(Ex).
4. From 3 by the truth conditions of E**s**: the sentence E**s** is true on I.
5. From 1.1–1.3 by the truth conditions of the universal quantifier: all basic substitutional instances of $(\forall x)(Ex\rightarrow Ax)$, $(\forall x)(Ax\rightarrow Gx)$, and $(\forall z)(Gz\rightarrow\neg Hz)$ are true on I; in particular, E**s**→A**s**, A**s**→G**s**, and G**s**→¬H**s** are true on I.
6. From 4 and 5 by the truth conditions of the conditional: A**s** is true on I.
7. From 5 and 6 by the truth conditions of the conditional: G**s** is true on I.
8. From 5 and 7 by the truth conditions of the conditional: ¬H**s** is true on I.
9. From 4 and 8 by the truth conditions of the conjunction: E**s**∧¬H**s** is true on I.
10. From 9 by the truth conditions of the existential quantifier: $(\exists z)(Ez\wedge\neg Hz)$ is true on I.
11. From 1 and 10: since I is arbitrary, all PL interpretations on which the premises are true make the conclusion true.
12. From 11 by the definition of logical consequence: $(\exists z)(Ez\wedge\neg Hz)$ is a logical consequence of the set $\{(\forall x)(Ex\rightarrow Ax), (\forall x)(Ax\rightarrow Gx), (\forall z)(Gz\rightarrow\neg Hz)\}$.

Observe that this proof would not be possible without the fact that the extension of E on I is not empty (Step 2). In fact a PL interpretation that assigns the empty extension to E, A, and G makes the premises true and the conclusion false. Thus while this relation of logical consequence holds in TL, its PL counterpart does not hold in PL. In other words, this is an example of a valid TL inference that is invalid in PL. We will employ this fact below to articulate the relation between TL and PL.

(b) As stated above, a TL argument Γ/W is deductively invalid if and only if PL(Γ/W) is deductively invalid in PL(TL). This follows from the facts that any TL diagram that is a counterexample to a TL argument can be converted into a PL(TL) interpretation that is a counterexample to all the PL counterparts of the TL argument, and that any PL(TL) counterexample to a PL counterpart of a TL argument can be converted into a TL counterexample to the TL argument. Here is an example.

S1 *s* is not *t*
S2 *t* is *K*
S3 some *K* are *B*

S4 *s* is not *B*

The TL diagram D below is a counterexample to this argument.

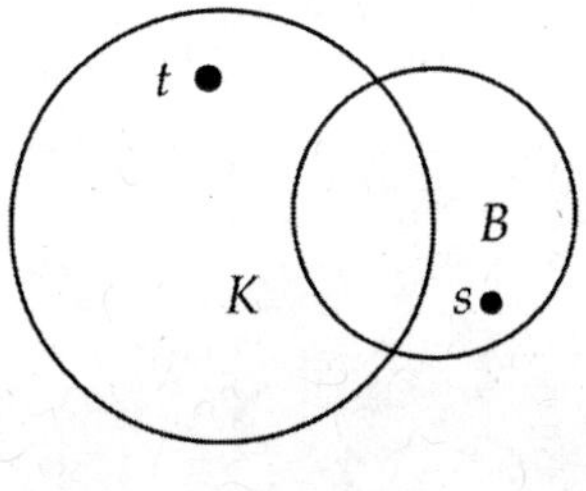

D: *s* is not *t* (S1, true); *t* is *K* (S2, true); some *K* are *B* (S3, true) / *s* is not *B* (S4, false)

A PL counterpart of this argument is this:

S1* $s \neq t$
S2* Kt
S3* $(\exists x)(Kx \wedge Bx)$

S4* $\neg Bs$

The PL(TL) interpretation I below *mimics* the TL diagram D. (The UD is generic.)

UD: $\{\alpha, \beta, \delta, \sigma, \tau\}$
LN: a, b, d, s, t

Semantical assignments

I(a): α; I(b): β; I(d): δ; I(s): σ; I(t): τ
I(Kx): $\{\alpha, \beta, \delta, \tau\}$; I($Bx$): $\{\beta, \delta, \sigma\}$

The reader can easily verify that S1*–S3* are true on I and that S4* is false on I. D shows that the TL argument is deductively invalid and I shows that the PL argument is deductively invalid. If we are given the PL(TL) interpretation I first, we simply draw a TL diagram such that I(Kx) and I(Bx) are represented by two circles, a K-circle and a B-circle, and I(s) and I(t) are represented by two dots, an s-dot and a t-dot. We then mirror the relations between the constituents of I. So we make the K-circle and the B-circle strictly overlap (because I(Kx) and I(Bx) are not disjoint and neither is a subset of the other), we make the s-dot and the t-dot not identical (because I(s) and I(t) are not identical), we place the t-dot inside the K-circle but outside the B-circle (because I(t) belongs to I(Kx) but not to I(Bx)), and we place the s-dot inside the B-circle but outside the K-circle (because I(s) belongs to I(Bx) but not to I(Kx)). This procedure generates the TL diagram D above.

(c) We previously stated that if Γ is a set of TL sentences, then Γ is consistent if and only if PL(Γ) is consistent in PL(TL). If Γ is consistent in TL, then there is a TL diagram that is a model of Γ. But this TL diagram can be converted into a PL(TL) interpretation that is a model of PL(Γ), which means that PL(Γ) is consistent in PL(TL). The same reasoning can be applied in reverse to show that if PL(Γ) is consistent in PL(TL), then Γ is consistent in TL. Note that since a PL(TL) interpretation is also a PL interpretation, if a set is consistent in PL(TL), then it is consistent in PL. This entails that if Γ is consistent in TL, then PL(Γ) is consistent in PL. The converse is not true: there are inconsistent TL sets that have PL counterparts that are consistent in PL. We will give an example of such a set in (d) below. The following is an example of a set of TL sentences and its PL counterpart. Both sets are consistent.

S5	e is A	S5*	Ae
S6	no A is B	S6*	$\neg(\exists z)(Az \wedge Bz)$
S7	some B are not C	S7*	$(\exists y)(By \wedge \neg Cy)$

The TL diagram D is a model of the set {S5, S6, S7}.

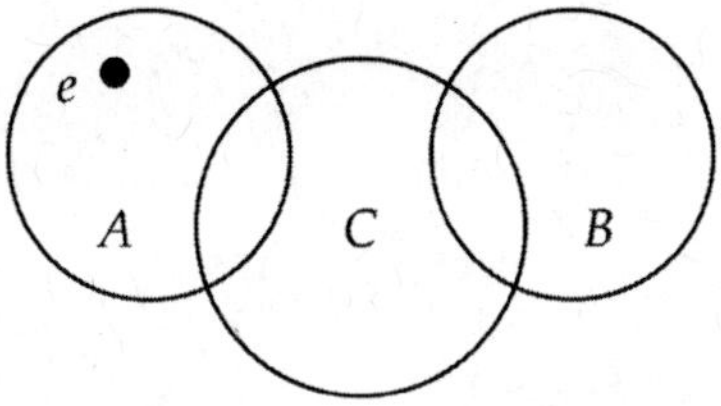

D: *e* is *A* (S5, true); no *A* is *B* (S6, true); some *B* are not *C* (S7, true)

The structure of the PL(TL) interpretation I below mirrors the structure of the TL diagram D.

UD: $\{\alpha, \beta, \delta, \sigma, \tau, \varepsilon\}$
LN: *a*, *b*, *d*, *s*, *t*, *e*

Semantical assignments

I(*a*): α; I(*b*): β; I(*d*): δ; I(*s*): σ; I(*t*): τ; I(*e*): ε
I(*Ax*): $\{\alpha, \beta, \varepsilon\}$; I(*Bx*): $\{\delta, \sigma\}$; I(*Cx*): $\{\alpha, \delta, \tau\}$

It is a simple exercise to verify that I is a model of the set {S5*, S6*, S7*}. The existence of I shows that this PL set is consistent. If we are given I first, we will be able to construct D on the basis of I. We simply assign circles to *A*, *B*, and *C*, and a dot to *e*. We place the dot inside the *A*-circle but outside the *C*-circle (because I(*e*) belongs to I(*Ax*) but not to I(*Cx*)), we make the *A*-circle and the *B*-circle disjoint (because I(*Ax*) and I(*Bx*) are disjoint), and we make the *C*-circle strictly overlap with both the *A*-circle and the *B*-circle (because I(*Cx*) share elements with I(*Bx*) and I(*Ax*), and has an element not shared with either I(*Ax*) or I(*Bx*)).

(d) As indicated above a set of TL sentences is inconsistent if and only if its PL counterparts are inconsistent in PL(TL). We said in (c) that if a TL set is consistent, then any of its PL counterparts is consistent in PL (and of course in PL(TL)). This statement is equivalent to saying that if a PL set that is a counterpart of a TL set is inconsistent in PL, then the TL set is inconsistent. The converse, however, is not true. Here is an example of an inconsistent TL set whose PL counterpart is consistent in PL.

S8	all *H* are *D*	S8*	$(\forall x)(Hx \rightarrow Dx)$
S9	all *D* are *K*	S9*	$(\forall x)(Dx \rightarrow Kx)$
S10	no *H* is *K*	S10*	$(\forall z)(Hx \rightarrow \neg Kz)$

The TL diagram below and its standard variants are, up to similarity, all the diagrams on which S8 and S9 are true. On all of them S10 is false. Hence there is no TL diagram on which S8, S9, and S10 are true. Thus the TL set is inconsistent.

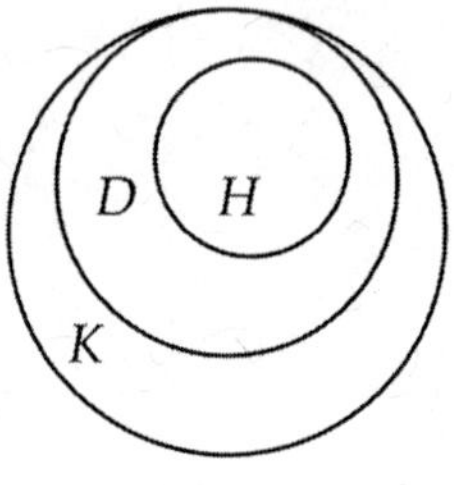

D: all *H* are *D* (S8, true); all *D* are *K* (S9, true); no *H* is *K* (S10, false)

We will use an indirect proof to show that the PL set is inconsistent in PL(TL).

Proof that {S8*, S9*, S10*} is inconsistent in PL(TL)

1. Reductio Assumption: there is a PL(TL) interpretation I such that:
    1.1. $(\forall x)(Hx \rightarrow Dx)$ is true on I.
    1.2. $(\forall x)(Dx \rightarrow Kx)$ is true on I.
    1.3. $(\forall z)(Hx \rightarrow \neg Kz)$ is true on I.
2. From 1.1–1.3 by the truth conditions of the universal quantifier: all the basic substitutional instances of $(\forall x)(Hx \rightarrow Dx)$, $(\forall x)(Dx \rightarrow Kx)$, and $(\forall z)(Hx \rightarrow \neg Kz)$ are true on I.
3. From 1 by the definition of a PL(TL) interpretation: the extension of *H* on I is not empty, that is, there is a name **s** in LN, such that I(**s**) belongs to I(*Hx*).
4. From 3 by the truth condition of *H***s**: the sentence *H***s** is true on I.
5. From 2 and 3 by the definition of a basic substitutional instance: *H***s**→*D***s**, *D***s**→*K***s**, and *H***s**→¬*K***s** are true on I.
6. From 4 and 5 by the truth condition of the conditional: *D***s** is true on I.
7. From 5 and 6 by the truth conditions of the conditional: *K***s** is true on I.
8. From 4 and 5 by the truth conditions of the conditional: ¬*K***s** is true on I, which contradicts 7.
9. From 1 through 8: the Reductio Assumption is false, and hence there is no PL(TL) interpretation on which S8*, S9*, and S10* are true.
10. From 9 by the definition of inconsistency: the set {S8*, S9*, S10*} is inconsistent.

This set is consistent in PL because a PL interpretation that assigns the empty extension to the predicates *H*, *D*, and *K* makes all the sentences true. Such an interpretation is not permitted by PL(TL) since no PL(TL) interpretation may assign the empty extension to any 1-place predicate.

Since every PL(TL) interpretation is also a PL interpretation (the converse is not true, though), any PL(TL) interpretation that is a counterexample to a PL argument is also a PL counterexample to the same argument. This entails that any PL argument that is invalid in PL(TL) is also invalid in PL. But as we explained above, a TL argument is invalid if and only if its PL counterparts are invalid in PL(TL). Hence if a TL argument is invalid, then its PL counterparts are also invalid in PL. This assertion is equivalent to saying that if a PL argument that is a counterpart of a TL argument is valid in PL, then the TL argument is also valid. It is incorrect to interpret the preceding assertion as stating that every inference that is valid in PL is also valid in TL. If this were true, TL would be an extension of PL, which is clearly false since PL is a much more powerful system than TL. The assertion does not say that every valid PL argument can be converted into a valid TL argument; rather, it states that if a PL argument *that is a counterpart of a TL argument* is valid, then the TL argument is valid too. There are many PL arguments that are not counterparts of any TL arguments. For example, the TL translations of the English arguments in Subsection 2.5.12:C4 (the argument about the biological mothers of the Chess Club member) and in exercise 2.6.10 (the argument about heads of horses) are both invalid, while, as we will see in Chapter Five, the PL translations of those English arguments are valid. So these are examples of valid PL arguments that cannot be translated into valid TL arguments. Also, PL is not an extension of TL. If it were, every valid TL argument would have a valid PL counterpart. But in (a) above, we gave an example of a valid TL argument that is invalid in PL. Thus none of TL and PL can be described as an extension of the other. The best way to describe the relation between these two systems is to say that PL *includes* TL, without being an extension of it. The language of *inclusion* is apt, since PL includes a restricted portion—namely PL(TL)—that is equivalent to TL.

4.5.12:C2 As we said at the outset of this section, PL is a proper extension of SL: every valid SL inference is valid in PL, but the converse is not true. As is the case with TL, we also have a notion of a PL counterpart of an SL sentence or of a set of SL sentences. We will use this notion to give a precise formulation of the previous claim. We describe below an effective procedure that transforms every SL sentence into a class of PL sentences and every SL set into a class of PL sets.

Let **X** be any SL sentence. We say that a PL sentence **Y** is a PL **counterpart** of the SL sentence **X** just in case **Y** is obtained from **X** by replacing every atomic component of **X** with a PL sentence such that all the occurrences of a single atomic sentence are replaced with the same PL sentence. It is permissible to replace two different atomic components of **X** with the same PL sentence. A PL counterpart of an SL sentence **X** is denoted as "PL(**X**)." It is clear that PL(**X**) does not designate a unique PL sentence. Here are examples of SL sentences and some of their PL counterparts.

SL sentence S1: $(A \wedge B) \rightarrow C$
Replace A with Je; B with $(\forall z)(Jz \rightarrow Kz)$; and C with Ke
A PL counterpart of S1: $(Je \wedge (\forall z)(Jz \rightarrow Kz)) \rightarrow Ke$

SL sentence S2: $A \rightarrow \neg(H \vee (G \leftrightarrow (C \wedge A)))$
Replace A with $(\exists v)(Rvt \rightarrow Dv)$; H with $\neg Ba$; and G and C with $\neg(\forall x)Qxs$
A PL counterpart of S2: $(\exists v)(Rvt \rightarrow Dv) \rightarrow \neg(\neg Ba \vee (\neg(\forall x)Qxs \leftrightarrow (\neg(\forall x)Qxs \wedge (\exists v)(Rvt \rightarrow Dv))))$

SL sentence S2: $A \rightarrow \neg(H \vee (G \leftrightarrow (C \wedge A)))$
Replace A and H with Pas; G with $(Ad \vee \neg Bd)$; and C with $(\exists x)(Kxe \leftrightarrow (\forall z)Jz)$
A PL counterpart of S2: $Pas \rightarrow \neg(Pas \vee ((Ad \vee \neg Bd) \leftrightarrow ((\exists x)(Kxe \leftrightarrow (\forall z)Jz) \wedge Pas)))$

SL sentence S3: $(E \rightarrow (D \rightarrow \neg B)) \rightarrow ((E \wedge D) \rightarrow \neg B)$
Replace E with $(Rs \wedge \neg Ac)$; D with $(\exists x)Gxj$; and B with $(\forall x)(Qx \leftrightarrow Px)$
A PL counterpart of S3: $((Rs \wedge \neg Ac) \rightarrow ((\exists x)Gxj \rightarrow \neg(\forall x)(Qx \leftrightarrow Px)))$
$\rightarrow (((Rs \wedge \neg Ac) \wedge (\exists x)Gxj) \rightarrow \neg(\forall x)(Qx \leftrightarrow Px))$

To produce PL counterparts of *sets* of SL sentences, we generalize the previous procedure. Precisely, if **Φ** is a set of SL sentences, a PL counterpart of **Φ** is a set **Ψ** that consists of PL counterparts of the members of **Φ** such that for every **X** in **Φ**, there is *exactly one* PL(**X**) in **Ψ**, and *all the occurrences* of a single atomic component of sentences in **Φ** are replaced with the same PL sentence. We denote a PL counterpart of **Φ** as "PL(**Φ**)." Again, it should be obvious that PL(**Φ**) does not designate a unique set of PL sentences. To illustrate this definition we consider a few examples of sets of SL sentences and some of their PL counterparts.

SL set Φ1: {(*A*∧¬*B*)→*C*, ¬*C*, ¬*B*, ¬*A*}
Replace *A* with (∀*y*)(*Jy*↔*Ky*); *B* with *Heg*; and *C* with *Jc*∧(∃z)*Kz*
A PL counterpart of Φ1: {((∀*y*)(*Jy*↔*Ky*)∧¬*Heg*)→(*Jc*∧(∃z)*Kz*), ¬(*Jc*∧(∃z)*Kz*), ¬*Heg*, ¬(∀*y*)(*Jy*↔*Ky*)}

SL set Φ2: {¬*H*→*D*, ¬*D*, ¬*E*}
Replace *H* with *Jc*∧¬(∃z)*Kz*; *D* with *Rab*; *E* with (∀*y*)(*Jy*↔*Ky*)
A PL counterpart of Φ2: {¬(*Jc*∧¬(∃z)*Kz*)→*Rab*, ¬*Rab*, ¬(∀*y*)(*Jy*↔*Ky*)}

SL set Φ2: {¬*H*→*D*, ¬*D*, ¬*E*}
Replace *H* with *Bst*; *D* with (∃z)(*Kz*∧*Jz*); and *E* with (∀*x*)(*Bxt*→*Kx*)
A PL counterpart of Φ2: {¬*Bst*→(∃z)(*Kz*∧*Jz*), ¬(∃z)(*Kz*∧*Jz*), ¬(∀*x*)(*Bxt*→*Kx*)}

Following the same convention we used for TL and PL(TL), we denote a PL counterpart of an SL argument **Φ**/**X** as "PL(**Φ**/**X**)" or "PL(**Φ**, **X**)." Now we are ready to state the relation between PL and SL in precise terms.

PL is a proper extension of SL. For every SL argument **Φ**/**X**, for all SL sentences **Y** and **Z**, and for every set Δ of SL sentences, the following four assertions are true. (1) If **Φ**/**X** is deductively valid, then PL(**Φ**/**X**) is deductively valid. This assertion is equivalent to the statement that if PL(**Φ**/**X**) is invalid, then **Φ**/**X** is invalid. (2) If **Y** is logically true (or logically false), then PL(**Y**) is logically true (or logically false). This assertion entails that if PL(**Y**) is contingent, then **Y** is contingent. (3) If **Y** and **Z** are logically equivalent, then PL(**X**) and PL(**Y**) are logically equivalent, where PL(**X**) and PL(**Y**) are the members of any PL counterpart of the SL set {**Y**, **Z**}. This assertion is equivalent to the statement that if PL(**X**) and PL(**Y**) are not logically equivalent, then **X** and **Y** are not logically equivalent. (4) If Δ is inconsistent, then PL(Δ) is inconsistent. The preceding assertion is equivalent to saying that if PL(Δ) is consistent, then Δ is consistent. All of these assertions are corollaries of a single theorem.

Theorem 4.4: for every SL sentence **Z** and for every set **Φ** of SL sentences, if **Φ** ⊨ **Z** in SL, then PL(**Φ**) ⊨ PL(**Z**) in PL, where PL(**Φ**) is a subset of PL(**Φ**, **Z**) and PL(**Z**) is a member of PL(**Φ**, **Z**).

The converse of this theorem and of each of the assertions above are not true. Using the term 'inference', we can express Theorem 4.4 compactly as follows: every inference that is valid in SL is also valid in PL. However, the converse is not true. The examples below illustrate some of these assertions and they *prove* the claim that there are valid PL inferences that are invalid in SL.

(a) The SL sentence $(E\to(D\to\neg B))\to((E\wedge D)\to\neg B)$ (S3 above) is logically true. Hence any PL counterpart of this sentence is logically true as well. The logical truth of S3 means that any arbitrary truth valuation that assigns any truth values to the atomic components *E*, *D*, and *B* makes S3 true. The reader can easily verify this claim by performing truth analysis. Let **X**, **Y**, and **Z** be any PL sentences (some of them could be the same sentence). If we replace *E* with **X**, *D* with **Y**, and *B* with **Z**, we obtain a PL counterpart of S3—namely, $(\mathbf{X}\to(\mathbf{Y}\to\neg\mathbf{Z}))\to((\mathbf{X}\wedge\mathbf{Y})\to\neg\mathbf{Z})$—which is also logically true in PL. Using our standard notation, we refer to this PL sentence as PL(S3). Any PL interpretation I for PL(S3) must assign (classical) truth values to **X**, **Y**, and **Z**. Since PL(S3) consists of **X**, **Y**, and **Z**, connected to each other by truth-functional sentential connectives, the truth value that I assigns to PL(S3) is determined fully by the truth values of **X**, **Y**, and **Z** on I and by the types of truth-functional connectives that connect **X**, **Y**, and **Z**. But these are the same connectives that occur in S3, and the relations between the constituents of S3 mirrors the relations between the *corresponding* constituents of PL(S3); hence regardless of the truth values that **X**, **Y**, and **Z** receive on I, PL(S3) is true on I. This shows that any PL(S3) is logically true. This is a very convenient conclusion. It entails that any PL sentence, no matter how complex it may be, that has the sentential form of S3 is logically true. For instance, the following two PL sentences are logically true. We know they are logically true because they exhibit the sentential form of S3; we do not need to engage in any complex semantical reasoning about PL interpretations.

$PL(S3)_1$: $((Rs\wedge\neg Ac)\to((\exists x)Gxj\to\neg(\forall x)(Qx\leftrightarrow Px)))\to(((Rs\wedge\neg Ac)\wedge(\exists x)Gxj)\to\neg(\forall x)(Qx\leftrightarrow Px))$
Replace *E* with $(Rs\wedge\neg Ac)$; *D* with $(\exists x)Gxj$; and *B* with $(\forall x)(Qx\leftrightarrow Px)$.

$PL(S3)_2$: $((\forall x)(Ax\vee Bx)\to((\exists z)\neg Bz\to\neg(\exists x)Ax))\to(((\forall x)(Ax\vee Bx)\wedge(\exists z)\neg Bz)\to\neg(\exists x)Ax)$
Replace *E* with $(\forall x)(Ax\vee Bx)$; *D* with $(\exists z)\neg Bz$; and *B* with $(\exists x)Ax$.

(b) The SL sentence $(A\wedge B)\to C$ (S1 above) is contingent: the truth valuation that assigns T to *A*, *B*, and *C* makes the sentence true and the truth valuation that assigns T to *A* and *B* and F to *C* makes the sentence false. However, if we replace *A* with *Je*, *B* with $(\forall z)(Jz\to Kz)$, and *C* with *Ke*, the resulting PL counterpart of S1, which is $(Je\wedge(\forall z)(Jz\to Kz))\to Ke$, is logically true. The antecedent of this conditional asserts that *e* is *J* and that for every individual *z*, if *z* is *J* then it is *K*. If the antecedent is true, then *e* is *K*, which means that the consequent of PL(S1) is true. Hence there is no PL interpretation for this sentence on which the antecedent is true and the consequent false. This entails that PL(S1) is true on every PL interpretation for it. So we have an example of a contingent SL sentence that has logically true PL counterparts. This shows that the converse of Assertion 2 above is not true: if PL(**X**) is logically true in PL, **X** need not be logically true in SL.

(c) The SL argument $\{(A\wedge\neg B)\to C, \neg C, \neg B\}/\neg A$ (Φ1 above) is deductively valid: there is no truth valuation on which the premises are true and the conclusion false. A simple truth analysis would demonstrate the validity of this argument. Assertion 1 above entails that any PL counterpart of this argument is also valid. Below are two PL counterparts of this SL argument. We denote the set of premises, $\{(A\wedge\neg B)\to C, \neg C, \neg B\}$, as Φ.

PL(Φ/¬*A*)$_1$: {((∀*y*)(*Jy*↔*Ky*) ∧ ¬*Heg*)→(*Jc* ∧ (∃z)*Kz*), ¬(*Jc* ∧ (∃z)*Kz*), ¬*Heg*} / ¬(∀*y*)(*Jy*↔*Ky*)
Replace *A* with (∀*y*)(*Jy*↔*Ky*); *B* with *Heg*; and *C* with *Jc* ∧ (∃z)*Kz*

PL(Φ/¬*A*)$_2$:
{((∃*x*)(∀*y*)(*Rxy*→*Ksy*) ∧ ¬(*Rab* ∧ *Hs*))→ ¬(∀z)(∀*v*)(*Kzv* ∨ *Hz*), ¬¬(∀z)(∀*v*)(*Kzv* ∨ *Hz*), ¬(*Rab* ∧ *Hs*)} / ¬(∃*x*)(∀*y*)(*Rxy*→*Ksy*)
Replace *A* with (∃*x*)(∀*y*)(*Rxy*→*Ksy*); *B* with *Rab* ∧ *Hs*; and *C* with ¬(∀z)(∀*v*)(*Kzv* ∨ *Hz*)

It is clear that we can generate extremely complex PL arguments that are counterparts to the simple SL argument Φ/¬*A*. We can prove the deductive validity of these PL arguments by giving semantical proofs, as we did in this chapter, or by constructing PL derivations, as we will do in the next chapter, or by simply showing that these PL arguments are all counterparts of Φ/¬*A*. There is no question that the last approach is the simplest of the three.

(d) The SL argument {¬*H*→*D*, ¬*D*}/¬*E* (Φ2 above) is deductively invalid: a truth valuation that assigns T to *H* and *E* and F to *D* makes the premises true and the conclusion false. This argument has infinitely many PL counterparts. Most of them are invalid but some are valid. The PL counterpart we gave above is deductively valid. We reproduce this PL argument below.

A PL counterpart of {¬*H*→*D*, ¬*D*}/¬*E*: {¬(*Jc* ∧ ¬(∃z)*Kz*)→*Rab*, ¬*Rab*} / ¬(∀*y*)(*Jy*↔*Ky*)
Replace *H* with *Jc* ∧ ¬(∃z)*Kz*; *D* with *Rab*; *E* with (∀*y*)(*Jy*↔*Ky*)

Proof that {¬(*Jc* ∧ ¬(∃z)*Kz*)→*Rab*, ¬*Rab*} / ¬(∀*y*)(*Jy*↔*Ky*) is valid

1. Let I be any PL interpretation for this PL argument such that
    1.1. ¬(*Jc* ∧ ¬(∃z)*Kz*)→*Rab* is true on I.
    1.2. ¬*Rab* is true on I.
2. From 1.2 by the truth conditions of the negation: *Rab* is false on I.
3. From 1.1 and 2 by the truth conditions of the conditional: ¬(*Jc* ∧ ¬(∃z)*Kz*) is false on I.
4. From 3 by the truth conditions of the negation: *Jc* ∧ ¬(∃z)*Kz* is true on I.
5. From 4 by the truth conditions of the conjunction: both *Jc* and ¬(∃z)*Kz* are true on I.
6. From 5 by the truth conditions of the negation: (∃z)*Kz* is false on I.
7. From 6 by the truth conditions of the existential quantifier: every basic substitutional instance of (∃z)*Kz* is false on I; in particular, *Kc* is false on I.
8. From 5 and 7 by the truth conditions of the biconditional: *Jc*↔*Kc* is false on I.
9. From 8 by the truth conditions of the universal quantifier: (∀*y*)(*Jy*↔*Ky*) is false on I.
10. From 9 by the truth conditions of the negation: ¬(∀*y*)(*Jy*↔*Ky*) is true on I.
11. From 1 and 10: every PL interpretation that makes the premises true makes the conclusion true.
12. From 11 by the definition of deductive validity: the PL argument {¬(*Jc* ∧ ¬(∃z)*Kz*)→*Rab*, ¬*Rab*} / ¬(∀*y*)(*Jy*↔*Ky*) is valid.

So this is an example of an invalid SL argument that has valid P counterparts. This shows that the converse of Assertion 1 above is not true: if a PL counterpart of an SL argument is valid, the SL argument need not be valid.

4.5.13 Because the PL **worldview** contains a relatively rich assortment of basic ingredients of reality, all TL diagrams and all SL truth valuations can be transformed into PL interpretations. Since there are superfluous TL diagrams and superfluous truth valuations, that is, TL diagrams and SL truth valuations that represent no *relevant* logical possibilities, the PL interpretations that correspond to these diagrams and truth valuations are superfluous too. We saw in Chapters Two and Three that the presence of superfluous TL diagrams and truth valuations was responsible for the existence of TL and SL arguments that are not faithful translations of natural-language arguments. However, the presence of superfluous PL interpretations does not entail that *every* PL translation of a natural-language argument whose TL or SL translation is not faithful is necessarily not faithful either. The reason is this: a natural-language argument typically has several PL **translations**, each of which offers a different level of analysis. Some of these translations are better than others in terms of capturing the logical relations that exist between the various parts of the natural-language argument. When we speak of *the* PL translation of a natural-language argument we suppose that this translation is as close to the original argument as the expressive power of PL permits. Such a PL translation might be faithful, even though the TL or SL translation is not faithful.

4.5.13:C COMMENTARY ON 4.5.13

4.5.13:C1 In 2.5.12:C4 and 3.5.12:C3 we gave examples of deductively valid English arguments whose TL and SL translations are invalid. We explained there that the TL diagrams and SL truth valuations that are counterexamples to the symbolic arguments are superfluous; they do not represent any relevant logical possibilities. In 2.5.12:C6 we gave controversial examples of deductively invalid English arguments whose TL translations are valid, and in 3.5.12:C4 less controversial examples of deductively invalid English arguments whose SL translations are valid. We also argued in 3.5.12:C4 that there are relevant logical possibilities that cannot be represented by *any* relevant SL truth valuations. In this section we will consider similar claims regarding PL. We will show that there are superfluous PL interpretations; we will give examples of English arguments that have multiple PL translations, some of which are more faithful than others; we will show that there are valid English arguments that have invalid PL translations; we will discuss the use of meaning-postulates in PL; we will study proposed examples of invalid English arguments whose PL translations are valid; we will discuss examples of relevant logical possibilities that cannot be represented by any relevant PL interpretations; and we will conclude with some general remarks about the degree to which the three logical systems developed in this book are successful in their attempts to capture the notion of logical possibility. However, following our usual practice, we will begin by discussing the relation of representation between PL interpretations and logical possibilities.

In 2.5.12:C1 and 3.5.12:C1, we gave a rough explanation of what we mean by 'relevant logical possibility.' and we defined the notion of a TL diagram and of an SL truth valuation representing a logical possibility. We need to do the same for PL. So let Σ be any set of natural-language declarative sentences and p be any logical possibility. We say that p is a logical possibility *for* Σ, or that p is *relevant to* Σ, if and only if every sentence in Σ makes a (true or false) assertion about some or all of the constituents of p. We repeat what we said in the previous three chapters: the expressions that occur in the members of Σ must have the same meanings across all the logical possibilities for Σ. The notion of a PL interpretation is defined in 4.4.1 relative to some set of PL sentences. Hence we have a precise understanding of the expression 'a PL interpretation for a set of PL sentences'.

As above, Σ is a set of declarative sentences of some natural language L. We take Σ^{PL} to be a set of PL sentences that are translations of the members of Σ. Voc(Σ^{PL}) is the PL extra-logical vocabulary of which the sentences in Σ^{PL} are composed, together with the logical vocabulary of PL. Voc(Σ) consists of the PL-L expressions that are translated into the symbols in Voc(Σ^{PL}), and L expressions that correspond to the five sentential connectives, the two quantifiers, and the identity predicate, together with all the variables of PL.[1] As we noted previously, PL-L is a wider language than L. PL-L is L augmented with PL variables and rules for using these variables in sentences. So every L sentence is a PL-L sentence, but the converse is not true. If X is a PL-L sentence composed of Voc(Σ) that has a PL translation composed of Voc(Σ^{PL}), then we denote this PL sentence as X^{PL}. We said 'PL-L' instead of 'L' because we want to leave the door open for the possibility of using variables in the sentences that are composed of Voc(Σ). Now we are ready to define the relation of representation.

Representation. A PL interpretation M for Σ^{PL} **represents** a logical possibility p for Σ if and only if for every PL sentence **Z** that is composed of Voc(Σ^{PL}) and for every PL-L sentence X that is composed of Voc(Σ), if **Z** is a translation of X (i.e., **Z** is X^{PL}), then **Z** is true (or false) on M if and only if X is true (or false) in p.

We saw in TL that the definition of representations entails the existence of a certain structural correspondence between constituents of M and constituents of p. We also said there that in most cases such a structural correspondence is partial and imperfect. The same is true for PL. We will see an example of a structural correspondence between a logical possibility and a PL interpretation that represents this possibility in the following subsection.

4.5.13:C2 In this subsection we illustrate through an example the notions and notation introduced above. Consider the following set of English sentences.

The Set Σ

S1 Nicholas and Stephanie have the same house.
S2 Stephanie's car is a Ford and Nicholas's car is also a Ford.
S3 Anyone who has a Ford likes all American cars.
S4 Nicholas does not like some American cars.

Translation Key

n: Nicholas
s: Stephanie
Ox: *x* is a house
Fv: *v* is a Ford
Ax: *x* is an American car
Hxy: *x* has *y*
Lvz: *v* likes *z*

1 If L is English, then PL-L is PL-English, which we have frequently used in this chapter.

The Set Σ^{PL}

S1^{PL} $(\exists x)(Ox \wedge (Hnx \wedge Hsx))$

S2^{PL} $(\exists y)(Fy \wedge Hsy) \wedge (\exists z)(Fz \wedge Hnz)$

S3^{PL} $(\forall x)(\forall w)((Fx \wedge Hwx) \rightarrow (\forall y)(Ay \rightarrow Lwy))$

S4^{PL} $(\exists z)(Az \wedge \neg Lnz)$

Voc(Σ^{TL}) = $\{n, s, O, F, A, H, L, \neg, \wedge, \vee, \rightarrow, \leftrightarrow, \forall, \exists, =, u, v, w, x, y, z, u_1, v_1, w_1, \ldots, (,)\}$

Voc(Σ) ={Nicholas; Stephanie; *x* is a house; *v* is a Ford; *x* is an American car; *x* has *y*; *v* likes *z*; it is not the case that; not; and; or; if-then; if and only if; for every; there is; *x* is identical with *y*; $u, v, w, x, y, z, u_1, v_1, w_1, \ldots$}

As is the case with SL and TL, we allow a certain degree of freedom in constructing PL-English sentences that are based on Voc(Σ). A quick review of S1–S4 shows that the expressions in Voc(Σ) are not followed strictly. For example, we wrote 'Nicholas and Stephanie have the same house', when the expression 'have the same house' is not in Voc(Σ). We noted previously that natural languages allow a great deal of flexibility that is absent from formal languages. However, in almost all cases PL-English sentences that are constructed on the basis of Voc(Σ) can be paraphrased using only the expressions in Voc(Σ) or close variants of them and, perhaps, some additional words. For instance, the sentence 'Nicholas and Stephanie have the same house' can be paraphrased as the PL-English sentence 'There is *x*, *x* is a house, and Nicholas has *x* and Stephanie has *x*'.

We now give an example of a PL interpretation for Σ^{PL} that represents a logical possibility p for Σ. J below is a PL interpretation for Σ^{PL}.

The PL Interpretation J

UD: {Right Triangle1, Right Triangle2, Square, Hexagon, Heptagon, Octagon}
LN: *n*, *s*, *q*, *r*, *t*, *g*

Semantical Assignment

J(*n*): Right Triangle1; J(*s*): Right Triangle2; J(*q*): Square ; J(*r*): Hexagon; J(*t*): Heptagon; J(*g*): Octagon
J(*Ox*): "*x* is a 4-sided geometric figure": {Square}
J(*Fv*): "*v* is a geometric figure that has 6 or 7 sides": {Hexagon, Heptagon}
J(*Ax*): "*x* is a geometric figure that contains an obtuse angle": {Hexagon, Heptagon, Octagon}
J(*Hxy*): "The sum of the interior angles of *x* is one half or one fourth the sum of the interior angles of *y*": {⟨Right Triangle1, Square⟩, ⟨Right Triangle2, Square⟩, ⟨Right Triangle1, Hexagon⟩, ⟨Right Triangle2, Hexagon⟩}
J(*Lvz*): "The number of the sides of *v* is less than or equal to half the number of the sides of *z*": {⟨Right Triangle1, Hexagon⟩, ⟨Right Triangle1, Heptagon⟩, ⟨Right Triangle1,

Octagon⟩, ⟨Right Triangle2, Hexagon⟩, ⟨Right Triangle2, Heptagon⟩, ⟨Right Triangle2, Octagon⟩}

Note: The sum of the interior angles of a triangle is 180°, of a square is 360°, of a hexagon is 720°, of a heptagon is 900°, and of an octagon is 1080°. A hexagon is 6-sided, a heptagon is 7-sided, and an octagon is 8-sided.

PL-English reading of ($S1^{PL}$) $(\exists x)(Ox \wedge (Hnx \wedge Hsx))$ on J:
There is a 4-sided geometric figure x such that the sum of the interior angles of Right Triangle1 and of Right Triangle2 is one half or one fourth the sum of x's interior angles.
$S1^{PL}$ is true on J.
Square (and only Square) meets these conditions.

Natural reading of ($S2^{PL}$) $(\exists y)(Fy \wedge Hsy) \wedge (\exists z)(Fz \wedge Hnz)$ on J:
There is a geometric figure that has 6 or 7 sides, such that the sum of the interior angles of Right Triangle1 is one half or one fourth the sum of its interior angles; and there is a geometric figure that has 6 or 7 sides, such that the sum of the interior angles of Right Triangle2 is one half or one fourth the sum of its interior angles.
$S2^{PL}$ is true on J.
Hexagon (and only Hexagon) meets both sets of conditions.

PL-English reading of ($S3^{PL}$) $(\forall x)(\forall w)((Fx \wedge Hwx) \rightarrow (\forall y)(Ay \rightarrow Lwy))$ on J:
For all x and w, if x is a geometric figure that has 6 or 7 sides, such that the sum of the interior angles of w is one half or one fourth the sum of interior angles of x, then the number of the sides of w is less than or equal to half the number of the sides of any geometric figure that contains an obtuse angle.
$S3^{PL}$ is true on J.
The only figures that make the antecedent true are Hexagon (for x) and Right Triangle1 and Right Triangle2 (for w); and the triangles make the consequent true too, since the number of sides of a triangle is less than or equal to half the sides of any figure that contains an obtuse angle (those are Hexagon, Heptagon, and Octagon).

Natural reading of ($S4^{PL}$) $(\exists z)(Az \wedge \neg Lnz)$ on J:
The number of the sides of Right Triangle1 is greater than half the number of the sides of some geometric figure that contains an obtuse angle.
$S4^{PL}$ is false on J.
Only Hexagon, Heptagon, and Octagon contain obtuse angles; and the number of the sides of a triangle is less than or equal to half the number of the sides of each one of them.

It is not a coincidence that $S1^{PL}$–$S3^{PL}$ are true, while $S4^{PL}$ is false on J. The set Σ^{PL} is inconsistent, which is a reflection of the inconsistency of Σ. The truth of $S2^{PL}$ and $S3^{PL}$ entails the falsity of $S4^{PL}$. We will offer a semantic proof of the inconsistency of Σ^{PL} (and hence, of Σ) later in this subsection.

We describe below a logical possibility p that is relevant to Σ and that is represented by the interpretation J. p is relevant to Σ because the sentences in Σ make assertions about constituents of p. We will then show that J and p satisfy the condition stated in the definition of representation (the Representation Condition). Informal logical possibilities that are represented by PL interpretations have structures that, at least partially, look a lot like PL interpretations. Just like PL interpretations, they typically comprise individuals, properties, and relations; some or all of their constituents correspond to some or all of the constituents of the PL interpretations that represent them; and at least some of their structural relations mirror structural relations of the representing interpretations. Here is a description of the logical possibility p.

Logical possibility p

p consists of the following constituents.

Six individuals: Nicholas Dowd, Stephanie Dowd, The Dowds' house, 2009 Ford Mustang, 2010 Ford Focus, 2010 Chrysler PT Cruiser
Property: being a house: {The Dowds' house}
Property: being a Ford: {2009 Ford Mustang, 2010 Ford Focus}
Property: being an American car: {2009 Ford Mustang, 2010 Ford Focus, 2010 Chrysler PT Cruiser}
2-place relation: ownership: {⟨Nicholas, the Dowds' house⟩, ⟨Stephanie, the Dowds' house⟩, ⟨Nicholas, 2009 Ford Mustang⟩, ⟨Stephanie, 2009 Ford Mustang⟩}
2-place relation: being fond of: {⟨Nicholas, 2009 Ford Mustang⟩, ⟨Nicholas, 2010 Ford Focus⟩, ⟨Nicholas, 2010 Chrysler PT Cruiser⟩, ⟨Stephanie, 2009 Ford Mustang⟩, ⟨Stephanie, 2010 Ford Focus⟩, ⟨Stephanie, 2010 Chrysler PT Cruiser⟩}
2-place relation: having a desire to own: {⟨Nicholas, 2010 Ford Focus⟩, ⟨Stephanie, 2010 Chrysler PT Cruiser⟩}[1]

According to the extensions above, Nicholas and Stephanie share one car, which is a 2009 Ford Mustang, they own the same house, and each of them is fond of all American cars. It is a simple exercise to verify that the English sentences S1–S3 are true and S4 is false in p. So we know that the truth values of $S1^{PL}$–$S4^{PL}$ on J are identical with the truth values of S1–S4 in p.

In order to show that J represents p, we need to argue that the Representation Condition is satisfied. Recall that the Representation Condition asserts that for every PL sentence **Z** that is composed of Voc(Σ^{TL}) and for every PL-English sentence X that is composed of Voc(Σ), if **Z** is a translation of X, then **Z** is true (or false) on J if and only if X is true (or false) in p. In order to give such an argument, we first need to describe some structural correspondence between J and p. These correspondence relations are implied partly by the translation key and partly by

1 I intentionally included a constituent of p that does not correspond to any constituent of J in order to stress the fact that a logical possibility may contain constituents that do not correspond to any constituents of the representing PL interpretations. The converse is true too: a PL interpretation that represents a logical possibility may contain constituents that do not correspond to any constituents of the logical possibility.

the descriptions of J and p. We will try to describe the properties and relations in a natural fashion, but occasionally we will use PL-English predicates when natural expressions are awkward. For the convenience of the reader, I reproduce the translation key below.

Translation key

n: Nicholas
s: Stephanie
Ox: *x* is a house
Fv: *v* is a Ford
Ax: *x* is an American car
Hxy: *x* has *y*
Lvz: *v* likes *z*

Individuals

Right Triangle1 corresponds to Nicholas Dowd
Right Triangle2 corresponds to Stephanie Dowd
Square corresponds to the Dowds' house
Hexagon corresponds to 2009 Ford Mustang
Heptagon corresponds to 2010 Ford Focus
Octagon corresponds to 2010 Chrysler PT Cruiser

Note: Right Triangle1 is named *n* on J, which is the translation of the name 'Nicholas', and Right Triangle2 is named *s* on J, which is the translation of the name 'Stephanie'.

Properties

The property J(*Ox*), "being a 4-sided geometric figure," corresponds to the property "being a house"
The property J(*Fv*), "being a geometric figure that has 6 or 7 sides," corresponds to the property "being a Ford"
The property J(*Ax*), "being a geometric figure that contains an obtuse angle," corresponds to the property "being an American car"

Relations

The 2-place relation J(*Hxy*), "the sum of the interior angles of *x* is one half or one fourth the sum of the interior angles of *y*," corresponds to the 2-place relation "ownership"
The 2-place relation J(*Lvz*), "the numbers of the sides of *v* is less than or equal to half the number of the sides of *z*," corresponds to the 2-place relation "being fond of"

An important observation to make is that the structural relations between the constituents of J mirror the structural relations between the corresponding constituents of p. This is the central fact that we need in order to argue that J and p satisfy the Representation Condition. Since there are infinitely many sentences that can be generated from Voc(Σ^{PL}), it is not obvious at all that J and p meet the requirement of the Representational Condition. However, if one assumes that the English connectives and quantifiers listed in Voc(Σ) have truth conditions similar to the truth conditions of their formal counterparts, then it is plausible to suppose that this Representation Condition holds for J and p. We can argue that since the expressions in Voc(Σ^{PL}) are translations of the expressions in Voc(Σ), and since the connectives and quantifiers in these sets have similar truth conditions, we can assume that X^{PL} and X make similar assertions about different sets of constituents. But there is a correspondence between the constituents about which X^{PL} makes an assertion and the constituents about which X makes an assertion. In addition, the structural relations between the constituents of the first set precisely mirrors the structural relations between the corresponding constituents of the second set. Given that the truth value of a sentence is, in general, determined by the assertion the sentence makes and the structure described by this assertion, and given that X^{PL} and X make similar assertions about corresponding constituents that have parallel structures, it is reasonable to expect that X^{PL} and X have the same truth value. We conclude that the PL interpretation J for Σ^{PL} represents the logical possibility p for Σ.

As we previously said, the set Σ is inconsistent. S2 and S3 entail that Nicholas and Stephanie, who own Fords, like all American cars. S4, however, states that Nicholas does not like some American cars. Any faithful PL translations of S2–S4 are also inconsistent. For instance, the set {$S2^{PL}$, $S3^{PL}$, $S4^{PL}$} is inconsistent. We will show this using an indirect proof.

Proof that {$S2^{PL}$, $S3^{PL}$, $S4^{PL}$} is inconsistent

1 Reductio Assumption: suppose that there is a PL interpretation H on which the following sentences are true.
   - 1.1. $(\exists x)(Fx \land Hnx) \land (\exists y)(Fy \land Hsy)$
   - 1.2. $(\forall x)(\forall w)((Fx \land Hwx) \rightarrow (\forall y)(Ay \rightarrow Lwy))$
   - 1.3. $(\exists z)(Az \land \neg Lnz)$
2. From 1.1 by the truth conditions of the conjunction: $(\exists x)(Fx \land Hnx)$ is true on H.
3. From 2 by the truth conditions of the existential quantifier: there is a basic substitutional instance of $(\exists x)(Fx \land Hnx)$ that is true on H. Hence there is a name **s** in LN such that $F\mathbf{s} \land Hn\mathbf{s}$ is true on H.
4. From 1.2 by the truth conditions of the universal quantifier: all basic substitutional instances of $(\forall x)(\forall w)((Fx \land Hwx) \rightarrow (\forall y)(Ay \rightarrow Lwy))$ are true on H. Hence $(F\mathbf{s} \land Hn\mathbf{s}) \rightarrow (\forall y)(Ay \rightarrow Lny)$ is true on H.
5. From 3 and 4 by the truth conditions of the conditional: $(\forall y)(Ay \rightarrow Lny)$ is true on H.
6. From 1.3 by the truth condition of the existential quantifier: there is a basic substitutional instance of $(\exists z)(Az \land \neg Lnz)$ that is true on H. Thus for some name **t** in LN, $A\mathbf{t} \land \neg Ln\mathbf{t}$ is true on H.
7. From 5 by the truth conditions of the universal quantifier: every basic substitutional instance of $(\forall y)(Ay \rightarrow Lny)$ is true on H. Hence, $A\mathbf{t} \rightarrow Ln\mathbf{t}$ is true on H.
8. From 6 by the truth conditions of the conjunction and the conditional: $\neg(A\mathbf{t} \rightarrow Ln\mathbf{t})$ is true on H.
9. From 7 and 8: $A\mathbf{t} \rightarrow Ln\mathbf{t}$ and $\neg(A\mathbf{t} \rightarrow Ln\mathbf{t})$ are true on H, which is a contradiction.
10. From 1 through 9: there is no PL interpretation that makes all of 1.1–1.3 true.
11. From 10 by the definition of inconsistency: the set {$S2^{PL}$, $S3^{PL}$, $S4^{PL}$} is inconsistent.

4.5.13:C3 Using the terminology introduced in 2.5.12:C4, we call an interpretation that represents a relevant logical possibility **genuine** and one that represents no relevant logical possibility **superfluous**. We know from Chapters Two and Three that there are superfluous TL diagrams and SL truth valuations. Since these TL diagrams and SL truth valuations can be transformed into PL interpretations, these interpretations are superfluous too. So it is certain that there are superfluous PL interpretations. We will give an example of such an interpretation in this subsection. In 3.5.12:C4, we gave a fairly convincing example of a relevant logical possibility (call it q) that is not represented by any SL truth valuation. It was the example of the two switches that are connected in series to a light bulb. Since every SL truth valuation can be transformed into a PL interpretation, the PL interpretations that correspond to the SL truth valuations that fail to represent q also fail to represent q. However, it does not follow from this that there are *no* PL interpretations that can represent q. A natural-language argument might have PL translations that are much more faithful than any SL translation. In this case, the logical possibility q might be represented by an interpretation for one of these more refined PL translations of the English argument. Later we will examine this issue more carefully. In this subsection we will give an example of a superfluous PL interpretation and an example of an English argument that has multiple PL translations, some of which are more faithful than others.

We showed in 2.5.12 and 3.5.12 that the presence of superfluous TL diagrams and SL truth valuations implies that a deductively valid natural-language argument might have a deductively invalid TL or SL translation. However, it is incorrect to conclude from this that if a valid natural-language argument has an invalid TL or SL translation, then *all* of its PL translations are also invalid. The relative richness of the PL worldview, which is reflected in its syntax and semantics, allows PL to offer multiple layers of analyses, descriptions, and translations. Some of these layers are shallow, and hence it might be incapable of capturing the various interdependencies that exist in a logical possibility or in a natural-language argument; and other layers are deeper allowing the PL interpretation or translation to mirror most, if not all, of the essential interdependencies in a relevant logical possibility or in the original argument. Let us consider an example similar to one discussed in 2.5.12:C4.

P1 All the Chess Club members are students in Symbolic Logic.
P2 Nora is a member of the Chess Club.

C Nora's mother is a mother of a student in Symbolic Logic.

The TL translation of this valid argument is invalid. We first give the TL translation of this argument and show that all the TL diagrams that are counterexamples to the TL translation are superfluous. We will then give a PL translation that parallels the TL translation.

TL Translation Key

n: Nora
m: Nora's mother
C: Members of the Chess Club
S: Students in Symbolic Logic
R: Mothers of students in Symbolic Logic

The TL Argument

S1 all *C* are *S*
S2 *n* is *C*

S3 *m* is *R*

The TL diagram D below shows that this TL argument is deductively invalid. S1 and S2 are true on D and S3 is false on D.

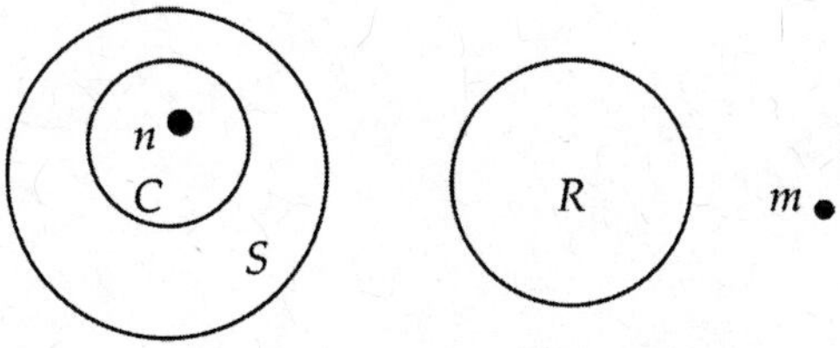

There are many other TL diagrams that are counterexamples to the TL argument. However, none of them represents a logical possibility that is a counterexample to the English argument. All of the TL counterexamples, regardless of their differences, make S1 and S2 true and S3 false. If we use the translation key to construct English stories based on these TL counterexamples, these stories would describe "situations" in which all the Chess Club members are students in Symbolic Logic and Nora is a member of the Chess Club, but her mother is not a mother of a student in Symbolic Logic. All these stories contain an implicit contradiction. The contradiction is easy to see. Since all the Chess Club members are students in Symbolic Logic and Nora is a member of the Chess Club, Nora would have to be a student in Symbolic Logic; hence her mother would be a mother of a student in Symbolic Logic, which contradicts the previous assertion that Nora's mother is not a mother of a student in Symbolic Logic. So all these stories describe impossible situations. Therefore none of these stories depicts a logical possibility. In other words, the TL diagrams that are counterexamples to the TL argument, and on the basis of which these stories are constructed, do not represent logical possibilities that are relevant to the English argument—they are superfluous. According to the procedure described in 2.5.12:C5, this is strong evidence that the original English argument is deductively valid. Of course, we do not need a sophisticated symbolic system to convince us of the deductive validity of the English argument. The validity of this argument is "near the surface"—we are intuitively convinced of its validity. But we are interested in examining the status of some of the PL translations of this argument. We would like to see how the logical resources of PL succeed in excluding the superfluous interpretations from the sphere of those interpretations that "matter" to the PL verdict regarding the logical status of the English argument.

We said that PL can offer multiple translations of a typical natural-language argument, some of which are more faithful than others. Let us study one of these PL translations. This translation parallels the TL translation.

PL Translation Key

n: Nora
m: Nora's mother
Cx: *x* is a member of the Chess Club
Sy: *y* is a student of Symbolic Logic
Rx: *x* is the mother of a student in Symbolic Logic

PL Argument

S4	$(\forall x)(Cx \rightarrow Sx)$
S5	Cn
S6	Rm

The PL argument is deductively invalid. There are infinitely many interpretations that make the premises, S4 and S5, true and the conclusion, S6, false. Here is one of these interpretations.

The PL Interpretation J

UD: {Whale1, Whale2, Tiger, Lion, Eagle}
LN: *n, h, t, c, m*

Semantical Assignments

J(*n*): Whale1; J(*h*): Whale2; J(*t*): Tiger; J(*c*): Lion; J(*m*): Eagle
Cx: "*x* is a whale": {Whale1, Whale2}
Sy: "*y* is a mammal": {Whale1, Whale2, Tiger, Lion}
Rz: "*z* is a feline": {Tiger, Lion}

Natural reading of (S4) $(\forall x)(Cx \rightarrow Sx)$ on J:
Every whale is a mammal.
S4 is true on J.

Natural reading of (S5) *Cn* on J:
Whale1 is a whale.
S5 is true on J.

Natural reading of (S6) *Rm* on J:
Eagle is a feline.
S6 is false on J.

J is a superfluous interpretation: it does not represent a logical possibility *that is relevant to the English argument*. The italicized clause is essential to the correctness of this claim. It is obvious that J represents *a* logical possibility. We explained in Chapters Two and Three that superfluous TL diagrams and SL truth valuations are not superfluous because they fail to represent any logical possibilities (in fact they do), but because they fail to represent any *relevant* logical possibilities. We will prove that J fails to represent any logical possibility that is relevant to the English argument {P1, P2}/C. We employ an indirect-proof strategy.

Assume (for reductio) that J represents a logical possibility q that is relevant to the English argument {P1, P2}/C. The definition of representation implies that a sentence that is composed of the vocabulary of the English argument is true (or false) in q if and only if its PL translation, according to the translation key above, is true (or false) on J. Since S4, S5, and S6 are the PL translations of P1, P2, and C, respectively, and since S4 and S5 are true and S6 is false on J, P1 and P2 must be true and C false in q. Thus in q the sentences 'All the Chess Club members are students in Symbolic Logic' and 'Nora is a member of the Chess Club' are true and the sentence 'Nora's mother is a mother of a student in Symbolic Logic' is false. As we stressed several times, the expressions of these sentences retain their meanings across all relevant logical possibilities. Given the basic meanings of these sentences, we know that Nora is a student in Symbolic Logic, since she is a member of the Chess Club and all members of the Chess Club are students in Symbolic Logic. But now if she is a student in Symbolic Logic, then her mother is a mother of a student in Symbolic Logic. Hence the sentence 'Nora's mother is a mother of a student in Symbolic Logic' is true in q. But we are given that this sentence is false in q. So q contains a contradiction. It follows that q is not a logical possibility after all, which allows us to reject the Reductio Assumption and conclude that J represents no logical possibility that is relevant to the English argument {P1, P2}/C. Therefore J is an example of a superfluous PL interpretation. This establishes our claim that there are PL interpretations that do not represent any relevant logical possibilities.

The presence of a deductively invalid PL translation of this argument does not entail that all possible PL translations of this argument are also invalid. Our goal is to produce a PL translation that is faithful to the original argument. A PL argument is a **faithful translation** of another argument if and only if the interdependencies that exist between the expressions of the PL argument are precisely those interdependencies that exist between the corresponding expressions of the translated argument. The previous PL argument {S4, S5}/S6 is not a faithful translation of the English argument {P1, P2}/C because the relational predicate 'is a mother of', which creates an interdependency between the singular terms 'Nora' and 'Nora's mother', is left untranslated in the PL argument. Hence, in the PL argument, there is no connection between the names '*n*' and '*m*' (the connection is that '*n*' translates a name that refers to Nora and '*m*' translates a singular term that refers to Nora's mother).

Since the PL syntax we described in this book contains the means for generating complex singular terms, we will be able to capture faithfully the English singular term 'the mother of Nora' as a PL term. We will do this near the end of this subsection. However in order to make an important point about "meaning-postulates," we will initially give a translation key that contains no function symbols. In this case we need to add a premise stating the relation between Nora and her mother.

Translation Key

n: Nora
m: Nora's mother
Cx: *x* is a member of the Chess Club
Sy: *y* is a student of Symbolic Logic
Rxy: *x* is the mother of *y*

PL Argument

S7 $(\forall x)(Cx \rightarrow Sx)$
S8 Cn
S9 Rmn

S10 $(\exists z)(Sz \wedge Rmz)$

We added the premise '*Rmn*' in order to capture that '*m*' stands for the mother of Nora. We said in previous chapters that a symbolic premise such as S9 is referred to as a **meaning-postulate**. Meaning-postulates are premises that are added to the symbolic argument in order to capture aspects of the meanings of the expressions used in the natural-language argument. The PL-English reading of the conclusion is that there is *z* such that *z* is a student of Symbolic Logic and Nora's mother is the mother of *z*. This PL argument is deductively valid. It would be easy to show its validity using the methods of Chapter Five. Until then, we have to rely on a semantical proof.

Proof that {S7, S8, S9}/S10 is deductively valid

1. Let I be any PL interpretation for the PL argument that makes the following sentences true:
    1.1. $(\forall x)(Cx \rightarrow Sx)$
    1.2. Cn
    1.3. Rmn
2. From 1.1 by the truth conditions of the universal quantifier: every basic substitutional instance of $(\forall x)(Cx \rightarrow Sx)$ is true on I.
3. From 2: $Cn \rightarrow Sn$ is true on I.
4. From 3 and 1.2 by the truth conditions of the conditional: Sn is true on I.
5. From 4 and 1.3 by the truth conditions of the conjunction: $Sn \wedge Rmn$ is true on I.
6. From 5 by the truth conditions of the existential quantifier: $(\exists z)(Sz \wedge Rmz)$ is true on I.
7. From 1 and 6: any PL interpretation that makes S7–S9 true also makes S10 true.
8. From 7 by the definition of deductive validity: the PL argument {S7, S8, S9}/S10 is valid.

As we promised earlier, we now give a PL translation of the English argument that is, perhaps, more faithful than the previous two translations and requires no meaning-postulates. This translation presupposes that everyone has one and only one mother. It is this presupposition that prevents us from saying that the translation below is totally faithful to the English.

Translation Key

n: Nora
hx: The mother of x
Cx: x is a member of the Chess Club
Sy: y is a student of Symbolic Logic

PL Argument

S11 $(\forall x)(Cx \rightarrow Sx)$
S12 Cn

S13 $(\exists z)(Sz \wedge gn = gz)$

The conclusion says in PL-English: "There is z, such that z is a student in Symbolic Logic, and the mother of Nora is the mother of z." This is a fair rendering of the English conclusion, C. Below we give a semantical proof showing that this argument is indeed valid.

Proof that {S11, S12}/S13 is deductively valid

1. Let I be any arbitrary interpretation for the argument {S11, S12}/S13 such that
   1.1 $(\forall x)(Cx \rightarrow Sx)$ is true on I
   1.2 Cn is true on I
2. From 1.1 by the truth conditions of the universal quantifier: every basic substitutional instance of $(\forall x)(Cx \rightarrow Sx)$ is true on I; in particular $Cn \rightarrow Sn$ is true on I.
3. From 1.2 and 2 by the truth conditions of the conditional: Sn is true on I.
4. By Theorem 4.1 and the truth conditions of the identity predicate: $gn = gn$ is true on I.
5. From 3 and 4 by the truth conditions of the conjunction: $Sn \wedge gn = gn$ is true on I.
6. From 5 by the truth conditions of the existential quantifier: $(\exists z)(Sz \wedge gn = gz)$ is true on I.[1]
7. From 1 and 6: since I is arbitrary, every PL interpretation that makes S11 and S12 true makes S13 true.
8. From 7 by the definition of deductive validity: {S11, S12}/S13 is deductively valid.

We did not give an example of a valid natural-language argument that has *no* valid PL translations. The previous example only shows that there are valid natural-language arguments that have invalid PL translations. However, we were able to produce two valid PL translations of the same argument. So at most we have shown that *some* of the PL translations of a valid natural-language argument might be invalid, and that a valid natural-language argument that has invalid PL translations might have valid PL translations. We will discuss in the following subsection the use of meaning-postulates in PL translations of natural-language arguments. We would like to know whether there are valid natural-language arguments that cannot be translated faithfully into a valid PL argument, even if we use *reasonable* meaning-postulates.

---

1 $Sn \wedge gn = gn$ is a substitutional instance of $(\exists z)(Sz \wedge gn = gz)$ and of $(\exists z)(Sz \wedge gz = gz)$.

We will return to this issue in Subsection 4.5.13:C5. But first we need to say a few things about meaning-postulates.

4.5.13:C4 The reader may recall that S9, which is '*Rmn*', is a meaning-postulate. We had to add it to the symbolic argument to capture the relation between '*m*' and '*n*'—namely, that '*m*' translates the term 'the mother of Nora' and '*n*' translates the name 'Nora'. Meaning-postulates are standardly added as premises to symbolic arguments. However, we can also express the conclusion as a conditional and add the meaning-postulates to the antecedent of the conclusion. For instance, we could have expressed the PL argument {S7, S8, S9}/S10 as {S7, S8}/(S9→S10). The second (nonstandard) approach gives the appearance that no additional premises are included. But this, of course, is a deceptive appearance; meaning-postulates are still added to the argument in the form of an antecedent for the conclusion. If meaning-postulates are disallowed, whether as additional premises or as an antecedent for the conclusion, then many natural-language arguments would not have any faithful PL translations. Here is an example of such an argument.

P1 Gregory is taller than Karen
P2 Karen is taller than Mariam

C1 Gregory is taller than Mariam

PL Translation Key

g: Gregory
k: Karen
m: Mariam
Txy: *x* is taller than *y*

PL Argument

S1 *Tgk*
S2 *Tkm*

S3 *Tgm*

The PL argument is obviously deductively invalid. The relation "taller than" is a *transitive* relation, that is, if α is taller than β and β is taller than δ, then α is taller than δ. This is part of the meaning of the expression 'taller than'. We know that the English argument is deductively valid because we know that the relation "taller than" is a transitive relation. If someone were to tell us that she does not know that "taller than" is transitive, we would think that she does not fully understand the meaning of the expression 'taller than'. Hence in order to capture this important aspect of the meaning of the English predicate 'taller than', we need to add a meaning-postulate to the PL argument. The following PL argument is deductively valid.

PL Argument

S4 Tgk
S5 Tkm
S6 $(\forall x)(\forall y)(\forall z)((Txy \wedge Tyz) \rightarrow Txz)$

S7 Tgm

Any interpretation that makes the premises of this argument true must assign a transitive relation to the predicate '*Txy*'. If S6 is dropped from the argument, then there are interpretations that make the remaining premises true while assigning intransitive relations to the predicate '*Txy*'. This might seem as a deficiency in PL. However, it does not seem fair to declare PL deficient because the predicate '*Txy*' designates in some PL interpretations intransitive relations. The meanings of natural-language expressions rely on large amounts of resources, which may include physical facts about the world, social practices, psychological states, and contextual elements. Humans typically make use of such resources to fix the meanings of the expressions of their languages. The problem with PL (and other logical systems) is not that these resources are not available to the semantics of PL expressions; rather in most cases they *cannot fix* the meanings of these expressions *across all* relevant PL interpretations.

Let us illustrate the point with an example. There are PL interpretations for the PL argument above that assign a meaning to '*Txy*' that shares central aspects with the meaning of the expression 'taller than'. The PL interpretation H described below is one of these interpretations.

The PL Interpretation H

UD: The set of all American citizens
LN: *g*, *k*, *m*, and the full names of all the American citizens (we assume that every American citizen has a full name, and if two citizens have the same full name, we disambiguate the name by adding numbers to it)

Semantical Assignments

H assigns to every name in LN the American citizen who has that full name.
H(*g*): Gregory Adam Johnson
H(*k*): Karen Elizabeth Weiss
H(*m*): Mariam Biwer Yaqub
H(*Txy*): *x* is taller than *y*

Facts of the interpretation H

We can make a blanket assumption that all the facts that are needed to give '*Txy*' the standard meaning of 'taller than' are presupposed. Usually such a blanket assumption is not needed. By identifying the relation assigned to '*Txy*' with the use of the PL-English predicate '*x* is taller than *y*', we presuppose that the predicate '*x* is taller than *y*' has its standard meaning. However, we may elect to specify certain aspects of the meaning of '*x* is taller than *y*' by explicitly describing features of the relation assigned to 'Txy' by H. For example, using the notion of extension, we may state:

1. H(Txy) is transitive: for all α, β, and δ in UD, if $\langle\alpha, \beta\rangle$ and $\langle\beta, \delta\rangle$ belong to the extension H(Txy), then $\langle\alpha, \delta\rangle$ also belongs to H(Txy).
2. H(Txy) is irreflexive: for every α in UD, $\langle\alpha, \alpha\rangle$ does not belong to H(Txy).
3. H(Txy) is asymmetric: for all α and β in UD, if $\langle\alpha, \beta\rangle$ belongs to H(Txy), then $\langle\beta, \alpha\rangle$ does not belong to H(Txy).
4. ⟨Gregory Adam Johnson, Karen Elizabeth Weiss⟩ and ⟨Karen Elizabeth Weiss, Mariam Biwer Yaqub⟩ belong to H(Txy).

If our interest is only to determine the truth values of S4–S7 on H, we have more than we need. S4–S7 are all true on H. (We know that ⟨Gregory Adam Johnson, Mariam Biwer Yaqub⟩ belongs to H(Txy) because we know that H(Txy) satisfies 1 and 4.)

It is clear from this example that a PL interpretation could be endowed with a large collection of facts, which enhance the semantical contents of the interpreted PL expressions. Thus the problem cannot be that PL interpretations are denied adequate resources needed to give "deep" semantics to the symbols they interpret. The problem is that there is little "uniformity" in the resources that can be made available to relevant PL interpretations. In natural-languages such resources are employed to give the meanings of the expressions that occur in a collection of sentences, and then these meanings are held fixed across all the logical possibilities that are relevant to this collection of sentences. We assume that every relevant logical possibility is endowed with adequate resources to give these expressions their *standard* meanings. This entails that there is a good deal of uniformity in these resources: they are all adequate for endowing the same expression with its standard meaning. For example, the expression 'taller than' has the same (standard) meaning in every logical possibility that is relevant to the argument {P1, P2}/C. It cannot be the case that the predicate 'taller than' designates a transitive relation in one logical possibility and an intransitive relation in another. This is not the case, however, in PL. We can describe a PL interpretation J for {S4, S5, S6}/S7 that is identical with H except that J assigns to 'Txy' the relation "a parent of," that is, for all American citizens α and β, $\langle\alpha, \beta\rangle$ belongs to the extension of J(Txy) if and only if α is a parent of β. This relation is intransitive in J (assuming that only fathers and mothers are parents). The resources of H and J vary significantly, so that the meaning of 'Txy' in H is vastly different from its meaning in J. PL imposes little semantical uniformity on relevant PL interpretations: all these interpretations are bivalent, they are all extensional (i.e., properties and relations are identified with their extensions), they all assign the relation of token identity to '=', the truth conditions of the quantifiers and the sentential connectives are the same in all PL interpretations, the variables range only over the UD of a PL interpretation, no UD is allowed to be empty, every name must be assigned a referent, and so on. This uniformity is not adequate to fix the meanings of all PL expressions across all relevant PL interpretations. Because of this great semantical freedom in PL, the use of meaning-postulates is essential for the production of faithful PL translations of many natural-language arguments.

It might appear that the lack of significant semantical uniformity between relevant PL interpretations is a great liability. However, one can argue that this semantical freedom is one of PL's great strengths. In fact, all logical systems permit degrees of semantical freedom that far exceed what is typically allowable in natural languages. The reason for this feature is that every logical system attempts to capture a certain notion (or notions) and, at the same time, to be as topic-neutral as possible in order to be applicable in as many contexts as possible. Of course, every logical system has its own presuppositions, which impose certain limitations on its neutrality. As we stated in this book, systems of classical deductive logic attempt to capture the notions of logical possibility and true-in-a-logical-possibility; classical proof theory is meant to capture the notion of demonstrative proof. Systems of classical logic presuppose, among other things, bivalence and non-relevance (i.e., they do not require a relation of relevance between the premises and the conclusion of a deductively valid argument). The great semantical freedom that PL allows is one of the reasons for its success. PL can be applied in a large variety of theories and contexts. Of course, there are other reasons for the success of PL. PL has an elegant and relatively powerful proof theory. Mathematicians believe that almost all of their proofs are formalizable as PL proofs (derivations). This is not a merely dogmatic belief. The modern foundation of nearly all of mathematics is set theory, and this theory is formalizable as a set of PL sentences. The semantics of PL is relatively simple and can be made to look quite natural. PL interpretations are reasonable candidates for representing logical possibilities. PL syntax is sufficiently rich and flexible to allow the constructions of very simple as well as very complex sentences and formulas. The metatheory of PL generates a large number of interesting, yet manageable, questions. (The *metatheory* of a logical system is the theory whose object of study is the logical system itself.) Almost every other system of deductive logic is either an extension of PL or of SL or a rival to these two systems, which attempts to correct their perceived deficiencies or failings. There are more reasons for the success of PL, but it is not our purpose here to account for all the reasons for the success of First-Order Predicate Logic, or to make a case for this logical system. It is important, nonetheless, to understand PL, since it is the modern foundation of many systems of deductive logic. So understanding PL is a prerequisite for understanding these systems as well as the systems that aim to replace it.

4.5.13:C5 In this subsection we investigate whether there are valid natural-language arguments that cannot be translated into valid PL arguments that are faithful to the original arguments. There are, of course, arguments that invoke semantical resources that far exceed the expressive power of PL. For instance, an argument that involves *second-order quantification,* that is, quantification over properties and relations, might be beyond the reach of PL. Here is an example of such an argument. (We are assuming that the Evening Star and the Morning Star are the planet Venus.)

P1 The Evening Star has every property that the Morning Star has.
P2 The Morning Star is a planet in the solar system.
P3 All space telescopes generated detailed images of every planet in the solar system.
P4 The Hubble Telescope is a space telescope.

C1 The Hubble Telescope generated a detailed image of the Evening Star.

Before presenting a symbolic translation of this argument, we need to discuss very briefly second-order quantification and predication. Logicians call sentences like P1 **second-order.** The predicate logic we study here is commonly called "First-Order Logic." It is first-order because its quantifiers range only over individuals, and its predicates apply only to names that designate individuals. Individuals are considered *first-order objects*. First-order quantifiers do not range over properties and relations and first-order predicates do not apply to other predicates. Properties and relations *of individuals* are considered *second-order objects*. Second-order quantifiers range over properties and relations of individuals and second-order predicates apply to first-order predicates. For example, the sentences 'This book is red' and 'There is a red book' are first-order sentences. The first may be translated into the PL sentence 'Rb' and the second into '$(\exists x)(Bx \wedge Rx)$', where '$b$', '$Bx$', and '$Rx$' translate, respectively, 'This book', 'x is a book', and 'x is Red'. On the other hand, the sentences 'Red is a color' and 'This book has a color' are appropriately considered second-order sentences, since color is a property of red, green, blue, and so on, and red, green, and blue are properties of individuals. These sentences may be translated into the second-order sentences '$\mathrm{C}R$' and '$(\exists X)(\mathrm{C}X \wedge Xb)$', where '$\mathrm{C}X$' is a second-order predicate that translates 'X is a color' and 'X' is a second-order variable that ranges over properties. Thus in PL-English '$(\exists X)(\mathrm{C}X \wedge Xb)$' says that there is X, such that X is a color and this book has X. P1 is a second-order sentence because it involves second-order quantification. To say that the Evening Star has every property that the Morning Star has is to say that for every property X, if the Morning Star has X, then the Evening Star has X.

Syntactically speaking, a first-order quantifier applies to a variable that occupies a "name-place," i.e., a place that could be occupied by a name if the quantifier were not present. For example, in the PL sentence $(\forall y)(Ay \rightarrow Lsy)$, the first-order universal quantifier applies to the second and third occurrences of the variable y. These occurrences occupy name-places. If we remove the quantifier, we obtain the proper formula $Ay \rightarrow Lsy$, in which the variable y occurs freely in two places. Each of these places could be occupied by a name. We could write $Aa \rightarrow Lsa$, $Ab \rightarrow Lsb$, $At \rightarrow Lst$, and so on. A second-order quantifier applies to a variable that occupies a "predicate-place," i.e., a place that could be occupied by a (first-order) predicate if the quantifier were not present. It is a standard practice to use a different type of letters to denote second-order variables, i.e., variables that occupy predicate-places. Let us use the uppercase forms of PL (first-order) variables to denote second-order variables. In the sentence $(\exists Z)(Za \wedge Zb)$, the second-order existential quantifier applies to the second and third occurrences of the variable Z. If we drop the quantifier, we obtain the formula $Za \wedge Zb$, which contains two free occurrences of the second-order variable Z. Both of these occurrences could be occupied by 1-place predicates. We could write $Pa \wedge Pb$, $Ga \wedge Gb$, $Sa \wedge Sb$, and so on. The semantical content of the first-order sentence $(\forall y)(Ay \rightarrow Lsy)$ is that for every individual y, if y has the property A, then the individual s bears the relation L to y; while the semantical content of the second-order sentence $(\exists Z)(Za \wedge Zb)$ is that there is a property Z that both individuals a and b have.

As its name suggests, a second-order predicate applies to second-order variables as well as to first-order predicates. For instance, the three-place second-order predicate D may be applied to the second-order variables X and Y and to the first-order predicate R to form the second-order formula $\mathrm{D}XYR$. On the other hand, a first-order predicate is used to form first-order formulas by appending to it a certain number of singular terms or first-order variables. For example, the three-place first-order predicate P may be applied to the names a and b and to the first-order variable z to form the first-order formula $Pabz$. If we augment First-Order Predicate Logic (PL) with second-order variables and predicates, the resulting logic is called **Second-Order Predicate Logic** (or simply, "Second-Order Logic") and is usually denoted as PL^2. Of course, we must modify the syntax and semantics of PL in order to accommodate the second-order quantification and predication.

The argument above is deductively valid. The most faithful symbolic translation of this argument employs second-order quantification. The following PL^2 translation is deductively valid.

PL^2 Translation Key

e: The Evening Star
m: The Morning Star
b: The Hubble Telescope
Pz: *z* is a planet in the solar system
Tx: *x* is a space telescope
Gvw: *v* generated a detailed image of *w*

PL^2 Argument

S1 (∀Z)(*Zm*→*Ze*)
S2 *Pm*
S3 (∀*x*)(*Px*→(∀*y*)(*Ty*→*Gyx*))
S4 *Tb*

S5 *Gbe*

The reason this argument is valid is because the predicate *P* can be substituted for the second-order variable Z in the formula *Zm*→*Ze*; we obtain the PL sentence *Pm*→*Pe*. This sentence and S2 give us *Pe*. The truth of S3 entails that *Pe*→(∀*y*)(*Ty*→*Gye*) is true. Since *Pe* is true, we have that (∀*y*)(*Ty*→*Gye*) is true. Again the truth of this universally quantified sentence entails that the substitutional instance *Th*→*Gbe* is true, which, together with S4, logically implies *Gbe*.

PL has no means for translating faithfully a premise such as P1. However, we can think of a PL translation that captures the deductive validity of the English argument.

PL Translation Key

e: The Evening Star
m: The Morning Star
b: The Hubble Telescope
p: The property of being a planet in the solar system
Tx: *x* is a space telescope
Rxy: *x* has the property *y*
Gvw: *v* generated a detailed image of *w*

PL Argument

S6 (∀*y*)(*Rmy*→*Rey*)
S7 *Rmp*
S8 (∀*x*)(*Rxp*→(∀*y*)(*Ty*→*Gyx*))
S9 *Tb*

S10 *Gbe*

This PL argument is deductively valid. It is not as faithful a translation of the original argument as the PL^2 one. The PL^2 argument correctly captures the second-order quantification of P1 and the predications of P2 and P3. S1 *explicitly* asserts that every property *m* has *e* has as well; we do not need to consult a translation key to know what S1 asserts about *m* and *e*. Of course, we do need to consult the translation key in order to determine what *m* and *e* stand for. In the PL argument, we need to consult the translation key in order to learn that S6 asserts something about the properties that *m* and *e* have. S2 and S3 also *explicitly* assert that *m* has the property *P* and that if anything has the property *P*, then it satisfies the consequent '$(\forall y)(Ty \rightarrow Gyx)$'. We do not need to refer to any translation key in order to learn that S2 attributes the property *P* to *m* and that S3 asserts something about all things that have the property *P*. In the PL argument, however, we need to consult the translation key in order to learn that S7 is attributing a property to *m* and that S8 asserts something about all things that have a certain property. PL has the resources to express faithfully the predications of P2 and P3. A natural PL translation of P2 is 'Pm' and of P3 is '$(\forall x)(Px \rightarrow (\forall y)(Ty \rightarrow Gyx))$'. S7 and S8 are contrived translations in order to cope with the second-order quantification of P1. In spite of this criticism, at the end of the day {S6, S7, S8, S9}/S10 is *a* PL translation of the English argument {P1, P2, P3, P4}/C1 that captures the deductive validity of the original argument, and which does not seem to exhibit logical features that are not present in the English argument. So this example fails to establish the existence of a valid natural-language argument that cannot be faithfully translated into a valid PL argument.

However, the previous example suggests a way of modifying the second-order argument so that the modified argument is deductively valid but cannot be faithfully translated into a valid PL argument, even if the use of *reasonable* meaning-postulates is allowed. We said "reasonable" because if there is no restriction on the use of meaning-postulates, every valid natural-language argument with finitely many premises could be translated into a valid PL argument by adding a specific conditional premise, which trivializes the logical force of the original argument. For instance, if {Q1, Q2, Q3}/Q4 is some deductively valid natural-language argument and R1, R2, R3, and R4 are PL sentences that translate, faithfully or not, the sentences Q1, Q2, Q3, and Q4 respectively, then we can add the conditional premise $(R1 \wedge R2 \wedge R3) \rightarrow R4$ as a meaning-postulate to the PL argument. The resulting symbolic argument $\{R1, R2, R3, (R1 \wedge R2 \wedge R3) \rightarrow R4\}/R4$ is trivially deductively valid. The deductive validity of the PL argument is not a reflection of the deductive validity of the original argument. It is merely a trivial consequence of adding the conditional premise. Every argument can be turned into a deductively valid argument by adding a conditional premise asserting that if the given premises are true, then the conclusion is true as well. This shows that it is essential that the added meaning-postulates be, in some sense, reasonable. They should be introduced only to capture aspects of the meanings of the original expressions that are presupposed by the use of these expressions and are relevant to the logical status of the original argument. Furthermore, the introduction of meaning-postulates should not make the PL translation deviate from the logical relations that exist in the original argument.

With this in mind, we can now give an example of a valid English argument whose best PL translations are invalid. Consider the following modification of the previous argument. ('Individual' is used in the philosophical sense of the word.)

P5 Whatever relation the Morning Star has to any individual the Evening Star has as well.
P6 All space telescopes generated detailed images of the Morning Star.
P7 The Hubble Telescope is a space telescope.

C2 The Hubble Telescope generated a detailed image of the Evening Star.

This is a deductively valid argument. To see this, observe that the second premise (P6) attributes a certain relation to the Morning Star—namely, that every space telescope generated a detailed image of it. Thus P5 and P6 entail that the Evening Star also has this relation. Hence, all space telescopes generated detailed images of the Evening Star. The preceding intermediary conclusion, together with P7, entails that the Hubble Telescope generated a detailed image of the Evening Star, which is C2.

Again, the most faithful symbolic translation of this argument is second-order. Let *W* be a second-order variable that ranges over binary relations, i.e., it occupies the place of a 2-place predicate. The following PL^2 translation is deductively valid.

PL^2 Translation Key

e: The Evening Star
m: The Morning Star
b: The Hubble Telescope
Tx: *x* is a space telescope
Gvw: *w* generated a detailed image of *v* (observe that we switched the order of the variables in this translation).

PL^2 Argument

S11 $(\forall W)(\forall x)(Wmx \rightarrow Wex)$
S12 $(\forall x)(Tx \rightarrow Gmx)$
S13 Tb

S14 Geb

It is easy to establish the validity of this argument. The second-order variable *W* in S11 can be replaced by the 2-place predicate G and the first-order variable *x* by the name *b*. We get the substitutional instance $Gmb \rightarrow Geb$. By replacing *x* with *b* in S12, we obtain the substitutional instance $Tb \rightarrow Gmb$, which, together with S13, gives us *Gmb*. The latter and $Gmb \rightarrow Geb$ logically imply *Geb*.

The best PL can do is something like the following.

PL Translation Key

e: The Evening Star
m: The Morning Star
b: The Hubble Telescope

p: The property "having detailed images of it generated by all space telescopes"
Tx: *x* is a space telescope
Rxy: *x* has the property *y*
Gvw: *w* generated a detailed image of *v*

PL Argument

S15 $(\forall y)(Rmy \rightarrow Rey)$
S16 *Rmp*
S17 *Tb*
S18 $(Rep \wedge Tb) \rightarrow Geb$

S19 *Geb*

This PL argument is deductively valid. '*Rep*' follows from S15 and S16. Thus given S17, we have '$Rep \wedge Tb$', which, together with S18, entails S19. S18 is intended as a meaning-postulate. The PL argument would be deductively invalid if this premise is dropped. There are good reasons to believe that S18 is not a reasonable meaning-postulate. It does not capture any aspects of the meanings of the extra-logical expressions in the original argument. Its only purpose is to capture the inference from the sentences 'All space telescopes generated detailed images of the Evening Star' and 'The Hubble Telescope is a space telescope' to the sentence 'The Hubble telescope generated a detailed image of the Evening Star'. This inference is a logical relation, which the English argument exhibits without the mediation of a conditional premise, and which the PL argument captures only by means of a conditional premise (S18). In this sense, the PL argument lacks some of the "logical force" of the original argument.

It is not hard to modify this example so that there would be many inferential relations between the premises and intermediary conclusions that the best PL translation would not be able to capture without the introduction of many additional conditional premises. Someone who wants to defend such a PL translation as sufficiently faithful to the original argument would be hard pressed to explain why the PL argument requires that all these referential relations, which the original argument exhibits without the mediation of any conditional premises, be encoded as additional premises of the PL translation: what grounds are there for saying that such a PL translation is "sufficiently faithful" when it lacks so much of the "logical force" of the original argument? If the previous reasoning is correct, then there are valid natural-language arguments that cannot be faithfully translated into valid PL arguments.

4.5.13:C6 In 3.5.12:C4 we gave examples to show that there are invalid natural-language arguments whose SL translations are valid, and to show that there are relevant logical possibilities that cannot be represented by any relevant SL truth valuations. We explained there that those examples were the product of two features of SL: the truth-functional semantics of the SL connectives, especially the material conditional, and the lack of any interdependency between the atomic sentences of SL. For the most part, atomic sentences of PL (i.e., atomic formulas that have no free variables) are independent in their truth values from each other; but this is not universally true. The use of the identity-predicate can create interdependencies between PL atomic sentences. For example, the atomic sentence *Ast* is not independent from the atomic sentences *Asr* and $r = t$. If the last two sentences are true, the first sentence must be true as well. In addition to the lack of interdependency between most PL atomic sentences, the sentential connectives of PL have the same truth-functional semantics they have in SL. This is a source of trouble for PL. In this subsection we will discuss the most serious of these problems: the presence of invalid natural-language arguments whose PL translations are all valid and the presence of relevant logical possibilities that cannot be represented by any relevant PL interpretations. To be sure, these are controversial issues. However, we will do our best to make a fairly convincing case for the existence of such arguments and of such logical possibilities.

Let us begin with the two-switches example. Here is the English argument.

P1 If we throw the first switch and we throw the second switch, the light bulb is turned on.

C1 Either if we throw the first switch then the light bulb is turned on, or if we throw the second switch then the light bulb is turned on.

If we assume that the two switches are connected in series to the light bulb, and that both are off, then we have a counterexample to this argument. Both switches must be thrown for the light to come on. So if only one switch is thrown, the light does not come on. We will refer to this logical possibility as q. The diagram of q is given below together with the truth values of the sentences in question.

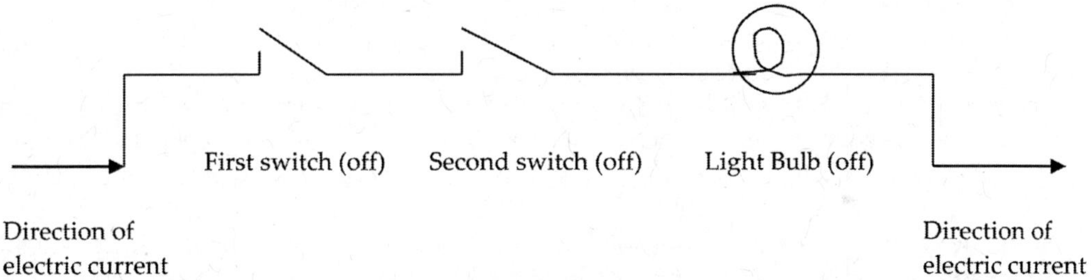

The logical possibility q:
(P1) If we throw the first switch and we throw the second switch, the light bulb is turned on (true)
(A) If we throw the first switch, the light bulb is turned on (false)
(B) If we throw the second switch, the light bulb is turned on (false)
(C1) Either if we throw the first switch then the light bulb is turned on, or if we throw the second switch then the light bulb is turned on (false)

Thus the English argument is deductively invalid. We saw in Chapter Three that the SL translation of this argument is deductively valid. The first PL translation that we will consider parallels the SL translation.

PL Translation Key

a: The first switch
b: The second switch
c: The light bulb
Tx: We throw *x*
Oz: *z* is turned on

PL argument

S1 $(Ta \wedge Tb) \rightarrow Oc$

S2 $(Ta \rightarrow Oc) \vee (Tb \rightarrow Oc)$

This is a valid PL argument. Theorem 4.4 of Subsection 4.5.12:C2 entails that any PL counterpart of a deductively valid SL argument is deductively valid. Since the argument {S1}/S2 is a PL counterpart of the deductively valid SL argument $\{(A \wedge B) \rightarrow C\}/(A \rightarrow C) \vee (B \rightarrow C)$ (see 3.5.12:C4), which is a translation of the English argument {P1}/C1, the PL argument above is deductively valid. Thus we have an invalid English argument whose PL translation is valid.

We will try to improve on this translation. The central problem with the previous translation is that the formal conditional is truth-functional while the English conditional is causal. The English 'if-then' in the premise P reports a causal connection between the switches and the light bulb: the throwing of the two switches *causes* the light to come on. It does not matter whether the switches are actually on or off, the causal fact does not change. On the other hand, whether throwing one of the switches causes the light to come on is dependent on the status of the other switch. If the other switch is off, then throwing the switch will not cause the light to come on; but if the other switch is on, then throwing the switch causes the light to come on. This explains the truth values of the sentences stated below the diagram.

Adding meaning-postulates will not help. Since S2 can be inferred from S1 alone, adding more premises will not invalidate this inference. We discussed this issue in Chapter One. We said there that a deductively valid argument could not be made invalid by adding new premises to it. What is required is to retranslate P and C in a manner that captures some of the causal meaning of the English conditional. Anyone who has studied the philosophy of causation knows how hard it is to define the relation of causation, and this is not the place to summarize some of the proposed definitions of causation. We only want to note that all those definitions make use of rich semantical resources that are available to natural languages. A typical formal language, such as the language of PL, lacks many of these resources. So we must set a more modest goal for ourselves: we are not aiming at capturing the causal sense of the English conditional as used in P and C, but only at capturing some aspects of this meaning—aspects that would allow us to convey a relation of dependency between *Oc*, on the one hand, and *Ta* and *Tb*, on the other hand.

Part of our understanding of the causal relations depicted in the diagram above is that *whenever* the first and second switches are on, the light comes on (we ignore the very short time it takes for the electric current to travel through the switches to the light bulb). This is not true of either switch considered on its own. For example, it is not true that whenever the first switch is thrown, the light comes on. Sometimes it does, when the second switch is also thrown, or is already on, and sometimes it doesn't, when the second switch is not thrown. Hence we need to incorporate a temporal parameter. We want to convey in PL that at any given moment of time, if both switches are thrown at that moment, then the light bulb is turned on. The claim that if we throw the first (or second) switch, the light bulb is turned on should be understood as implying that at any given moment in time, if the first (or second) switch is thrown at that moment, the light bulb is turned on. According to the circuit described in the diagram above, the first claim is true while the second is false. These implications can be captured to a certain degree in PL. Here is the new translation.[1]

PL Translation Key

a: The first switch
b: The second switch
c: The light bulb
Txv: We throw *x* at time *v*
Ozv: *z* is turned on at time *v*

PL Argument

S3 $(\forall x)((Tax \wedge Tbx) \rightarrow Ocx)$

S4 $(\forall z)(Taz \rightarrow Ocz) \vee (\forall y)(Tby \rightarrow Ocy)$

The PL-English readings of S3 and S4, given the translation key above, are as follows:
(S3) *For every x, if we throw the first switch at time x and we throw the second switch at time x, then the light bulb is turned on at time x.*
(S4) *Either for every z, if we throw the first switch at time z, the light bulb is turned on at time z, or for every y, if we throw the second switch at time y, the light bulb is turned on at time y.*

This translation captures some aspects of the meaning of the causal conditionals employed in P and C above. Of course, the formal conditionals are still material conditionals, but the introduction of variables and quantifiers creates some interdependencies that were absent from the previous translation. The important thing about this new translation is that it is, like the English argument, deductively invalid. The following PL interpretation demonstrates the invalidity of this argument: it makes the premise true and the conclusion false.

The PL Interpretation J

UD: ℕ: {0, 1, 2, 3, …}
LN: $a, b, c, n_0, n_1, n_2, n_3, \ldots, n_k, \ldots$

Semantical Assignments

J(a): 5; J(b): 7; J(c): 35; J(n_0): 0; J(n_1): 1; J(n_2): 2; J(n_3): 3; …; J(n_k): k; …
J(*Txv*): "*v* is a multiple of *x*" (observe the order of the variables)
J(*Ozv*): "*v* is divisible by *z*" (observe the order of the variables)

1 I am indebted to Martin Barrett for this PL analysis of the original English argument.

The natural reading of (S3) $(\forall x)((Tax \wedge Tbx) \rightarrow Ocx)$ on J:
Every number that is both a multiple of 5 *and a multiple of* 7 *is divisible by* 35.
S3 is true on J.

The natural reading of (S4) $(\forall z)(Taz \rightarrow Ocz) \vee (\forall y)(Tby \rightarrow Ocy)$ on J:
Either every multiple of 5 *is divisible by* 35, *or every multiple of* 7 *is divisible by* 35.
S4 is false on J.
Both disjuncts are false. The first one is false because there are numbers, such as 20, that are multiples of 5 but not divisible by 35. Similarly, the second disjunct is false since there are numbers, such as 21, that are multiples of 7 but not divisible by 35.

Thus one could argue that since the PL translation above is deductively invalid, this example fails to establish the existence of an invalid natural-language argument whose PL translations are all valid. One could further argue that since the second PL argument is a more sophisticated and a more faithful translation of the original English argument, PL cannot be faulted for the existence of valid PL translations of this argument, for these PL arguments are not sufficiently faithful to the original argument.

Even if this reasoning is correct, it does not show that the logical possibility q can be represented by some PL interpretation for the PL argument {S3}/S4. The individuals in the logical possibility q are the two switches, the light bulb, and moments of time. For the sake of simplicity, we will assume that these moments of time are intervals each of which is one minute; so 0 stands for the present time, 1 for the first minute in the future, 2 for the second minute in the future, 3 for the third minute in the future, and so on. q also contains as constituents the relations "we throw *x* at time *v*" and "*z* is turned on at time *v*." Since in q the switches and the light are off *all* the time, the extensions of these relations in q are empty: we never throw any switch and the light never comes on. However, this does not prevent P1 from being true in q and A and B from being false in q. (See the circuit diagram above for the sentences P1, A, and B.) As we previously said, the English 'if-then' is used in these sentences as a causal conditional, and not as a truth-functional one. Hence the truth values of P1, A, and B are not determined by the standard truth conditions of the material conditional '→'. P1 is true in q regardless of the status of the two switches: throwing both switches *causes* the light to come on. The fact that the extension of the relation "we throw *x* at time *v*" is empty means that both switches are off. Thus throwing the first switch *does not cause* the light to come on, because the second switch is off; and similarly throwing the second switch *does not cause* the light to come on, because the first switch is off. This explains why the sentences 'If we throw the first switch, the light bulb is turned on' (A) and 'If we throw the second switch, the light bulb is turned on' (B) are false in q. We assume that the disjunction 'A or B' (which is C1) has the same truth conditions as the truth-functional connective '∨'. Since both A and B are false, C1 is false in q.

No PL interpretation that is relevant to the PL argument {S3}/S4 can represent the logical possibility q (which is depicted by the circuit diagram above). The reason is that no such interpretation can satisfy the Representation Condition. We will prove this claim using an indirect strategy. But, first, we need to modify the original English argument to make it suitable for the revised PL argument {S3}/S4. Consider the following PL-English argument.

P1* For every x, if we throw the first switch at time x and we throw the second switch at time x, then the light bulb is turned on at time x.

C1* Either for every z, if we throw the first switch at time z then the light bulb is turned on at time z, or for every y, if we throw the second switch at time y then the light bulb is turned on at time y.

Let Σ be the set {P1*, C1*} and Σ^{PL} be the set {S3, S4}. Given the translation key we used to produce S3 and S4, we have the following sets of vocabulary.

Voc(Σ^{PL}) = {a, b, c, Txv, Ozv, $\neg$, $\wedge$, $\vee$, $\rightarrow$, $\leftrightarrow$, $\forall$, $\exists$, u, v, w, x, y, z, u_1, v_1, …, (,)}

Voc(Σ) = {The first switch, The second switch, The light bulb, We throw x at time v, z is turned on at time v, It is not the case that, not, and, or, if-then, if and only if, for all, there is, u, v, w, x, y, z, u_1, v_1, …}

Now we state our Reductio Assumption. Assume that there is a PL interpretation M for the PL argument {S3}/S4 that represents the logical possibility q for the PL-English argument {P1*}/C1*. The Representation Condition must hold for M and q. Therefore for every PL sentence **Z** that is composed of Voc(Σ^{PL}) and for every PL-English sentence X that is composed of Voc(Σ), if **Z** is a translation of X (i.e., if **Z** is X^{PL}), then **Z** is true (or false) on M if and only if X is true (or false) in q. The table below lists various PL-English sentences and their PL translations with their truth values in q and on M, respectively, and the reasons for these truth values. It is important to remember that in q we never throw any switch and the light never comes on. The 'if-then' in Voc(Σ) is a causal conditional: it is used in the argument {P1*}/C1* to describe causal relations that obtain in q.

	PL-English sentence X composed of Voc(Σ)	**PL sentence X^{PL} composed of Voc(Σ^{PL})**
1	*There is x, such that we throw the first switch at time x.* Truth value in q: False Reason: The first switch is never thrown.	$(\exists x)Tax$ Truth value on M: False Reason: Representation Condition
2	*There is x, such that we throw the second switch at time x.* Truth value in q: False Reason: The second switch is never thrown.	$(\exists x)Tbx$ Truth value on M: False Reason: Representation Condition
3	*There is x, such that the light bulb is turned on at time x.* Truth value in q: False Reason: The light bulb is never turned on.	$(\exists x)Ocx$ Truth value on M: False Reason: Representation Condition
4	*For every x, if we throw the first switch at time x and we throw the second switch at time x, then the light bulb is turned on at time x.* Truth value in q: True Reason: Assuming that the electric circuit depicted in the diagram above is functional, the sentence describes the causal relation between the light bulb and the two switches. Since the switches and the light	$(\forall x)((Tax \wedge Tbx) \rightarrow Ocx)$ Truth value on M: True Reason: From 1 and 2, by the truth conditions of the existential quantifier, all the basic substitutional instances of $(\exists x)Tax$ and $(\exists x)Tbx$ are false on M. This means that for every name **s**, both T**a**s and T**b**s are false on M. By the truth conditions of the conjunc-

	PL-English sentence X composed of Voc(Σ)	**PL sentence X^{PL} composed of Voc(Σ^{PL})**
	bulb are connected in series, throwing both switches at the same time turns the light on.	tion, for every name **s**, $Ta\mathbf{s} \wedge Tb\mathbf{s}$ is false on M. By the truth conditions of the conditional, for every name **s**, $(Ta\mathbf{s} \wedge Tb\mathbf{s}) \rightarrow Oc\mathbf{s}$ is true on M. Thus all the basic substitutional instances of $(\forall x)((Tax \wedge Tbx) \rightarrow Ocx)$ are true on M. Therefore, by the truth conditions of the universal quantifier, $(\forall x)((Tax \wedge Tbx) \rightarrow Ocx)$ is true on M.
5	*For every z, if we throw the first switch at time z, the light bulb is turned on at time z.* Truth value in q: False Reason: Throwing the first switch at any time does not turn the light on since the second switch is off; both switches must be thrown in order for the light to come on.	$(\forall z)(Taz \rightarrow Ocz)$ Truth value on M: True Reason: From 1 by the truth conditions of the existential quantifier, every basic substitutional instance of $(\exists x)Tax$ is false on M. Hence for every name **s**, $Ta\mathbf{s}$ is false on M. By the truth conditions of the conditional, for every name **s**, the conditional $Ta\mathbf{s} \rightarrow Oc\mathbf{s}$ is true on M. This means that all the basic substitutional instances of $(\forall z)(Taz \rightarrow Ocz)$ are true on M. Therefore, by the truth conditions of the universal quantifier, $(\forall z)(Taz \rightarrow Ocz)$ is also true on M.
6	*For every y, if we throw the second switch at time y, the light bulb is turned on at time y.* Truth value in q: False Reason: Throwing the second switch at any time does not turn the light on since the first switch is off; both switches must be thrown in order for the light to come on.	$(\forall y)(Tby \rightarrow Ocy)$ Truth value on M: True Reason: From 2 by the truth conditions of the existential quantifier, every basic substitutional instance of $(\exists x)Tbx$ is false on M. Hence for every name **s**, $Tb\mathbf{s}$ is false on M. By the truth conditions of the conditional, for every name **s**, the conditional $Tb\mathbf{s} \rightarrow Oc\mathbf{s}$ is true on M. This means that all the basic substitutional instances of $(\forall y)(Tby \rightarrow Ocy)$ are true on M. Therefore, by the truth conditions of the universal quantifier, $(\forall y)(Tby \rightarrow Ocy)$ is also true on M.
7	*Either for every z, if we throw the first switch at time z then the light bulb is turned on at time z, or for every y, if we throw the second switch at time y then the light bulb is turned on at time y.* Truth value in q: False Reason: From 5 and 6, both disjuncts of this sentence are false in q. Hence the sentence is also false in q.	$(\forall z)(Taz \rightarrow Ocz) \vee (\forall y)(Tby \rightarrow Ocy)$ Truth value on M: True Reason: From 5 and 6, both disjuncts of this sentence are true. Thus the sentence is true on M.

The sentences in the last three pairs have different truth values: the PL sentences are all true on M and the English sentences are all false in q. Hence q and M do not satisfy the Representation Condition, which entails that M does not represent q. Since M was chosen arbitrarily, no PL interpretation that is relevant to the PL argument {S3}/S4 can represent q.

We have established so far that there are relevant logical possibilities that cannot be represented by any relevant PL interpretations. But we did not succeed in giving an example of an invalid natural-language argument whose PL translations are all valid. However, the refined analysis we gave of the two-switches argument suggests a way of modifying that argument, so that its "best" PL translations are all deductively valid. Here is the modified argument. As we explained earlier, the PL-English argument below is deductively invalid.

P2 For any time *x*, if we throw the first switch at *x* and we throw the second switch at *x*, then the light bulb is turned on at *x*.

C2 Either there is a time *z* such that if we throw the first switch at *z* then the light bulb is turned on at *z*, or there is a time *y* such that if we throw the second switch at *y* then the light bulb is turned on at *y*.

Using the same translation key above, we obtain the following PL translation of the preceding argument.

S5 $(\forall x)((Tax \land Tbx) \rightarrow Ocx)$

S6 $(\exists z)(Taz \rightarrow Ocz) \lor (\exists y)(Tby \rightarrow Ocy)$

This PL argument is deductively valid. Below is a proof of this claim. We employ an indirect-proof strategy.

Proof that {S5}/S6 is valid

1. Reductio Assumption: there is a PL interpretation I that makes S5 true and S6 false, that is:
    1.1. $(\forall x)((Tax \land Tbx) \rightarrow Ocx)$ is true on I.
    1.2. $(\exists z)(Taz \rightarrow Ocz) \lor (\exists y)(Tby \rightarrow Ocy)$ is false on I.
2. From 1.2 by the truth conditions of the disjunction: both $(\exists z)(Taz \rightarrow Ocz)$ and $(\exists y)(Tby \rightarrow Ocy)$ are false on I.
3. From 1.1 by the truth conditions of the universal quantifier: every basic substitutional instance of the sentence $(\forall x)((Tax \land Tbx) \rightarrow Ocx)$ is true on I.
4. By the definition of interpretation: The UD of I is not empty; and hence there is at least one name **t** in LN.
5. From 3 and 4 by the definitions of basic substitutional instance: $(Ta\mathbf{t} \land Tb\mathbf{t}) \rightarrow Oc\mathbf{t}$ is true on I.
6. From 2 by the truth conditions of the existential quantifier: all basic substitutional instances of $(\exists z)(Taz \rightarrow Ocz)$ and of $(\exists y)(Tby \rightarrow Ocy)$ are false on I; in particular, $Ta\mathbf{t} \rightarrow Oc\mathbf{t}$ and $Tb\mathbf{t} \rightarrow Oc\mathbf{t}$ are false on I.

7. From 6 by the truth conditions of the conditional: *Ta*t and *Tb*t are true on I and *Oc*t is false on I.
8. From 7 by the truth conditions of the conjunction and the conditional: (*Ta*t∧*Tb*t)→*Oc*t is false on I, which contradicts 5.
9. From 1 through 8: there is no PL interpretation on which S5 is true and S6 is false.
10. From 9 by the definition of deductive validity: the argument {S5}/S6 is deductively valid.

Adding meaning-postulates will not help matters. As we explained previously, adding premises to a deductively valid argument cannot transform it into an invalid argument. We conclude that the preceding English argument is an example of an invalid argument whose best PL translation is valid.

4.5.13:C7 As we approach the end of Chapter Four, it is important to examine how far we traveled in answering the central questions we posed in Chapter One: What is the nature of logical possibility? And what does it mean to say that a declarative sentence is true (or false) in a logical possibility? Since the notion of a declarative sentence is part of the last question, we also inquired about the class of these sentences. In Chapter One we did not attempt to give any explanation of the notion of true-in-a-logical-possibility. We simply took truth as an intuitively clear, primitive notion. We used truth to define a declarative sentence as a sentence for which it is meaningful to say that it is true or that it is false. We also gave an intuitive explanation of the notion of logical possibility. We said that a logical possibility is a story that contains no explicit or implicit contradictions. An explicit contradiction is a sentence of the form "Z and not-Z" that occurs in the story. The difficulty we encountered was how to define an implicit contradiction. We observed there that the only meaning we could attribute to saying that a contradiction is implicit in some story is that a sentence of the form "Z and not-Z" can be *inferred* from the story. But to infer something from a story is to show that it *logically follows* from the story, i.e., it is a *logical consequence* of the story. We noted there that this analysis is circular, since logical consequence is defined in terms of logical possibility: to say that X is a logical consequence of Y is to say that there is no logical possibility in which Y is true and X is false, or, equivalently, that in every logical possibility in which Y is true X is true as well. Thus the circle was this: logical possibility was defined as a story that contains no explicit or implicit contradictions; an implicit contradiction was defined as a contradiction that is a logical consequence of the story; and logical consequence (or deductive validity) was defined in terms of logical possibility. Thus we start with logical possibility and end with logical possibility. Although circular definitions can sometimes be explanatory, they are not characterizations. To characterize a notion is to define it in terms of other (preferably, simpler) notions that do not presuppose the notion that is being defined.

Thus our task was threefold: identify the class of all declarative sentences, give a characterization of the notion of logical possibility, and define the notion of true-in-a-logical-possibility. Given that the eight logical concepts—deductive validity (or logical consequence), deductive invalidity, logical truth, logical falsehood, contingency, logical equivalence, consistency, and inconsistency—introduced in Chapter One were all defined in terms of declarative sentences being true or false in some or all logical possibilities, rigorous accounts of these three notions would immediately yield rigorous definitions of those eight logical concepts.

We carried out this task three times with varying degrees of success. In Chapter Two we developed a modern version of an ancient logical system: Term Logic (TL). TL is based on a worldview that takes individuals (in the philosophical sense of 'individuals') and properties as the basic ingredients of reality. Logical possibilities, according to this worldview, are any combinations of these ingredients. As with any system of classical deductive logic, TL consists of a syntax that generates all the declarative sentences of this system, a semantics that offers a characterization of the notion of *relevant* logical possibility, and a definition of what it means to say that a sentence is true (or false) in some logical possibility. In TL, logical possibilities are characterized as TL diagrams for some sets of TL sentences, and truth or falsehood on TL diagrams is defined by specifying truth conditions for all the possible forms of TL sentences. In Section 2.5.12, we examined the extent to which TL diagrams can be said to represent logical possibilities. We gave a relatively rigorous definition of what it means to say a TL diagram represents a logical possibility. Based on that definition we demonstrated the existence of superfluous TL diagrams, that is, TL diagrams that represent no relevant logical possibilities. This fact is responsible for there being valid English arguments whose TL translations are invalid.

In 2.5.12:C6, we also examined proposed examples of invalid English arguments whose TL translations are valid. Our goal was to see whether there are relevant logical possibilities that cannot be represented by any TL diagrams. Our analysis there did not demonstrate the existence of such logical possibilities. This in itself does not show that every relevant logical possibility can be represented by a TL diagram or that every invalid natural-language argument can be faithfully translated into an invalid TL argument. Our discussion there only shows that some popular types of counterexamples to TL do not work. However, there are at least two reasons why it is not easy to think of relevant logical possibilities that fail to be represented by any TL diagrams. First, the TL worldview is very "basic": almost every imaginable situation can be decomposed into individuals and properties. Second, the TL worldview does not allow properties to have empty extensions. Most of the counterexamples to TL involve properties whose extensions are empty. Thus it is customary to dismiss such counterexamples as inapplicable to TL because of TL's prohibition against properties with empty extensions. Regardless of whether there are relevant logical possibilities that cannot be represented by any TL diagrams, the fact remains that there are TL diagrams that do not represent any relevant logical possibilities. This is enough to show that TL diagrams are not a successful surrogate for the notion of logical possibility.

In Chapter Three we considered another system of classical logic: Sentence Logic. The worldview of SL dispenses with individuals, properties, and relations in favor of states of affairs. It was hoped that states of affairs are more "fundamental" ingredients of reality. In fact, many philosophers believe that the world is not the totality of individuals and their properties and relations, but rather of states of affairs (or facts).[1] The SL worldview makes the further presupposition that there are no interdependencies between states of affairs: each may obtain or fail to obtain independently of the status of any other state of affairs. This worldview

1 The most important philosopher of the 20th century to hold this position is the Austrian philosopher Ludwig Wittgenstein (1889–1951), whose book *Tractatus Logico-Philosophicus* begins with the famous proposition that the world is the totality of facts, not of things. A more recent philosopher who subscribes to a similar position is the Australian philosopher David Armstrong (b. 1926).

is reflected in the syntax and semantics of SL. SL syntax consists of sentence letters (atomic sentences), which are intended to represent states of affairs, and logical connectives, which, intuitively speaking, allow the system to make complex assertions about states of affairs. Those logical connectives are truth-functional. As defined in Chapter Three, a truth-functional connective is a connective such that the truth value of any compound sentence that is generated by the use of this connective is fully determined by the truth values of its immediate components. Logical possibilities, according to SL, are characterized as truth valuations, which are distributions of truth values to the atomic sentences. The intuitive idea is that if an atomic sentence represents a state of affairs that obtains, then the atomic sentence is true, and if it represents a state of affairs that does not obtain, then the atomic sentence is false. The notion of true-in-a-logical possibility becomes in SL true-on-a-truth-valuation.

As we have seen in Chapter Three, the logical independence of SL atomic sentences and its truth-functional semantics were a source of trouble for SL. Even if we accept that states of affairs are the most fundamental ingredients of reality (or, at least, part of the basic ingredients of reality), we may seriously doubt the presupposition of the logical independence of all states of affairs. Does it sound at all reasonable to suppose that the facts that Ricardo is Maria's brother and that Maria is Ricardo's sister are logically independent of each other if we are talking about the same Ricardo and Maria? What sort of a logical possibility is it in which Ricardo is Maria's brother but she is not his sister? Recall that one of our central requirements is that the meanings of the expressions in question be held fixed across all relevant logical possibilities. So it is not a permissible move to say that he is her brother in logical possibility p but she is not his sister there because 'being his sister' means in p what 'being his mother' means in our world. It should be obvious that if we allow meanings to alter across logical possibilities, then our logical concepts would be trivialized: many logically true and logically false sentences would be contingent, many deductively valid arguments would be deductively invalid, and many pairs of logically equivalent sentences would not be equivalent. For instance, the sentence 'A triangle is a three-sided figure' would be contingent (and not logically true), since in some logical possibility 'triangle' could mean 'square'; the sentence 'Some bachelors are married' would also be contingent (and not logically false), since 'bachelor' could mean 'married male' in some logical possibility; and the claim that the sentence 'Gregory is taller than Mariam' is a logical consequence of the sentences 'Gregory is taller than Karen' and 'Karen is taller than Mariam' would be false since in some logical possibility 'taller than' could mean 'a parent of'. Thus the presupposition that all the states of affairs are logically independent of each other appears to be indefensible and problematic.[1] The other feature, which is the truth-functional semantics of SL, limits the system's ability to express many extra-logical relations, such as causal, spatial, and temporal relations, that might exist between many states of affairs.

1 Some philosophers would argue that the fact that Maria is Ricardo's sister and that Ricardo is Maria's brother are the same fact: it consists of the same two individuals and a specific relation, which, if viewed from Ricardo's "direction," is the relation of brotherhood and, if viewed from Maria's "direction," is the relation of sisterhood. The standard example that is usually given for dependent states of affairs is about colors: the facts that this dog is red and that this dog is blue; because it is not logically possible for anything to be both red and blue (all over). I think the relations of sisterhood and brotherhood are not matters of "direction" (this is Russell's term). These relations designate different ordered pairs; for example, Leonardo could be Ricardo's brother, but not his sister, and Maria could be Juanita's sister, but not her brother.

The most problematic SL truth-functional connective is the material conditional. As we have seen, English conditionals are employed to express extra-logical relations between states of affairs, which might be beyond the reach of the material conditional. We explained in 3.5.12 that these two features of SL are responsible for there being superfluous truth valuations and for there being relevant logical possibilities that cannot be represented by any truth valuations. Just as is the case with TL diagrams, SL truth valuations fall short of capturing the sphere of logical possibilities.

In this chapter, we introduced a much more powerful system: Predicate Logic. As explained in a previous subsection, PL includes TL and extends SL. Unlike the SL worldview, PL worldview does not include states of affairs as basic ingredients of reality. The PL worldview specifies individuals, properties, and relations (including the relation of token identity) as the basic ingredients of reality. The loss of states of affairs does not seem to hinder PL's expressive power. Intuitively speaking, the PL worldview replaces states of affairs with individuals' having certain properties or standing in certain relations to each other. This worldview is reflected in the syntax and semantics of PL. PL syntax comprises names (for individuals) and predicates (for properties and relations). But it also comprises logical devices, which are sentential connectives and quantifiers, that allow it to make complex assertions about individuals, properties, and relations. The syntax of PL generates the class of all the PL declarative sentences. The semantics of PL sentences is given by PL interpretations. Intuitively speaking, PL interpretations are meant to capture the notion of logical possibility. PL interpretations have a certain natural feel to them; many of them read as stories. A PL interpretation typically consists of a set of individuals together with properties and relations that are instantiated by these individuals, as well as semantical assignments that correlate PL expressions with the constituents of the interpretations. The semantics of PL also includes truth conditions for all the sentence forms of PL. These truth conditions offer a definition of the notion of true-on-an-interpretation, which is intended to capture the informal notion of true-in-a-logical-possibility.

In 4.5.13, we gave examples of superfluous PL interpretations and of relevant logical possibilities that cannot be represented by any PL interpretations. This demonstrates that, like TL diagrams and SL truth valuations, the notion of a PL interpretation is not a successful characterization of the informal notion of a logical possibility. Of course, we did not cover all systems of classical deductive logic. For instance, there are higher-order logical systems. We discussed briefly in this chapter Second-Order Logic. It is doubtful that any known system of classical logic actually succeeds in capturing perfectly the notion of logical possibility. There are, however, many systems of non-classical logics, some of which could be considered as attempting to capture the notion of logical possibility. Some non-classical systems are arguably more successful in this task than any known system of classical logic. Other systems of non-classical logic cannot be considered as attempts at capturing the notion of logical possibility; rather they are concerned with other notions. For instance, epistemic logic concerns knowledge and belief, and deontic logic concerns obligation and other related notions such as permissibility. The study of other systems of classical and non-classical logic is beyond the aims of this book. The three systems developed in this book are the "central" systems, in the sense that almost every other logical system is either an extension of one of them or a rival to one of them. Hence the study of these systems is a prerequisite for the study of many other logical systems.

If, as we said above, all of these three systems fall short of capturing the sphere of logical possibilities, then what have we gained from studying them? An area of study that aims at capturing a certain notion might not be fully successful in characterizing that notion, yet it can be very illuminating and informative in offering an elucidation and partially successful characterization of the notion. If we always require complete success before we consider the area worthy of further study and investigation, we would have to dismiss many important and fruitful areas of human knowledge. Consider for instance the area of Normative Decision Theory. It is an interdisciplinary area with practitioners from economics, philosophy, political science, psychology, and statistics. This theory studies the nature and conditions of rational decision-making. The theory contains many competing proposed conditions and characterizations of rational decision-making. Some of these proposals are incompatible with each other; so we know that they cannot all be correct. In fact there is near universal agreement among the practitioners of this theory that no proposed characterization of rational decision-making is completely successful. However, this area of study has been very fruitful, with many applications in different fields. It is also conceptually illuminating and informative. It has uncovered a good deal about the conditions and nature of rational decision-making. It would be foolish to dismiss the achievements of this theory on the basis that none of them is fully successful in accomplishing the stated aims of the theory.

Similar reasoning can be given for the study of these three logical systems and many other systems of classical and non-classical logic. Most logicians would agree that the various logical systems that they develop and investigate are not completely successful in achieving what they aim to achieve. However, we continue to study and investigate these systems as well as develop new ones. Logical Theory has been very illuminating and informative. We learned a great deal about the scope and limitations of these systems and with that we learned a great deal about the nature of the notions we hoped to characterize. Term, Sentence, and Predicate Logic taught us a lot about the nature of logical possibility, about the nature of demonstrative proof (Chapter Five), about the concept of truth, and about various logical concepts, such as deductive validity and consistency. Predicate Logic specifically was very successful in supplying mathematics with a medium for expressing mathematical concepts, theories, and proofs. It is true that hardly any mathematicians (including mathematical logicians) actually use strict PL formalism in stating their definitions, axioms, theorems, and proofs. But it is agreed upon that almost all of these aspects of mathematics can, in principle, be expressed as PL sentences and PL derivations. We said "in principle" because most of the time it is practically impossible to formalize fully mathematical proofs. Mathematicians actually use what we called PL-English, augmented with a large number of symbols, in expressing their theories and proofs. Hence we learned a lot from our hard work in studying these three systems. We are conceptually more enlightened and informed and we have at our disposal powerful methodologies for determining the applicability of many logical concepts, even though some of these methodologies, as we explained earlier, are not effective decision procedures.

4.6 Exercises

Notes: All answers must be adequately justified. As stated previously, the quantifiers range over the universe of discourse only.

4.6.1 Draw the construction tree of each of the following PL expressions. Determine which of them are PL formulas and which are PL sentences.

4.6.1a* $(\exists y)((T^2yy \wedge \neg R^3yac) \rightarrow (\forall x)(T^2xy \vee \neg(\exists w)R^3wex))$

4.6.1b* $(\forall z)(\forall y)(R^2zy \rightarrow \neg(z = y))$

4.6.1c $((H^2ad \wedge (\exists v)H^2vb) \leftrightarrow (\forall v)(H^2vv))$

4.6.1d $(\exists y)(((\forall x)\neg x = e \rightarrow (\exists u)u = x) \vee J^3eyt)$

4.6.1e $(\exists x)(\exists z)(\exists y)(G^3xyz \wedge (\neg x = y \vee (x = z \vee D^2zy)))$

4.6.1f $(\forall x)(\neg(\exists w)P^1w \rightarrow ((\exists y)(L^2xk \wedge L^2yk) \vee L^2h^2abx))$

4.6.1g* $(\neg(\exists z)(\forall y)R^2zy \leftrightarrow (M^2sj \vee (\forall y)(G^1a \wedge \neg M^2yz)))$

4.6.1h $(\forall v)(((\exists u)A^2ug^1e \wedge \neg A^2g^1eg^1e) \leftrightarrow (\exists u)(G^1v \wedge \neg A^2yu))$

4.6.1i $(\forall x)((\exists z)(\neg z = a \rightarrow (\exists z)G^3xze) \leftrightarrow \neg(J^3zbx \wedge K^1s))$

4.6.2 Use the following translation key and the supplied universe of discourse to translate the PL sentences below into PL-English and into idiomatic English wherever possible.

b: Bob
n: Nathan
Sx: *x* is a smoker
Ey: *y* exercises
Ox: *x* is overweight
Hz: *z* is at high risk for heart disease
Dv: *v* is vindictive
Az: *z* is happy
Cu: *u* is a college course
Pxw: *x* does poorly in *w*
Rzx: *z* seeks revenge against *x*
Ixy: *x* is an instructor of *y*
UD: The set of all living people and all college courses

Assume that the predicates *S*, *E*, *O*, *H*, *D*, *A*, and *R* can only apply to people, that *C* only applies to college courses, and that *P* and *I* can only apply to ordered pairs of the form $\langle\alpha, \beta\rangle$ where α is a person and β is a course.

4.6.2a $(Sb \wedge \neg Eb) \wedge Ob$
4.6.2b $(\neg Sn \wedge \neg On) \wedge Dn$

4.6.2c*	$(\exists y)((\neg Hy \wedge Sy) \wedge Ey)$
4.6.2d	$(\forall x)(((Sx \wedge \neg Ex) \wedge Ox) \rightarrow Hx)$
4.6.2e	$((\exists x)(Cx \wedge Pbx) \vee Hb) \rightarrow An$
4.6.2f*	$(\forall v)((Cv \wedge Pnv) \rightarrow (\forall y)(Iyv \rightarrow Rny))$
4.6.2g	$(\forall x)(\forall w)(((Cx \wedge Pwx) \wedge Dw) \rightarrow (\exists z)Rwz)$
4.6.2h	$(\exists x)(\exists y)(\exists z)(((Cx \wedge Iyx) \wedge Pzx) \rightarrow (Dz \leftrightarrow Rzy))$
4.6.2i	$An \rightarrow (\exists v)(\exists w)((Cw \wedge Pvw) \vee Hv)$
4.6.2j*	$(\forall x)(((Sx \wedge \neg Ex) \wedge Ox) \rightarrow (\neg(\exists y)Hy \rightarrow Ax))$

4.6.3 Provide appropriate translation keys and translate the following sentences into PL. Do not specify any universe of discourse.

4.6.3a* If anyone can do this problem, Sarah can.
4.6.3b Robert hates all those who do not hate themselves.
4.6.3c Seniors date only juniors.
4.6.3d* Johanna and William are siblings who have different mothers.
4.6.3e Every man who believes that he is the King of China is either mad or is the King of China.
4.6.3f* Every woman who adopts a child loves him or her unconditionally and deserves our respect.
4.6.3g* All people are impressed by those who say what they want to hear, not what they ought to hear.
4.6.3h If you stand on the shoulders of giants, you will see great things.
4.6.3i* If you believe that your children are capable of achieving great things and you tell them this, then either they will, indeed, achieve great things or they will share your illusion.
4.6.3j If you deny the existence of the external world, then either you live inside nothing or you live inside your mind.
4.6.3k* All people are created equal and are endowed with certain natural rights.
4.6.3l Every person who is aware of her humanity and is not subject to the oppression of others or of needs will never allow someone to treat her as merely a means to an end.
4.6.3m* Each one of us desires happiness; except some people are happy only if they make others unhappy, and some are happy only if they make others happy.
4.6.3n* Someone entered the room and searched for the keys; she wasn't my sister, Barbara.
4.6.3o Someone, whose name I don't know, went to my instructor and complained to her about me.
4.6.3p I met someone who went to a huge university; she didn't like it at all.
4.6.3q All land animals evolved from marine animals.
4.6.3r* All land animals evolved from a single marine animal.
4.6.3s Either birds evolved from reptiles or they both evolved from a common ancestor.

4.6.4* Construct two PL interpretations of the set below such that they are substantively different from each other, each one of them is a model of the set, neither assigns the empty extension to any predicate, and one is finite and the other is infinite.

$Hm \wedge Sb$
$(\forall x)(Gx \rightarrow Hx)$
$(\forall x)(Rx \rightarrow Gx)$
$\neg(\exists z)(Sz \wedge Rz)$

4.6.5 Give three PL interpretations of the following set of PL sentences such that the first satisfies the set and is of size 3, the second also satisfies the set but it is of an infinite size, and the third is of size 4 and satisfies only the first four sentences. Is the last sentence a logical consequence of the other sentences?

$(\forall x)((x = a \vee x = b) \rightarrow Lx)$
$a \neq b$
Pe
$Oea \wedge Oeb$
$(\forall z)((Pz \wedge (Oza \wedge Ozb)) \rightarrow z = e)$

4.6.6 Demonstrate that '$s = t$' is not a logical consequence of the set Γ whose members are the following PL sentences.

$(\forall x)(Ax \rightarrow \neg Gx)$
$(\exists x)Ax \leftrightarrow Gs$
$(\exists w)\neg Gw \rightarrow Kt$
$(\exists x)(Gx \wedge Kx)$

4.6.7* Let Γ be the set consisting of the following three PL sentences.

$(\forall x)\neg Sox$
$(\forall x)(\exists u)((u \neq x \wedge Sux) \wedge (\forall z)(Szx \rightarrow z = u))$
$(\forall x)(x \neq o \rightarrow (\exists u)Sxu)$

4.6.7a Construct a model of Γ that is of size of 3.
4.6.7b Construct a model of Γ that is of size $\aleph_0$.
4.6.7c Is it possible for a PL interpretation of size 1 or 2 to satisfy Γ? Explain.

4.6.8 Show that the following sets of PL sentences are consistent.

4.6.8a* $(\forall y)(Py \rightarrow (\exists x)(\exists z)Bxyz)$
$(\forall x)(\forall y)(\forall z)((Bxyz \wedge Px) \rightarrow (Py \wedge Pz))$
$\neg Pn$
$(\exists x)(\exists z)Bxnz$

4.6.8b $(\neg Ha \vee \neg Kb) \to a = gb$
$a \neq gb$
$(\exists x)(a = gx \wedge x = gb)$
$(\forall x)(\forall y)(\forall z)((x = gy \wedge a = gz) \to (Sxz \wedge Syb))$
$(\forall u)(Sau \to Hu)$
$(\forall z)(Sbz \to Kz)$

4.6.8c $(\forall x)Gox$
$(\forall x)(x \neq o \to ((\exists y)Gxy \wedge (\exists v)Lxv))$
$(\forall x)(\forall y)(\forall z)((Gxy \wedge Lzy) \to x \neq z)$
$(\forall x)(\forall y)((\exists z)(Gxz \wedge Lyz) \to Gxy)$

4.6.9 Establish that each of the sets below is consistent. Explain why all the models of these sets are infinite.

4.6.9a* $(\forall x)(\forall y)(\forall z)((Kxy \wedge Kyz) \to Kxz)$
$(\forall x)(\exists y)Kxy$
$(\forall x)(\forall y)(Kxy \to \neg Kyx)$[1]

4.6.9b $(\forall x)(\forall y)(\forall z)((Gxz \wedge Gyz) \to x = y)$
$(\forall x)(\exists y)Gxy$
$(\forall x)\neg Gxe$

4.6.10 Construct appropriate PL interpretations to establish that the PL sentences in each of the pairs below are not logically equivalent.

4.6.10a* $(\exists v)(Hv \wedge Fv)$ and $(\exists v)Hv \wedge (\exists v)Fv$
4.6.10b $(\forall u)(Nu \to Ca)$ and $(\forall u)Nu \to Ca$
4.6.10c $(\forall w)(Dw \vee Sw)$ and $(\forall w)Dw \vee (\forall w)Sw$

4.6.11 Show that the PL sentences below are contingent.

4.6.11a* $(Em \wedge Pn) \to (\exists y)(Ey \wedge Py)$
4.6.11b $(\forall x)(\exists y)Bxy \wedge \neg(\exists y)(\forall x)Bxy$

---

1 This sentence could be replaced with a simpler one: $(\forall x)\neg Kxx$. The resulting set is also consistent and all its models are infinite.

Solutions to the Starred Exercises

SOLUTIONS TO 4.6.1

4.6.1a

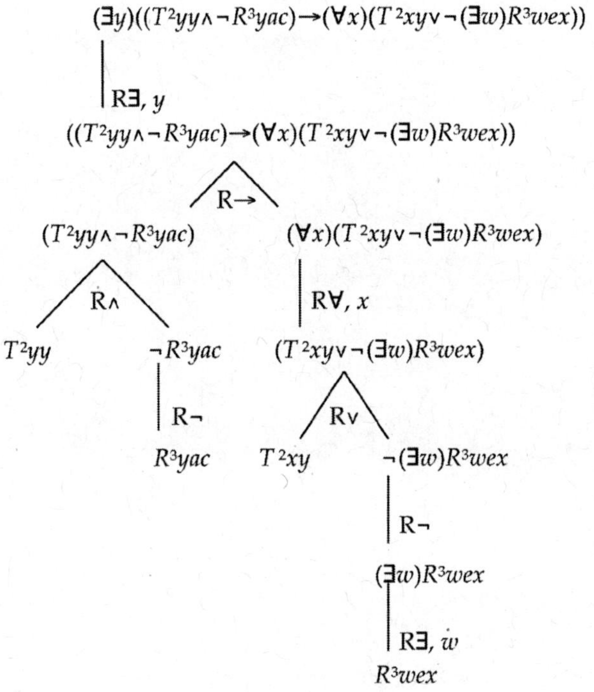

Since the tree terminates with atomic formulas, this PL expression has a complete construction tree. Hence it is a PL formula. Every variable that occurs in an atomic formula is quantified over; therefore, this formula is a PL sentence.

4.6.1b

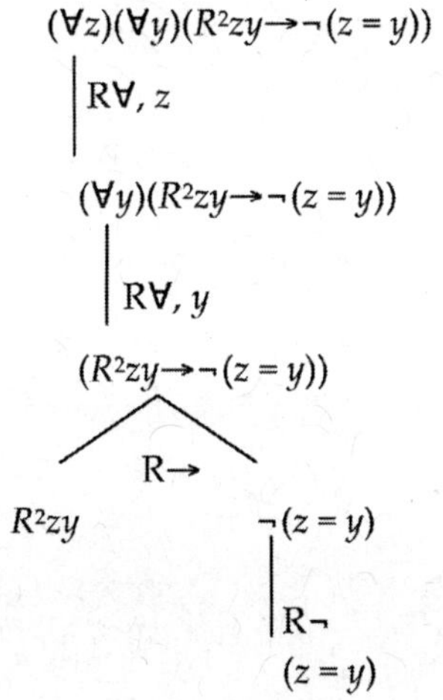

The expression '(z = y)' is not an atomic formula, and it cannot be decomposed. None of the formation rules of PL can generate this expression. Atomic formulas that are formed by the use of the identity predicate do not have outermost parentheses. The original expression does not have a complete expression tree; and hence it is not a PL formula.

4.6.1g

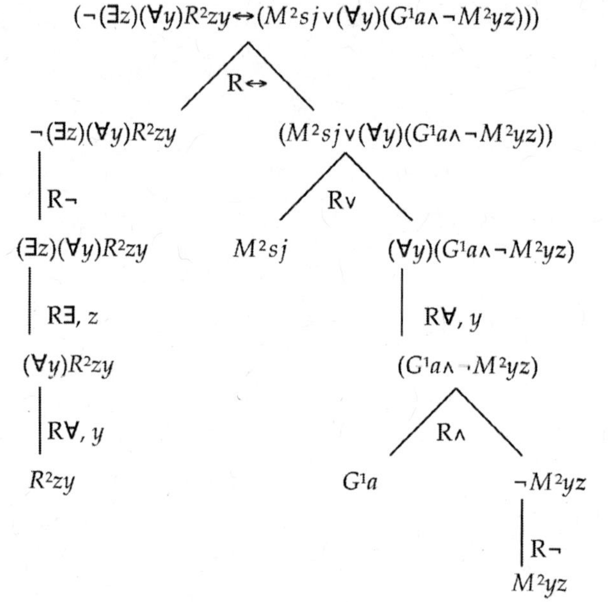

This is a complete construction tree because all branches terminate with atomic formulas. All the variables in the atomic formulas, except 'z' in 'M^2yz', are quantified over. The last occurrence of 'z', however, is free since it receives no quantification. Hence this PL expression is a *proper* formula, i.e., it is a PL formula but not a sentence.

SOLUTIONS TO 4.6.2

4.6.2c

PL-English translation of (∃y)((¬Hy∧Sy)∧Ey): *There is y, such that y is not at high risk for heart disease, y smokes, and y exercises.*
English translation: *There is someone who exercises and who is not at high risk for heart disease, but he is a smoker.*

Note: We used 'but' instead of 'and' because the sentence sounds more natural with 'but'.

4.6.2f

PL-English translation of (∀v)((Cv∧Pnv)→(∀y)(Iyv→Rny)): *For every v, if v is a college course and Nathan does poorly in v, then he seeks revenge against all the instructors of v.*
English translation: *If Nathan does poorly in a college course, then he seeks revenge against all the instructors of that course.*

4.6.2j

PL-English translation of $(\forall x)(((Sx \wedge \neg Ex) \wedge Ox) \rightarrow (\neg(\exists y)Hy \rightarrow Ax))$: *For every x, if x smokes, does not exercise, and is overweight, then x is happy if there is no y such that y is at high risk for heart disease.*
English translation: *Every overweight smoker who doesn't exercise is happy if no one is at high risk for heart disease.*

Note: It might seem that our English translation of this PL sentence is not faithful, for our translation could be interpreted as follows: *If no one is at high risk for heart disease, then every overweight smoker who doesn't exercise is happy.* The PL translation of the last sentence is $\neg(\exists y)Hy \rightarrow (\forall x)(((Sx \wedge \neg Ex) \wedge Ox) \rightarrow Ax)$, which is surely different from the original PL sentence, $(\forall x)(((Sx \wedge \neg Ex) \wedge Ox) \rightarrow (\neg(\exists y)Hy \rightarrow Ax))$. In fact, our translation is correct. Both English readings convey the same meaning. As evidence for this, we note that the two PL translations are logically equivalent. It is not hard to show that they are logically equivalent using the methods of Chapter Five, but here we can give a semantical proof. To simplify matters, we replace the conjunction $((Sx \wedge \neg Ex) \wedge Ox)$ with one metalinguistic variable, $\mathbf{G}x$, and the sentence $\neg(\exists y)Hy$ with one metalinguistic variable, $\mathbf{Z}$. These simplifications do not affect our conclusion since these formulas occur, with all their components intact, in both sentences. So our goal is to show that the following two PL sentences are logically equivalent: (S1) $(\forall x)(\mathbf{G}x \rightarrow (\mathbf{Z} \rightarrow Ax))$ and (S2) $\mathbf{Z} \rightarrow (\forall x)(\mathbf{G}x \rightarrow Ax)$.

Proof that S1 and S2 are logically equivalent

1. Reductio Assumption: there is a PL interpretation I on which $(\forall x)(\mathbf{G}x \rightarrow (\mathbf{Z} \rightarrow Ax))$ and $\mathbf{Z} \rightarrow (\forall x)(\mathbf{G}x \rightarrow Ax)$ have different truth values.
2. Case One: assume that
    - 2.1. $(\forall x)(\mathbf{G}x \rightarrow (\mathbf{Z} \rightarrow Ax))$ is true on I, and
    - 2.2. $\mathbf{Z} \rightarrow (\forall x)(\mathbf{G}x \rightarrow Ax)$ is false on I.
3. From 2.2 by the truth conditions of the conditional: $\mathbf{Z}$ is true and $(\forall x)(\mathbf{G}x \rightarrow Ax)$ is false on I.
4. From 3 by the truth conditions of the universal quantifier: there is a name $\mathbf{s}$ in LN such that the substitutional instance $\mathbf{Gs} \rightarrow A\mathbf{s}$ is false on I.
5. From 4 by the truth conditions of the conditional: $\mathbf{Gs}$ is true and $A\mathbf{s}$ is false on I.
6. From 2.1 by the truth conditions of the universal quantifier: all basic substitutional instances of $(\forall x)(\mathbf{G}x \rightarrow (\mathbf{Z} \rightarrow Ax))$ are true on I; in particular, the sentence $\mathbf{Gs} \rightarrow (\mathbf{Z} \rightarrow A\mathbf{s})$ is true on I.
7. From 5 and 6 by the truth conditions of the conditional: $\mathbf{Z} \rightarrow A\mathbf{s}$ is true on I.
8. From 3 and 7 by the truth conditions of the conditional: $A\mathbf{s}$ is true on I, which contradicts 5.
9. From 2 through 8: Case One leads to a contradiction; therefore, the PL interpretation I cannot make $(\forall x)(\mathbf{G}x \rightarrow (\mathbf{Z} \rightarrow Ax))$ true and $\mathbf{Z} \rightarrow (\forall x)(\mathbf{G}x \rightarrow Ax)$ false.
10. Case Two: assume that
    - 2.1. $(\forall x)(\mathbf{G}x \rightarrow (\mathbf{Z} \rightarrow Ax))$ is false on I, and
    - 2.2. $\mathbf{Z} \rightarrow (\forall x)(\mathbf{G}x \rightarrow Ax)$ is true on I.
11. From 2.1 by the truth conditions of the universal quantifier: There is a name $\mathbf{s}$ in LN, such that the substitutional instance $\mathbf{Gs} \rightarrow (\mathbf{Z} \rightarrow A\mathbf{s})$ is false on I.

12. From 11 by the truth conditions of the conditional: **Gs** is true and **Z**→*A***s** is false on I.
13. From 12 by the truth conditions of the conditional: **Z** is true and *A***s** is false on I.
14. From 2.2 and 13 by the truth conditions of the conditional: (∀*x*)(**G***x*→*Ax*) is true on I.
15. From 14 by the truth conditions of the universal quantifier: all basic substitutional instances of (∀*x*)(**G***x*→*Ax*) are true on I; in particular, the sentence **Gs**→*A***s** is true on I.
16. From 12 and 15 by the truth conditions of the conditional: *A***s** is true on I, which contradicts 13.
17. From 10 through 16: Case Two also leads to a contradiction; therefore, the PL interpretation I cannot make (∀*x*)(**G***x*→(**Z**→*Ax*)) false and **Z**→ (∀*x*)(**G***x*→*Ax*) true.
18. From 1, 9, and 17: there is no PL interpretation that assigns different truth values to S1 and S2.
19. From 18 by the definition of logical equivalence: S1 and S2 are logically equivalent.

Solutions to 4.6.3

4.6.3a

Px: *x* is a person; *Cy*: *y* can do this problem; *s*: Sarah
$(\exists y)(Py \wedge Cy) \rightarrow Cs$

4.6.3d

Szw: *z* and *w* are siblings; *gy*: The mother of *y*; *j*: Johanna; *l*: William
$Sjl \wedge gj \neq gl$

4.6.3f

Wx: *x* is a woman; *Axy*: *x* adopts *y*; *Cz*: *z* is a child; *Lzv*: *z* loves *v* unconditionally; *Dxw*: *x* deserves *w*'s respect; *Px*: *x* is a person
$(\forall x)(\forall y)((Wx \wedge (Cy \wedge Axy)) \rightarrow (Lxy \wedge (\forall z)(Pz \rightarrow Dxz)))$

Note: We treated 'our' in 'deserves our respect' as a universal quantifier ranging over all people. The meaning is that such a woman deserves the respect of all people. 'We' and 'you', when used without specifying who "we" and "you" are, function as universal quantifiers ranging over all people. For example, when one says that we ought to respect people's privacy, one means that every person ought to respect the privacy of others; and when one says that if you work hard, you will attain your goals, one means that anyone who works hard will attain his goals.

4.6.3g

Py: *y* is a person; *Izx*: *z* is impressed by *x*; *Sxv*: *x* says *v*, *Wxy*: *x* wants to hear *y*; *Oxy*: *x* ought to hear *y*
$(\forall z)(\forall x)(((Pz \wedge Px) \wedge ((\exists v)(Wxv \wedge Szv) \wedge (\forall w)(Oxw \rightarrow \neg Szw))) \rightarrow Ixz)$; or
$(\forall z)(\forall x)(((Pz \wedge Px) \wedge ((\forall v)(Wxv \rightarrow Szv) \wedge (\forall w)(Oxw \rightarrow \neg Szw))) \rightarrow Ixz)$

Note: Translating the last clause of the English sentence as $(\forall w)(Oxw \rightarrow \neg Szw)$ is defensible. The English sentence seems to imply that the speaker does not say *anything* that the listener ought to hear. On the other hand, to say that there is *something* the listener ought to hear, yet the speaker does not say, does not rule out that the speaker may say many other things that the listener does not want to hear but he ought to hear. Also, if one were to state the second conjunct in the second translation as $(\forall v)(Szv \rightarrow Wxv)$, then one would imply that the English clause asserts that whatever z says is something x wants to hear. Although this is a possible interpretation of the English clause, it does not seem to be the best interpretation. z may say a lot of things that are irrelevant to x. For instance, z may say all sorts of things to her family and relatives that x does not care about. So it is better to interpret the clause as saying that if the listener wants to hear v, the speaker says v.

4.6.3i

> Px: x is a person; Cxz: x is the child of z; $Bxyz$: x believes that y is capable of achieving z; $Txyz$: x tells y that y will achieve z; Gw: w is a great thing; Azv: z will achieve v; Sxy: x will share y's illusion
> $(\forall w)((Pw \wedge (\forall x)(Cxw \rightarrow (\exists z)(Gz \wedge (Bwxz \wedge Twxz)))) \rightarrow (\forall v)(Cvw \rightarrow ((\exists u)(Gu \wedge Avu) \vee Svw)))$

Note: The translation above represents one possible reading of the English sentence: a person must believe that *all* his children are capable of achieving great things and he must tell this to *all* of them in order for them to achieve great things or share his illusion. There is another possible reading of the English sentence: *only those* children who are believed to be able of achieving great things and who are told about this belief will actually succeed in achieving great things or will share their parent's illusion. The second reading can be translated as follows:

> $(\forall w)(Pw \rightarrow ((\forall x)(Cxw \wedge (\exists z)(Gz \wedge (Bwxz \wedge Twxz))) \rightarrow ((\exists u)(Gu \wedge Axu) \vee Sxw)))$

4.6.3k

> Px: x is a person; Cvw: v and w are created equal; Exy: x is endowed with y; Nx: x is a natural right
> $(\forall v)(\forall w)((Pv \wedge Pw) \rightarrow (Cvw \wedge (\exists x)(Nx \wedge Evx)))$

Note: Representing "All people" by only two PL variables v and w is adequate, since these variables range over the set of *all* individuals.

4.6.3m

> Px: x is a person; Dx: x desires happiness; Hw: w is happy; Mzv: z makes v happy; Uzv: z makes v unhappy
> $(\forall w)(Pw \rightarrow Dw) \wedge ((\exists y)(Py \wedge (Hy \rightarrow (\exists z)(Pz \wedge Uyz))) \wedge (\exists v)(Pv \wedge (Hv \rightarrow (\exists u)(Pu \wedge Mvu))))$

4.6.3n

Pz: *z* is a person; *Ex*: *x* entered the room; *Cv*: *v* searched for the keys; *Sxy*: *x* is the sister of *y*; *i:* I (the speaker); *b*: Barbara
$(\exists x)((Px \wedge (Ex \wedge Cx)) \wedge (x \neq b \wedge (Sbi \wedge (\forall w)(Swi \rightarrow w = b))))$

Note: The comma before the name 'Barbara' indicates that this name is *not restrictive*, that is, it is not used to identify the sister of the speaker. This means that the speaker has only one sister. So if the speaker says "my sister," then he has fully specified the identity of the person—namely, his sister—and the name 'Barbara' is additional information about the person. On the other hand, had the speaker written that the intruder was not his sister Barbara (without a comma), then the name 'Barbara' would be restrictive, which means that it is needed in order to specify the identity of the sister. This would imply that the speaker has more than one sister and he needs to identify her by stating her name. Since the speaker indicates that he has only one sister, who is Barbara, we added the standard *uniqueness clause* $(\forall w)(Swi \rightarrow w = b)$.

4.6.3r

Lx: *x* is a land animal; *Mx*: *x* is a marine animal; *Exy*: *x* evolved from *y*
$(\exists x)((Mx \wedge (\forall w)(Lw \rightarrow Ewx)) \wedge (\forall z)((Mz \wedge (\forall w)(Lw \rightarrow Ewz)) \rightarrow z = x))$

Note: The first conjunct says that there is an *x*, such that *x* is a marine animal and all land animals evolved from *x*; the second clause says that this *x* is unique. This is the standard PL uniqueness clause. In PL-English it says: "if *z* is also a marine animal from which all land animals evolved, then *z* is identical with *x*."

SOLUTIONS TO 4.6.4

The PL Interpretation I

UD: {Amber, Deborah, Rubin, Jolene, Augustine, Javier}
LN: *b, d, r, j, a, i*

Semantical Assignments

I(*b*): Amber; I(*d*): Deborah; I(*r*): Rubin; I(*j*): Jolene; I(*a*): Augustine; I(*i*): Javier
I(*Hx*): "*x* is a physics major": {Amber, Rubin, Jolene, Augustine}
I(*Sy*): "*y* plays basketball": {Amber, Javier, Deborah}}
I(*Gy*): "*y* is interested in astronomy": {Rubin, Jolene, Augustine}
I(*Rz*): "*z* owns a telescope": {Jolene}

Natural reading of $(Hm \wedge Sb)$ on I: *Amber is a physics major and plays basketball.*
True on I.

Natural reading of $(\forall x)(Gx \rightarrow Hx)$ on I: *All those who are interested in astronomy are physics majors.*
True on I.

Natural reading of $(\forall x)(Rx \rightarrow Gx)$ on I: *Everyone who owns a telescope is interested in astronomy.*
True on I.

Natural reading of $\neg(\exists z)(Sz \wedge Rz)$ on I: *No one who plays basketball owns a telescope.*
True on I.

Since UD consists of six individuals, I is finite. Also, I is a model of the set.

The PL Interpretation J

UD: The set of all even natural numbers: {0, 2, 4, 6, …}
LN: $m, b, c_0, c_1, c_2, c_3, \ldots, c_k, \ldots$

Semantical Assignments

I(m): 18; I(b): 12; I(c_0): 0; I(c_1): 2; I(c_2): 4; I(c_3): 6; …; I(c_k): 2k; …
I(Hx): "x is a multiple of 6": {6, 12, 18, 24, 30, …}
I(Sy): "y is a divisor of 60": {2, 4, 6, 10, 12, 20, 30, 60}
I(Gy): "y is a multiple of 12": {12, 24, 36, 48, 60, …}
I(Rz): "z is a multiple of 24": {24, 48, 72, 96, 120, …}

Natural reading of $(Hm \wedge Sb)$ on J: *18 is a multiple of 6 and 12 is a divisor of 60.*
True on J.

Natural reading of $(\forall x)(Gx \rightarrow Hx)$ on J: *All multiples of 12 are multiples of 6.*
True on J.

Natural reading of $(\forall x)(Rx \rightarrow Gx)$ on J: *All multiples of 24 are multiples of 12.*
True on J.

Natural reading of $\neg(\exists z)(Sz \wedge Rz)$ on J: *No divisor of 60 is a multiple of 24.*
True on J.

Since UD consists of infinitely many numbers, J is infinite; and J is a model of the set.

SOLUTIONS TO 4.6.7

4.6.7a Here is a PL interpretation I of size 3 that satisfies the set Γ.

UD: {Sarah, John, Robert}
LN: o, j, r

Semantical Assignments

I(o) = Sarah; I(j) = John; I(r) = Robert
I(Sxy) = x is a half-brother of y

Facts of I

1. John and Robert are half-brothers (they share the same father but have different mothers).
2. John is a half-brother of Sarah (they share the some mother but have different fathers).
3. Robert is not a brother of Sarah (they have different parents).

The natural reading of $(\forall x)\neg Sox$ on I: *Sara is not a half-brother of anyone.*
True on I.

The natural reading of $(\forall x)(\exists u)((u \neq x \wedge Sux) \wedge (\forall z)(Szx \rightarrow z = u))$ on I: *Every person has exactly one half-brother, who is different from him or her.*
True on I.

The natural reading of $(\forall x)(x \neq o \rightarrow (\exists u)Sxu)$ on I: *Everyone who is not Sarah is a half-brother of someone.*
True on I.

4.6.7b The following PL interpretation J satisfies the set Γ and is of size $\aleph_0$.

UD: $\mathbb{N}$: {0, 1, 2, 3, …}
LN: $o, a_1, a_2, a_3, \ldots, a_k, \ldots$

Semantical Assignments

J(o): 0; J(a_k): k, for every k = 1, 2, 3, …
J(Sxy): "x is the successor of y" (i.e., $x = y+1$)

The natural reading of $(\forall x)\neg Sox$ on J: *0 is not the successor of any natural number.*
True on J.

The natural reading of $(\forall x)(\exists u)((u \neq x \wedge Sux) \wedge (\forall z)(Szx \rightarrow z = u))$ on J: *Every natural number has exactly one successor, which is different from it.*
True on J.

The natural reading of $(\forall x)(x \neq o \rightarrow (\exists u)Sxu)$ on J: *Every natural number that is not 0 is the successor of some natural number.*
True on J.

4.6.7c Let K be any PL interpretation of **size 1** that is relevant to the set Γ. This means that there is only one individual α in UD. Since every name has a referent and o is a name in LN, α must be named o. On K, the sentence $(\forall x)(\exists u)((u \neq x \wedge Sux) \wedge (\forall z)(Szx \rightarrow z = u))$ is false since the following basic substitutional instance is false: $(\exists u)((u \neq o \wedge Suo) \wedge (\forall z)(Szo \rightarrow z = u))$. The latter is false on K because it implies that there is an individual *other than* α that bears the relation K(S) to α, and of course there is no such individual. Therefore no interpretation of size 1 can satisfy the sentence $(\forall x)(\exists u)((u \neq x \wedge Sux) \wedge (\forall z)(Szx \rightarrow z = u))$, and hence no interpretation of size 1 can satisfy the set Γ.

Now let H be any PL interpretation of **size 2** that is relevant to the set Γ. The UD of H consists of only two individuals α and β. Since o is a name in LN, one of these individuals must be named o; say α is named o. β must also have a name; let us call it e (the actual name is irrelevant as long as it is not o). Suppose that H satisfies the sentence $(\forall x)(\exists u)((u \neq x \wedge Sux) \wedge (\forall z)(Szx \rightarrow z = u))$. Thus all the basic substitutional instances of this sentence are true on H. There are at least two basic substitutional instances: $(\exists u)((u \neq o \wedge Suo) \wedge (\forall z)(Szo \rightarrow z = u))$ and $(\exists u)((u \neq e \wedge Sue) \wedge (\forall z)(Sze \rightarrow z = u))$. The truth of these instances entails that there is an individual other than α that bears the relation H(S) to α and that there is an individual other than β that bears the relation H(S) to β. Given that we only have the two individuals α and β, it follows that β bears the relation H(S) to α and that α bears the relation H(S) to β. In other words, the ordered pairs $\langle\beta, \alpha\rangle$ and $\langle\alpha, \beta\rangle$ belong to the extension H(S). By the truth conditions of the atomic sentences, the sentences *Seo* and *Soe* must be true on H. Now assume that H also satisfies the sentence $(\forall x)\neg Sox$. This implies that the substitutional instance $\neg Soe$ is true on H. Therefore, both *Soe* and $\neg Soe$ are true on H, which is impossible. It follows that if H makes the sentence $(\forall x)(\exists u)((u \neq x \wedge Sux) \wedge (\forall z)(Szx \rightarrow z = u))$ true, it cannot make $(\forall x)\neg Sox$ true as well. In other words, H cannot satisfy the set Γ. This establishes that no PL interpretation of size 2 can satisfy Γ.

SOLUTION TO 4.6.8

4.6.8a

The PL Interpretation I

UD: $\{-1, 0, 1, 2, 3, \ldots\}$
LN: $a, n, c_1, c_2, c_3, \ldots, c_k, \ldots$

Semantical Assignments

I(a): –1; I(n): 0; I(c_k): k, for every k = 1, 2, 3, …
I(Pz): "z is positive"
I($Bxyz$): "y is between x and z" (i.e., $x < y < z$)

Natural Reading $(\forall y)(Py \rightarrow (\exists x)(\exists z)Bxyz)$ on I: *Every positive number is between two numbers.*
True on I.

PL-English reading of $(\forall x)(\forall y)(\forall z)((Bxyz \wedge Px) \rightarrow (Py \wedge Pz))$ on I: *For all numbers x, y, and z, if y is between x and z and x is positive, then y and z are positive as well.*
True on I.

Natural reading of $\neg Pn$ on I: *Zero is not positive.*
True on I.

Natural reading of $(\exists x)(\exists z)Bxnz$ on I: *Zero is between two numbers.*
True on I.

Since all the sentences are true on I, the set is consistent.

SOLUTION TO 4.6.9

4.6.9a We will prove that this set is consistent by describing a PL interpretation J on which every member of the set is true.

> UD: The set that consists of infinitely many planets that orbit a black hole. These planets revolve around the black hole in circular orbits such that no two planets share the same orbit, and for any two planets one of them must be inside the orbit of the other. The picture below depicts the relations between these heavenly bodies.[1] (We are assuming, of course, that no planet is inside its own orbit.)

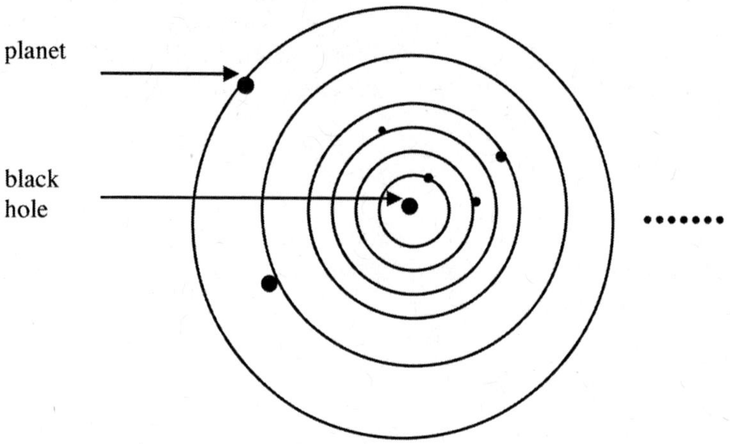

> LN: $p_1, p_2, p_3, \ldots, p_k, \ldots$

1 It might be objected that this PL interpretation is not physically possible since no black hole has a gravitational field that is so powerful as to keep infinitely many planets revolving in orbits around it. Besides, it is very likely that the universe contains only finitely many planets. However, physical possibility is not the same as logical possibility. There is no reason to suppose that logical possibilities must be constrained by the laws of physics or by physical facts. The PL interpretation described above might not be physically possible but it appears to be perfectly consistent.

Semantical Assignments

J(p_1): The first planet from the black hole; J(p_2): The second planet from the black hole; J(p_3): The third planet from the black hole; I(p_k): The kth planet from the black hole
I(Kxy): "x is inside the orbit of y"

Natural reading of $(\forall x)(\forall y)(\forall z)((Kxy \wedge Kyz) \rightarrow Kxz)$ on J:
For any three planets, if the first is inside the orbit of the second and the second is inside the orbit of the third, then the first is inside the orbit of the third.
PL-English: *For all planets x, y, and z, if x is inside the orbit of y and y is inside the orbit of z, then x is inside the orbit of z.*
True on J.

Natural reading of $(\forall x)(\exists y)Kxy$ on J: *Every planet is inside the orbit of some planet.*
True on J.

Natural reading of $(\forall x)(\forall y)(Kxy \rightarrow \neg Kyx)$ on J:
If a planet is inside the orbit of another planet, then the latter is not inside the orbit of the former.
PL-English: *For all planets x and y, if x is inside the orbit of y, then y is not inside the orbit of x.*
True on J.

Let Φ be the set of the three PL sentences above. A very interesting fact about Φ is that every model of it is infinite. The set Φ consists of three sentences that involve a 2-place predicate K. Any PL interpretation that is relevant to Φ must assign a binary (i.e., 2-place) relation to K. Let M be any arbitrary PL interpretation that is a *model* of Φ. Thus M(K) is a binary relation on the UD of M. To simplify our notation, let us call this relation R. Since PL is an extensional system, we can think of M(K) as an extension whose members are ordered pairs. Intuitively speaking, an ordered pair $\langle \alpha, \beta \rangle$, where α and β are members of UD, belongs to the extension M(K) if and only if α bears the relation R to β. Rather than saying that $\langle \alpha, \beta \rangle$ belongs to the extension M(K), we will simply write "α Rβ." We are really talking about the same thing in two different ways. R is the binary relation that M assigns to the 2-place predicate K, that is, R is M(K), and since PL is an extensional system, R (and hence M(K)) is identified with its extension on M. However, in order to stay close to common usage, we chose to speak of R as the binary relation whose extension on M is M(K).

At this stage, we do not know whether the extension M(K) is empty or not. What we know is that M is a model of Φ and that R is the binary relation that M assigns to the 2-place predicate K. Our goal is to show that the extension of R on M (i.e., M(K)) is an infinite set. This immediately entails that the UD of M is infinite; and hence M is an infinite PL interpretation. Once we accomplish this, we will be in a position to discuss a significant philosophical implication. We end this solution with a digression into common properties of relations.

Since M is a model of Φ, the sentences in Φ are all true on M. The truth of these sentences entails that R has certain properties. To be precise, there are three properties R has, each of which is entailed by one of the sentences in Φ. We discuss each one in turn.

Transitivity: For all α, β, and δ in UD, if α R β and β R δ, then α R δ. (R is said to be *transitive.*)
Extendability: For every α in UD, there is β in UD, such that α R β. (R is said to be *extendable.*)
Asymmetry: For all α and β in UD, if α R β, then not-(β R α) (i.e., "it is not the case that β R α").[1] (R is said to be *asymmetric.*)

Transitivity follows from the truth of the sentence $(\forall x)(\forall y)(\forall z)((Kxy \land Kyz) \rightarrow Kxz)$ on M, extendability from the truth of $(\forall x)(\exists y)Kxy$ on M, and asymmetry from the truth of $(\forall x)(\forall y)(Kxy \rightarrow \neg Kyx)$ on M. So we conclude that since M is a model of the set Φ, the relation that M assigns to Φ—namely R—is transitive, extendable, and asymmetric.

We will now prove that R has an infinite extension. We begin by making a helpful observation. We can describe an asymmetric relation graphically by saying that such a relation does not allow "loops." To see this, we diagram α R β as two dots—one is α and the other is β—that are connected by an arrow, representing R and its direction from α to β. If we want, instead, to represent β R α, we draw the arrow from β to α. If we have α R β and β R α, then we obtain two arrows, one from α to β and the other from β to α, forming a loop. To say that R is asymmetric is to say that there are never such loops. Of course, there is also no loop between an individual and itself if R is asymmetric; otherwise, we would have α R α and not-(α R α), which is impossible. Here are diagrams of a relation Z that exhibits two loops: δ Z δ; and α Z β and β Z α.

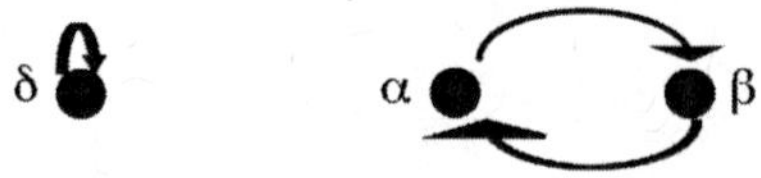

δ Z δ α Z β and β Z α

As indicated above, an asymmetric relation cannot have any such loops. So it is clear that Z is not asymmetric.

As is the case with any PL interpretation, the UD of M cannot be empty. So there is at least one individual in UD; we call it ε_1. Since R is extendable, there is an individual in UD such that ε_1 bears the relation R to this individual. This individual cannot be ε_1 itself; otherwise we end up with the loop ε_1 R ε_1. Thus this individual must be something other than ε_1; call it ε_2. So far we have ε_1 R ε_2. Again, since R is extendable, ε_2 must relate to an individual in UD. It cannot relate to itself or to ε_1; for either case generates a loop. Hence it must relate to a third individual ε_3. We now have the following situation: ε_1 R ε_2 and ε_2 R ε_3. To simplify matters, we will write this sequence as "ε_1 R ε_2 R ε_3". Given that R is transitive, ε_1 bears the relation R to ε_3. i.e., ε_1 R ε_3. The diagram below describes the current situation

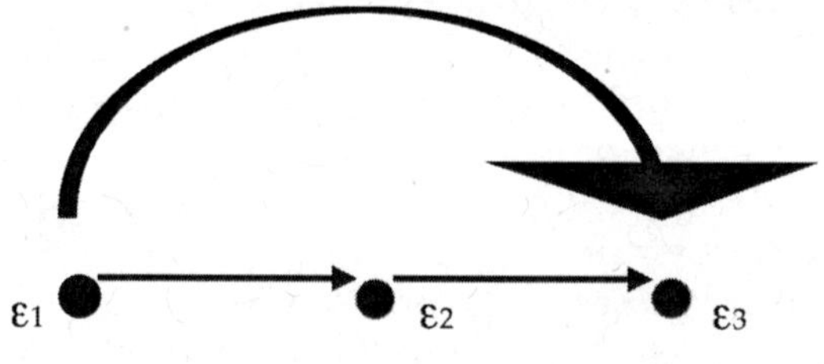

1 We included the clause 'in UD' for emphasis. It is understood that all quantifiers range over the UD.

Since R is extendable, ε_3 must relate to some individual. It cannot relate to itself, ε_2, or ε_1, because any such relation would create a loop and, as we explained above, the asymmetry of R prevents the formation of loops. The only option we have is that ε_3 relates to a fourth individual ε_4. Thus we obtain that $\varepsilon_1 \mathrm{R} \varepsilon_2 \mathrm{R} \varepsilon_3 \mathrm{R} \varepsilon_4$. By the transitivity of R, we have the following additional relations: $\varepsilon_1 \mathrm{R} \varepsilon_3$, $\varepsilon_1 \mathrm{R} \varepsilon_4$, $\varepsilon_2 \mathrm{R} \varepsilon_4$. The diagram below describes all of the relations we have so far.

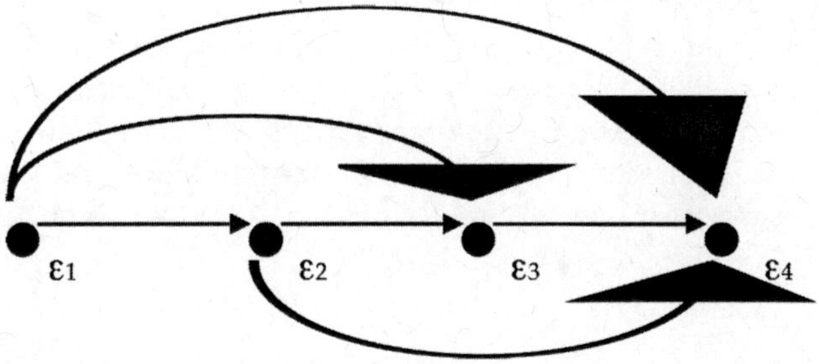

The same reasoning applies to ε_4. The property of extendability forces ε_4 to bear R to some individual. This individual cannot be ε_4 or any of the previous individuals, since each one of these individuals relates to ε_4, and hence if ε_4 bears R to any one of them, a loop would be formed, which is ruled out by the asymmetry of R.

This pattern of reasoning applies to any finite sequence $\varepsilon_1 \mathrm{R} \varepsilon_2 \mathrm{R} \varepsilon_3 \mathrm{R} \varepsilon_4 \mathrm{R} \ldots \mathrm{R} \varepsilon_{n-1} \mathrm{R} \varepsilon_n$. The transitivity of R ensures that every ε_k bears the relation R to ε_n, where k is less than n-1. The extendability of R requires ε_n to bear the relation R to some individual. But this individual cannot be $\varepsilon_1, \varepsilon_2, \varepsilon_3, \varepsilon_4, \ldots,$ or ε_n; for otherwise a loop would form. In other words, there are infinitely many individuals in UD that form an unending sequence "connected" by the binary relation R: $\varepsilon_1 \mathrm{R} \varepsilon_2 \mathrm{R} \varepsilon_3 \mathrm{R} \varepsilon_4 \mathrm{R} \ldots \mathrm{R} \varepsilon_n \mathrm{R} \ldots$. For every $\varepsilon_j \mathrm{R} \varepsilon_k$, the ordered pair $\langle \varepsilon_j, \varepsilon_k \rangle$ belongs to the extension of R on M. The previous reasoning shows that the extension of R on M is infinite. However, R is the relation that M assigns to the 2-place predicate *K*, that is, R is precisely M(*K*). The extensionality of M reduces R to its extension on M. We can say, therefore, that M(*K*) is an infinite set of ordered pairs, each of which is of the form $\langle \varepsilon_j, \varepsilon_k \rangle$, where j is less than k. This allows us to conclude that UD is infinite, and to say that R is an infinite relation (rather than, R is a relation whose extension is infinite). We summarize the conclusions we arrived at: (1) any model of the set Φ is infinite (i.e., it has an infinite universe of discourse); (2) any such PL interpretation assigns to *K* a transitive extendable asymmetric relation (a TEA relation, for short), and (3) every TEA relation is infinite in any PL interpretation.

The concept of a TEA relation gives rise to the concept of the infinite. It is not identical with it. It is incorrect to say that the concept of the infinite is the concept of a transitive extendable asymmetric relation. Branch (b) of this exercise shows that the concept of the infinite can arise from a different type of relation. We have three concepts here: transitivity, extendibility, and asymmetry. As we will see later each of these concepts is finite, in the sense that we can arrive at a full understanding of it using finite resources only. Each one of these concepts describes a certain structure. Again these structures could be exhibited by finitely many elements. Furthermore, any partial combination of these three concepts is finite. It is only when these three concepts are put together that a new structure *emerges* that can only be exhibited by infinitely many elements. The infinite size of a TEA relation is an outcome of the interaction of three structures. In other words, it is an "organizational" outcome of the interaction of three structures. Many philosophers call such new organizational outcomes *emergent properties.* Very roughly speaking, an emergent property is a property of a certain system that is due to the organization of several components none of which has this property and no partial combination of which has this property. This is not meant as a definition of emergence. The concept of "emergent property" is philosophically complex, and philosophers do not agree on a general characterization of it.

In what follows we show that none of the components of a TEA relation is infinite nor is any partial combination of these components. Consider first the concept of a transitive relation. As defined above, a relation R is transitive if and only if for all objects α, β, and δ, if α R β and β R δ, then α R δ. We can give an example of finitely many objects that are related to each other via a transitive relation. Consider four people, John, Sarah, Mary, and William, and the relation "taller than." Assume that they are ordered in terms of their height as follows:

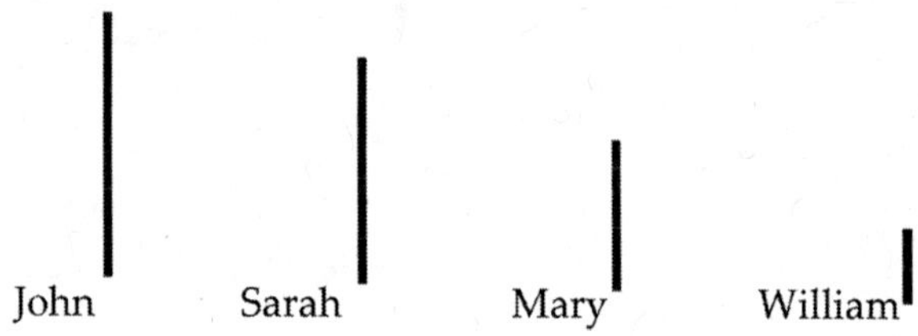

It is obvious from the diagram above that "taller than" is transitive: for any three individuals *a*, *b*, and *c*, if *a* is taller than *b* and *b* is taller than *c*, then *a* is taller *c*. So we can fully understand what a transitive relation is by relying only on finite resources. The concept of a transitive relation, therefore, is a finite concept.

Second, a relation R is asymmetric if and only if for any objects α and β, if α R β, then not-(β R α). Said differently, an asymmetric relation is not "reversible." The relation "taller than" above is an example of an asymmetric relation. The order of any two individuals related by "taller than" is not reversible. Again, we explained asymmetry using only a finite universe of discourse; hence the concept of an asymmetric relation is also a finite concept.

Third, a relation R is extendable if and only if for every object α, there is an object β, such that α R β. In other words, R is extendable if and only if every object bears the relation R to some object. Consider three people, Arthur, Barbara, and Sandy, such that Arthur likes Barbara, Barbara likes Sandy, and Sandy likes Arthur. This is all the information we have; we are not told, for instance, whether Arthur likes Sarah. Consider the relation "likes." It is clear from the preceding example that everyone likes someone. This shows that "likes" in this example is extendable. Our universe of discourse is finite; so the concept of an extendable relation is a finite concept.

In order to claim that a concept *emerges* from the combination of three other concepts, it is imperative that this concept is not located in any of these three concepts *or in any partial combination of them,* that is, in any combination consisting of only two of these concepts. So we must show that if we combine any two of these three concepts, no infinite resources are needed to understand the combined concept. The relation "being taller than" in the example above is an instance of a transitive and asymmetric relation. Since the example consists of finitely many individuals, we can understand this combined concept—namely, the concept of a transitive asymmetric relation—by relying only on finite resources. This combined concept, therefore, is a finite concept.

Consider now the previous example about Arthur, Barbara, and Sandy. Add the following information: Barbara does not like Arthur, Sandy does not like Barbara, Arthur does not like

Sandy, and no one likes himself or herself. We can describe the whole situation by indicating the ordered pairs that have the relation "likes" and the ones that do not. For the sake of simplicity we use *a*, *b*, and *s* to refer to Arthur, Barbara, and Sandy, respectively.

> Here are the ordered pairs that have the relation "likes": $\langle a, b\rangle$, $\langle b, s\rangle$, and $\langle s, a\rangle$.
> And here are the pairs that do *not* have the relation "likes": $\langle b, a\rangle$, $\langle s, b\rangle$, $\langle a, s\rangle$, $\langle a, a\rangle$, $\langle b, d\rangle$, and $\langle s, s\rangle$.

It is obvious that "likes" is extendable: everyone likes someone. And it is obvious that "likes" is asymmetric: no pair is reversible. Thus we managed to understand this combined concept—namely, the concept of an asymmetric extendable relation—by relying on finite resources only. So this combined concept is also finite.

Let us return to our original example about Arthur, Barbara, and Sandy before the modification we introduce in the preceding paragraph. We have "Arthur likes Barbara," "Barbara likes Sandy," and "Sandy likes Arthur." Let us add the following facts to this situation: Arthur likes Sandy, Barbara likes Arthur, Sandy likes Barbara, Arthur likes himself, Barbara likes herself, and Sandy likes herself. If we let *a*, *b*, and *s* stand for Arthur, Barbara, and Sandy, respectively, we can describe the extension of the relation "likes" as follows.

$$\{\langle a, b\rangle, \langle b, s\rangle, \langle s, a\rangle, \langle a, s\rangle, \langle b, a\rangle, \langle s, b\rangle, \langle a, a\rangle, \langle b, b\rangle, \langle s, s\rangle\}$$

In this example, the relation "likes" is clearly extendable—everyone likes someone—but it is also transitive. For instance, Arthur likes Barbara and Barbara likes Sandy, so Arthur likes Sandy; Barbara likes Sandy and Sandy likes Arthur, so Barbara likes Arthur; Arthur likes Barbara and Barbara likes Arthur, so Arthur likes himself; and so on. The combined concept of a transitive extendable relation is, therefore, a finite concept. These are all the possible partial combinations.

The concept of a TEA relation gives rise to the concept of the infinite and it emerges from the combination of three finite concepts. When these three structures, each of which can be fully exhibited by a finite set of elements, are combined together, a new structure emerges, which can only be exhibited by an infinite set of elements. As we have seen, no partial combination of these structures is infinite. So the concept of the infinite cannot be reduced to the component concepts that collectively give rise to it or to any partial combination of them. A TEA relation has the property of being infinite, and this property arises from the "organization" of combining three finite structures. Hence, this property seems to be a genuine case of an emergent property. Another point to note is that no infinite conceptual resources are needed to understand the concept of the infinite. Some philosophers have claimed that the concept of the infinite can only arise from an infinite source. As we have seen this is not true. There are only three concepts involved—namely, transitivity, extendability, and asymmetry—each of which can be applied to a finite structure. Furthermore, the proof that a TEA relation is infinite makes no mention of the concept of the infinite. The proof proceeds by showing that in order for a finite sequence of individuals to be ordered by a TEA relation, the sequence must be "expanded indefinitely," that is, new individuals must constantly be added to the sequence. We *inferred* from this fact that a sequence whose elements are related to each other by a TEA relation must be infinite. We inferred the existence of an *actually infinite* sequence from the existence of an *indefinitely expandable* sequence. In philosophical lingo, we inferred the existence of an *actual infinity* from the existence of a *potential infinity*. The inference is justified since in order to say that this sequence exhibits the structure of a TEA relation, the process of "generating" this sequence must be, in a certain sense, "completed," that is, all the new individuals that need to be added to the sequence have been added. Without this "completion" of the indefinitely expandable sequence, the structure of a TEA relation is not exhibited by the sequence at any stage of its expansion. Only when the expansion is "completed" and all the new individuals are added to the sequence, would the sequence exhibit the structure of a TEA relation. Such a sequence is *actually* infinite.

We conclude the solution to this exercise with a brief digression in which we define some common properties of relations. Let J be any PL interpretation and R any binary relation that J assigns to some 2-place predicate *K* (i.e., R is J(*K*)). Previously we defined the properties of transitivity, asymmetry, and extendibility. Now we define a few more properties, which are quite useful in the study of logic in general. We state first the PL sentence that might be true on J and then describe in PL-English the corresponding property of the relation R. It is important to note that these properties are such that R *might* have *some* of them in J. R cannot have all of them in a single PL interpretation. It is obvious that not all of these properties are independent of each other. For example, a reflexive relation cannot be irreflexive; a symmetric relation cannot be asymmetric; a total relation must be reflexive; a reflexive relation is extendable; an asymmetric relation must be irreflexive; and so on.

PL Sentence: $(\forall x)Kxx$
Reflexivity: For all α in UD, α R α. (R is said to be *reflexive*.)

PL Sentence: $(\forall x)\neg Kxx$
Irreflexivity: For all α in UD, not-(α R α). (R is said to be *irreflexive*.)

PL Sentence: $(\forall x)(\forall y)(Kxy \rightarrow Kyx)$
Symmetry: For all α and β in UD, if α R β, then β R α. (R is said to be *symmetric*.)

PL Sentence: $(\forall x)(\forall y)((Kxy \wedge Kyx) \rightarrow x = y)$
Antisymmetry: For all α and β in UD, if α R β and β R α, then $\alpha = \beta$. (R is said to be *antisymmetric*.)

PL Sentence: $(\forall x)(\forall y)(Kxy \vee Kyx)$
Totality: For all α and β in UD, either α R β or β R α. (R is said to be *total*.)

PL Sentence: $(\forall x)(\forall y)((Kxy \vee Kyx) \vee x = y)$
Connexity: For all α and β in UD, either α R β, β R α, or $\alpha = \beta$. (R is said to be *connex*.)

PL Sentence: $(\forall x)(\forall y)(\forall z)((Kxz \vee Kyz) \rightarrow x = y)$
R is injective: For all α, β, and δ in UD, if α R δ and β R δ, then $\alpha = \beta$.

PL Sentence: $(\exists z)(\forall y)\neg Kyz$
R has a minimal element: There is δ in UD, such that for every β in UD, not-(β R δ). (δ is called "an R-minimal element in UD.")

PL Sentence:	$(\exists z)(\forall y)\neg Kzy$
R has a maximal element:	There is δ in UD, such that for every β in UD, not-(δ R β). (δ is called "an R-maximal element in UD.")

There are many examples of each type of these relations. The relation of monogamous marriage among people is irreflexive (no one is married to himself or herself), symmetric (if *a* is married to *b*, then *b* is married to *a*), injective (no two people are married to the same person), and has minimal as well as maximal elements (there are many people who have no spouses). The relation "less than" among the natural numbers is irreflexive (no number is less than itself), transitive (if *n* is less than *m* and *m* is less than *k*, then *n* is less than *k*), asymmetric (if *n* is less than *m*, then *m* is not less than *n*), connex (any two numbers are either identical or one of them is less than the other), has a minimal element—namely, 0 (no natural number is less than 0), has no maximal element (every number is less than some number), and is not injective (two different numbers could be less than the same number). The relation "has the same biological father as" among people is reflexive (everyone has the same biological father as himself or herself), symmetric (if *a* has the same biological father as *b*, then *b* has the same biological father as *a*), transitive (if *a* and *b* have the same biological father and *b* and *c* also have the same biological father, then *a* and *c* have the same biological father), and has minimal as well as maximal elements (there are people who have no biological siblings). The relation "a divisor of" between positive integers is reflexive (every numbers is a divisor of itself), anti-symmetric (if two numbers are divisors of each other, then they must be the same number), transitive (if *a* is a divisor of *b* and *b* is a divisor of *c*, then *a* is a divisor of *c*), has no minimal element (1 is divisible by 1),[1] has no maximal element (every number is a divisor of some number), and is neither total nor connex (there are numbers none of which is a divisor of the others).

SOLUTION TO 4.6.10

4.6.10a The following PL interpretation I makes the first sentence false and the second true. This proves that these sentences are not logically equivalent.

The PL Interpretation I

UD: {Cow, Pig, Whale, Manatee}
LN: *c*, *p*, *h*, *m*

Semantical Assignments

I(*c*): Cow; I(*p*): Pig; I(*h*): Whale; I(*m*): Manatee
I(*Hx*): "*x* is a farm animal": {Cow, Pig}
I(*Fv*): "*v* is a marine mammal": {Whale, Manatee}

1 Some textbooks define a minimal element δ as an element such that for every β, if β is not identical with δ, then not-(β R δ). According to this definition 1 would be a minimal element of the relation "a divisor of" in the set of all positive integers.

Natural reading of $(\exists v)(Hv \wedge Fv)$ on I: *There is a farm animal that is a marine mammal.*
False on I.

Natural reading of $(\exists v)Hv \wedge (\exists v)Fv$ on I: *There is a farm animal and there is a marine mammal.*
True on I.

SOLUTION TO 4.6.11

4.6.11a Consider the following two PL interpretations.

The PL Interpretation K

UD: {Minnesota, New Mexico, Arizona, Iowa}
LN: *m, n, a, i*

Semantical Assignments

K(*m*): Minnesota; K(*n*): New Mexico; K(*a*): Arizona; K(*i*): Iowa
K(*Ez*): "*z* is a Midwestern state": {Minnesota, Iowa}
K(*Py*): "*y* is a Southwestern state": {New Mexico, Arizona}

Natural reading of $(Em \wedge Pn) \rightarrow (\exists y)(Ey \wedge Py)$ on K:

If Minnesota is a Midwestern state and New Mexico is a Southwestern state, then there is a Midwestern state that is also Southwestern.
The sentence is false on K since its antecedent is true and its consequent is false on K.

PL Interpretation H

UD: {Minnesota, Wisconsin, Iowa, Arizona}
LN: *m, n, i, a*

Semantical Assignments

H(*m*): Minnesota; H(*n*): Wisconsin; H(*i*): Iowa; H(*a*): Arizona
H(*Ez*): "*z* is an agricultural state": {Minnesota, Wisconsin, Iowa}
H(*Py*): "*y* borders one of the Great Lakes": {Minnesota, Wisconsin}

Natural reading of $(Em \wedge Pn) \rightarrow (\exists y)(Ey \wedge Py)$ on H:
If Minnesota is an agricultural state and Wisconsin borders one of the Great Lakes, then there is an agricultural state that borders one of the Great Lakes.
This sentence is true on H since its antecedent and consequent are both true on H.

Given that $(Em \wedge Pn) \rightarrow (\exists y)(Ey \wedge Py)$ is false on one interpretation and true on another, it is contingent.

Chapter Five

Classical Proof Theory

5.1 The Notion of Demonstrative Proof

In Section 4.5.2:C, we distinguished between "universal" and "existential" concepts. In general, a universal concept is a concept whose applicability to any object requires that *all* objects of a certain type meet certain conditions. On the other hand, an existential concept is a concept whose applicability to any object requires the *existence* of an object of a certain type. The denial of a universal concept is an existential concept, and the denial of an existential concept is a universal concept. For instance, the logical concepts of deductive validity, logical truth, logical falsehood, and inconsistency are universal concepts, while the logical concepts of deductive invalidity, contingency, and consistency are existential concepts. In Chapters Two, Three, and Four, we discussed many examples of deductively invalid arguments, of contingent sentences, and of consistent sets of sentences. In all these examples, we established the applicability of these concepts by establishing the existence of TL diagrams, SL truth valuations, and PL interpretations that fulfill certain conditions. We typically established the existence of those diagrams, truth valuations, and interpretations by presenting examples of them: we "constructed" PL interpretations, we "recovered" SL truth valuations, and we "drew" TL diagrams. On the other hand, when we wanted to show that a universal concept is applicable to some set of PL sentences, we reasoned about all PL interpretations of certain types. For instance, to show that some set of PL sentences is inconsistent or that some PL argument is deductively valid, we *demonstrated* that *all* the relevant PL interpretations fulfill certain conditions. Such a **demonstration** is typically a **proof** that proceeds from a certain set of premises and terminates with the desired conclusion. Between the initial and terminal stages, a proof comprises consecutive steps in which intermediate conclusions are inferred according to specific modes of reasoning. We call such proofs **demonstrative proofs** or simply **demonstrations**.[1] **Classical proof theory** is introduced precisely for the purpose of developing the necessary understanding and tools for constructing **formal representations** of demonstrative proofs. These formal representations are referred to as **derivations**. In this Chapter, we will study SL as well as PL derivations.

5.1:C COMMENTARY ON 5.1

5.1:C1 Demonstrations are not used only in logic, but also in many other fields of knowledge, most notably in the mathematical and the theoretical sciences, such as geometry, mathematical economics, and theoretical physics. However, demonstrative proofs are also common

1 It is very common to use the word 'proof' in the sense of demonstrative proof. In this book, we follow this common practice: almost always we refer to demonstrative proofs as simply "proofs."

in philosophy, jurisprudence, and even in everyday reasoning. For instance, when Donasia consults an income-tax table to find how much tax she owes on her last year's personal income, she most likely performs a bit of demonstrative reasoning. Donasia might reason as follows: (1) last year I made an adjusted gross income of $80,000; (2) according to the applicable income-tax table, anyone whose adjusted gross income falls between $45,000 and $82,000 owes 18% of this income in taxes;[1] (3) since my adjusted gross income is more than $45,000 but less than $82,000, I owe 18% of my adjusted gross income in taxes; (4) 18% of $80,000 is $14,400, therefore I owe this amount in taxes. We can present this reasoning as a demonstration carried out in steps proceeding from the premises (1) and (2) and terminating with the conclusion: "I owe $14,400 in taxes." To simplify matters, let 'AGI' stand for 'last year's adjusted gross income'. We will replace 'I' and 'my' with 'Donasia' and 'Donasia's'.

Demonstration 1

0.1 Premise: Donasia's AGI = $80,000.
0.2 Premise: All those whose AGI is greater than $45,000 and less than $82,000 owe 18% of their AGI in taxes.
0.3 Premise (arithmetical fact): $80,000 is greater than $45,000 and less than $82,000.
0.4 Premise (arithmetical fact): 18% of $80,000 = $14,400.
1 From 0.1 and 0.3: Donasia's AGI is greater than $45,000 and less than $82,000.
2 From 0.2 and 1: Donasia owes 18% of her AGI in taxes.
3 From 0.1 and 2: Donasia owes 18% of $80,000 in taxes.
4 From 0.4 and 3: Donasia owes $14,400 in taxes.

Perhaps, no one actually mentally entertains such a detailed demonstration in order to determine how much one owes in taxes. But this is not the point. The point is that Donasia's ordinary reasoning *can be represented* as Demonstration 1. People are semantical creatures. They understand the semantical relations between the various premises, including some implicit premises, and then they "leap" to the appropriate conclusion. If this reasoning is demonstrative, then we can fill in this "leap" with a sequence of "very short leaps," each of which is licensed by a specific rule of inference. We call these "short leaps" **inferential steps**. Almost all inferential steps consist of three components: (1) the **antecedents**, (2) the **conclusion**, and (3) the **inferential license**. The term 'antecedent' here has different meaning from the term 'antecedent' that we used in Chapters Three and Four to refer to the first clause of a conditional, i.e., to the **X** of the conditional **X→Y**. To avoid confusion, we will refer to the antecedents of an inferential step as *the inference's antecedents*; and in order to distinguish the conclusion of an inferential step from the (final) conclusion of the demonstration, we will refer to it as *the inference's conclusion*. The inferential license explains how this conclusion is inferred. It comprises references to the inference's antecedents and the name of the rule of inference that is used to infer this conclusion from those antecedents.

Let us analyze Demonstration 1 by making explicit the inferential steps. First we begin by listing all the premises that are invoked in the demonstration. Premises do not require an-

1 This is a hypothetical example. Actual tax tables are more complicated than what is described in this example.

tecedents or inference rules. They are the "given." We can introduce any of the given premises at any stage of the demonstration. However, it is a common practice *in logic* to collect all the premises needed in a demonstration at the beginning of the sequence of inferential steps. This is only a practice and it is only common in logic. Usually (in practice outside logic) premises are entered into a demonstration "as needed": wherever the proof calls for a certain premise, the premise is introduced at that stage. But in this book, we will follow the common logical practice of listing all the relevant premises at the initial stage. We call this stage the **zero stage** and it consists of as many lines as there are premises.

Lines 0.1–0.4 are the premises of the demonstration. Since we do not "infer" any of these sentences, the zero stage is not really an inferential step: it is a list of the "given." The first inferential step occurs at line 1. Line 1 contains the conclusion of this inferential step, i.e., the inference's conclusion. It is inferred from two antecedents 0.1 and 0.3 by invoking a rule of inference called **Substitution**. The rule allows the inference of a sentence of the form "... *t* ..." from two sentences: (1) $s = t$ and (2) "... *s* ...", where *s* and *t* are any singular terms. In Demonstration 1, the first inference's antecedent, 0.1, asserts that Donasia's AGI = $80,000, and the second inference's antecedent, 0.3, makes an arithmetical claim about the number $80,000. The rule Substitution allows us to substitute 'Donasia's AGI' for the numeral '$80,000' in 0.3. These two antecedents together with the rule Substitution license the inference of the conclusion listed on line 1. If we were to specify the inferential license of this step, we would write "From 0.1 and 0.3 by the rule Substitution" or more concisely "0.1, 0.3, Sub." This inferential step gives rise to a simple deductively valid argument:

0.1 Donasia's AGI = $80,000.
0.3 $80,000 is greater than $45,000 and less than $82,000.
————————————————————: Substitution
1 Donasia's AGI is greater than 45,000 and less than $82,000.

As usual the line represents the logical link between the premises and the conclusion; in this case this link is the rule Substitution. Since this argument is deductively valid, we can assert that in every logical possibility in which the premises are true the conclusion is true as well. In other words, the rule Substitution is **truth-preserving**. A conclusion that is inferred by means of this rule must be true *if* its antecedents are true. The rule Substitution, therefore, if applied to true antecedents, *preserves* the truth of the antecedents by generating a true conclusion. It is not hard to see why Substitution is truth-preserving: if two things are identical (i.e., they are really one thing), then whatever is true of one of them is true of the other.[1] For

1 Many philosophers believe that Substitution is not applicable to what is termed "intensional contexts." These are contexts in which propositional attitudes, such as belief and desire, or modalities, such as necessity and possibility, are invoked. The classic example for propositional-attitude context is the sentence 'Oedipus wants to marry Jocasta'. But Jocasta = Oedipus's mother; by Substitution we have 'Oedipus wants to marry his mother'. It seems clear that the former sentence is true while the latter sentence is false. The standard example for modal context is the sentence 'It is necessary that the number 8 is even'. Since 8 = the number of the solar planets, by Substitutions we have 'It is necessary that the number of the solar planets is even'. Again it seems clear that the former sentence is true while the latter sentence is false. If these judgments are correct, then Substitution is not applicable to such contexts. It is worth noting that there are philosophers who dispute these judgments.

instance, since George Washington is identical with the first President of the United States (they are the same person) and since George Washington was married to Martha, it follows that the first President of the United States was married to Martha.

The second inferential step occurs on line 2. Here we have two inference's antecedents 0.2 and 1. The second antecedent is the conclusion of the first inferential step. This is perfectly all right. With some qualification, which we will note later when we discuss "blocks," anything we infer in a demonstration can be used as an antecedent for a later inference. The first inference's antecedent, 0.2, asserts that all those whose AGI is greater than \$45,000 and less than \$82,000 owe 18% of their AGI in taxes. The second inference's antecedent implies that Donasia is one of those people whose AGI is greater than \$45,000 and less than \$82,000. Therefore the consequent of the conditional in 0.2 applies to her: Donasia owes 18% of her AGI in taxes. This inferential step also constitutes a short valid argument:

0.2 All those whose AGI is greater than \$45,000 and less than \$82,000 owe 18% of their AGI in taxes.
1 Donasia's AGI is greater than 45,000 and less than \$82,000.

---

2 Donasia owes 18% of her AGI in taxes.

There is no traditional single name for this inference rule.[1] We will see later that this inferential step can be decomposed into two inferential steps, each of which invokes a single rule of inference. However, the truth-preserving nature of this inference is intuitively obvious. There is no compelling reason why we do not have a single rule that licenses this inference. So let us invent a temporary rule: we call this rule of inference "Conditional Instantiation." It allows the inference of a sentence of the form "*t* is *B*" from two antecedents of the forms "All *A*s are *B*s" and "*t* is *A*," where '*t*' is a singular term and '*A*' and '*B*' are general terms. For this specific argument, we may take '*A*', '*B*', and '*t*' as representing, respectively, 'Those whose AGI is greater than \$45,000 and less than \$82,000', 'Those who owe 18% of their AGI in taxes', and 'Donasia'.

Each of the conclusions of the third and fourth inferential steps is inferred by the rule Substitution. The conclusion 'Donasia owes 18% of \$80,000 in taxes' is obtained by substituting '\$80,000' for 'her AGI' in the conclusion of the second inferential step—namely 'Donasia owes 18% of her AGI in taxes'. The final conclusion, 'Donasia owes \$14,400 in taxes', results from substituting '\$14,400' for '18% of \$80,000' in the conclusion of the third inferential step—namely 'Donasia owes 18% of \$80,000 in taxes'. These two inferential steps generate two simple valid arguments.

0.1 Donasia's AGI = \$80,000.
2 Donasia owes 18% of her AGI in taxes.

——————————————: Substitution

3 Donasia owes 18% of \$80,000 in taxes.

---

1 This inference has the valid form: *all A are B, d is A*; therefore *d is B*. In traditional syllogistic logic, this inference is reformulated as: *all A are B, all those that are identical with d are A*; therefore *all those that are identical with d are B*. The latter has the standard form: *all A are B, all D are A*; therefore *all D are B*. This inference has a traditional name—Barbara.

0.4 18% of \$80,000 = \$14,400.
3 Donasia owes 18% of \$80,000 in taxes.
———————————————: Substitution
4 Donasia owes \$14,400 in taxes

5.1:C2 We now consider a mathematical example of a demonstrative proof. Let *a*, *b*, and *d* be positive integers such that *d* is a divisor of *a* and of *a*+*b* (i.e., *a* and *a*+*b* are divisible by *d*). It follows that *d* is also a divisor of *b*. The standard mathematical proof of this fact is as follows. Since *d* is a divisor of *a* and of *a*+*b*, there are two (positive) integers *p* and *q* such that *a* = *pd* and *a*+*b* = *qd*. By substituting *pd* for *a* in *a*+*b* = *qd*, we obtain *pd*+*b* = *qd*. Subtracting *pd* from both sides of the equality produces *b* = *qd*–*pd*. Factoring out *d* gives *b* = (*q*–*p*)*d*. It follows that *b* is divisible by *d*. As with almost all mathematical proofs, this proof is demonstrative. We can represent it as a precise demonstration consisting of a sequence of inferential steps. We use 'iff' as an abbreviation for 'if and only if'.

Demonstration 2

0.1 Premise: *a*, *b*, and *d* are positive integers such that *d* is a divisor of *a* and of *a*+*b*.
0.2 Premise (definition of a divisor): For all positive integers *m* and *n*, *n* is a divisor of *m* iff there is an integer *k* such that *m* = *kn*.
0.3 Premise (arithmetical fact): For all integers *m*, *n*, and *k*, if *m*+*n* = *k*, then *n* = *k*–*m*.
0.4 Premise (arithmetical fact): For all integers *n*, *k*, *h*, and *g*, if *n* = *kg*–*hg*, then *n* = (*k*–*h*)*g*.
0.5 Premise (arithmetical fact): For all integers *m* and *n*, *m*–*n* is an integer.
1 From 0.1 and 0.2: There is an integer *p* such that *a* = *pd*.
2 From 0.1 and 0.2: There is an integer *q* such that *a*+*b* = *qd*.
3 From 1 and 2: *pd*+*b* = *qd*.
4 From 0.3 and 3: *b* = *qd*–*pd*.
5 From 0.4 and 4: *b* = (*q*–*p*)*d*.
6 From 0.5, 1, and 2: *q*–*p* is an integer.
7 From 0.2, 5, and 6: *d* is a divisor of *b*.

College students know a lot of arithmetical facts, because they have all studied arithmetic and algebra. Thus most people who are told the original reasoning presuppose the arithmetical premises listed at the zero stage. But when we write a *precise* and *complete* demonstrative proof, we have to list all the premises needed in the proof.

Every non-zero line represents an inferential step. The conclusions on lines 1 and 2 are inferred from the same antecedents, 0.1 and 0.2, by a rule that is similar to our temporary rule Conditional Instantiation. The difference is that here we have 'if and only if' instead of 'if-then'. Conditional Instantiation can be paraphrased as follows: from (1) *t* is *A* and (2) for all *x*, if *x* is *A*, then *x* is *B*, we may infer that *t* is *B*. But it is clear that replacing 'if-then' with 'if and only if' does not make the new rule non-truth-preserving. Let us call this new (also temporary) rule of inference "Biconditional Instantiation." This rule licenses the inference of a sentence of the form "*t* is *B*" from two antecedents of the forms "*t* is *A*" and "For all *x*, *x* is *A* iff *x* is *B*." Thus the inferences at lines 1 and 2 are truth-preserving. As usual, this gives us two simple valid arguments:

0.1 a, b, and d are positive integers such that d is a divisor of a and of $a+b$.
0.2 For all positive integers m and n, n is a divisor of m iff there is an integer k such that $m = kn$.

---

1 There is an integer p such that $a = pd$.

0.1 a, b, and d are positive integers such that d is a divisor of a and of $a+b$.
0.2 For all positive integers m and n, n is a divisor of m iff there is an integer k such that $m = kn$.

---

2 There is an integer q such that $a+b = qd$.

The inferential step at line 3 invokes the familiar rule Substitution. This step gives rise to the valid argument:

1 There is an integer p such that $a = pd$.
2 There is an integer q such that $a+b = qd$.

———————————————: Substitution

3 $pd+b = qd$.

Inferential steps 4, 5, and 6 employ our temporary rule of Conditional Instantiation. Hence we have the following valid arguments:

0.3 For all integers m, n, and k, if $m+n = k$, then $n = k-m$.
3 $pd+b = qd$.

———————————————: Conditional Instantiation

4 $b = qd-pd$.

0.4 For all integers n, k, h, and g, if $n = kg-hg$, then $n = (k-h)g$.
4 $b = qd-pd$.

———————————————: Conditional Instantiation

5 $b = (q-p)d$.

0.5 For all integers m and n, $m-n$ is an integer. (Paraphrase: for all m and n, if m and n are integers, then so is $m-n$.)
1 There is an integer p such that $a = pd$.
2 There is an integer q such that $a+b = qd$.

———————————————: Conditional Instantiation

6 $q-p$ is an integer.

The last inferential step has three antecedents and it invokes our second temporary rule, Biconditional Instantiation. It constitutes the valid argument below:

0.2 For all positive integers m and n, n is a divisor of m if and only if there is an integer k such that $m = kn$.
5 $b = (q-p)d$
6 $q-p$ is an integer.
———————————————————: Biconditional Instantiation
7 d is a divisor of b.

The original reasoning moved quickly from the given premise 0.1, the definition of a divisor, and a few implicit arithmetical premises to the conclusion that b is divisible by d. We transformed that "natural" reasoning into a precise demonstration. A formal representation of this demonstration is longer and more involved.

5.1:C3 People ordinarily do not reason formally. Even experts, such as mathematicians, who are in the business of discovering and constructing demonstrative proofs hardly ever reason formally. So it is natural to ask, What is the point of developing formal representations of demonstrative proofs? The answer is threefold. First, formal derivations are extremely precise and maximally explicit in identifying the rules of inference invoked in them. As we will see later, these rules of inference are demonstratively truth-preserving. Hence to know that there is a (classical) formal derivation that represents a certain proof is conclusive evidence that this proof is a (classical) demonstrative proof. In the next section, we will state a theorem that establishes that all the deductively valid arguments of PL can be demonstrated to be valid by means of formal PL derivation. It is widely believed that almost all of mathematical theories are reducible to set theory, which is a (first-order) PL theory. This fact entails that almost all mathematical demonstrations are formalizable as PL derivations. Hence if a mathematical proof is shown to be non-formalizable as a PL derivation, then it is established that such a proof cannot be formulated in set theory. Second, sometimes it is possible to clarify certain errors in informal proofs by formalizing them and showing which steps of the proofs are not licensed by any known rule of inference. Finally, specifying explicit rules of inference focuses the debate between competing logical systems as to which logical system is a better candidate for representing certain modes of demonstrative reasoning.

These are the contributions that formal derivations have for our understanding and employment of demonstrative proofs. But *classical* proof theory has a further advantage that is peculiar (but not unique) to PL (and SL). We will explain later that the applicability of all the universal logical concepts of PL, such as deductive validity and logical truth, can be established by demonstrating that PL derivations of certain types exist. This fact shows that those *universal* concepts can be transformed into *existential* concepts. For instance, to show semantically that a PL argument $\Gamma/\mathbf{X}$ is deductively valid, we need to show that *all* the PL interpretations that satisfy Γ make $\mathbf{X}$ true. On the other hand, to show that $\Gamma/\mathbf{X}$ is deductively valid proof-theoretically is to show that *there exists* a PL derivation of $\mathbf{X}$ from the set Γ. While the semantical concept is universal, the proof-theoretic concept is existential.

5.2 The Notion of Formal Derivation

5.2.1 Assume that **L** is a symbolic logical system and **DS** (for "Deduction System") is a collection of *formal* rules of inference. Let Γ be a set of **L** sentences and **X** an **L** sentence. An **L derivation of X from** Γ is a finite sequence of **L** sentences such that the last sentence of the sequence is **X** and every sentence in the sequence is either a member of Γor is licensed by one of the rules in **DS**. If there is an **L** derivation of **X** from Γ, we say that **X** is **derivable** from Γ in **L** or that **X** is a **theorem** of Γ in **L**. Symbolically, we write "$\Gamma \vdash_{\mathbf{L}} \mathbf{X}$." In this chapter, **L** is either SL or PL. In most cases, we will not specify whether the logical system is SL or PL; we will rely on the context to make clear which logical system is employed. If the context is neutral as to which logical system is invoked, then the statements made apply equally to SL and PL. The collection of inference rules that we will use in this chapter is called **the Natural Deduction System** and is described in 5.3.

5.2.1:C COMMENTARY ON 5.2.1

A formal derivation is a sequence of sentences that typically begins with a list of premises and terminates with the desired conclusion. Following the standard practice in logic, the premises are listed at the start of the derivation, which we also call the **zero stage**. The sentences at the non-zero stages of a derivation are licensed by specific rules of inference. These rules are formal, in the sense that their applicability is determined by the *forms* of the inference's antecedents or the inference's conclusion. The form of a sentence may be defined as the *syntactical structure* of the sentence. This is not a precise definition. According to this definition, a sentence may have several forms. For instance, we can recognize four syntactical structures for the SL sentence $(A \rightarrow B) \vee \neg C$: (1) $\mathbf{X} \vee \mathbf{Y}$, (2) $\mathbf{X} \vee \neg \mathbf{Y}$, (3) $(\mathbf{X} \rightarrow \mathbf{Y}) \vee \mathbf{Z}$, and (4) $(\mathbf{X} \rightarrow \mathbf{Y}) \vee \neg \mathbf{Z}$. The first syntactical structure of this sentence is the coarsest and the last is the most refined. In applying formal inference rules, in most cases 'the form of a sentence or a formula' refers to its coarsest syntactical structure: the main connective or operator together with the immediate component(s).

We said above that a demonstrative proof comprises inferential steps, and that a typical inferential step consists of antecedents, a conclusion, and an inferential license. Formal derivations, on the other hand, consist solely of symbolic sentences. In practice, we will write a formal derivation as a series of inferential steps. These inferential steps, however, are for our consumption; they are not proper parts of the derivation. The derivation itself consists only of symbolic sentences. These sentences are either premises, or assumptions and conclusions that are licensed by specific formal rules of inference. All the additional information required to recognize an inferential step—information such as the number associated with a stage or its inferential license—is extraneous to the derivation: it is added for the benefit of the agent who is constructing or reading a derivation. This information tells the agent the place of a sentence in a derivation and the inferential license used to justify the introduction of this sentence at that stage of the derivation.

In this chapter, we will construct many SL and PL derivations. But to give a feel for how formal derivations look, we will formalize Demonstration 1 in PL. As is the case with arguments, we begin by translating the expressions that occur in a demonstration into PL.

Translation Key (5 names, 3 function symbols, and 2 predicates)

d: Donasia
a: $80,000
b: $45,000
c: $82,000
e: $14,400
gx: The AGI of *x*
fy: 18% of *y*
hz: The tax *z* owes
Rxy: *x* is greater than *y*
Lvw: *v* is less than *w*

We now use this translation key to formalize Demonstration 1 in PL. We indicate the step's number to the left of the inference's conclusion and the interferential license to its right. We will give an approximation of a PL derivation and then a standard PL derivation.

An Approximation of a PL Derivation

0.1	$gd = a$	Premise
0.2	$(\forall x)((Rgxb \wedge Lgxc) \rightarrow hx = fgx)$	Premise
0.3	$Rab \wedge Lac$	Premise
0.4	$fa = e$	Premise
1	$Rgdb \wedge Lgdc$	From 0.1 and 0.3 by Substitution
2	$hd = fgd$	From 0.2 and 1 by Conditional Instantiation
3	$hd = fa$	From 0.1 and 2 by Substitution
4	$hd = e$	From 0.4 and 3 by Substitution

The Natural Deduction System (NDS) described in 5.3 does not contain an inference rule that corresponds to Conditional Instantiation. Thus in order to transform this derivation into a standard PL derivation, we need either to introduce a PL inference rule that corresponds to Conditional Instantiation or find a way of deriving the conclusion of step 2 from 0.2 and 1 by means of the rules of NDS. Since we do not want to change our set of formal rules every time we encounter a rule of inference that has no counterpart in NDS, we will adopt the second approach. But first we discuss the rule of Substitution. This rule has a formal counterpart that is also called "Substitution" (Sub). If we have an antecedent of the form **t** = **s**, where **t** and **s** are any PL singular terms, and another antecedent **X** that contains occurrences of the term **t**, then Sub permits the inference of a PL sentence **Y** that is obtained from **X** by replacing **t** in one or more of its occurrences in **X** with **s**. We usually denote **Y** as **X[s, t]** in order to indicate how **Y** is obtained from **X**. Using this notation we can express the formal rule Sub as follows: If **t** occurs in **X**, then from **t** = **s** and **X**, **X[s, t]** may be inferred. Observe that Sub is purely formal: its applicability is fully determined by the forms of its antecedents. No semantical understanding is required in order to determine whether Sub is applicable to certain PL sentences or not. We only need to see a PL sentence of the form

t = **s** and another PL sentence in which the PL term **t** occurs in order to determine that Sub is applicable to these two sentences.

Of course, Sub is justified semantically. As with the case with demonstrative proofs, our formal rules of inference are truth-preserving. Sub is truth-preserving: any PL interpretation on which **t** = **s** and **X** are true makes **X**[**s**, **t**] true as well. This justification of Sub requires a proof of its own. It would be sufficient to show that if **t** and **s** have the same referent on any relevant PL interpretation I, then **X** (in which **t** occurs) and **X**[**s**, **t**] have, in some sense, the "same" truth conditions on I. This proof would actually show more than we need in order to justify Sub; it would show that **X** and **X**[**s**, **t**] have the same truth value on I: they are either both true or both false. Such a proof is not trivial, since we must show that Sub is truth-preserving no matter how complex the PL terms **t** and **s** and the PL sentence **X** are. Since this proof is relatively too involved, we will not describe it here.

Logicians describe a formal inference rule that is truth-preserving as **sound**. This is a different sense of the word 'sound' from the sense defined in Chapter One. In Chapter One, we defined an argument to be sound just in case it is deductively valid and its premises are all true. On the other hand, a sound formal rule of inference is a rule that is truth-preserving. Thus a proof that shows that Sub is truth-preserving is a proof that establishes the **soundness** of Sub. Not every soundness proof is substantive. Some of these proofs are trivial and we will be able to give some of them in 5.3 below.

Although NDS does not contain a rule that corresponds to what we termed Conditional Instantiation, it can capture this rule as a combination of two formal rules. These are Universal Instantiation (UI) and Modus Ponens (MP). The first rule licenses the inference from a universally quantified sentence of any of its substitutional instances. In other words, UI permits the inference of **X**[**t**] from (∀**z**)**X**, where **t** is any PL singular term. Recall that we defined in Chapter Four the substitutional instance **X**[**t**] as the sentence that is obtained from (∀**z**)**X** by removing the universal quantifier (∀**z**) and replacing *all* the occurrences of **z** in **X** with the singular term **t**. The second rule permits the inference of the consequent of a conditional sentence from the conditional sentence and its antecedent. That is, MP licenses the inference of **Y** if both **X**→**Y** and **X** are given. These two rules are sound and their soundness proofs are trivial. We will give these proofs in Section 5.3. Now we have all the rules we need to give a PL derivation that represent Demonstration 1. The square brackets at the left of the derivation indicate the *level* of the derivation that is enclosed between two brackets with the same numeric designation. Bracketed derivations are called **blocks**. We will explain the role of blocks and how to use them in Section 5.3. For now all that we are required to know is that we begin and terminate every derivation with brackets that are designated with the numeral '0'. We call this block the **0-block** and it represents the **main derivation**. Later we will encounter derivations that consist of main derivations and subderivations. Here is the PL derivation that represents Demonstration 1. (We use 'P' to abbreviate 'premise' and we drop the use of 'from', 'and', and 'by' from the descriptions of the inferential licenses.)

PL Derivation

[0	0.1	$gd = a$	P
	0.2	$(\forall x)((Rgxb \wedge Lgxc) \rightarrow hx = fgx)$	P
	0.3	$Rab \wedge Lac$	P
	0.4	$fa = e$	P

	1	$Rgdb \wedge Lgdc$	0.1, 0.3, Sub
	2	$(Rgdb \wedge Lgdc) \rightarrow hd = fgd$	0.2, UI
	3	$hd = fgd$	1, 2, MP
	4	$hd = fa$	0.1, 3, Sub
0]	5	$hd = e$	0.4, 4, Sub

Let 'D' stand for the sequence of nine PL sentences above. It is clear from the definition of a PL derivation given in 5.2.1 that D is a PL derivation of the conclusion '$hd = e$' from the set of premises, which consists of the sentences 0.1–0.4. D fulfills this definition: it is a finite sequence of PL sentences such that its last sentence is '$hd = e$' and each of its sentences is either a premise or is licensed by an NDS rule. As stated previously, D consists solely of the nine PL sentences; the numbers and brackets to the left of these sentences and the inferential licenses to their right are not part of D. They are part of the metalanguage of PL, and their purpose is to make clear how the sentences of D are introduced.

5.2.2 The **Soundness Theorem for PL** asserts that for every set **Γ** of PL sentences and every sentence **X** of PL, if **X** is a theorem of **Γ**, then **X** is a logical consequence of **Γ**. Symbolically, if **Γ** ⊢ **X**, then **Γ** ⊨ **X**. The **Soundness Theorem for SL** makes a similar assertion about SL: for every SL set **Γ** and every SL sentence **X**, if **Γ** ⊢ **X**, then **Γ** ⊨ **X**.

5.2.2:C COMMENTARY ON 5.2.2

The Soundness Theorems establish not only that every NDS rule of inference is truth-preserving (i.e., sound) but also that every PL and SL derivation is truth-preserving. Proving that every rule is truth-preserving is only part of proving that every derivation is truth-preserving. Every derivation involves a combination of inference rules applied in some order. To show that the proof theory of PL and SL are sound is to show that no matter how these rules are combined the resulting derivations are truth-preserving. It is clear that such a proof is complex. Thus we will not attempt to present this proof in this book. However, we will present pieces of this proof in Section 5.3. We will establish the soundness of some NDS rules. Given the scope of this book, we will not be able to establish the soundness of every NDS rule, since some of these proofs are beyond the level of this book. Books on the metatheory of PL and SL contain complete demonstrations of the Soundness and Completeness Theorems of these logical systems.[1] In this book we will make use of these theorems without giving their proofs.

Let us consider closely the statement made by the Soundness Theorem for PL in order to see that this theorem indeed affirms that every PL derivation is truth-preserving (a similar explanation applies to SL). The Soundness Theorem for PL asserts that every theorem of a set of PL sentences is a logical consequence of that set. We claim that this assertion is equivalent to saying that every PL derivation is truth-preserving. We first explain what we mean by saying that a PL derivation D is truth-preserving. According to the definition of a PL derivation given in 5.2.1, D is a finite sequence of PL sentences whose members are either premises or are introduced by NDS rules. Let Σ_D be the PL set that consists of the premises invoked in D and let **X** be the conclusion of D, that is, **X** is the last sentence of D. It is clear that Σ_D is finite. As we

1 For instance, see Yaqub, *An Introduction to Metalogic* (Broadview Press, forthcoming).

will see later, Σ_D could be the empty set; in other words, there are derivations that invoke no premises. To say that D is truth-preserving is to say that if the premises invoked in D are true, then the conclusion of D is true as well. Precisely speaking, D is truth-preserving if and only if every PL interpretation that makes the members of Σ_D all true also makes **X** true. Given the definition of logical consequence, the preceding statement is simply the assertion that **X** is a logical consequence of Σ_D. To simplify our expression we shall refer to the Soundness Theorem for PL as stated in 5.2.2 above as ST and to the statement that every PL derivation is truth-preserving as TP. We claimed that ST and TP are equivalent.

We first show that if ST is true, then TP is true. Hence we assume that ST is true and let D be any PL derivation. It is clear that D is a derivation of **X** from Σ_D, where Σ_D is defined as above and **X** is the conclusion of D. Thus **X** is a theorem of Σ_D. By ST, **X** is a logical consequence of Σ_D. In other words, D is truth-preserving, which means that TP is true. Now we show that if TP is true, then ST is true. Suppose that TP is true, i.e., every PL derivation is truth-preserving. We want to show that ST is true. Let Γ be any PL set and **X** any PL sentence such that **X** is a theorem of Γ. By definition, this means that there is a PL derivation D from **X** to Γ. Now, D might make use of all the members of Γ or of only some of them. In the case of an infinite Γ, we know for sure that D cannot invoke all the members of Γ, since D is a *finite* sequence of sentences. In general, we know that Σ_D is a subset of Γ. This is an immediate consequence of the definitions of a PL derivation and of Σ_D (the premises that D invokes must all come from Γ). Since TP is assumed to be true, we know that D is truth-preserving, that is, we know that the conclusion of D, which is **X**, is a logical consequence of Σ_D. Our goal is to show that **X** is a logical consequence of Γ as well. To show this we need to establish that every PL interpretation I for Γ and **X** on which the members of Γ are all true makes **X** true. Thus let I be any PL interpretation for Γ and **X** that satisfies the set Γ. Since Σ_D is a subset of Γ, I must also satisfy the set Σ_D. But **X** is a logical consequence of Σ_D; therefore **X** is true on I. This shows that **X** is a logical consequence of Γ. We have proved that if TP is true, then ST is true, and that if ST is true, then TP is true. Hence ST and TP are equivalent statements.

5.2.3 The **Completeness Theorem for PL** asserts that for every set Γ of PL sentences and every sentence **X** of PL, if **X** is a logical consequence of Γ, then **X** is a theorem of Γ. Symbolically, if $\Gamma \vDash \mathbf{X}$, then $\Gamma \vdash \mathbf{X}$. Similarly, the **Completeness Theorem for SL** asserts that for every SL set Γ and every SL sentence **X**, if $\Gamma \vDash \mathbf{X}$, then $\Gamma \vdash \mathbf{X}$.

5.2.3:C COMMENTARY ON 5.2.3

The Completeness Theorems establish that the NDS rules of inference are **adequate**. The adequacy of these rules means that if a sentence of these logical systems is a logical consequence of a set of premises, then it is possible to produce a demonstrative proof that establishes this fact and that can be represented by a formal derivation. A completeness theorem for any logical system has no significance without an accompanying soundness theorem. To see the point, suppose that we have only one rule of inference: for any sentence **X**, infer **X**. It is a silly rule, but it is complete; it allows us to derive any logical consequence since it allows us to derive any sentence whatsoever. Any proof theory that contains this rule is totally insignificant because it is unsound. It is clear that this rule is not truth-preserving. For instance, it licenses the inference of $A \wedge \neg A$, which is logically false. The difficult task is to construct a proof theory that is both sound and complete. The Soundness and Completeness Theorems for PL and SL demonstrate that NDS accomplishes this task. They show that the rules of NDS are truth-preserving and adequate.

There is a very significant implication of these theorems: the relation of logical consequence in PL and SL is **formalizable**. In order to explain this fact, we first combine these two theorems below.

The Soundness and Completeness Theorems for PL (SL): For every set Γ of PL (SL) sentences and every PL (SL) sentence **X**, $\Gamma \vDash \mathbf{X}$ if and only if $\Gamma \vdash \mathbf{X}$.

The 'if' direction is the Soundness Theorem, and the 'only-if' direction is the Completeness Theorem. The combined statement above shows that the notions of logical consequence and derivability are equivalent in PL and SL: every logical consequence of Γ is derivable from it, and every sentence that is derivable from Γ is a logical consequence of it. In other words, the logical consequences of Γ are precisely its theorems. The notion of derivability is a formal notion. Whether **X** is derivable from Γ or not depends solely on the forms of the members of Γ and **X**. To say that a certain logical notion is formalizable is to say that it is equivalent to a formal notion. Since the notion of logical consequence in PL and SL is equivalent to the notion of derivability and since the latter is formal, the notion of logical consequence in PL and SL is formalizable.

This fact justifies our description of these logical systems as formal logics. Not every logical system is formal in this sense. Second-Order Predicate Logic (PL^2), which we described briefly in Chapter Four, is not a formal system since its notion of logical consequence is not formalizable. It is known that no "reasonable" collection of sound rules of inference for PL^2 can be complete. This fact is sometimes described by saying that PL^2 is an "incomplete" logic or that it is an "essentially semantical" logic. The idea is that reasoning about logical consequence in PL^2 must make use of semantical notions, such as PL^2 interpretation and truth; this reasoning cannot be reduced in general to PL^2 derivations or facts about PL^2 derivations.

5.2.4 The Soundness and Completeness Theorems for PL and SL have many important **corollaries**. Some of these corollaries link certain logical concepts to the notion of derivability. We list five corollaries below.

5.2.4a A PL (SL) argument is **deductively valid** if and only if its conclusion is derivable from the set of its premises.

5.2.4b A set of PL (SL) sentences is **inconsistent** if and only if a sentence of the form '$\mathbf{X} \wedge \neg \mathbf{X}$' is derivable from it.

5.2.4c Two PL (SL) sentences are **logically equivalent** if and only if each one is derivable from the other (i.e., they are **co-derivable**).

5.2.4d A PL (SL) sentence is **logically true** if and only if it is derivable from the empty set.

5.2.4e A PL (SL) sentence is **logically false** if and only if a sentence of the form '$\mathbf{X} \wedge \neg \mathbf{X}$' is derivable from it.

5.2.4:C Commentary on 5.2.4

We will prove that all of these statements follow from the Soundness and Completeness Theorems. We will prove them for PL and the same proofs can straightforwardly be converted to proofs about SL logical concepts. Since all of these statements are biconditionals, we need to prove two parts for each one: an 'if' direction and an 'only-if' direction.

Proof of 5.2.4a: the 'if' direction

1 Suppose that Γ is a PL set and **X** is a PL sentence such that $\Gamma \vdash \mathbf{X}$.
2 From 1 by the Soundness Theorem: $\Gamma \vDash \mathbf{X}$.
3 From 2: the PL argument $\Gamma/\mathbf{X}$ is deductively valid (since the notions of logical consequence and deductive validity are identical).

Proof of 5.2.4a: the 'only-if' direction

1 Suppose that the PL argument $\Gamma/\mathbf{X}$ is deductively valid.
2 From 1: $\Gamma \vDash \mathbf{X}$.
3 From 2 by the Completeness Theorem: $\Gamma \vdash \mathbf{X}$.

Proof of 5.2.4b: the 'if' direction

1 Let Γ be a PL set such that $\Gamma \vdash \mathbf{X} \wedge \neg \mathbf{X}$, for some PL sentence **X**.
2 From 1 by the Soundness Theorem: $\Gamma \vDash \mathbf{X} \wedge \neg \mathbf{X}$.
3 Reductio Assumption: Γ is consistent.
4 From 3 by the definition of consistency: there is a PL interpretation I that is a model of Γ.
5 If I is not relevant to **X**, we expand I into I* such that I* interprets all the extra vocabulary in **X** without changing any of the semantical assignments made by I. Since I is a model of Γ, I* is also a model of Γ.
6 From 2 and 5 by the definition of logical consequence: $\mathbf{X} \wedge \neg \mathbf{X}$ is true on I*, which is impossible since $\mathbf{X} \wedge \neg \mathbf{X}$ is logically false.
7 From 3 and 6: the Reductio Assumption is false; hence Γ is inconsistent.

Proof of 5.2.4b: the 'only-if' direction

1 Let Γ be an inconsistent set of PL sentences, and let **X** be any PL sentence.
2 From 1 by the definition of inconsistency: there is no PL interpretation that satisfies Γ; and hence there is no interpretation that satisfies Γ and makes $\mathbf{X} \wedge \neg \mathbf{X}$ false.
3 From 2 by the definition of logical consequence: $\Gamma \vDash \mathbf{X} \wedge \neg \mathbf{X}$.
4 From 3 by the Completeness Theorem: $\Gamma \vdash \mathbf{X} \wedge \neg \mathbf{X}$.

Proof of 5.2.4c: the 'if' direction

1 Suppose that **X** and **Y** are any two PL sentences such that $\{\mathbf{X}\} \vdash \mathbf{Y}$ and $\{\mathbf{Y}\} \vdash \mathbf{X}$.
2 From 1 by the Soundness Theorem: $\{\mathbf{X}\} \vDash \mathbf{Y}$ and $\{\mathbf{Y}\} \vDash \mathbf{X}$.

3 Let I be any PL interpretation for **X** and **Y** on which **X** is true.
4 From 2 ($\{\mathbf{X}\} \vDash \mathbf{Y}$) and 3 by the definition of logical consequence: **Y** is true on I too.
5 Now let I be a PL interpretation for **X** and **Y** on which **X** is false.
6 From 2 ($\{\mathbf{Y}\} \vDash \mathbf{X}$) and 5 by the definition of logical consequence: **Y** is false on I.
7 From 3 through 6: **X** and **Y** have identical truth values on every PL interpretation for them.
8 From 7 by the definition of logical equivalence: **X** and **Y** are logically equivalent.

Proof of 5.2.4c: the 'only-if' direction

1 Suppose that **X** and **Y** are logically equivalent PL sentences.
2 From 1 by the definition of logical equivalence: there is no PL interpretation on which **X** is true and **Y** is false, or **Y** is true and **X** is false.
3 From 2: any PL interpretation for **X** and **Y** on which **X** is true makes **Y** true, and any PL interpretation for **X** and **Y** on which **Y** is true makes **X** true.
4 From 3 by the definition of logical consequence: $\{\mathbf{X}\} \vDash \mathbf{Y}$ and $\{\mathbf{Y}\} \vDash \mathbf{X}$.
5 From 4 by the Completeness Theorem: $\{\mathbf{X}\} \vdash \mathbf{Y}$ and $\{\mathbf{Y}\} \vdash \mathbf{X}$.

Proof of 5.2.4d: the 'if' direction

1 Let **X** be any PL sentence that is derivable from the empty set, i.e., $\varnothing \vdash \mathbf{X}$.
2 From 1 by the Soundness Theorem: $\varnothing \vDash \mathbf{X}$.
3 Let I be any PL interpretation for **X**. Since there is no sentence in $\varnothing$ that is false on I (because there are no sentences in $\varnothing$), I satisfies $\varnothing$.
4 From 2 and 3 by the definition of logical consequence: **X** is true on I.
5 From 3 and 4: since I is arbitrary, **X** is true on every PL interpretation for it.
6 From 5 by the definition of logical truth: **X** is logically true.

Proof of 5.2.4d: the 'only-if' direction

1 Let **X** be any logically true PL sentence.
2 From 1 by the definition of logical truth: **X** is true on every PL interpretation for it.
3 By the reasoning in step 3 of the preceding proof: every PL interpretation satisfies $\varnothing$.
4 From 1 and 2 by the definition of logical consequence: $\varnothing \vDash \mathbf{X}$ (since every PL interpretation for **X** that satisfies $\varnothing$ makes **X** true).
5 From 4 by the Completeness Theorem: $\varnothing \vdash \mathbf{X}$.

Proof of 5.2.4e: the 'if' direction

1 Let **Y** be a PL sentence such that $\{\mathbf{Y}\} \vdash \mathbf{X} \wedge \neg \mathbf{X}$.
2 From 1 by the Soundness Theorem: $\{\mathbf{Y}\} \vDash \mathbf{X} \wedge \neg \mathbf{X}$.
3 Reductio Assumption: there is a PL interpretation I on which **Y** is true.
4 If I is not relevant to **X**, we expand I into I* such that I* interprets all the extra vocabulary in **X** without changing any of the semantical assignments made by I. Since **Y** is true on I, it is true on I*.
5 From 2 and 4 by the definition of logical consequence: $\mathbf{X} \wedge \neg \mathbf{X}$ is true on I*, which is impossible since $\mathbf{X} \wedge \neg \mathbf{X}$ is logically false.

6 From 3 and 5: the Reductio Assumption is false; and hence there is no PL interpretation on which **Y** is true.
7 From 6 by the definition of logical falsehood: **Y** is logically false.

Proof of 5.2.4e: the 'only-if' direction

1 Let **Y** be any logically false PL sentence and **X** any PL sentence.
2 From 1 by the definition of logical falsehood: there is no PL interpretation on which **Y** is true; and hence there is no PL interpretation on which **Y** is true and $\mathbf{X} \wedge \neg \mathbf{X}$ is false.
3 From 2 by the definition of logical consequence: $\{\mathbf{Y}\} \vDash \mathbf{X} \wedge \neg \mathbf{X}$.
4 From 3 by the Completeness Theorem: $\{\mathbf{Y}\} \vdash \mathbf{X} \wedge \neg \mathbf{X}$.

5.2.5 The **Compactness Theorem** asserts that for every PL (SL) set Γ and every PL (SL) sentence **X**, if **X** is a logical consequence of Γ, then **X** is a logical consequence of a **finite** subset of Γ. The Compactness Theorem is a corollary of the Soundness and Completeness Theorems.

5.2.5:C Commentary on 5.2.5

5.2.5:C1 We will give a generic proof of the Compactness Theorem that works for any logical system **L** that has a sound and complete proof theory whose derivations are finite sequences. It is understood that **L** has a well-defined notion of logical consequence (if it did not, its Soundness and Completeness theorems would not make much sense). We denote the relations of logical consequence and derivability in **L** as $\vDash_{\mathbf{L}}$ and $\vdash_{\mathbf{L}}$, respectively.

Proof of the Compactness Theorem for **L**

1 Suppose that Γ is any set of **L** sentences and **X** is any **L** sentence such that $\Gamma \vDash_{\mathbf{L}} \mathbf{X}$.
2 From 1 by the Completeness Theorem for **L**: $\Gamma \vdash_{\mathbf{L}} \mathbf{X}$.
3 From 2 by the definition of derivability: there is an **L** derivation D of **X** from Γ.
4 Let Σ be the subset of Γ whose members are all the sentences in Γ that D invokes.
5 From 4: since D is a finite sequence of **L** sentences, Σ must be a finite subset of Γ.
6 From 3 and 4 by the definition of a derivation: $\Sigma \vdash_{\mathbf{L}} \mathbf{X}$.
7 From 6 by the Soundness Theorem for **L**: $\Sigma \vDash_{\mathbf{L}} \mathbf{X}$.
8 From 5 and 7: **X** is a logical consequence of a finite subset of Γ.

5.2.5:C2 The Compactness Theorem has an interesting philosophical implication about the expressive powers of PL and SL. Suppose that Δ/Z is a natural-language argument whose set of premises, Δ, is infinite. Suppose further that all the premises are needed for the deductive validity of the argument, that is, if a single premise is removed from Δ, the resulting argument would be invalid. We considered an example of such an argument in Chapter One, Subsection 1.1.1:C2 (the argument about the Arabic word *thalathah* and its English translation). The Compactness Theorem implies that such an argument cannot be translated faithfully into PL or SL (or into any logical system for which the Compactness Theorem holds). For if it were possible to translate Δ/Z faithfully into, say, PL, then the resulting PL argument, by the Compactness Theorem, would require only finitely many premises for its deductive validity. So there are two possibilities: either the PL translation of Δ/Z is valid or it is invalid. If it is invalid, then the PL translation fails to capture the logical relation between the premises and the conclusion of the original argument; and hence the PL translation would not be faithful. On the other hand, if it is valid, then the PL translation fails to capture the fact that the conclusion of the original argument is not a logical consequence of any finite collection of the premises (we assumed that all the members of Δ are needed for the argument to be deductively valid); therefore the PL argument would also fail to be a faithful translation of the original argument, Δ/Z.

This implication supplies further evidence that the notions of SL truth valuations and PL interpretations do not correctly (completely) capture the intuitive notion of logical possibility. We will explain this claim for the case of PL. Similar explanation can be given for SL. Suppose that the best PL translation of Δ/Z is **Γ/X**. We first assume that **Γ/X** is deductively invalid. Thus there is a PL interpretation that is a counterexample to the PL argument. Since the original argument Δ/Z is deductively valid, there are no counterexamples to this argument. It follows that the PL interpretations that are counterexamples to the PL argument **Γ/X** are all superfluous (i.e., they do not represent any relevant logical possibilities). Now we assume the second case: **Γ/X** is deductively valid. By the Compactness Theorem, there is a finite subset **Σ** of **Γ** such that **Σ/X** is deductively valid. Let L be the finite subset of Δ whose PL translation is **Σ**. So now we obtain two arguments: a natural language argument L/Z and its best PL translation **Σ/X**. We assumed that the PL argument **Σ/X** is deductively valid. But the natural language argument L/Z cannot be valid because it is given that Z does not logically follow from any finite subset of Δ. Therefore, there is a logical possibility that is a counterexample to the argument L/Z. Since its PL translation **Σ/X** is deductively valid, there are no PL interpretations that are counterexamples to the PL argument. This means that the logical possibilities that are counterexamples to the natural-language arguments L/Z cannot be represented by any PL interpretations for the argument **Σ/X**. Hence either there are PL interpretations that fail to represent any relevant logical possibilities, or there are relevant logical possibilities that cannot be represented by any PL interpretations. Either way, the notion of PL interpretation fails to capture the intuitive notion of logical possibility.

5.2.5:C3 Sometimes the Compactness Theorem is given a different formulation. We will give this version here and show that it is equivalent to the original formulation. We call this second formulation the "Finite-Satisfiability Theorem." We restrict our discussion to PL, though it is a trivial exercise to reformulate everything we say here for SL. If there is a PL interpretation that satisfies some set of PL sentences (i.e., it makes every member of the set true), we say that this set is **satisfiable**. In other words, a set of PL sentences is satisfiable if and only if it has a model. If every finite subset of some set **Γ** of PL sentences has a model, we say that this set is **finitely satisfiable**. Now we are ready to state our theorem.

The Finite-Satisfiability Theorem for PL: Every PL set that is finitely satisfiable is satisfiable (or, if every finite subset of a PL set has a model, then the set has a model too).

We give two proofs. The first shows that the Compactness Theorem entails the Finite-Satisfiability Theorem, and the second shows that the Finite-Satisfiability Theorem entails the Compactness Theorem. These two proofs establish that the two theorems are equivalent.

Proof that the Compactness Theorem entails the Finite-Satisfiability Theorem

1 Let Γ be a finitely satisfiable set of PL sentences and **X** be any PL sentence.
2 Reductio Assumption: Γ is not satisfiable.
3 From 2 by the definition of satisfiability: there is no PL interpretation that satisfies Γ; and hence there is no PL interpretation that satisfies Γ and makes **X**∧¬**X** false.
4 From 3 by the definition of logical consequence: Γ ⊨ **X**∧¬**X**.
5 From 4 by the Compactness Theorem: there is a finite subset Σ of Γ such that Σ ⊨ **X**∧¬**X**.
6 From 1 and 5 by the definition of finite satisfiability: there is a PL interpretation I that is a model of Σ.
7 If I is not relevant to **X**, we expand I into I* such that I* interprets all the additional vocabulary in **X** without changing any of the semantical assignments made by I. Since I is a model of Σ, I* is also a model of Σ.
8 From 5 and 7 by the definition of logical consequence: **X**∧¬**X** is true of I*, which is impossible.
9 From 2 through 8: the Reductio Assumption is false; and hence Γ is satisfiable.

Proof that the Finite-Satisfiability Theorem entails the Compactness Theorem

1 Let Γ be any PL set and **X** any PL sentence such that Γ⊨ **X**.
2 Reductio Assumption: **X** is not a logical consequence of any finite subset of Γ.
3 From 2 by the definition of logical consequence: for any finite subset Σ of Γ, there is a PL interpretation that satisfies Σ and makes **X** false.
4 From 3 by the principle of bivalence: for every finite subset Σ of Γ, there is a PL interpretation that satisfies Σ and makes ¬**X** true.
5 From 4 by the definition of satisfiability: for every finite subset Σ of Γ, Σ is satisfiable and Σ∪{¬**X**} is satisfiable (recall that Σ∪{¬**X**} is the union of Σ and {¬**X**}, that is, the set that is obtained by adding ¬**X** to the members of Σ).
6 From 5: every finite subset of Γ∪{¬**X**} is satisfiable (a finite subset of Γ∪{¬**X**} is either a finite subset of Γ alone or a finite subset of Γ with the addition of ¬**X**).
7 From 6 by the definition of finite satisfiability: the set Γ∪{¬**X**} is finitely satisfiable.
8 From 7 by the Finite-Satisfiability Theorem: Γ∪{¬**X**} is satisfiable.
9 From 8 by the definition of satisfiability: there is a PL interpretation that satisfies Γ∪{¬**X**}, that is, there is a PL interpretation that satisfies Γ and makes **X** false.
10 From 9 by the definition of logical consequence: **X** is not a logical consequence of Γ, which contradicts 1.
11 From 2 through 10: the Reductio Assumption is false; and hence **X** is a logical consequence of some finite subset of Γ.

The Finite-Satisfiability Theorem describes an interesting feature of PL and SL. We may describe this feature compactly by saying that no set can be made inconsistent by adding to it infinitely many new sentences none of which contradicts the existing sentences. Let us explain this compact statement. Say we start with the empty set, and we have at our disposal an infinite supply of PL or SL sentences none of which is logically false. Suppose we order these sentences in some fashion, obtaining an infinite sequence of sentences: $\mathbf{X}_1, \mathbf{X}_2, \mathbf{X}_3, \ldots, \mathbf{X}_n, \ldots$. We add the first sentence to the empty set; we get the set $\{\mathbf{X}_1\}$. This set is clearly consistent since we assumed that none of these sentences is logically false. Now we add the second sentence. The set now is $\{\mathbf{X}_1, \mathbf{X}_2\}$. Although none of these sentences is logically false, a contradiction might ensue: these sentences might conflict with each other; for instance $\mathbf{X}_1$ might be $(\forall x)\neg Ax$ and $\mathbf{X}_2$ might be Ae. Suppose that no contradiction ensues, and hence this expanded set is consistent. Again we expand the set by adding $\mathbf{X}_3$ obtaining the set $\{\mathbf{X}_1, \mathbf{X}_2, \mathbf{X}_3\}$. Assume that this addition does not result in any contradiction, which means that the new set is also consistent. We repeat this process indefinitely adding one sentence at a time and expanding the set further and further. The Finite-Satisfiability Theorem entails that if this process does not generate a contradiction at any stage, that is, every finite set $\{\mathbf{X}_1, \mathbf{X}_2, \mathbf{X}_3, \ldots, \mathbf{X}_n\}$ is consistent, consistency will be preserved even if we add *all* the sentences in our infinite sequence. The resulting infinite set will be consistent too. This is an interesting result because it seems intuitively reasonable to expect the resulting infinite set to be consistent. If no contradiction ensues at any stage of the expansion, then it seems plausible to expect that no contradiction would ensue even if we add all the sentences in our infinite sequence. On the other hand, the Compactness Theorem does not seem intuitive at all. If infinitely many premises logically imply a sentence, then why do finitely many premises suffice for the logical implication? However, as we proved above, these two theorems are equivalent.

The Compactness Theorem is surprising, and, as the Swedish logician Per Lindström (1936–2009) proved, it is one of two features of First-Order Predicate Logic (PL) that jointly characterize this logic uniquely. Informally, the Compactness Theorem says that if there is a deductively valid PL argument whose set of premises Σ is infinite, then all but finitely many premises are needed for the argument to be deductively valid. This is because PL allows us to construct a demonstrative proof showing that any PL interpretation that is a model of Σ also satisfies the conclusion of the argument. Such a demonstrative proof is a finite sequence of sentences. Hence this demonstrative proof can only invoke finitely many members of Σ. This entails that only finitely many members of Σ are needed for deriving the conclusion of the argument. Thus if we replace Σ with these finitely many premises, the resulting argument is still deductively valid.

5.3 The Natural Deduction System (NDS)

5.3.1 The **rules of NDS** are divided into three groups: **normal rules**, **hypothetical rules**, and **replacement rules**.

5.3.1:C COMMENTARY ON 5.3.1

In the next section we will describe a relatively large collection of rules. Many of these rules can be derived from other NDS rules. We could have elected to give a "minimal" collection of rules, in the sense that these rules are adequate and none of them could be derived from the others. While such a strategy gives us fewer rules, in practice it is more difficult to construct derivations using very few rules. We would like to have rules that correspond to our natural modes of reasoning; and the fewer rules we have, the fewer modes of reasoning we can represent in our proof theory. On the other hand, if we have too many rules, it would become difficult to manage and remember all of these rules. The rules included in our NDS are all traditional ones, which can be found in many logic books. In Subsection 5.3.3, however, we will describe a deduction system that has a relatively small collection of rules, from which all the NDS rules can be derived. That system has philosophical significance.

The rules described in Subsection 5.3.2a are called "normal rules," in Subsection 5.3.2b "hypothetical rules," and in Subsection 5.3.2c "replacement rules." Each rule is presented in a schematic form as it may appear in an actual derivation. Before delving into the rules in the following Subsection, we will illustrate how to interpret these schematic descriptions with a few examples. We begin with two normal rules.

Constructive Dilemma (CD)

[n
⋮
h $\mathbf{X} \vee \mathbf{Y}$
⋮
i $\mathbf{X} \rightarrow \mathbf{Z}_1$
⋮
j $\mathbf{Y} \rightarrow \mathbf{Z}_2$
⋮
k $\mathbf{Z}_1 \vee \mathbf{Z}_2$ h, i, j, CD
⋮
n]

Disjunctive Syllogism (DS)

[n
⋮
h $\mathbf{X} \vee \mathbf{Y}$
⋮
i $\neg\mathbf{Y}$
⋮
k $\mathbf{X}$ h, i, DS
⋮
n]

The letters 'n', 'h', 'i', 'j', and 'k' are replaced with numerals in actual derivations. The lines h, i, and j are the inference's antecedents and the line k is always the inference's conclusion. The vertical ellipses '⋮' between two lines indicates that some lines might occur in-between. To the right of the inference's conclusion is the inferential license. Thus the license "h, i, j, CD" abbreviates "From lines h, i, and j by Constructive Dilemma," and "h, i, DS" says "From lines h and i by Disjunctive Syllogism." The metalinguistic variables '$\mathbf{X}$', '$\mathbf{Y}$', '$\mathbf{Z}_1$', and '$\mathbf{Z}_2$' stand for any PL or SL sentences. The order of the lines h, i, and j is irrelevant. The fact that line k comes after the other lines is essential to the correctness of the rule. The brackets to the left represent the block within which the rule is applied. Normal rules must be applied within open blocks only. The meaning of this and the use of blocks will be explained later. We only note here that a left bracket followed by a numeral n represents the first line of the nth block, and a numeral n followed by a right bracket represents the last line of the nth block. Prior to '[n' the nth block has not been opened yet and after 'n]' the nth block is closed.

Below is an example of how CD and DS may be applied in an actual derivation.

[2

⋮

6	$Ps \rightarrow (\exists y)By$	
7	$\neg(\forall x)\neg Dx$	
8	$Ps \vee (Rab \wedge Qc)$	
9	$(Rab \wedge Qc) \rightarrow (\forall x)\neg Dx$	
10	$(\exists y)By \vee (\forall x)\neg Dx$	6, 8, 9, CD
11	$(\exists y)By$	7, 10, DS

⋮

2]

In this example, the portion of the derivation displayed is part of the second block. Observe that the rules are applied fully within an open block: this block is opened at some point prior to the 6th line and is closed after the 11th line. CD is applied at line 10. It has three antecedents: two conditionals of the form $\mathbf{X} \rightarrow \mathbf{Z}_1$ and $\mathbf{Y} \rightarrow \mathbf{Z}_2$ on lines 6 and 9 and a disjunction of the form $\mathbf{X} \vee \mathbf{Y}$ on line 8. These are the forms of the antecedents required for an application of CD. The conclusion of CD is a sentence of the form $\mathbf{Z}_1 \vee \mathbf{Z}_2$, which is the sentence that occurs on line 10. On line 11, DS is applied. It has two antecedents: a disjunction on line 10 and the negation of one of the disjuncts on line 7. The conclusion of CD is the other disjunct, which is the sentence that occurs on line 11. It is clear from this example that any conclusion we infer may be used as an antecedent for a later inference *provided that they all occur in an open block.*

We now consider an example of the following replacement rules as they may be applied in an actual derivation.

De Morgan's Laws (DeM):[1]	$\neg(\mathbf{X} \wedge \mathbf{Y})$	$\Leftrightarrow$	$\neg\mathbf{X} \vee \neg\mathbf{Y}$
	$\neg(\mathbf{X} \vee \mathbf{Y})$	$\Leftrightarrow$	$\neg\mathbf{X} \wedge \neg\mathbf{Y}$
Material Conditional (MC):	$\mathbf{X} \rightarrow \mathbf{Y}$	$\Leftrightarrow$	$\neg\mathbf{X} \vee \mathbf{Y}$

[1

⋮

6	$Ps \rightarrow (\forall x)\neg(Rab \wedge Dx)$	
7	$Ps \rightarrow (\forall x)(\neg Rab \vee \neg Dx)$	6, DeM
8	$Ps \rightarrow (\forall x)(Rab \rightarrow \neg Dx)$	7, MC
9	$\neg Ps \vee (\forall x)(Rab \rightarrow \neg Dx)$	8, MC

⋮

1]

The double arrow '⇔' in the descriptions of the replacement rules indicates that each side might be replaced by the other. Replacement rules are powerful inference rules because, un-

1 These laws are named after their discoverer, the British mathematician and logician Augustus De Morgan (1806–1871).

like normal and hypothetical rules, they can be applied not only to sentences, to whole lines, but also to formulas which are parts of sentences. Normal and hypothetical rules can only be applied to whole lines (and hence they can only be applied to sentences). We will explain these requirements with more elaboration later. For now we make the observation that DeM was applied not to the whole conditional on line 6 but only to part of its consequent, which is a proper formula. We replaced $\neg(Rab \wedge Dx)$ with $(\neg Rab \vee \neg Dx)$, and copied the rest of the sentence as is, obtaining the sentence on line 7. On line 8 we made an inference using MC by applying the rule to $(\neg Rab \vee \neg Dx)$, which is only part of line 7. The rule allows us to replace $(\neg Rab \vee \neg Dx)$ with $(Rab \rightarrow \neg Dx)$. Note that we applied the reverse direction of MC (i.e., $\mathbf{X} \rightarrow \mathbf{Y} \Leftarrow \neg\mathbf{X} \vee \mathbf{Y}$). Finally, we applied MC to the whole sentence on line 8, replacing the conditional $Ps \rightarrow (\forall x)(Rab \rightarrow \neg Dx)$ with the disjunction $\neg Ps \vee (\forall x)(\neg Rab \rightarrow Dx)$, which is the sentence on line 9. As the case with normal rules, replacement rules must be applied within open blocks.

The last example we discuss is of a hypothetical rule. Below are the rule of Conditional Proof and an application of it.

Conditional Proof (CP): 'CPA' stands for 'Conditional Proof Assumption'.

[n			
	⋮		
[n+1	h	$\mathbf{X}$	CPA
	⋮		
n+1]	k	$\mathbf{Y}$	
	k+1	$\mathbf{X} \rightarrow \mathbf{Y}$	h–k, CP
	⋮		
n]			

[1			
	⋮		
[2	i	Ps	CPA
	⋮		
2]	k	$(\forall x)\neg(Rab \wedge Dx)$	
	k+1	$Ps \rightarrow (\forall x)\neg(Rab \wedge Dx)$	i–k, CP
	⋮		
1]			

Blocks are opened at two stages only: the zero stage and a stage at which the assumption of a hypothetical rule occurs. The block that is opened at the zero stage is the 0-block and it encloses the main derivation. The 0-block closes at the conclusion of the derivation. A block that is opened due to the application of some hypothetical rule closes at a line permitted by that specific rule. The inference's conclusion of a hypothetical rule is stated immediately after the line at which its block is closed. The inference's antecedent of a hypothetical rule is the derivation enclosed in its block. Blocks are numbered according to the order in which they are opened. Thus if the last block opened carries the numeral n, the next block to be opened carries the numeral n+1. A block B* is a **subblock** of a block B if and only if B* is opened after B is opened and before B is closed. Hence all the blocks of a derivation, other than the 0-block itself, are subblocks of the 0-block. Blocks that are ordered in a series B_1, B_2, B_3, …, B_n such that B_{k+1} is a subblock of B_k, for all k = 1, 2, 3, …, n-1, are called **nested blocks**. *Nested blocks are stacks*, that is, the last block to be opened is the first to be closed. Hence, a block can be closed only if all its subblocks are closed. A very important fact to remember about subblocks is that once a subblock is exited, the hypothetical rule that initiated that block permits a *one-time* reference to the subderivation enclosed in that subblock; after that none of the lines of this subderivation can serve as an antecedent for any subsequent rule. In other words, once a subblock is exited and a reference is made to its lines as dictated by the hypothetical rule that initiated the subblock, the lines of the subblock become *inaccessible* to all subsequent stages of the derivation. The assumption of a hypothetical rule is "active" only within the block of its rule. After the block is exited the assumption becomes "inactive"; we say in this case that the assumption has been **discharged**; so an "active" assumption is an **undischarged** assumption. A discharged assumption no longer has a role in the derivation. Since all the subblocks of a block must be closed before the block can be closed, all the assumptions of the hypothetical rules invoked in some derivation must be discharged before inferring the conclusion of the main derivation.

The rule CP allows us to start a new block with any assumption we may want. There is no restriction on the form of the CP Assumption. A CP block may be closed at any line after its initial line. There is no restriction on the form of the last line of a CP block. However, the inference of CP is highly restricted: it is the conditional whose antecedent is the CP Assumption and whose consequent is the last line of the CP block. In the example above, a CP block was opened at line i with the CP Assumption '*Ps*'. The CP block is the 2nd block. It closes at line k with the sentence '$(\forall x)\neg(Rab \wedge Dx)$'. The CP inference is the conditional $Ps \rightarrow (\forall x)\neg(Rab \wedge Dx)$ on line k+1. The license "i–k, CP" abbreviates "From the derivation on lines i–k by Conditional Proof." Of course, the derivation on lines i–k is the derivation enclosed in the 2nd subblock.

5.3.2 The NDS we adopt in this book consists of seventeen **normal rules**, three **hypothetical rules**, and thirteen **replacement rules.**

5.3.2a **Normal Rules**: Every normal rule must be applied fully within an open block. **X, Y, Z, $\mathbf{Z_1}$**, and **$\mathbf{Z_2}$** are SL or PL sentences (or, where noted, PL formulas). In each of the rules of this category the order of the lines h, i, and j is irrelevant; the only order that matters is that line k comes after lines h, i, and j. The last five normal rules in this subsection do not apply to SL derivations.

Reiteration (**Reit**):

[n
 ⋮
 h X
 ⋮
 k X h, Reit
 ⋮
n]

Conjunction (**Conj**):

```
[n
    ⋮
    h   X
    ⋮
    i   Y
    ⋮
    k   X∧Y        h, i, Conj
    ⋮
n]
```

Simplification (**Simp**):

```
[n
    ⋮
    h   X∧Y
    ⋮
    k   X (or Y)   h, Simp¹
    ⋮
n]
```

Addition (**Add**):

```
[n
    ⋮
    h   X
    ⋮
    k   X∨Y (or Y∨X)           h, Add
    ⋮
n]
```

Disjunctive Syllogism (**DS**):

```
[n
    ⋮
    h   X∨Y
    ⋮
    i   ¬X
    ⋮
    k   Y          h, i, DS
    ⋮
n]
```

---

1 A sentence enclosed in parentheses and preceded by 'or' indicates an alternative conclusion or an alternative antecedent for the rule.

```
[n
      ⋮
      h    X∨Y
      ⋮
      i    ¬Y
      ⋮
      k    X          h, i, DS
      ⋮
n]
```

Modus Ponens (**MP**):

```
[n
      ⋮
      h    X→Y
      ⋮
      i    X
      ⋮
      k    Y          h, i, MP
      ⋮
n]
```

Modus Tollens (**MT**):

```
[n
      ⋮
      h    X→Y
      ⋮
      i    ¬Y
      ⋮
      k    ¬X         h, i, MT
      ⋮
n]
```

Biconditional Modus Ponens (**BcMP**):

```
[n
      ⋮
      h    X↔Y
      ⋮
      i    X
      ⋮
      k    Y          h, i, BcMP
      ⋮
n]
```

```
[n
   ⋮
   h   X⇔Y
   ⋮
   i   Y
   ⋮
   k   X          h, i, BcMP
   ⋮
n]
```

Biconditional Modus Tollens (**BcMT**):

```
[n
   ⋮
   h   X⇔Y
   ⋮
   i   ¬X
   ⋮
   k   ¬Y         h, i, BcMT
   ⋮
n]
[n
   ⋮
   h   X⇔Y
   ⋮
   i   ¬Y
   ⋮
   k   ¬X         h, i, BcMT
   ⋮
n]
```

Hypothetical Syllogism (**HS**):

```
[n
   ⋮
   h   X→Y
   ⋮
   i   Y→Z
   ⋮
   k   X→Z        h, i, HS
   ⋮
n]
```

Constructive Dilemma (**CD**):

```
[n
  ⋮
  h    X∨Y
  ⋮
  i    X→Z1
  ⋮
  j    Y→ Z2
  ⋮
  k    Z1∨Z2   h, i, j, CD
  ⋮
n]
```

Explosion (**Expl**): **Y** is any PL or SL sentence.

```
[n
  ⋮
  h    X∧¬X
  ⋮
  k    Y           h, Expl
  ⋮
n]
```

Universal Instantiation (**UI**): **s** is any PL *singular term*, **z** is a PL variable, and **X** is a PL formula that contains occurrences of **z** but no **z**-quantifiers. **X[s]** is the PL sentence formed by replacing *all* the occurrences of **z** in **X** with **s**.

```
[n
  ⋮
  h    (∀z)X
  ⋮
  k    X[s]        h, UI
  ⋮
n]
```

Universal Generalization (**UG**): **X** is a PL sentence, **s** is a PL *name* that occurs in **X**, and **z** is a PL variable that does not occur in **X**. **s** is **arbitrary** at line h, that is, **s** does not occur in any premise or undischarged assumption listed on line h or prior to it. **X[z]** is the PL formula formed by replacing *all* the occurrences of **s** in **X** with **z**.

```
[n
  ⋮
  h    X
  ⋮
  k    (∀z)X[z] h, UG
  ⋮
n]
```

Existential Generalization (**EG**): **X** is a PL sentence, **s** is a PL *singular term* that occurs in **X**, and **z** is a PL variable that does not occur in **X**. **X[z, s]** is a PL formula formed by replacing *one or more* of the occurrences of **s** in **X** with **z**.

```
[n
    ⋮
    h   X
    ⋮
    k   (∃z)X[z, s]  h, EG
    ⋮
n]
```

Identity (**Id**): **r** is any PL *singular term*. Id does not require an antecedent.

```
[n
    ⋮
    k   r = r          Id
    ⋮
n]
```

Substitution (**Sub**): **s** and **t** are PL *singular terms* and **X** is a PL sentence that contains occurrences of **s**. **X[t, s]** is a PL sentence formed by replacing *one or more* of the occurrences of **s** in **X** with **t**.

```
[n
    ⋮
    h   s = t (or t = s)
    ⋮
    i   X
    ⋮
    k   X[t, s]        h, i, Sub
    ⋮
n]
```

5.3.2b **Hypothetical Rules**: Every hypothetical rule starts a new block and adds an assumption, and it terminates with exiting the block and discharging the assumption. **X** and **Y** are SL or PL sentences (or, where noted, PL formulas). In these rules the order of the lines h and k is relevant and, therefore, must be observed. The last rule is not part of SL inference rules.

Reductio Ad Absurdum (**RAA**): 'RA' stands for 'Reductio Assumption'.

[n			
	⋮		
[n+l	h	$\mathbf{X}$	RA
	⋮		
n+l]	k	$\mathbf{Y}\wedge\neg\mathbf{Y}$	
	k+1	$\neg\mathbf{X}$	h–k, RAA
	⋮		
n]			

[n			
	⋮		
[n+l	h	$\neg\mathbf{X}$	RA
	⋮		
n+l]	k	$\mathbf{Y}\wedge\neg\mathbf{Y}$	
	k+1	$\mathbf{X}$	h–k, RAA
	⋮		
n]			

Conditional Proof (**CP**): 'CPA' stands for 'Conditional Proof Assumption'.

[n			
	⋮		
[n+l	h	$\mathbf{X}$	CPA
	⋮		
n+l]	k	$\mathbf{Y}$	
	k+1	$\mathbf{X}\rightarrow\mathbf{Y}$	h–k, CP
	⋮		
n]			

Existential Instantiation (**EI**): **s** is a PL *name*, **z** is a PL variable, and **X** is a PL formula that contains occurrences of **z** but no **z**-quantifiers. **s** satisfies three conditions: (1) it does not occur in any premise or undischarged assumption prior to line h, (2) it does not occur in (∃**z**)**X**, and (3) it does not occur in **Y**. **X**[**s**] is the PL sentence formed by replacing *all* the occurrences of **z** in **X** with **s**. 'EIA' stands for 'Existential Instantiation Assumption'.

[n			
	⋮		
	h-1	(∃**z**)**X**	
[n+l	h	**X**[**s**]	EIA, **s**
	⋮		
n+l]	k	**Y**	
	k+1	**Y**	h–k, EI
	⋮		
n]			

5.3.2c **Replacement Rules**: Every one of these rules must be applied fully within an open block. Replacement rules may be applied to a complete sentence or to a component of a sentence. All replacements may be performed in the forward or reverse direction. **X**, **Y**, and **Z** are SL sentences or PL formulas. The last rule is not part of SL inference rules.

Double Negation (**DN**):	$\neg\neg X$	$\Leftrightarrow$	X
Idempotence (**Idem**):	$X \wedge X$	$\Leftrightarrow$	X
	$X \vee X$	$\Leftrightarrow$	X
Commutation (**Com**):	$X \wedge Y$	$\Leftrightarrow$	$Y \wedge X$
	$X \vee Y$	$\Leftrightarrow$	$Y \vee X$
Association (**Assoc**):	$X \wedge (Y \wedge Z)$	$\Leftrightarrow$	$(X \wedge Y) \wedge Z$
	$X \vee (Y \vee Z)$	$\Leftrightarrow$	$(X \vee Y) \vee Z$
Distribution (**Dist**):	$X \wedge (Y \vee Z)$	$\Leftrightarrow$	$(X \wedge Y) \vee (X \wedge Z)$
	$X \vee (Y \wedge Z)$	$\Leftrightarrow$	$(X \vee Y) \wedge (X \vee Z)$
De Morgan's Laws (**DeM**):	$\neg(X \wedge Y)$	$\Leftrightarrow$	$\neg X \vee \neg Y$
	$\neg(X \vee Y)$	$\Leftrightarrow$	$\neg X \wedge \neg Y$
Material Conditional (**MC**):	$X \rightarrow Y$	$\Leftrightarrow$	$\neg X \vee Y$
Negated Conditional (**NC**):	$\neg(X \rightarrow Y)$	$\Leftrightarrow$	$X \wedge \neg Y$
Contraposition (**Cont**):	$X \rightarrow Y$	$\Leftrightarrow$	$\neg Y \rightarrow \neg X$
Exportation (**Expr**):	$X \rightarrow (Y \rightarrow Z)$	$\Leftrightarrow$	$(X \wedge Y) \rightarrow Z$
Biconditional (**Bc**)	$X \leftrightarrow Y$	$\Leftrightarrow$	$(X \rightarrow Y) \wedge (Y \rightarrow X)$
Negated Biconditional (**NBc**):	$\neg(X \leftrightarrow Y)$	$\Leftrightarrow$	$\neg X \leftrightarrow Y$
	$\neg(X \leftrightarrow Y)$	$\Leftrightarrow$	$X \leftrightarrow \neg Y$
Negated Quantifiers (**NQ**):	$\neg(\forall z)X$	$\Leftrightarrow$	$(\exists z)\neg X$
	$\neg(\exists z)X$	$\Leftrightarrow$	$(\forall z)\neg X$

5.3.2:C Commentary on 5.3.2

All of these rules can be justified on a semantical basis. Since not all of them are applicable to SL, we will only consider justifications that utilize the semantics of PL. However, any justification we give for an inference rule that is applicable also to SL can be trivially transformed to one that utilizes only SL semantics. The transformation is trivial since the truth conditions of the sentential connectives are the same in SL and PL; all that is required is to replace any mention of a PL interpretation with that of an SL truth valuation. Given the scope of this book, we will not be able to give complete justifications for all these rules, but we can make a convincing case for some of them.

For each normal rule of inference we need to show that any PL interpretation that makes the inference's antecedents true also makes the inference's conclusion true. This would prove that the rule is sound (i.e., truth-preserving). We will use the letter 'J' to stand for any PL interpretation. The rule of Reiteration is obvious. The soundness of Conjunction and Simplification follows immediately from the truth conditions of the conjunction: the conjunction $\mathbf{X} \wedge \mathbf{Y}$ is true on J if and only if both conjuncts **X** and **Y** are true on J. Addition is sound, since if a sentence **X** is true on J, then $\mathbf{X} \vee \mathbf{Y}$ or $\mathbf{Y} \vee \mathbf{X}$ is also true on J no matter what **Y** is. The soundness of Disjunctive Syllogism follows from the fact that if $\mathbf{X} \vee \mathbf{Y}$ is true on J and one of the disjuncts is false on J, then the other disjunct must be true on J; for if it were not, $\mathbf{X} \vee \mathbf{Y}$ would be false. The soundness of Modus Ponens and Modus Tollens is an immediate consequence of the truth conditions of the conditional. If $\mathbf{X} \rightarrow \mathbf{Y}$ is true on J, then either **X** is false or **Y** is true on J. It follows that if **X** is true (i.e., not false) on J, **Y** must be true on J, and if **Y** is false (i.e., not true) on J, **X** must be false on J. The truth conditions of the biconditional justify the rules of Biconditional Modus Ponens and Biconditional Modus Tollens. If $\mathbf{X} \leftrightarrow \mathbf{Y}$ is true on J, then J assigns identical truth values to **X** and **Y**; hence if one of them is true on J, the other is true as well, and if one of them is false on J, the other must also be false. Hypothetical Syllogism is justified on the basis of the truth conditions of the Conditional. Assume that $\mathbf{X} \rightarrow \mathbf{Y}$ and $\mathbf{Y} \rightarrow \mathbf{Z}$ are true on J. There are two cases: **X** is true on J or **X** is false on J. If **X** is false on J, $\mathbf{X} \rightarrow \mathbf{Y}$ is true on J. On the other hand, if **X** is true on J, then **Y** is true on J (since $\mathbf{X} \rightarrow \mathbf{Y}$ is true on J), and hence **Z** is true on J (since $\mathbf{Y} \rightarrow \mathbf{Z}$ is true on J); this shows that $\mathbf{X} \rightarrow \mathbf{Z}$ is true on J. The soundness of Constructive Dilemma is a consequence of the truth conditions of the disjunction and the conditional. Assume that $\mathbf{X} \vee \mathbf{Y}$, $\mathbf{X} \rightarrow \mathbf{Z}_1$, and $\mathbf{X} \rightarrow \mathbf{Z}_2$ are all true on J. By the truth conditions of the disjunction either **X** is true on J or **Y** is true on J (or both). Given the truth conditions of the conditional, it follows that if **X** is true on J, $\mathbf{Z}_1$ must be true on J, and if **Y** is true on J, then $\mathbf{Z}_2$ must be true on J; hence the disjunction $\mathbf{Z}_1 \vee \mathbf{Z}_2$ is true on J.

The justification for Explosion is less straightforward and less intuitive than the rest of the normal rules. We said that these rules are truth-preserving, and we also said that this means that any PL interpretation on which the inference's antecedents are true makes the inference's conclusion true. But the inference's antecedent of Explosion is a sentence of the form $\mathbf{X} \wedge \neg \mathbf{X}$, which is logically false. So it is unclear what it means to say that if the antecedent is true, then the conclusion is true as well. I will give two justifications for this rule: the first justification is the standard one, but the second is more intuitive even though it is a form of "logic fiction." Let us examine carefully the definition of "truth-preserving." To simplify matters, we will use the notation "J(Σ) = T" to mean that every member of the set Σ is true on J. We assume that Σ consists of the antecedents of some inference. We have the following definition:

> An inference $\Sigma \vdash \mathbf{Y}$ is truth-preserving if and only if for every PL interpretation J for Σ and **Y**, if J(Σ) = T, then J(**Y**) = T.

This definition will need to be modified to be applicable to more complex rules of inference; but it will do for now. In the case of Explosion, Σ consists of only one member, which is a sentence of the form $\mathbf{X} \wedge \neg \mathbf{X}$. All PL interpretations make this sentence false, that is, for every PL interpretation J for **X**, J($\mathbf{X} \wedge \neg \mathbf{X}$) = F. In other words, the antecedent of the conditional 'if-then' above is false for any given PL interpretation. If the English conditional is read as a material conditional, then this conditional is true for any given PL interpretation. It follows that the following universal conditional is true: for every PL interpretation J for **X** and **Y**, if J($\mathbf{X} \wedge \neg \mathbf{X}$) = T, J(**Y**) = T, no matter what **Y** is. The truth of the conditional is guaranteed by the fact that its antecedent is logically false. But this conditional is the right-hand side of the definition above; hence the inference $\{\mathbf{X} \wedge \neg \mathbf{X}\} \vdash \mathbf{Y}$ is a truth-preserving inference according to that definition. An important observation to make is that the correctness of this reasoning depends on attributing truth conditions to the English conditional above, which is the right-hand side of the definition, similar to the truth conditions of the material conditional '→'. This in fact is a standard feature of the metatheory of classical logic. The connectives of the metatheory, such as 'if-then' and 'or', have the same truth conditions as their formal counterparts, such as '→' and '∨'.

The second justification presupposes an impossible state of affairs: there is a PL interpretation that makes a sentence of the form $\mathbf{X} \wedge \neg \mathbf{X}$ true. This, of course, is logic fiction, for the semantics of PL does not allow for the existence of such a PL interpretation. But we will indulge ourselves in assuming that such a PL interpretation exists. We call it O. If Explosion is a sound rule, then *every* sentence **Y** that is interpretable by O must be true on O. This is because Explosion licenses the inference of any sentence from a contradiction. We will show that this is the case by invoking a famous argument that is due to the American philosopher and logician C. I. Lewis (1883–1964).

Lewis's Argument for Explosion

1 Suppose that a sentence of the form $\mathbf{X} \wedge \neg \mathbf{X}$ is true on some PL interpretation O.
2 From 1: since the inference rule Simplification is sound, **X** is true on O.
3 From 2: since the inference rule Addition is sound, $\mathbf{X} \vee \mathbf{Y}$ is true on O, for any sentence **Y** that is interpretable by O.
4 From 1: since the inference rule Simplification is sound: $\neg \mathbf{X}$ is true on O.
5 From 3 and 4: since the inference rule Disjunctive Syllogism is sound, **Y** is true on O, which is the desired conclusion.

This argument is standardly presented without the logic fiction. It is employed to show that the inference rule of Explosion is derivable from the rules of Simplification, Addition, and Disjunctive Syllogism. The argument shows that if these three rules are justified, then Explosion is equally justified. We modified the setting of the argument to show that if these three rules are truth-preserving, then Explosion is also truth-preserving without assuming that the English conditional has the same truth conditions as the material conditional. Lewis's original argument is stronger: it aims to show that if we have reason to believe that Simplification, Addition, and Disjunctive Syllogism are correct rules of inference, then we have no option but to accept the correctness of Explosion. Counterarguments against Lewis's argument have traditionally focused on showing that Disjunctive Syllogism is unjustifiable. We will not be able to consider any of these counterarguments here, for they rely on logical resources that are beyond the scope of this book.

Universal Instantiation is easy to justify. The rule licenses the inference of any substitutional instance **X**[**t**] from a universally quantified sentence $(\forall \mathbf{z})\mathbf{X}$. The truth conditions of the universal quantifier state that if $(\forall \mathbf{z})\mathbf{X}$ is true on J, then every *basic* substitutional instance of it is true on J. Theorem 4.2 (Chapter Four) asserts that if all basic substitutional instances of $(\forall \mathbf{z})\mathbf{X}$ are true on J, then all substitutional instances of $(\forall \mathbf{z})\mathbf{X}$ are also true on J. Recall that a basic substitutional instance of $(\forall \mathbf{z})\mathbf{X}$ is obtained when a PL *name* is substituted for the variable **z** in all its occurrences in **X**, and a substitutional instance is obtained when a *singular term* is substituted for **z** in all its occurrence in **X**. Existential Generalization has a similar justification. The rule allows us to infer an existentially quantified sentence (∃**z**)**X**[**z**, **s**] from a substitutional instance **X** in which the singular term **s** occurs but the variable **z** does not occur. According to Theorem 4.2, if a substitutional instance of (∃**z**)**X**[**z**, **s**] is true on J, then a basic substitutional instance of the existentially quantified sentence is true on J. But now according to the truth conditions of the existential quantifier, if a basic substitutional instance of (∃**z**)**X**[**z**, **s**] is true on J, then (∃**z**)**X**[**z**, **s**] is true on J as well. This shows that Existential Generalization is sound.

Universal Generalization (UG) is more involved. This formal rule is supposed to correspond to the informal mode of reasoning that we previously called arbitrary-object proof. The idea behind this kind of proof is that if a statement is proved for some arbitrary object of a certain type, then it is proved for *all* objects of that type. An arbitrary object of type T is an object such that no assumption about it is permitted in the proof other than its membership in type T. The same idea is behind UG. If we prove some **X** in which the name **s** occurs but the variable **z** does not occur, and **s** does not appear in any premise of the proof or any undischarged assumption through the line at which **X** occurs, then **s** is an arbitrary name at that line, that is, **s** designates an arbitrary object. Thus we should be entitled to infer (∀**z**)**X**[**z**], where **X**[**z**] is obtained from **X** by substituting **z** for **s** in all its occurrences. Loosely speaking, we have proved that **X** is true of all objects.

Let us state precisely what it means to say that UG is truth-preserving. Suppose that we have a derivation D whose line h contains the sentence **X** in which a name **s** occurs and a variable **z** does not occur. Let Σ be the set whose members are all the premises of the derivation D, which are listed at the zero stage, and all the undischarged assumptions through line h. Without loss of generality, further suppose that the sentence (∀**z**)**X**[**z**] occurs at line h+1. To show that UG is sound we must show that for any PL interpretation J, if J(Σ) = T, then J((∀**z**)**X**[**z**]) = T. Unfortunately, we will not be able to prove the soundness of UG here. The proof invokes a principle called the "Principle of Mathematical Induction," which we will not discuss in this book.

It remains to show that the inference rules Identity and Substitution are sound. Identity is trivial: for any given singular term **t**, the referent of **t** is a single object on all PL interpretations for **t**. Hence **t** = **t** is logically true. The truth of **t** = **t** does not follow from any antecedent; it is true on every PL interpretation for **t**. Substitution is sound too. If two singular terms **t** and **s** have the same referent, then substituting one of them for the other in any sentence would not change the assertions made by that sentence on J. Thus these sentences would have the same truth value. In other words if J(**t** = **s**) = T, J(**X**) = T, and **s** occurs in **X**, then replacing **s** with **t** in **X** would not change the assertion that **X** makes on J. Therefore the resulting sentence **Y** has the same truth value on J as **X** does, which is true.

Hypothetical rules present us with a conceptual challenge. We defined a rule of inference to be sound if and only if every PL interpretation that makes the inference's antecedents true makes the inference's conclusion true as well. But the "antecedent" of a hypothetical rule is a *derivation*. Since it makes no sense to say that a PL interpretation makes a derivation true, we must revise our conception of soundness for these rules. We should not lose sight of the dictum that to be sound is to be truth-preserving. Let R be a hypothetical rule and let RD be the subderivation that is required by R. RD relies on a set of antecedents. Call this set Σ. Σ consists of the premises of the main derivation, that is, the sentences that are listed at the zero stage of the derivation, and all the undischarged assumptions that occur prior to RD, *together* with the assumption made by R (recall that every hypothetical rule begins with an assumption). The inference's conclusion of R is made immediately after the last sentence of RD. In other words, the conclusion of R is inferred when the subblock that encloses RD is exited. But once this subblock is exited, the assumption introduced by R becomes "inactive," i.e., discharged. Let the conclusion of R be **X** and assume that it occurs on line m. The set of the premises of the derivation together with the undischarged assumptions made when line m is reached is Σ minus the assumption introduced by R. This is the same set of premises and undischarged assumption that we had before invoking RD. Call this set Δ. Now we are ready to define what we mean by saying that R is sound. R is sound if and only if any interpretation that makes the members of Δ all true makes **X** true as well. To prove the soundness of any of the hypothetical rules, we need to employ the Principle of Mathematical Induction; we said above that we do not plan to discuss this principle in this book. So we will not be able to present proofs for the soundness of these rules here.

A full justification of a replacement rule R also requires the Principle of Mathematical Induction. One must prove that if a PL interpretation I makes some sentence **X** true, then any sentence that is obtained from **X** by replacing a component of it with another formula according to the rule R is also true on I. Such a proof must work for any PL sentence no matter how complex it may be. Thus in order to prove the soundness of the replacement rule Material Conditional, it is not enough to show that any sentence of the form **X→Y** may be replaced with a sentence of the form ¬**X**∨**Y**, and vice versa, but we must show that any PL sentence **Z** that contains a component formula of the form **X→Y** may be replaced with the sentence **Z*** that is obtained from **Z** by replacing the component **X→Y** with the component formula ¬**X**∨**Y**, and vice versa. We need this general proof to justify, for example, inferring (∀*x*)(*Rxc*∧(∃z)(*Pz*⇔(¬*Qza* ∨*Dxz*))) from (∀*x*)(*Rxc*∧(∃z)(*Pz*⇔(*Qza*→*Dxz*))), where the component formula (*Qza*→*Dxz*) is replaced with the formula (¬*Qza*∨*Dxz*). It is easy to demonstrate the soundness of each replacement rule when applied to whole sentences. All that is required is to show that the two sentences on both sides of the double arrow '⇔' are logically equivalent. This would demonstrate that inferring each side from the other is truth-preserving. The difficulty is to prove that the inference is truth-preserving even when the replacement rule is applied to components of sentences and formulas. To be sure, such proofs are not pretty; but they are straightforward once one learns how to carry out one of them. We will omit the proofs of the soundness of replacement rules, because these proofs rely on the Principle of Mathematical Induction.

The reader is invited to prove these rules for whole sentences. These proofs are trivial. One may use either the method of truth analysis or of truth tables to show that every pair of sentences that occur on both sides of '⇔' is a pair of logically equivalent sentences. These proofs would be sufficient for SL and PL, since the sentential connectives have the same truth conditions in SL and PL. The only exception is the last rule: Negated Quantifiers. To show that this rule is sound when applied to whole sentences, we need to prove that both sides of the double arrow are logically equivalent in PL, that is, they have the same truth value on every PL interpretation for them. The logical equivalence between sentences of the forms ¬(∀**z**)**X** and (∃**z**)¬**X** and between sentences of the forms ¬(∃**z**)**X** and (∀**z**)¬**X** are immediate consequences of the truth conditions of the quantifiers and the negation. For instance, any PL interpretation I that makes ¬(∀**z**)**X** true makes at least one basic substitutional instance of (∀**z**)**X** false. In other words, there is a name **t** in the LN of I such that **X**[**t**] is false on I. But this means that ¬**X**[**t**] is true on I. Since ¬**X**[**t**] is a basic substitutional instance of the sentence (∃**z**)¬**X**, the latter is true on I. Of course, these simple proofs do not show the soundness of the replacement rules. Applying these rules to whole sentences is but a very small part of the full range of their applicability, as we explained above.

Replacement rules are unique in their applicability to components of sentences and formulas. The rest of the NDS rules must be applied only to whole sentences that occupy complete lines. Applications that violate this condition might not be sound. Consider, for instance, the following three inferences using normal rules.

1. From $(A \wedge B) \rightarrow C$ by Simplification (applied to $A \wedge B$): $A \rightarrow C$.
2. From $Pa \wedge \neg(\forall x)Rx$ by Universal Instantiation (applied to $(\forall x)Rx$): $Pa \wedge \neg Ra$.
3. From $\neg(Pa \rightarrow Ra)$ by Existential Generalization (applied to Ra): $\neg(Pa \rightarrow (\exists x)Rx)$.

All of these inferences are unsound. If we assign T to A and F to B and C, $(A \wedge B) \rightarrow C$ is true and $A \rightarrow C$ is false. Hence the first inference is not truth-preserving. For the second inference, let a stand for John, P for being tall, and R for being taller than 6 feet. Suppose that John is actually taller than 6 feet. The first sentence says that John is tall and not everyone is taller than 6 feet, which is true, and the second says that John is tall but he is not taller than 6 feet, which is false. So the inference is not truth-preserving. For the last inference, let a stand for number 7, P for being odd, and R for being divisible by 2. The first sentence asserts that it is not the case that if 7 is odd then it is divisible by 2, which is true, and the second asserts that it is not the case that if 7 is odd then there is a number that is divisible by 2, which is clearly false. Again, this inference is not truth-preserving. These examples are sufficient to show that if a normal (or a hypothetical) rule is applied to a component of a sentence, the inference might not be sound. We state these conditions compactly in the following dictum: normal and hypothetical rules apply to whole sentences but not to their parts, while replacement rules apply to both whole sentences and their parts.

5.3.3 The Natural Deduction System (NDS) we described in 5.3.2 is not the **standard deduction system** that is given in many logic textbooks. The standard deduction system for PL consists of seventeen rules only. Our system contains thirty-three rules (some of which are two-part rules). All these rules are traditional rules of inference. The standard deduction system consists of Reiteration and **introduction** and **elimination rules**: for every PL logical symbol, there is an inference rule that introduces the symbol into the conclusion of the rule and there is an inference rule that eliminates the logical symbol from the conclusion of the rule. Since there are eight PL logical symbols—five sentential connectives, two quantifiers, and the identity predicate—there are seventeen rules of inference in the standard deduction system.

5.3.3:C Commentary on 5.3.3

Our purpose for giving so many rules in NDS is to allow us to formalize natural demonstrations in a relatively straightforward fashion. Nevertheless, natural demonstrations can be analyzed with the aid of many rules of inference, many of which have no formal counterparts in our Natural Deduction System. For instance, the rule that we called "Conditional Instantiation," and which we discussed in 5.1:C1, has no formal counterpart among our NDS rules. There is no compelling reason why we did not include in our system a formal representation of Conditional Instantiation. It is a sound rule of inference and commonly invoked in natural reasoning. Furthermore, it is not hard to formalize. A formal counterpart of this rule is an inference rule that has two antecedents of the forms $(\forall \mathbf{z})(\mathbf{X} \rightarrow \mathbf{Y})$ and $\mathbf{X}[\mathbf{t}]$, where $\mathbf{t}$

is any singular term, and whose conclusion is of the form **Y[t]**. But we elected not to include such a rule. If we invoke Conditional Instantiation in some demonstration and want to formalize this demonstration, we can always capture it formally by applying two rules, Universal Instantiation and Modus Ponens. This choice sounds arbitrary, and, in fact, it is mostly arbitrary. This rule is not one of the traditional rules of inference of *modern* classical logic. However, as we explained in note 1, p. 330, the traditional system of Aristotelian Syllogistic Logic has a rule of inference that is called "Barbara," which is very similar to Conditional Instantiation. So it is really a matter of tradition. On the other hand, there is a good reason for not including in NDS formal counterparts of all the natural modes of demonstrative reasoning that we can identify as sound rules of inference. The reason is that there is a large number of such natural modes of reasoning. Including a formal rule for each one of them is not feasible: it would render the deduction system unmanageable. So it seems that the best approach is to include a relatively large number of formal rules that mirror common modes of demonstrative reasoning, while keeping the size and complexity of the deduction system within reasonable limits. This goal guided our selection of the natural deduction system that we described in the previous subsection.

There is also a philosophical reason for including almost all the traditional rules of formal classical logic in our NDS. Different logical systems might question the justification of some of these rules. Not all of these rules are included in the standard deduction system, which we will describe in this subsection. To be sure, many logical systems focus their disagreement with classical logic on some of the rules of the standard deduction system. But these rules do not exhaust all the debates. For instance, there are non-classical logical systems that challenge the correctness of Disjunctive Syllogism. As we saw, Disjunctive Syllogism is employed in an argument that is meant to justify Explosion. There are many non-classical systems that reject Explosion. Thus these systems must address the correctness of Disjunctive Syllogism. Disjunctive Syllogism, however, is not part of the standard deduction system. A possible reaction to this point is that although Disjunctive Syllogism is not part of the standard rules of inference, it can be derived from them. Hence anyone who rejects Disjunctive Syllogism must identify the standard rules that give rise to Disjunctive Syllogism and explain what is wrong with them. This is a reasonable response, as all the rules we included in our NDS are derivable from the standard rules, which we will describe in this subsection. However, there is still value to NDS that is not fully exhibited by the standard rules: Let us continue with our example of DS. One might be willing to reject some of the standard rules that give rise to DS but still accept DS on independent grounds. Hence someone who rejects Lewis's argument for Explosion would have to address the correctness of DS independently of the standard rules of inference.

The standard deduction system consisting of seventeen rules was discovered by the German mathematician and logician Gerhard Gentzen (1909–45). We will refer to it as the **Gentzen Deduction System** (GDS). The reader, however, must be warned that many logic textbooks use the title "natural deduction system" to describe this standard system instead of the one we introduced in the preceding subsection. As indicated above, GDS consists of Reiteration plus eight pairs of introduction and elimination rules, abbreviated as #I and #E, respectively, where # is replaced by a sentential connective, a quantifier, or the identity predicate. We have already studied several of these rules in our NDS.

We will begin by listing the ones we already introduced. (1) Reiteration, (2–3) **Conjunction Introduction** ($\wedge$I) is the NDS rule Conjunction, and **Conjunction Elimination** ($\wedge$E) is the NDS rule Simplification. It is clear why these are introduction and elimination rules. The first takes two antecedents, none of which needs to be a conjunction, and infers a conclusion that is a conjunction; hence the conjunction sign is *introduced* by this rule into its conclusion. The second takes an antecedent that is a conjunction and infers a conclusion that is not necessarily a conjunction; hence the conjunction sign is *eliminated* by this rule from its conclusion. The same explanation applies to all these rules. (4–5) **Conditional Introduction** (→I) is the NDS rule Conditional Proof, and **Conditional Elimination** (→E) is the NDS rule Modus Ponens. (6–7) **Universal Introduction** (∀I) is the NDS rule Universal Generalization, and **Universal Elimination** (∀E) is the NDS rule Universal Instantiation. (8–9) **Existential Introduction** (∃I) is the NDS rule Existential Generalization, and **Existential Elimination** (∃E) is the NDS rule Existential Insanitation. (10–11) **Identity Introduction** (=I) is the NDS rule Identity, and **Identity Elimination** (=E) is the NDS rule Substitution.

The rest of the standard rules are either part of our NDS rules, modifications of them, or new ones. (12–13) **Disjunction Introduction** (∨I) is the NDS rule Addition, and **Disjunction Elimination** (∨E) is the following hypothetical rule (j is an integer greater than 1):

[n			
	⋮		
	h	**X∨Y**	
[n+1	h+1	**X**	∨E Assumption
	⋮		
n+1]	m	**Z**	
[n+j	m+1	**Y**	∨E Assumption
	⋮		
n+j]	k	**Z**	
	k+1	**Z**	h, (h+1)–m, (m+1)–k, ∨E
	⋮		
n]			

(14–15) **Biconditional Elimination** (↔E) is the NDS rule Biconditional Modus Ponens, and **Biconditional Introduction** (↔I) is the following hypothetical rule (j is an integer greater than 1):

[n			
	⋮		
[n+1	h	**X**	↔I Assumption
	⋮		
n+1]	m	**Y**	
[n+j	m+1	**Y**	↔I Assumption
	⋮		
n+j]	k	**X**	
	k+1	**X↔Y (or, Y↔X)**	h–m, (m+1)–k, ↔I
	⋮		
n]			

(16) **Negation Introduction** ($\neg$I) is a modification of the first part of the NDS rule Reductio Ad Absurdum (the order of the sentences **Y** and $\neg$**Y** is irrelevant).

[n			
	⋮		
[n+1	h	**X**	RA
	⋮		
	k–1	**Y**	
n+1]	k	$\neg$**Y**	
	k+1	$\neg$**X**	h–k, RAA
	⋮		
n]			

(17) **Negation Elimination** is a modification of the second part of the NDS rule Reductio Ad Absurdum (the order of the sentences **Y** and $\neg$**Y** is irrelevant).

[n			
	⋮		
[n+1	h	$\neg$**X**	RA
	⋮		
	k–1	**Y**	
n+1]	k	$\neg$**Y**	
	k+1	**X**	h–k, RAA
	⋮		
n]			

These are the seventeen rules of GDS. All the NDS rules introduced in Subsection 5.3.2 can be derived from these seventeen rules. The reverse is also true. All the GDS rules are either NDS rules or can be derived from NDS rules. Thus from a logical point of view the two systems, NDS and GDS, are equivalent. Because of this fact, any debate that concerns one of the NDS rules can be reduced to a debate about the correctness or the scope of some of the GDS rules. The first philosophical significance of GDS, therefore, is that it focuses the debate between systems of non-classical logic and classical logic on some of these introduction and elimination rules. The second philosophical significance is that the GDS rules may be considered as "defining" the sentential connectives, quantifiers, and the identity predicate. We defined all the logical symbols of PL semantically by giving their truth conditions. We then invoked these truth conditions to justify many of the NDS rules. In fact, the Soundness Theorem for PL offers complete semantical justifications for all the NDS and GDS rules. There is, however, a different approach to the "nature" of these logical symbols. Rather than defining them semantically by giving their truth conditions, this approach defines the logical symbols of PL by giving their proof-theoretic roles, that is, their introduction and elimination rules. For example, the conditional may be defined as the binary sentential connective that can be introduced into any conclusion by Conditional Proof (i.e., Conditional Introduction) and that can be eliminated from any conclusion by Modus Ponens (i.e., Conditional Elimination). This formal approach to the logical symbols of PL reverses the direction of justification of the semantical approach. The main point is that the meaning of these logical symbols can be "fixed" by their introduction and elimination rules.[1] In a certain sense, the role of introduction and elimination rules in this approach is analogous to the role of truth conditions in the semantical approach. We explained in Chapter Four, that we presuppose a certain theory of meaning that is called the "truth conditional theory of meaning." This theory affirms that the meaning of a sentence is captured by its truth conditions. Hence the meaning of the logical symbols is given by their truth conditions. The formal approach affirms that the meaning of these logical symbols is given by their "provability conditions"—namely, their introduction and elimination rules.

In fact, the formal approach to the PL logical symbols might be considered as a consequence of the truth conditional theory of meaning. The formal approach is preferred by advocates of a certain theory of truth and meaning that is called "verificationism." Modern verificationism about the meaning of declarative sentences is associated with the British philosopher Michael Dummett (1925-2011); it is an heir of an important school of philosophy of mathematics that is called "intuitionism," which was founded by the Dutch mathematician L.E.J. Brouwer (1881-1966). At the risk of oversimplifying, intuitionism may be defined as the philosophy of mathematics that affirms that the truth of a mathematical proposition is reduced to the existence of a "constructive" proof for it. It is not easy to explain precisely what is meant by a "constructive proof." Roughly speaking, however, we may define a constructive proof as a proof in which an object of a certain type is shown to exist by actually constructing this object, or in which it is shown that no object of a certain type can exist by deriving a contradiction from the supposition that it is possible to construct such an object. In very general terms, intuitionism implies that the truth of a mathematical proposition is cashed out in terms of its provability conditions. We can now see that if the truth conditions of a proposition are reduced to its provability conditions, then the truth conditional theory of meaning implies that the meaning of a proposition must be captured by its provability conditions. If we take the introduction and elimination rules for a logical symbol as its provability conditions, then the preceding position entails that the meaning of a logical symbol is given by its introduction and elimination rules. This is not the place to evaluate philosophically any of these theories; such an evaluation is philosophically involved, and hence it is beyond the scope of this book. Our goal is only to explain the philosophical significance of the Gentzen Deduction System. The preceding discussion is adequate for giving a general idea about the philosophical issues that relate to this deduction system.

1 We say "fixed" with caution, since it has been pointed out by many logicians and philosophers that the introduction and elimination rules are not strong enough to fix the semantics of these logical symbols uniquely. However, a defender of the formal approach most likely would not be moved by this objection, since she might insist that whatever semantical significance these introduction and elimination rules attribute to the logical symbols is the whole of their semantical content. If this content is not fixed uniquely, then this fact is part of the semantical significance of these logical symbols.

5.4 Strategies for Constructing Formal Derivations

There are **three strategies** for constructing PL derivations: one semantical and two formal. The semantical strategy is **semantical reasoning**. This strategy relies on the truth conditions of the premises and the conclusion to show either that if the premises are true, then the conclusion is true, or that the truth of the premises together with the *falsity* of the conclusion generates a contradiction. We should be methodical in our reasoning, progressing through short steps. Since the rules of NDS are based on the truth conditions of the logical connectives and quantifiers, we should be able to capture formally the successive steps of our reasoning. In most cases, there will be NDS rules that correspond to the steps of the semantical reasoning, but occasionally we will have to follow a roundabout path in order to arrive *formally* at a conclusion of some reasoning step. This annoying divergence between semantical reasoning and PL derivations is due to the fact that PL proof theory is a *formal* enterprise: no inference is allowed unless it is licensed by an NDS rule, and each one of these rules can only be applied if the given sentences have exactly the same forms that appear in the definition of the rule. The first formal strategy is **goal analysis**. This strategy proceeds in reverse order, and hence it is useful only when the conclusion is known. One examines the main connective or quantifier of the conclusion and then asks how to "generate" a sentence with this form. For example, if the conclusion is a conditional, then we might try Conditional Proof; if it is a universally quantified sentence, then we might try Universal Generalization. Once we decide what sort of sentence we need in order to derive the conclusion, we apply the same strategy to this (discovered) sentence and continue in reverse fashion. If the strategy works well, we should be able to proceed from the conclusion to the premises. The second formal strategy is **simplification and detachment**. The terms 'simplification' and 'detachment' have technical meaning in logic. 'Simplification' refers to the NDS rule that allows us to infer the conjuncts of a conjunction, if the conjunction is given, and 'detachment' is usually used to refer to the rule Modus Ponens. We use them here *not* in their technical senses. In this strategy we try to *simplify* sentences by making them more amenable to the NDS rules and we try to *detach* as many component sentences as possible. For instance, if we have a conjunction, we use Simplification to break it up; if we have a universally quantified sentence, we use Universal Instantiation to get rid of the universal quantifier. There is no precise definition of what we mean by 'simplification and detachment' here. It is a matter of simplifying the sentences, removing their quantifiers, transforming negations into non-negations, breaking up the components, and so on. This strategy is sometimes the least useful because it is a trial-and-error approach. We manipulate the sentences, hoping to make it easier to see how to proceed. There are three important **reminders**: (1) all the sentences listed in a PL derivation are supposed to be true unless the sentence is a contradiction or it entails a contradiction; (2) when all else fails, use Reductio Ad Absurdum; and (3) if one strategy is not adequate, use multiple strategies.

5.4:C Commentary on 5.4

5.4:C1 We will illustrate the use of these strategies by working out several exercises. Here is a practice exercise demonstrating the **strategy of semantical reasoning**. We will construct a derivation to show that

$$\{(B \wedge D) \leftrightarrow \neg P,\ \neg(B \rightarrow \neg P)\} \vdash \neg D$$

1. We begin with the second premise. It says that the conditional $B \rightarrow \neg P$ is false. But the conditional is false if and only if its antecedent is true and consequent is false. Thus we conclude that B is true and $\neg P$ is false (that is, P is true). The NDS rule that corresponds to this reasoning is Negated Conditional (NC), which is one of the replacement rules.
2. Given that the biconditional $(B \wedge D) \leftrightarrow \neg P$ is true and $\neg P$ is false, it follows that $B \wedge D$ is also false. The relevant NDS rule is Biconditional Modus Tollens (BcMT).
3. If $B \wedge D$ is false, then at least one of the conjuncts must be false. But from 1 above we know that B is true. Hence D must be false, which is the desired conclusion ($\neg D$). The applicable NDS rule here is Disjunctive Syllogism (DS).

As we will see, some of these steps cannot be formalized verbatim; we will need to fill some gaps using other rules. Let us summarize what we learned from the analysis we performed. First, we apply NC to the second premise to get B and $\neg\neg P$. It is semantically obvious that if the conditional $B \rightarrow \neg P$ is false, then P must be true. The proof theory, since it is a formal enterprise, does not allow this "jump." According to the rule NC when we are given a sentence of the form $\neg(\mathbf{X} \rightarrow \mathbf{Y})$, we are permitted to infer $\mathbf{X}$ and $\neg\mathbf{Y}$. Therefore if $\mathbf{Y}$ is $\neg P$, we ought to infer $\neg\neg P$, not P; if we need P, we can always infer P from $\neg\neg P$ by the rule Double Negation (DN).[1] Second, we apply BcMT to the first premise and $\neg\neg P$ to obtain $\neg(B \wedge D)$. Third, we apply DS to get $\neg D$. However, DS cannot be applied directly to the negation of a conjunction. It can only be applied to a disjunction. So we must transform $\neg(B \wedge D)$ to a disjunction. The NDS rule that allows us to do this is De Morgan's Laws (DeM). Here is the complete derivation. ('P' stands for 'Premise'.)

[0	0.1	$(B \wedge D) \leftrightarrow \neg P$	P
	0.2	$\neg(B \rightarrow \neg P)$	P
	1	$B \wedge \neg\neg P$	0.2, NC
	2	$\neg\neg P$	1, Simp
	3	$\neg(B \wedge D)$	0.1, 2, BcMT
	4	$\neg B \vee \neg D$	3, DeM
	5	B	1, Simp
	6	$\neg\neg B$	5, DN
0]	7	$\neg D$	4, 6, DS

The following practice exercise also demonstrates the **strategy of semantical reasoning**. We prove the following claim: S8 is a theorem of the set consisting of S1–S7. S1–S8 are given below.

S1 $(E \wedge N) \rightarrow H$
S2 $H \rightarrow (\neg D \vee V)$
S3 R
S4 $(R \wedge V) \rightarrow M$

---

1 We should warn that a sentence of the form $(\mathbf{X} \wedge \mathbf{Y})$ cannot be inferred from $\neg(\neg\mathbf{X} \wedge \mathbf{Y})$ by DN. These negation signs negate different sentences: the first negates $(\mathbf{X} \wedge \mathbf{Y})$ and the second negates only $\mathbf{X}$.

S5 $M \rightarrow B$
S6 $B \rightarrow \neg H$
S7 D
S8 $E \rightarrow \neg N$

The conclusion is a conditional, so one naturally thinks of Conditional Proof (CP). In this case we assume that the antecedent of the conclusion, E, is true, and then try to derive the consequent of the conclusion, $\neg N$, from this assumption, together with the seven premises. However, this approach is not the easiest, because once we assume that E is true, there isn't much we can do with this assumption on its own, as a careful examination of the premises readily reveals. It seems we would be in a much stronger position if we assume that the conclusion is false and try to derive a contradiction. If the conditional $E \rightarrow \neg N$ is false, E is true and $\neg N$ is false, that is, N is true. So a Reductio Assumption gives us two other sentences, in addition to the premises, to work with. This should make things much easier. Thus we will follow a Reductio Ad Absurdum (RAA) approach.

1. We begin by assuming that the conclusion, $E \rightarrow \neg N$, is false. This is the Reductio Assumption (RA). It follows that E and N are true (because $\neg N$ is false). The NDS rule that corresponds to this reasoning is Negated Conditional (NC).
2. S1 asserts that the conditional $(E \wedge N) \rightarrow H$ is true. Since E and N are true, H is true as well. The relevant NDS rule is Modus Ponens (MP).
3. S2 asserts that the conditional $H \rightarrow (\neg D \vee V)$ is true. Given that H is true, it follows that $\neg D \vee V$ is also true. The rule again is MP.
4. Looking down the list of premises, we see S6, which says that the conditional $B \rightarrow \neg H$ is true. But $\neg H$ is false (because H is true). Therefore B is false. The applicable rule here is Modus Tollens (MT).
5. S5 asserts that the conditional $M \rightarrow B$ is true. Given that B is false, we infer that M is false as well. The rule is again MT.
6. The same reasoning applies to S4. Hence we obtain that $R \wedge V$ is false. This is another application of MT.
7. But if the conjunction $R \wedge V$ is false, then one of the conjuncts is false. The rule that captures this reasoning is De Morgan's Laws (DeM).
8. We know from S3 that R is true. This implies that V is false. The relevant rule here is Disjunctive Syllogism (DS).
9. Now we need to look at our previous reasoning to see what to do with the fact that V is false. We notice that 3 above says that $\neg D \vee V$ is true. But if V is false, $\neg D$ must be true. This is another application of DS.
10. Thus we obtain a contradiction, because S7 asserts that D is true. Our reductio proof is completed. We showed that the premises, together with the negation of the conclusion, entail a contradiction. So we reject the Reductio Assumption and conclude that $E \rightarrow \neg N$ is true.

We formalize the previous steps, using the rules mentioned, and filling gaps when necessary.

[0	0.1	$(E \wedge N) \rightarrow H$	P
	0.2	$H \rightarrow (\neg D \vee V)$	P
	0.3	R	P
	0.4	$(R \wedge V) \rightarrow M$	P
	0.5	$M \rightarrow B$	P
	0.6	$B \rightarrow \neg H$	P
	0.7	D	P
[1	1	$\neg(E \rightarrow \neg N)$	RA
	2	$E \wedge \neg\neg N$	1, NC
	3	$E \wedge N$	2, DN
	4	H	0.1, 3, MP
	5	$\neg D \vee V$	0.2, 4, MP
	6	$\neg\neg H$	4, DN
	7	$\neg B$	0.6, 6, MT
	8	$\neg M$	0.5, 7, MT
	9	$\neg(R \wedge V)$	0.4, 8, MT
	10	$\neg R \vee \neg V$	9, DeM
	11	$\neg\neg R$	0.3, DN
	12	$\neg V$	10, 11, DS
	13	$\neg D$	5, 12, DS
1]	14	$D \wedge \neg D$	0.7, 13, Conj
0]	15	$E \rightarrow \neg N$	1–14, RAA

The practice exercise below demonstrates the **strategy of goal analysis**. We chose this strategy in order to illustrate the use of goal analysis, but it is equally feasible to use semantical reasoning. We prove that S4 is derivable from S1, S2, and S3. Here are the sentences.

S1 $(\forall z)(Tz \rightarrow Hz)$

S2 $(\exists x)Tx$

S3 $(\forall v)(Tv \rightarrow Mv)$

S4 $(\exists y)(Hy \wedge My)$

1. We want to derive $(\exists y)(Hy \wedge My)$. An existentially quantified sentence suggests that we use Existential Generalization (EG). In order to apply this rule, we must have a sentence with a name in place of the variable. EG is a liberal rule: no restriction is placed on the name. Thus we should have a sentence like $Ha \wedge Ma$.
2. In order to get the conjunction $Ha \wedge Ma$, we need to apply the rule Conjunction (Conj) to Ha and Ma.
3. Ha could be obtained from $(\forall z)(Tz \rightarrow Hz)$, which is S1, by Modus Ponens (MP), if we have Ta as well. So we need to get Ta.
4. If we have Ta, we can also get Ma from S3, which is $(\forall v)(Tv \rightarrow Mv)$, by MP.

5. *Ta* can be obtained from $(\exists x)Tx$, which is S2, by Existential Instantiation (EI).

Now we put these steps together in reverse order to obtain the desired conclusion.

[0	0.1	$(\forall z)(Tz \rightarrow Hz)$	P
	0.2	$(\exists x)Tx$	P
	0.3	$(\forall v)(Tv \rightarrow Mv)$	P
	1	$(\exists x)Tx$	0.2, Reit
[1	2	Ta	EIA, *a*
	3	$Ta \rightarrow Ha$	0.1, UI
	4	Ha	2, 3, MP
	5	$Ta \rightarrow Ma$	0.3, UI
	6	Ma	2, 5, MP
	7	$Ha \wedge Ma$	4, 6, Conj
1]	8	$(\exists y)(Hy \wedge My)$	7, EG
0]	9	$(\exists y)(Hy \wedge My)$	2–8, EI

A note about the use of EI: In order to initiate an EI block, the sentence that immediately precedes the block must be an existentially quantified sentence, such as $(\exists x)Tx$. The first line of the block, which is the assumption of the EI rule (EIA), is a basic substitutional instance of this sentence, such as *Ta*. The name of the existential instantiation, such as '*a*', is restricted: it must not occur in any premise or undischarged assumption prior to the line of EIA, it must not occur in the existentially quantified sentence that immediately precedes the EI block, and it must not occur in the last line of the EI block. Thus the EI block can be terminated at any sentence that does not contain the name of the instantiation. In this case, we ended the block at the sentence $(\exists y)(Hy \wedge My)$, because this sentence does not contain the name '*a*' and because it is the conclusion we want to derive. After the EI block is closed, the last sentence in the block is repeated at the line that immediately follows the EI block. We repeat the sentence outside the block to indicate that the sentence no longer depends on EIA, because after closing a subblock the assumption of that subblock is *discharged*, that is, it is no longer relevant to the lines after the subblock.

The next practice exercise demonstrates a **combined strategy of simplification and detachment and semantical reasoning**. We will show that the set of PL sentences

S1 $(\forall u)Du \vee (\forall u)Eu$
S2 $(\exists v)Qvv$
S3 $(\forall x)(\forall y)(Qxy \rightarrow (\neg Dx \wedge \neg Ey))$

is inconsistent. Thus we need to derive a contradiction from this set. Since we do not know what this contradiction is, the strategy of goal analysis is not helpful at this point. We could employ only semantical reasoning (and in fact that approach would be the easiest), but we will use a combined strategy of simplification and detachment and semantical reasoning in order to illustrate this approach.

1. We first simplify $(\exists v)Qvv$ by using Existential Instantiation (EI) to obtain Qaa.
2. Now we simplify $(\forall x)(\forall y)(Qxy \rightarrow (\neg Dx \wedge \neg Ey))$ by applying Universal Instantiation (UI) twice. We get $Qaa \rightarrow (\neg Da \wedge \neg Ea)$. Notice that we instantiated the sentence with the name '*a*' for the variables '*x*' and '*y*'. We did this because we want to use Qaa to detach the consequent $\neg Da \wedge \neg Ea$.
3. We detach $\neg Da \wedge \neg Ea$ from $Qaa \rightarrow (\neg Da \wedge \neg Ea)$ by Modus Ponens (MP) since we have Qaa.
4. We simplify the conjunction $\neg Da \wedge \neg Ea$ by breaking it up into two sentences, using the NDS rule Simplification (Simp). We now have $\neg Da$ and $\neg Ea$.
5. Since it is not clear how to proceed further with this strategy of simplification and detachment, we revert to semantical reasoning at this stage. So far we have $\neg Da$, $\neg Ea$, and $(\forall u)Du \vee (\forall u)Eu$ (we have not invoked S1 yet). Observe that we cannot instantiate the universal quantifiers in S1 because these quantifiers are "inside" the sentence, and we are not permitted to apply NDS normal rules to components of sentences. Only replacement rules can be so applied. $\neg Da$ asserts that *a* is not *D*. This implies that not everything is *D*. Hence $(\forall u)Du$, which says that everything is *D*, is false. Semantically, it is obvious that $\neg Da$ entails $\neg(\forall u)Du$, but there is no NDS rule that allows us to infer $\neg(\forall u)Du$ from $\neg Da$ directly. We need to apply two rules. First we use Existential Generalization (EG) to get $(\exists u)\neg Du$, and second we use Negated Quantifier (NQ) to derive $\neg(\forall u)Du$.
6. Given that $(\forall u)Du \vee (\forall u)Eu$ is true and $(\forall u)Du$ is false, it follows that $(\forall u)Eu$ must be true. The relevant NDS rule is Disjunctive Syllogism (DS).
7. But $(\forall u)Eu$ asserts that everything is *E* and we know, from 5 above, that *a* is not *E*. This is contradictory. The contradiction is clear semantically, but formally we capture this reasoning by employing two rules. First, we instantiate $(\forall u)Eu$ with *a*, obtaining Ea, and, second, we conjoin Ea and $\neg Ea$, using the rule Conjunction (Conj).

We put these steps together to form a standard PL derivation.

[0	0.1	$(\forall u)Du \vee (\forall u)Eu$	P
	0.2	$(\exists v)Qvv$	P
	0.3	$(\forall x)(\forall y)(Qxy \rightarrow (\neg Dx \wedge \neg Ey))$	P
	1	$(\exists v)Qvv$	0.2, Reit
[1	2	Qaa	EIA, *a*
	3	$(\forall y)(Qay \rightarrow (\neg Da \wedge \neg Ey))$	0.3, UI
	4	$Qaa \rightarrow (\neg Da \wedge \neg Ea)$	3, UI
	5	$\neg Da \wedge \neg Ea$	2, 4, MP
	6	$\neg Da$	5, Simp
	7	$\neg Ea$	5, Simp
	8	$(\exists u)\neg Du$	6, EG
	9	$\neg(\forall u)Du$	8, NQ
	10	$(\forall u)Eu$	0.1, 9, DS
	11	Ea	10, UI
	12	$Ea \wedge \neg Ea$	7, 11, Conj
1]	13	$Eb \wedge \neg Eb$	12, Expl
0]	14	$Eb \wedge \neg Eb$	2–13, EI

A note on the use of Explosion (Expl): If we end the EI block with the contradiction $Ea \wedge \neg Ea$, we would not be able to take this sentence outside the block, because the name of the existential instantiation, 'a', cannot occur in the conclusion of the EI block. So we used the NDS rule Explosion to introduce a new contradiction in which the name 'a' does not occur. Expl is an extremely liberal rule: it allows us to infer *any* sentence from a contradiction. We inferred $Eb \wedge \neg Eb$ because this is sufficient for our need. Since a contradiction is derivable from the set, it is inconsistent according to Corollary 5.2.4b.

Here is a practice exercise, demonstrating a **combined strategy of goal analysis and simplification and detachment**. We will prove the following claim:

$\{(\forall v)(Mv \rightarrow (\forall y)(Ey \rightarrow Gvy)), (\exists x)(Mx \wedge (\exists y)\neg Gxy)\} \vdash \neg(\forall v)Ev$

1. We first observe that the conclusion $\neg(\forall v)Ev$ is a negated universally quantified sentence. This suggests that we get this sentence from an existentially quantified sentence by the rule Negated Quantifier (NQ). So we should have $(\exists v)\neg Ev$.
2. $(\exists v)\neg Ev$ can be obtained from $\neg Eb$ by Existential Generalization (EG). Thus our goal now is to derive $\neg Eb$.
3. At this *stage* we employ the strategy of simplification and detachment. We simplify $(\exists x)(Mx \wedge (\exists y)\neg Gxy)$ using Existential Instantiation (EI) to obtain the Existential Instantiation Assumption (EIA) $Ma \wedge (\exists y)\neg Gay$.
4. The next natural step is to break up the conjunction $Ma \wedge (\exists y)\neg Gay$ into two sentences, using Simplification (Simp). We get Ma and $(\exists y)\neg Gay$.
5. Again we simplify $(\exists y)\neg Gay$ by instantiating the sentence with a *new* name, using EI. We obtain $\neg Gab$.
6. Now we instantiate $(\forall v)(Mv \rightarrow (\forall y)(Ey \rightarrow Gvy))$. Since we have Ma, we should replace the variable 'v' with the name 'a'. Using Universal Instantiation (UI), we infer $Ma \rightarrow (\forall y)(Ey \rightarrow Gay)$.
7. Given Ma, we detach the consequent of the conditional $Ma \rightarrow (\forall y)(Ey \rightarrow Gay)$ using Modus Ponens (MP). Hence we obtain $(\forall y)(Ey \rightarrow Gay)$.
8. We simplify $(\forall y)(Ey \rightarrow Gay)$ by instantiating it with the name 'b' since we have $\neg Gab$. Universal Instantiation (UI) gives us $Eb \rightarrow Gab$.
9. Given $\neg Gab$ and $Eb \rightarrow Gab$, we derive $\neg Eb$ using Modus Tollens (MT).
10. We have reached the *limit* of simplification and detachment. We should not lose sight of our goal. According to 1 and 2 above, our purpose is to derive $\neg Eb$ (which we already did), and then to use EG and NQ to infer the desired conclusion—namely, $\neg(\forall v)Ev$. So our analysis is completed.

The PL derivation below is constructed on the basis of the steps above.

[0	0.1	$(\forall v)(Mv \rightarrow (\forall y)(Ey \rightarrow Gvy))$	P
	0.2	$(\exists x)(Mx \wedge (\exists y)\neg Gxy)$	P
[1	1	$Ma \wedge (\exists y)\neg Gay$	EIA, *a*
	2	Ma	1, Simp
	3	$(\exists y)\neg Gay$	1, Simp
[2	4	$\neg Gab$	EIA, *b*

	5	$Ma \rightarrow (\forall y)(Ey \rightarrow Gay)$	0.1, UI
	6	$(\forall y)(Ey \rightarrow Gay)$	2, 5, MP
	7	$Eb \rightarrow Gab$	6, UI
	8	$\neg Eb$	4, 7, MT
2]	9	$(\exists v)\neg Ev$	8, EG
1]	10	$(\exists v)\neg Ev$	4–9, EI
	11	$(\exists v)\neg Ev$	1–10, EI
0]	12	$\neg(\forall v)Ev$	11, NQ

A note on lines 9–11: The reason the sentence on line 9 is repeated twice is that we have two *nested* EI blocks. It is obvious that ending the second EI block [2] at line 8 would not be useful since we would not be able to take the sentence $\neg Eb$ outside the block. $\neg Eb$ contains the name *b*, which cannot occur in the conclusion of the EI block [2]. The sentence on line 9 is suitable for terminating the block: it contains neither *a* nor *b*. So we closed the second EI block at line 9 and reiterated the sentence $(\exists v)\neg Ev$ at line 10, effectively discharging the second EI assumption. We closed the first EI block [1] at line 10, and again reiterated the sentence $(\exists v)\neg Ev$ at line 11. This means that the first EI assumption has been discharged as well. Thus the lines 11 and 12 are independent of the EI assumptions on lines 1 and 4.

5.4:C2 We made some promises in Chapters Two and Three, which we will fulfill in this subsection. These will serve as further practice exercises. The first promise was made in Subsection 2.5.12:C4. It concerns the following argument.

P All the Chess Club members are students in Symbolic Logic.

C All biological mothers of the Chess Club members are biological mothers of students in Symbolic Logic.

We explained there that TL does not have the means to capture this argument correctly. We will show that this argument can be translated faithfully into PL.

Translation Key

Cx: *x* is a member of the Chess Club
Sy: *y* is a student in Symbolic Logic
gz: The biological mother of *z*

PL Argument

S1 $(\forall x)(Cx \rightarrow Sx)$

S2 $(\forall z)((\exists x)(Cx \wedge z = gx) \rightarrow (\exists y)(Sy \wedge z = gy))$

We will prove that this PL argument, like the English argument, is deductively valid. By Corollary 5.2.4a, it suffices to derive S2 from S1. We will follow a **goal analysis strategy**.

1. Since S2 is a universally quantified sentence, we use Universal Generalization (UG) to infer S2. Thus we need to have a substitutional instance of S2 in which the name is arbitrary. Any name will do since the premise contains no names. A sentence such as $(\exists x)(Cx \wedge a = gx) \rightarrow (\exists y)(Sy \wedge a = gy)$ would be adequate for the inference.
2. $(\exists x)(Cx \wedge a = gx) \rightarrow (\exists y)(Sy \wedge a = gy)$ is a conditional. This suggests that we use Conditional Proof (CP) to infer this sentence. The assumption (CPA) of the CP subderivation is $(\exists x)(Cx \wedge a = gx)$, and the last sentence of the subderivation is $(\exists y)(Sy \wedge a = gy)$.
3. Our goal now is to derive $(\exists y)(Cy \wedge a = gy)$ from S1 and CPA. Since $(\exists y)(Cy \wedge a = gy)$ is an existentially quantified sentence, we can derive it from a substitutional instance $Sb \wedge a = gb$. The relevant rule is Existential Generalization (EG).
4. The conjunction $Sb \wedge a = gb$ can be derived by the rule Conjunction (Conj) from Sb and $a = gb$.
5. $a = gb$ and Cb may be obtained from CPA, which is $(\exists x)(Cx \wedge a = gx)$, if we instantiate it with the name b. The rule here is Existential Instantiation (EI).
6. Sb can be inferred by Modus Ponens (MP) from $Cb \rightarrow Sb$ if we have Cb as well. Cb is available to us from the preceding step.
7. $Cb \rightarrow Sb$ can be obtained by applying Universal Instantiation (UI) to S1, which is $(\forall x)(Cx \rightarrow Sx)$.

We put these steps together in reverse order to construct the following PL derivation.

PL Derivation

[0	0.1	$(\forall x)(Cx \rightarrow Sx)$	P
[1	1	$(\exists x)(Cx \wedge a = gx)$	CPA
[2	2	$Cb \wedge a = gb$	EIA, *b*
	3	Cb	2, Simp
	4	$a = gb$	2, Simp
	5	$Cb \rightarrow Sb$	0.1, UI
	6	Sb	3, 5, MP
	7	$Sb \wedge a = gb$	4, 6, Conj
2]	8	$(\exists y)(Sy \wedge a = gy)$	7, EG
1]	9	$(\exists y)(Sy \wedge a = gy)$	2–8, EI
	10	$(\exists x)(Cx \wedge a = gx) \rightarrow (\exists y)(Sy \wedge a = gy)$	1–9, CP
0]	11	$(\forall z)((\exists x)(Cx \wedge z = gx) \rightarrow (\exists y)(Sy \wedge z = gy))$	10, UG

The second promise was made in Subsection 3.5.12:C3 regarding the following argument.

P1 If Uranus is the last planet and if Newtonian physics is correct, then the theoretically calculated orbit of Uranus describes the motion of the planet correctly.
P2 The theoretically calculated orbit of Uranus differs significantly from the observational orbit.
P3 Newtonian physics is correct.
P4 The observational orbit of Uranus is the correct one.

C Uranus is not the last planet.

We were unable to capture the deductive validity of this argument in SL without the use of a meaning-postulate. However, PL is adequate for the task without the addition of any premises.

Translation Key

r: The planet Uranus
n: Newtonian Physics
t: The theoretically calculated orbit of Uranus
o: The observational orbit of Uranus
a: The actual orbit of Uranus
Lx: *x* is the last planet
Cy: *y* is correct

PL Argument

S1	$(Lr \wedge Cn) \rightarrow t = a$
S2	$t \neq o$
S3	Cn
S4	$o = a$
S5	$\neg Lr$

We employ a **strategy of simplification and detachment**. (1) Substituting a for o in $t \neq o$ gives us the negation of the consequent of $(Lr \wedge Cn) \rightarrow t = a$. (2) By Modus Tollens we detach the negation of the antecedent, that is, $\neg(Lr \wedge Cn)$. (3) The last sentence can be simplified by applying one of De Morgan's Laws to it; we get $\neg Lr \vee \neg Cn$. (4) By Disjunctive Syllogism, we can detach $\neg Lr$ from $\neg Lr \vee \neg Cn$ since we have Cn (S3). Here is the complete PL derivation.

[0	0.1	$(Lr \wedge Cn) \rightarrow t = a$	P
	0.2	$t \neq o$	P
	0.3	Cn	P
	0.4	$o = a$	P
	1	$t \neq a$	0.2, 1, Sub
	2	$\neg(Lr \wedge Cn)$	0.1, 1, MT
	3	$\neg Lr \vee \neg Cn$	2, DeM
	4	$\neg\neg Cn$	0.3, DN
0]	5	$\neg Lr$	3, 4 DS

Since the conclusion is derivable from the premises, the PL argument is deductively valid by Corollary 5.2.4a.

The last promise was made in the Solution to Exercise 3.6.13a. We said then that some of the premises of the argument in the passage were logically true, and hence they could be removed from the argument without affecting its deductive validity. We also said that the SL translation of the argument could not be valid without those premises. PL is an adequate system to capture the deductive validity of the English argument without the aid of those logically true premises. However, PL also requires the addition of one meaning-postulate asserting that the belief that God exists is a belief. In English, this is a logically true statement, and hence it is extraneous to the deductive validity of the argument. The PL argument, on the other hand, needs to be "informed" of the logical relation between the specific belief that God exists and the property of being a belief. Here is the "economical" English argument.

P1 If beliefs are processes that occur in the brain, then they are physical processes.
P2 All physical processes can be duplicated by certain machines.
P3 It is possible for a machine to duplicate the belief that God exists only if it is possible for a machine to believe in God.
P4 It is impossible for a machine to believe in God.

C It is not the case that beliefs are processes that occur in the brain.

We will show that this argument is deductively valid by showing that its PL translation is deductively valid. The PL argument below is deductively valid because its conclusion is derivable from its premises (see Corollary 5.2.4a).

Translation Key

e: The belief that God exists
Bx: x is a belief
Rx: x is a process that occurs in the brain
Pz: z is a physical process
My: y is a machine
Dzv: z can duplicate v
Gw: It is possible for w to believe in God.

PL Argument

S1 $(\forall x)(Bx \rightarrow Rx) \rightarrow (\forall x)(Bx \rightarrow Px)$
S2 $(\forall y)(Py \rightarrow (\exists x)(Mx \wedge Dxy))$
S3 $(\forall z)((Mz \wedge Dze) \rightarrow Gz)$
S4 $(\forall z)(Mz \rightarrow \neg Gz)$
S5 Be [meaning-postulate]

S6 $\neg(\forall x)(Bx \rightarrow Rx)$

We will use a **semantical reasoning strategy**. Since the conclusion is a negation and since Reductio Ad Absurdum (RAA) gives us an additional assumption, which increases our inferential resources, we will employ RAA.

1. Our Reductio Assumption (RA) is $(\forall x)(Bx \rightarrow Rx)$.
2. The first premise is a conditional whose antecedent is RA. Hence we can infer the consequent of the conditional, which is $(\forall x)(Bx \rightarrow Px)$. The relevant rule is Modus Ponens (MP)

3. By the truth conditions of the universal quantifier, all the basic substitutional instances of $(\forall x)(Bx \rightarrow Px)$ are true; in particular $Be \rightarrow Pe$ is true. The rule invoked here is Universal Instantiation (UI).
4. Another application of MP to $Be \rightarrow Pe$ and Be yields Pe.
5. The truth of S2 entails that the following substitutional instance is true: $Pe \rightarrow (\exists x)(Mx \wedge Dxe)$. This is another application of UI.
6. By MP, Pe and $Pe \rightarrow (\exists x)(Mx \wedge Dxe)$ give us $(\exists x)(Mx \wedge Dxe)$.
7. Since $(\exists x)(Mx \wedge Dxe)$ is true, a basic substitutional instance of it is true. Let this instance be $Mc \wedge Dce$. The rule invoked here is Existential Instantiation (EI).
8. The truth of S3 implies that the substitutional instance $(Mc \wedge Dce) \rightarrow Gc$ is true. The relevant rule is UI again.
9. From 7 and 8, by MP, we obtain Gc.
10. The truth conditions of $(\forall z)(Mz \rightarrow \neg Gz)$ entail that the substitutional instance $Mc \rightarrow \neg Gc$ is true. This is another application of UI.
11. Applying Simplification to $Mc \wedge Dce$ (step 7 above) generates Mc.
12. A final application of MP to 10 and 11 gives us $\neg Gc$, which contradicts 9.
13. RAA is complete; hence RA is false.

We transform these inferential steps into a standard PL derivation.

PL Derivation

[0	0.1	$(\forall x)(Bx \rightarrow Rx) \rightarrow (\forall x)(Bx \rightarrow Px)$	P
	0.2	$(\forall y)(Py \rightarrow (\exists x)(Mx \wedge Dxy))$	P
	0.3	$(\forall z)((Mz \wedge Dze) \rightarrow Gz)$	P
	0.4	$(\forall z)(Mz \rightarrow \neg Gz)$	P
	0.5	Be	P
[1	1	$(\forall x)(Bx \rightarrow Rx)$	RA
	2	$(\forall x)(Bx \rightarrow Px)$	0.1, 1, MP
	3	$Be \rightarrow Pe$	2, UI
	4	Pe	0.5, 3, MP
	5	$Pe \rightarrow (\exists x)(Mx \wedge Dxe)$	0.2, UI
	6	$(\exists x)(Mx \wedge Dxe)$	4, 5, MP
[2	7	$Mc \wedge Dce$	EIA, *c*
	8	$(Mc \wedge Dce) \rightarrow Gc$	0.3, UI
	9	Gc	7, 8, MP
	10	$Mc \rightarrow \neg Gc$	0.4, UI
	11	Mc	7, Simp
	12	$\neg Gc$	10, 11, MP
	13	$Gc \wedge \neg Gc$	9, 12, Conj
2]	14	$Ga \wedge \neg Ga$	13, Expl
1]	15	$Ga \wedge \neg Ga$	7–14, EI
0]	16	$\neg(\forall x)(Bx \rightarrow Rx)$	1–15, RAA

5.5 Exercises

Note: All answers must be adequately justified.

5.5.1 Give SL derivations to establish the following.

5.5.1a $\{\neg B \wedge (A \leftrightarrow B), D \rightarrow A\} \vdash \neg D$

5.5.1b* $\{(\neg A \vee B) \rightarrow D, A \rightarrow E, \neg E\} \vdash D$

5.5.1c $\{\neg A \rightarrow (D \vee H), A \rightarrow E, \neg E, \neg H\} \vdash D$

5.5.1d* $\{\neg A \wedge (D \vee H), A \leftrightarrow E, H \rightarrow E\} \vdash D \wedge \neg A$

5.5.1e $\{A \wedge (D \rightarrow H), A \leftrightarrow \neg E, H \rightarrow E\} \vdash \neg D \wedge A$

5.5.1f* $\{A \vee (D \leftrightarrow \neg H), S \wedge T, S \rightarrow (T \rightarrow \neg A), (S \vee R) \rightarrow \neg\neg H\} \vdash \neg D \wedge (T \rightarrow \neg A)$

5.5.1g $\{A \vee D, S \wedge T, S \rightarrow (A \rightarrow (F \wedge R)), T \rightarrow (D \rightarrow \neg L)\} \vdash (F \wedge R) \vee \neg L$

5.5.1h $\{\neg A \vee B, B \rightarrow \neg(C \rightarrow C)\} \vdash \neg A$

5.5.1i* $\{(R \wedge \neg L) \rightarrow \neg N, N \leftrightarrow R, (M \wedge \neg L) \rightarrow N, M \vee N\} \vdash L$

5.5.2 Give PL derivations to establish the following.

5.5.2a $\{\neg Ae, (\forall v)(Bv \rightarrow Av), (\forall x)(Bx \vee Dx)\} \vdash (\exists z)Dz$

5.5.2b* $\{Rst, (\forall u)(\forall w)((Ruw \wedge Gwu) \rightarrow u = w), (\forall x)(\forall z)(Rxz \leftrightarrow Gzx)\} \vdash s = t$

5.5.2c $\{Rst, (\forall u)(\forall w)((Ruw \wedge Gwu) \rightarrow u = w), (\forall x)(\forall z)(Rxz \leftrightarrow Gzx)\} \vdash (\exists y)Gyy$

5.5.2d $\{(\forall x)\neg Kxx, (\forall x)(\forall w)(Gwx \rightarrow Kwx), (\forall z)(Gzz \vee Dz)\} \vdash Db$

5.5.2e $\{(\forall x)\neg Kxx, (\forall x)(\forall w)(Gwx \rightarrow Kwx), (\forall z)(Gzz \vee Dz)\} \vdash (\forall x)Dx$

5.5.2f* $\{(\forall x)(\forall y)(x = y \rightarrow \neg Sxy), Pe, (\forall w)((Pw \wedge Lw) \rightarrow Swe)\} \vdash \neg Le$

5.5.2g* $\{(\forall x)(x = q \vee Nxq), (\forall x)(\forall z)(Nxz \rightarrow x = z)\} \vdash (\forall x)(\forall y)x = y$

5.5.2h $\{(\forall z)(Jz \vee \neg Hz), (\forall z)(Jz \rightarrow z \neq d)\} \vdash \neg Hd$

5.5.2i* $\{(\exists u)Au \rightarrow (\exists u)Bu, \neg(\exists z)(Az \wedge Bz), Ac\} \vdash (\exists x)(\exists y)x \neq y$

5.5.3 Give PL derivations to demonstrate the truth of the following statements.

5.5.3a* The PL sentence '$(\forall u)(Rub \wedge \neg Qu) \wedge (\neg Qa \rightarrow \neg Rab)$' is logically false.

5.5.3b The PL sentence '$(\forall z)(Lz \rightarrow (\exists x)(Mx \wedge Ezx)) \wedge ((\forall x)(\forall y)\neg Exy \wedge (\exists x)Lx)$' is logically false.

5.5.3c The PL sentence '$(\forall x)(\forall y)((Dxgy \rightarrow Agx) \wedge Drgs) \rightarrow (\exists z)(z = gr \wedge Az)$ is logically true.

5.5.3d* The PL sentence '$\neg(\exists v)Mvv \rightarrow (\forall x)(\forall y)(\neg Mxy \vee x \neq y)$' is logically true.

5.5.3.e The PL sentences '$(\forall z)(Ezz \vee (\exists v)(Ezv \wedge z \neq v))$' and '$(\forall z)(\exists v)Ezv$' are logically equivalent.

5.5.4 Give PL derivations to show that the following sets of PL sentences are inconsistent.

5.5.4a $\{(\forall x)(Ux \rightarrow (\forall z)Cxz), (\forall x)(\forall y)(Cxy \rightarrow (x \neq y \vee \neg Uy)), (\exists w)Uw\}$

5.5.4b $\{(\forall x)(\forall y)(Kxy \rightarrow \neg Kyx), (\exists w)(\forall y)Kwy\}$

5.5.4c* $\{(\forall u)(Du \vee Eu), (\exists v)(\exists x)Qvx, (\forall x)(\forall y)(Qxy \rightarrow (Ax \vee By)), (\forall z)(Az \rightarrow (\neg Dz \wedge \neg Ez)), (\forall z)(Bz \rightarrow (\neg Dz \wedge \neg Ez))\}$

5.5.5 Construct SL derivations to show that the SL translations of the arguments found in the exercises listed below are deductively valid.

5.5.5a 3.6.11a
5.5.5b* 3.6.11b
5.5.5c 3.6.12
5.5.5d* 3.6.13a
5.5.5e 3.6.13b
5.5.5f 3.6.13c
5.5.5g 3.6.14a
5.5.5h* 3.6.14b
5.5.5i 3.6.15

5.5.6 Since every TL argument is expressible as a PL argument, convert the TL translations of the arguments found in the exercises listed below into PL arguments, and then use PL derivations to show that the PL arguments are deductively valid.

5.5.6a* 2.6.6a
5.5.6b 2.6.6b
5.5.6c* 2.6.7a
5.5.6d* 2.6.7d
5.5.6e 2.6.7e
5.5.6f* 2.6.8a
5.5.6g 2.6.8b
5.5.6h 2.6.8c
5.5.6i* 2.6.8d
5.5.6j 2.6.8e
5.5.6k* 2.6.9a (first argument only)
5.5.6l 2.6.9b (first argument only)

5.5.7 Schematize the arguments in the passages below, translate the schematizations into PL, and use PL derivations to demonstrate that the arguments are deductively valid.

5.5.7a* Not everyone in this room is Julie. No person in this room is a teaching assistant but Julie. Therefore, someone in this room is not a teaching assistant.

5.5.7b My friend Salim believes that his mind and his brain are two different things. He argues as follows. It is clear that my mind is capable of having beliefs. My brain, on the other hand, is simply a machine (albeit a very complex biological machine). Now, whatever is capable of having beliefs could believe in God. Since nothing could believe in God unless it could be religious, and since nothing that is simply a machine could be religious, my mind isn't the same thing as my brain.

5.5.7c* All horses are animals. Therefore, all heads of horses are heads of animals.

5.5.7d Philosophers have long worried about a skeptical hypothesis that is called "the brain-in-a-vat hypothesis" (BVH, for short). BVH says that each one of us is a brain in a vat filled with appropriate nutrients (in order to keep the brain alive) and is connected to a supercomputer, which generates in the brain all the ordinary thoughts about an external world, where in fact there is no external world and all these thoughts are mere illusions. Some philosophers use BVH to argue that since BVH is logically possible and since we can never have a conclusive refutation of it, we should be skeptical about the existence of the external world. The American philosopher Hilary Putnam (b. 1926) presented a famous argument against this skeptical hypothesis. What follows is a simplified version of this argument. Every coherent thought I have about a non-abstract thing is causally related to that thing. The external world is not abstract. If a thought I have is causally related to a thing, then this thing exists. If I believe that BVH is true, then I have a coherent thought about BVH. But I have a coherent thought about BVH only if I have a coherent thought about the external world (because BVH mentions the external world). If the external world exists, BVH is false. Therefore if I believe that BVH is true, then BVH is false.

5.5.7e The British logician and philosopher Bertrand Russell (1872–1970) discovered a devastating paradox in an earlier version of set theory. After its discoverer, the paradox was named 'Russell's Paradox'. Here is an argument that is based on a version of the paradox. A Russellian barber is one who shaves *all and only* those males living in Russell Town who don't shave themselves. All Russellian barbers are males. All Russellian barbers live in Russell Town. Therefore there are no Russellian barbers.

5.5.7f* As explained in 3.6.13c, the adjective 'coreferential' describes terms that have the same referent and 'synonymous' describes terms that have the same meaning. Gottlob Frege (1848–1925) believed that the meaning (sense) of a term is a *different entity* from its referent. He gave several arguments for this thesis. Here is one of them. The terms 'The Evening Star' and 'The Morning Star' are coreferential. If these terms are synonymous, then the sentences 'The Evening Star is the Morning Star' and 'The Evening Star is the Evening Star' are expressing exactly the same thought. But sentences express the same thought only if they have the same epistemic value (i.e., they convey the same information). Given the obvious fact that the sentences above have different epistemic values, it follows that at least some coreferential terms are not synonymous.

5.5.7g Here is a simplified version of an argument by the British scholar C.S. Lewis (1898–1963) that attempts to show the incoherence of the *therapeutic theory of justice,* which is the thesis that therapy, and not punishment, is the appropriate legal response to a criminal act. It is clear that a theory that does not render a concept meaningful is not really a theory of that concept. A theory renders the concept of justice meaningful only if for every criminal act the theory assigns legal consequences that are fairly proportionate to the act itself. It is clear, however, that there are criminal acts for which the therapeutic theory of justice would impose an undeservedly painful therapy or an undeservedly merciful therapy. If it is the former, then there is a criminal act for which the therapeutic theory does not assign legal consequences that are fairly proportionate to the act itself. If it is the latter, similarly we reach the same negative conclusion. Hence we conclude that the therapeutic theory of justice is not *really* a theory of the concept of justice.

5.5.7h* The Spanish-Dutch philosopher Baruch Spinoza (1632–77) gave an important argument for *monism*, which is the thesis that reality consists of only one substance and its modes. Here is a part of that argument. Everything exists either in itself or in something else. A substance is that which exists in itself. A mode is that which exists in something else. Every mode exists in some substance. Given that I exist, it follows that there is at least one substance.

5.5.7i The same argument as in the preceding exercise with the following modification: the first premise is replaced by the statement: 'Everything exists in something'.

5.5.7j There are two main philosophical schools about the nature of mathematical objects:[1] realism and antirealism. Although adherents of each school subscribe to a variety of positions, they share a common core. Mathematical realists are unified in their conviction that it is rational (for us) to believe in the (literal) truth of at least some existential assertions about mathematical objects.[2] Antirealists, on the other hand, do not accept the (literal) truth of such assertions. Perhaps, the most prominent argument for mathematical realism is called *the indispensability argument*.[3] Here is a simplified version of this argument. There are two principles of rationality that seem self-evident. First, it is rational to believe in the truth of any assertion that forms an indispensable component of a highly confirmed theory. Second, it is rational to believe in the truth of a theory only if it is rational to believe in the truth of any assertion logically implied by the theory. Now, it is clear that each of our best scientific theories (such as general relativity and quantum mechanics) incorporates at least one mathematical theory as an indispensable part of it. It is also clear that all mathematical theories logically imply existential assertions about mathematical objects. Given the obvious fact that our best scientific theories are highly confirmed, it follows that it is rational to believe in the truth of at least some existential assertions about mathematical objects.

5.5.7k* This is an argument based on one originally given by the Austrian philosopher Ludwig Wittgenstein (1889–1951) in *Philosophical Investigations*. It can be shown

---

1 Roughly speaking, mathematical objects are either entities (such as numbers, lines, and vectors) or structures (such as number systems and vector spaces) that are described by mathematical theories.

2 'There are infinitely many prime numbers' and 'There exists a four-dimensional vector space' are examples of such assertions.

3 Outside philosophy of mathematics, this argument is sometimes called "the no-miracle argument," because one of its champions, the American philosopher Hilary Putnam (b. 1926), expressed the main point of this argument, when it is applied to philosophy of science, as follows: "Scientific realism is the only philosophy of science that does not render the success of modern science a miracle."

that the following claim is false: the referent of a name is also its meaning. The sentence that Plato was a philosopher is true. It is obvious that the name 'Plato' is the subject of the preceding sentence and that the object to which the name refers is the philosopher Plato. But Plato is dead. These two principles require no argument: (1) if the meaning of a name is dead, then the name is meaningless; (2) no sentence whose subject is meaningless can be true.

5.5.71 Elizabeth is not the mother of anyone. Any two people who have the same mother are identical. Each tall person is the mother of someone who is tall. If Elizabeth is not tall, and if the mother of any person who is not tall is also not tall, then no one is tall. It follows that no one is tall.

Solutions to the Starred Exercises

SOLUTIONS TO 5.5.1

5.5.1b

[0	0.1	$(\neg A \vee B) \rightarrow D$	P
	0.2	$A \rightarrow E$	P
	0.3	$\neg E$	P
	1	$\neg A$	0.2, 0.3, MT
	2	$\neg A \vee B$	1, Add
0]	3	D	0.1, 2, MP

5.5.1d

[0	0.1	$\neg A \wedge (D \vee H)$	P
	0.2	$A \leftrightarrow E$	P
	0.3	$H \rightarrow E$	P
	1	$\neg A$	0.1, Simp
	2	$\neg E$	0.2, 1, BcMT
	3	$\neg H$	0.3, 2, MT
	4	$D \vee H$	0.1, Simp
	5	D	3, 4, DS
0]	6	$D \wedge \neg A$	1, 5, Conj

5.5.1f

[0	0.1	$A \vee (D \leftrightarrow \neg H)$	P
	0.2	$S \wedge T$	P
	0.3	$S \rightarrow (T \rightarrow \neg A)$	P
	0.4	$(S \vee R) \rightarrow \neg\neg H$	P
	1	S	0.2, Simp
	2	T	0.2, Simp
	3	$T \rightarrow \neg A$	0.3, 1, MP

	4	$\neg A$	2, 3, MP
	5	$S \vee R$	1, Add
	6	$\neg\neg H$	0.4, 5, MP
	7	$D \leftrightarrow \neg H$	0.1, 4, DS
	8	$\neg D$	6, 7, BcMT
0]	10	$\neg D \wedge (T \rightarrow \neg A)$	3, 8, Conj

5.5.1i

[0	0.1	$(R \wedge \neg L) \rightarrow \neg N$	P
	0.2	$N \leftrightarrow R$	P
	0.3	$(M \wedge \neg L) \rightarrow N$	P
	0.4	$M \vee N$	P
[1	1	$\neg L$	RA
[2	2	R	RA
	3	$R \wedge \neg L$	1, 2, Conj
	4	$\neg N$	0.1, 3, MP
	5	$\neg R$	0.2, 4, BcMT
2]	6	$R \wedge \neg R$	2, 5, Conj
	7	$\neg R$	2–6, RAA
	8	$\neg N$	0.2, 7, BcMT
	9	M	0.4, 8, DS
	10	$M \wedge \neg L$	1, 9, Conj
	11	N	0.3, 10, MP
1]	12	$N \wedge \neg N$	8, 11, Conj
0]	13	L	1–12, RAA

SOLUTIONS TO 5.5.2

5.5.2b

[0	0.1	Rst	P
	0.2	$(\forall u)(\forall w)((Ruw \wedge Gwu) \rightarrow u = w)$	P
	0.3	$(\forall x)(\forall z)(Rxz \leftrightarrow Gzx)$	P
	1	$(\forall w)((Rsw \wedge Gws) \rightarrow s = w)$	0.2, UI
	2	$(Rst \wedge Gts) \rightarrow s = t$	1, UI
	3	$(\forall z)(Rsz \leftrightarrow Gzs)$	0.3, UI
	4	$Rst \leftrightarrow Gts$	3, UI
	5	Gts	0.1, 4, BcMP
	6	$Rst \wedge Gts$	0.1, 5, Conj
0]	7	$s = t$	2, 6, MP

5.5.2f

[0	0.1	$(\forall x)(\forall y)(x = y \rightarrow \neg Sxy)$	P
	0.2	Pe	P

	0.3	$(\forall w)((Pw \wedge Lw) \rightarrow Swe)$	P
	1	$(\forall y)(e = y \rightarrow \neg Sey)$	0.1, UI
	2	$e = e \rightarrow \neg See$	1, UI
	3	$e = e$	Id
	4	$\neg See$	2, 3, MP
	5	$(Pe \wedge Le) \rightarrow See$	0.3, UI
	6	$\neg(Pe \wedge Le)$	4, 5, MT
	7	$\neg Pe \vee \neg Le$	6, DeM
	8	$\neg\neg Pe$	0.2, DN
0]	9	$\neg Le$	7, 8, DS

5.5.2g

[0	0.1	$(\forall x)(x = q \vee Nxq)$	P
	0.2	$(\forall x)(\forall z)(Nxz \rightarrow x = z)$	P
[1	1	$a \neq q$	RA
	2	$a = q \vee Naq$	01, UI
	3	$(\forall z)(Naz \rightarrow a = z)$	0.2, UI
	4	$Naq \rightarrow a = q$	3, UI
	5	$\neg Naq$	1, 4, MT
	6	$a = q$	2, 5, DS
1]	7	$a = q \wedge a \neq q$	1, 6, Conj
	8	$a = q$	1–7, RAA
	9	$(\forall z)z = q$	8, UG
	10	$b = q$	9, UI
	11	$a = b$	8, 10, Sub
	12	$(\forall y)a = y$	11, UG
0]	13	$(\forall x)(\forall y)x = y$	12, UG

Note on lines 8–13: Since the name '*q*' occurs in premise 0.1, it can never be arbitrary in the derivation above. Hence we could not apply UG to '*q*'. We applied it to '*a*' because '*a*' is arbitrary. After we obtained $(\forall z)z = q$, we instantiated this sentence to get $b = q$. We used Sub to get rid of '*q*' and obtain the identity '$a = b$', in which both names are arbitrary. We then applied UG twice to '*a*' and '*b*'.

5.5.2i

[0	0.1	$(\exists u)Au \rightarrow (\exists u)Bu$	P
	0.2	$\neg(\exists z)(Az \wedge Bz)$	P
	0.3	Ac	P
	1	$(\exists u)Au$	0.3, EG
	2	$(\exists u)Bu$	0.1, 1, MP
[1	3	Be	EIA, *e*
	4	$(\forall z)\neg(Az \wedge Bz)$	0.2, NQ
	5	$\neg(Ac \wedge Bc)$	4, UI
	6	$\neg Ac \vee \neg Bc$	5, DeM

	7	$\neg\neg Ac$	0.3, DN
	8	$\neg Bc$	6, 7, DS
[2	9	$c = e$	RA
	10	$\neg Be$	8, 9, Sub
2]	11	$Be \land \neg Be$	3, 10, Conj
	12	$c \neq e$	9–11, RAA
	13	$(\exists y)c \neq y$	12, EG
1]	14	$(\exists x)(\exists y)x \neq y$	13, EG
0]	15	$(\exists x)(\exists y)x \neq y$	3–14, EI

SOLUTIONS TO 5.5.3

5.5.3a

[0	0.1	$(\forall u)(Rub \land \neg Qu) \land (\neg Qa \to \neg Rab)$	P
	1	$(\forall u)(Rub \land \neg Qu)$	0.1, Simp
	2	$\neg Qa \to \neg Rab$	0.1, Simp
	3	$Rab \land \neg Qb$	1, UI
	4	$Rab \to Qa$	2, Cont
	5	$\neg(Rab \to Qa)$	3, NC
0]	6	$(Rab \to Qa) \land \neg(Rab \to Qa)$	4, 5, Conj

Since a contradiction is derivable from the sentence '$(\forall u)(Rub \land \neg Qu) \land (\neg Qa \to \neg Rab)$', it is logically false by Corollary 5.2.4e.

5.5.3d

[0	0	$\varnothing$	
[1	1	$\neg(\exists v)Mvv$	CPA
	2	$(\forall v)\neg Mvv$	1, NQ
	3	$\neg Maa$	2, UI
[2	4	Mab	CPA
[3	5	$a = b$	RA
	6	Maa	4, 5, Sub
3]	7	$Maa \land \neg Maa$	3, 6, Conj
2]	8	$a \neq b$	5–7, RAA
	9	$Mab \to a \neq b$	4–8, CP
	10	$\neg Mab \lor a \neq b$	9, NC
	11	$(\forall y)(\neg May \lor a \neq y)$	10, UG
1]	12	$(\forall x)(\forall y)(\neg Mxy \lor x \neq y)$	11, UG
0]	13	$\neg(\exists v)Mvv \to (\forall x)(\forall y)(\neg Mxy \lor x \neq y)$	1–12, CP

Note on the zero stage of a derivation without premises: To derive **X** from the empty set is to derive it without premises. In such derivations, we indicate this fact by writing the empty set $\varnothing$ at the zero stage. We still start the main block with the line 0 but the actual first line of the derivation is line 1. Derivations without premises can only start by invoking the NDS rules

Identity, Conditional Proof, or Reductio Ad Absurdum. All other rules require antecedents, and hence they are not suitable to initiate a derivation without premises. Since the sentence '$\neg(\exists v)Mvv \rightarrow (\forall x)(\forall y)(\neg Mxy \vee x \neq y)$' is derivable from the empty set, it is logically true by Corollary 5.2.4d.

SOLUTION TO 5.5.4

5.5.4c

[0	0.1	$(\forall u)(Du \vee Eu)$	P
	0.2	$(\exists v)(\exists x)Qvx$	P
	0.3	$(\forall x)(\forall y)(Qxy \rightarrow (Ax \vee By))$	P
	0.4	$(\forall z)(Az \rightarrow (\neg Dz \wedge \neg Ez))$	P
	0.5	$(\forall z)(Bz \rightarrow (\neg Dz \wedge \neg Ez))$	P
	1	$(\exists v)(\exists x)Qvx$	0.2, Reit
[1	2	$(\exists x)Qsx$	EI, *s*
[2	3	Qst	EI, *t*
	4	$(\forall y)(Qsy \rightarrow (As \vee By))$	0.3, UI
	5	$Qst \rightarrow (As \vee Bt)$	4, UI
	6	$As \vee Bt$	3, 5, MP
	7	$As \rightarrow (\neg Ds \wedge \neg Es)$	0.4, UI
	8	$As \rightarrow \neg(Ds \vee Es)$	7, DeM
	9	$Ds \vee Es$	0.1, UI
	10	$\neg\neg(Ds \vee Es)$	9, DN
	11	$\neg As$	8, 10, MT
	12	Bt	6, 11, DS
	13	$Bt \rightarrow (\neg Dt \wedge \neg Et)$	0.5, UI
	14	$\neg Dt \wedge \neg Et$	12, 13, MP
	15	$\neg(Dt \vee Et)$	14, DeM
2]	16	$(\exists u)\neg(Du \vee Eu)$	15, EG
1]	17	$(\exists u)\neg(Du \vee Eu)$	3–16, EI
	18	$(\exists u)\neg(Du \vee Eu)$	2–17, EI
	19	$\neg(\forall u)(Du \vee Eu)$	18, NQ
0]	20	$(\forall u)(Du \vee Eu) \wedge \neg(\forall u)(Du \vee Eu)$	0.1, 19, Conj

Since a contradiction is derivable from the set, it is inconsistent by Corollary 5.2.4b.

SOLUTIONS TO 5.5.5

Note: All the problems of this exercise have the same justification: since the conclusion is derivable from the premises, the argument is deductively valid by Corollary 5.2.4a.

5.5.5b

S1	$\neg P \rightarrow (K \vee \neg K)$
S2	$K \rightarrow S$
S3	$S \rightarrow P$
S4	$\neg P \rightarrow \neg K$

SL Derivation

[0	0.1	$\neg P \rightarrow (K \vee \neg K)$	P
	0.2	$K \rightarrow S$	P
	0.3	$S \rightarrow P$	P
[1	1	$\neg P$	CPA
	2	$\neg S$	0.3, 1, MT
1]	3	$\neg K$	0.2, 2, MT
0]	4	$\neg P \rightarrow \neg K$	1–3, CP

As explained in Chapter Three, the first premise is logically true, and hence it is superfluous. Observe that we did not make use of it in the derivation above.

5.5.5d

S1	$B \rightarrow P$
S2	$P \rightarrow G$
S3	M
S4	$(G \wedge M) \rightarrow R$
S5	$R \rightarrow L$
S6	$\neg L$
S7	$\neg B$

SL Derivation

[0	0.1	$B \rightarrow P$	P
	0.2	$P \rightarrow G$	P
	0.3	M	P
	0.4	$(G \wedge M) \rightarrow R$	P
	0.5	$R \rightarrow L$	P
	0.6	$\neg L$	P
	1	$\neg R$	0.5, 0.6, MT
	2	$\neg (G \wedge M)$	0.4, 1, MT
	3	$\neg G \vee \neg M$	2, DeM
	4	$\neg\neg M$	0.3, DN
	5	$\neg G$	3, 4, DS
	6	$\neg P$	0.2, 5, MT
0]	7	$\neg B$	0.1, 6, MT

5.5.5h

S1	B
S2	$B \rightarrow N$
S3	G
S4	$(G \wedge N) \rightarrow L$
S5	$K \vee S$
S6	$(L \wedge S) \rightarrow D$
S7	$\neg R \rightarrow \neg D$
S8	$\neg R$
S9	K

SL Derivation

[0	0.1	B	P
	0.2	$B \rightarrow N$	P
	0.3	G	P
	0.4	$(G \wedge N) \rightarrow L$	P
	0.5	$K \vee S$	P
	0.6	$(L \wedge S) \rightarrow D$	P
	0.7	$\neg R \rightarrow \neg D$	P
	0.8	$\neg R$	P
	1	$\neg D$	0.7, 0.8, MP
	2	$\neg(L \wedge S)$	0.6, 1, MT
	3	N	0.1, 0.2, MP
	4	$G \wedge N$	0.3, 3, Conj
	5	L	0.4, 4, MP
	6	$\neg L \vee \neg S$	2, DeM
	7	$\neg\neg L$	5, DN
	8	$\neg S$	6, 7, DS
0]	9	K	0.5, 8, DS

SOLUTIONS TO 5.5.6

Note: All the problems of this exercise have the same justification: since the conclusion is derivable from the premises, the argument is deductively valid by Corollary 5.2.4a.

5.5.6a

The TL argument

S1	all *M* are *I*
S2	no *R* is *I*
S3	no *M* is *R*

The PL Argument

S1*	$(\forall x)(Mx \to Ix)$
S2*	$(\forall x)(Rx \to \neg Ix)$
S3*	$(\forall x)(Mx \to \neg Rx)$

PL Derivation

[0	0.1	$(\forall x)(Mx \to Ix)$	P
	0.2	$(\forall x)(Rx \to \neg Ix)$	P
[1	1	Ma	CPA
	2	$Ma \to Ia$	0.1, UI
	3	Ia	1, 2, MP
	4	$(\forall x)(\neg\neg Rx \to \neg Ix)$	0.2, DN
	5	$(\forall x)(Ix \to \neg Rx)$	4, Cont
	6	$Ia \to \neg Ra$	5, UI
1]	7	$\neg Ra$	3, 6, UI
	8	$Ma \to \neg Ra$	1–7, CP
0]	9	$(\forall x)(Mx \to \neg Rx)$	8, UG

5.5.6c

The TL argument

S1	*b* is *I*
S2	all *I* are *D*
S3	all *D* are *M*
S4	*b* is *M*

The PL Argument

S1*	Ib
S2*	$(\forall x)(Ix \to Dx)$
S3*	$(\forall x)(Dx \to Mx)$
S4*	Mb

[0	0.1	Ib	P
	0.2	$(\forall x)(Ix \to Dx)$	P
	0.3	$(\forall x)(Dx \to Mx)$	P
	1	$Ib \to Db$	0.2, UI
	2	$Db \to Mb$	0.3, UI
	3	$Ib \to Mb$	1, 2, HS
0]	4	Mb	0.1, 3, MP

5.5.6d

The TL argument

S1	all *K* are *M*
S2	all *M* are *E*
S3	no *G* is *E*
S4	some *T* are *G*
S5	some *T* are not *K*

The PL Argument

S1*	$(\forall z)(Kz \rightarrow Mz)$
S2*	$(\forall x)(Mx \rightarrow Ex)$
S3*	$(\forall y)(Gy \rightarrow \neg Ey)$
S4*	$(\exists x)(Tx \wedge Gx)$
S5*	$(\exists x)(Tx \wedge \neg Kx)$

PL Derivation

[0	0.1	$(\forall z)(Kz \rightarrow Mz)$	P
	0.2	$(\forall x)(Mx \rightarrow Ex)$	P
	0.3	$(\forall y)(Gy \rightarrow \neg Ey)$	P
	0.4	$(\exists x)(Tx \wedge Gx)$	P
[1	1	$Tc \wedge Gc$	EIA, *c*
	2	$Gc \rightarrow \neg Ec$	0.3, UI
	3	Gc	1, Simp
	4	$\neg Ec$	2, 3, MP
	5	$Mc \rightarrow Ec$	0.2, UI
	6	$\neg Mc$	4, 5, MT
	7	$Kc \rightarrow Mc$	0.1, UI
	8	$\neg Kc$	6, 7, MP
	9	Tc	1, Simp
	10	$Tc \wedge \neg Kc$	8, 9, Conj
1]	11	$(\exists x)(Tx \wedge \neg Kx)$	10, EG
0]	12	$(\exists x)(Tx \wedge \neg Kx)$	1–11, EI

5.5.6f

The TL argument

S1	s_1 is *E*
S2	s_2 is *M*
S3	no *E* is *P*
S4	all *M* are *P*
	———
S5	s_1 is not s_2

The PL Argument

S1*	Es_1
S2*	Ms_2
S3*	$(\forall z)(Ez \rightarrow \neg Pz)$
S4*	$(\forall z)(Mz \rightarrow Pz)$
	———
S5*	$s_1 \neq s_2$

PL Derivation

[0	0.1	Es_1	P
	0.2	Ms_2	P
	0.3	$(\forall z)(Ez \rightarrow \neg Pz)$	P
	0.4	$(\forall z)(Mz \rightarrow Pz)$	P
[1	1	$s_1 = s_2$	RA
	2	$Es_1 \rightarrow \neg Ps_1$	0.3, UI
	3	$\neg Ps_1$	0.1, 2, MP
	4	$Ms_2 \rightarrow Ps_2$	0.4, UI
	5	Ps_2	0.2, 4, MP
	6	Ps_1	1, 5, Sub
1]	7	$Ps_1 \wedge \neg Ps_1$	3, 6, Conj
0]	8	$s_1 \neq s_2$	1–7, RAA

5.5.6i

The TL argument

S1	*d* is *U*
S2	all *U* are *B*
S3	all *B* are *T*
S4	all *T* are *K*
	———
S5	some *U* are *K*

The PL Argument

S1*	Ud
S2*	$(\forall y)(Uy \rightarrow By)$
S3*	$(\forall z)(Bz \rightarrow Tz)$
S4*	$(\forall y)(Ty \rightarrow Ky)$
S5*	$(\exists x)(Ux \wedge Kx)$

PL Derivation

[0	0.1	Ud	P
	0.2	$(\forall y)(Uy \rightarrow By)$	P
	0.3	$(\forall z)(Bz \rightarrow Tz)$	P
	0.4	$(\forall y)(Ty \rightarrow Ky)$	P
	1	$Ud \rightarrow Bd$	0.2, UI
	2	$Bd \rightarrow Td$	0.3, UI
	3	$Td \rightarrow Kd$	0.4, UI
	4	$Ud \rightarrow Td$	1, 2, HS
	5	$Ud \rightarrow Kd$	3, 4, HS
	6	Kd	0.1, 5, MP
	7	$Ud \wedge Kd$	0.1, 6, Conj
0]	8	$(\exists x)(Ux \wedge Kx)$	7, EG

Note on the first premise: In TL, the first premise is superfluous because the general terms in TL cannot have empty extensions. In PL, predicates are allowed to have empty extensions, and hence the first premise is needed in order to establish that the extension of *U* is not empty.

5.5.6k

The TL Argument

S1	*c* is *N*
S2	*c* is *K*
S3	all *K* are *G*
S4	some *N* are *G*

The PL Argument

S1*	Nc
S2*	Kc
S3*	$(\forall x)(Kx \rightarrow Gx)$
S4*	$(\exists x)(Nx \wedge Gx)$

[0	0.1	Nc	P
	0.2	Kc	P
	0.3	$(\forall x)(Kx \rightarrow Gx)$	P
	1	$Kc \rightarrow Gc$	0.3, UI
	2	Gc	0.2, 1, MP
	3	$Nc \wedge Gc$	0.1, 2, Conj
0]	4	$(\exists x)(Nx \wedge Gx)$	3, EG

SOLUTIONS TO 5.5.7

Note: All the problems of this exercise have the same justification: by Corollary 5.2.4a the PL argument is deductively valid because its conclusion is derivable from its premises. Since the schematization is accurate and the translation is faithful, the original English argument is valid as well.

5.5.7a

Schematization of the Argument

P1 Not everyone in this room is Julie.
P2 No person in this room is a teaching assistant but Julie.

C Someone in this room is not a teaching assistant.

Translation Key

j: Julie
Rx: *x* is in this room
Tx: *x* is a teaching assistant

PL Argument

S1 $\neg(\forall x)(Rx \rightarrow x = j)$
S2 $(\forall x)((Rx \wedge Tx) \leftrightarrow x = j)$

S3 $(\exists x)(Rx \wedge \neg Tx)$

PL Derivation

[0	0.1	$\neg(\forall x)(Rx \rightarrow x = j)$	P
	0.2	$(\forall x)((Rx \wedge Tx) \leftrightarrow x = j)$	P
	1	$(\exists x)\neg(Rx \rightarrow x = j)$	0.1, NQ
	2	$(\exists x)(Rx \wedge x \neq j)$	1, NC
[1	3	$Ra \wedge a \neq j$	EI, *a*
	4	Ra	3, Simp
	5	$a \neq j$	3, Simp

	6	$(Ra \wedge Ta) \leftrightarrow a = j$	0.2, UI
	7	$\neg(Ra \wedge Ta)$	5, 6, MT
	8	$\neg Ra \vee \neg Ta$	7, DeM
	9	$\neg\neg Ra$	4, DN
	10	$\neg Ta$	8, 9, DS
	11	$Ra \wedge \neg Ta$	4, 10, Conj
1]	12	$(\exists x)(Rx \wedge \neg Tx)$	11, EG
0]	13	$(\exists x)(Rx \wedge \neg Tx)$	3–12, EI

5.5.7c

Schematization of the Argument

P All horses are animals.

C All heads of horses are heads of animals.

Translation Key

Rz: *z* is an a horse
Av: *v* is an animal
Hxy: *x* is the head of *y*

PL Argument

S1 $(\forall z)(Rz \rightarrow Az)$

S2 $(\forall x)(\forall z)((Rx \wedge Hzx) \rightarrow (\exists y)(Ay \wedge Hzy))$

PL Derivation

[0	0	$(\forall z)(Rz \rightarrow Az)$	P
[1	1	$Rt \wedge Hst$	CPA
	2	$Rt \rightarrow At$	0, UI
	3	Rt	1, Simp
	4	At	2, 3, MP
	5	Hst	1, Simp
	6	$At \wedge Hst$	4, 5, Conj
1]	7	$(\exists y)(Ay \wedge Hsy)$	6, EG
	8	$(Rt \wedge Hst) \rightarrow (\exists y)(Ay \wedge Hsy)$	1–7, CP
	9	$(\forall z)((Rt \wedge Hzt) \rightarrow (\exists y)(Ay \wedge Hzy))$	8, UG
0]	10	$(\forall x)(\forall z)((Rx \wedge Hzx) \rightarrow (\exists y)(Ay \wedge Hzy))$	9, UG

5.5.7f

Schematization of the Argument

P1	The terms 'The Evening Star' and 'The Morning Star' are coreferential.
P2	If the terms 'The Evening Star' and 'The Morning Star' are synonymous, then the sentences 'The Evening Star is the Morning Star' and 'The Evening Star is the Evening Star' are expressing exactly the same thought.
P3	Sentences express the same thought only if they have the same epistemic value.
P4	The sentences 'The Evening Star is the Morning Star' and 'The Evening Star is the Evening Star' have different epistemic values.
C	At least some coreferential terms are not synonymous.

Translation Key

e: The term 'The Evening Star'.
m: The term 'The Morning Star'.
s: The sentence 'The Evening Star is the Morning Star'.
r: The sentence 'The Evening Star is the Evening Star'.
Cxy: *x* and *y* are coreferential.
Sxy: *x* and *y* are synonymous.
Tzy: *z* and *y* express the same thought.
Ezv: *z* and *v* have the same epistemic value.

PL Argument

S1	Cem
S2	$Sem \rightarrow Tsr$
S3	$(\forall x)(\forall y)(Txy \rightarrow Exy)$
S4	$\neg Esr$
S5	$(\exists z)(\exists v)(Czv \wedge \neg Szv)$

PL Derivation

[0	0.1	Cem	P
	0.2	$Sem \rightarrow Tsr$	P
	0.3	$(\forall x)(\forall y)(Txy \rightarrow Exy)$	P
	0.4	$\neg Esr$	P
	1	$(\forall y)(Tsy \rightarrow Esy)$	0.3, UI
	2	$Tsr \rightarrow Esr$	1, UI
	3	$\neg Tsr$	0.4, 2, MT
	4	$\neg Sem$	0.2, 3, MT
	5	$Cem \wedge \neg Sem$	0.1, 4, Conj
	6	$(\exists v)(Cev \wedge \neg Sev)$	5, EG
0]	7	$(\exists z)(\exists v)(Czv \wedge \neg Szv)$	6, EG

5.5.7h

Schematization of the Argument

P1	Everything exists either in itself or in something else.
P2	A substance is that which exists in itself.
P3	A mode is that which exists in something else.
P4	Every mode exists in some substance.
P5	I exist.
C	There is at least one substance.

Translation Key

Exy: *x* exists in *y*
Sz: *z* is a substance
Mw: *w* in a mode
i: I (the speaker)

PL Argument

S1	$(\forall x)(Exx \vee (\exists y)(Exy \wedge x \neq y))$
S2	$(\forall z)(Sz \leftrightarrow Ezz)$
S3	$(\forall z)(Mz \leftrightarrow (\exists v)(Ezv \wedge z \neq v))$
S4	$(\forall x)(Mx \rightarrow (\exists y)(Sy \wedge Exy))$
S5	$(\exists x)x = i$
S6	$(\exists w)Sw$

PL Derivation

[0	0.1	$(\forall x)(Exx \vee (\exists y)(Exy \wedge x \neq y))$	P
	0.2	$(\forall z)(Sz \leftrightarrow Ezz)$	P
	0.3	$(\forall z)(Mz \leftrightarrow (\exists v)(Ezv \wedge z \neq v))$	P
	0.4	$(\forall x)(Mx \rightarrow (\exists y)(Sy \wedge Exy))$	P
	0.5	$(\exists x)x = i$	P
	1	$Eii \vee (\exists y)(Eiy \wedge i \neq y)$	0.1, UI
[1	2	Eii	CPA
	3	$Si \leftrightarrow Eii$	0.2, UI
	4	Si	2, 3, BcMP
1]	5	$(\exists w)Sw$	4, EG
	6	$Eii \rightarrow (\exists x)Sx$	2–5, CP
[2	7	$(\exists y)(Eiy \wedge i \neq y)$	CPA
	8	$Mi \leftrightarrow (\exists v)(Eiv \wedge i \neq v)$	0.3, UI
	9	Mi	7, 8, BcMP
	10	$Mi \rightarrow (\exists y)(Sy \wedge Eiy)$	0.4, UI
	11	$(\exists y)(Sy \wedge Eiy)$	9, 10, MP

[3	12	$Sb \wedge Eib$	EIA, b
	13	Sb	12, Simp
3]	14	$(\exists w)Sw$	13, EG
2]	15	$(\exists w)Sw$	12–14, EI
	16	$(\exists y)(Eiy \wedge i \neq y) \rightarrow (\exists w)Sw$	7–15, CP
	17	$(\exists w)Sw \vee (\exists w)Sw$	1, 6, 16, CD
0]	18	$(\exists w)Sw$	17, Idem

Note on premise S5: Our derivation did not make use of this premise. The English argument employs its English counterpart to establish that there exists at least one thing in the world (namely the thinker who is entertaining this argument). PL does not need this premise because the PL worldview presupposes that every logically possibility contains at least one individual. This requirement is reflected in the PL semantics by defining the universe of discourse of any PL interpretation as a nonempty set. Furthermore, every singular term in PL must have a referent; so the sentence $(\exists x)x = i$ is logically true in PL; and we know from previous discussion that a logically true premise is superfluous for the deductive validity of an argument. The same is true in proof theory, if a derivation of **X** from Γ contains a logically true premise, then there exists a derivation also of **X** from Γ that does not contain that premise. However, it is not true that if a derivation invokes a logically true sentence (not as a premise), then it can always be reformulated *without* the use of that sentence. Many PL derivations make *essential* use of the rule of Identity, yet this rule permits the inference (without any antecedent) of a sentence of the form **t** = **t**, which is logically true.

5.5.7k

Schematization of the Argument

P1 The sentence that Plato was a philosopher is true.
P2 The term 'Plato' is the subject of the sentence 'Plato was a philosopher', and the object to which the term 'Plato' refers is the philosopher Plato.
P3 Plato is dead.
P4 If the meaning of a term is dead, then the term is meaningless.
P5 No sentence whose subject is meaningless can be true.

C It is false that the referent of a name is also its meaning.

Translation Key

n: The sentence 'Plato was a philosopher'
p: The philosopher Plato
e: The term 'Plato'
gy: The subject of the sentence *y* (we are assuming that every declarative sentence has a unique subject)
Tx: *x* is true
Rxy: *x* is the object to which the term *y* refers
Mxy: *x* is the meaning of the term *y*
Dx: *x* is dead
Lx: *x* is meaningless

PL Argument

S1	Tn
S2	$e = gn \wedge Rpe$
S3	Dp
S4	$(\forall x)(\forall y)((Mxy \wedge Dx) \rightarrow Ly)$
S5	$(\forall x)(\forall y)((x = gy \wedge Lx) \rightarrow \neg Ty)$
S6	$\neg(\forall x)(\forall y)(Rxy \leftrightarrow Mxy)$

PL Derivation

[0	0.1	Tn	P
	0.2	$e = gn \wedge Rpe$	P
	0.3	Dp	P
	0.4	$(\forall x)(\forall y)((Mxy \wedge Dx) \rightarrow Ly)$	P
	0.5	$(\forall x)(\forall y)((x = gy \wedge Lx) \rightarrow \neg Ty)$	P
[1	1	$(\forall x)(\forall y)(Rxy \leftrightarrow Mxy)$	RA
	2	$(\forall y)(Mpy \leftrightarrow Rpy)$	1, UI
	3	$Mpe \leftrightarrow Rpe$	2, UI
	4	Rpe	0.2, Simp
	5	Mpe	3, 4, BcMP
	6	$(\forall y)((Mpy \wedge Dp) \rightarrow Ly)$	0.4, UI
	7	$(Mpe \wedge Dp) \rightarrow Le$	6, UI
	8	$Mpe \wedge Dp$	0.3, 5, Conj
	9	Le	7, 8, MP
	10	$(\forall y)((e = gy \wedge Le) \rightarrow \neg Ty)$	0.5, UI
	11	$(e = gn \wedge Le) \rightarrow \neg Tn$	10, UI
	12	$e = gn$	0.2, Simp
	13	$e = gn \wedge Le$	9, 12, Conj
	14	$\neg Tn$	11, 13, MP
1]	15	$Tn \wedge \neg Tn$	0.1, 14, Conj
0]	16	$\neg(\forall x)(\forall y)(Rxy \leftrightarrow Mxy)$	1-15, RAA

Index

from the publisher

A name never says it all, but the word "broadview" expresses a good deal of the philosophy behind our company. We are open to a broad range of academic approaches and political viewpoints. We pay attention to the broad impact book publishing and book printing has in the wider world; we began using recycled stock more than a decade ago, and for some years now we have used 100% recycled paper for most titles. As a Canadian-based company we naturally publish a number of titles with a Canadian emphasis, but our publishing program overall is internationally oriented and broad-ranging. Our individual titles often appeal to a broad readership too; many are of interest as much to general readers as to academics and students.

Founded in 1985, Broadview remains a fully independent company owned by its shareholders—not an imprint or subsidiary of a larger multinational.

If you would like to find out more about Broadview and about the books we publish, please visit us at **www.broadviewpress.com**. And if you'd like to place an order through the site, we'd like to show our appreciation by extending a special discount to you: by entering the code below you will receive a 20% discount on purchases made through the Broadview website.

Discount code: **broadview20%**

Thank you for choosing Broadview.

Please note: this offer applies only to sales of bound books within the United States or Canada.

RECYCLED
Paper made from recycled material
FSC® C103567

LIST
of products used:

1,560 lb(s) of Rolland Enviro100 Print
100% post-consumer

RESULTS
Based on the Cascades products you selected compared to products in the industry made with 100% virgin fiber, your savings are:

13 trees

12,906 gal. US of water
140 days of water consumption

1,631 lbs of waste
15 waste containers

4,241 lbs CO_2
8,042 miles driven

21 MMBTU
100,579 60W light bulbs for one hour

13 lbs NOX
emissions of one truck during 18 days